ALBUM OF SCIENCE

Antiquity and the Middle Ages

Primum mobile
Spera Lune
Spera mercurij
Spera Veneris
Spera Solis
Spera Martis
Spera Jouis

ALBUM OF SCIENCE

Antiquity and the Middle Ages

JOHN E. MURDOCH

I. B. COHEN, GENERAL EDITOR, ALBUMS OF SCIENCE

CHARLES SCRIBNER'S SONS
New York

To My Parents

(Frontispiece) Scipio's Dream. Taken from an Italian manuscript written in 1383, this picture serves as a frontispiece to the commentary on Cicero's *Somnium Scipionis* composed by the Roman scholar Macrobius (fourth to fifth centuries). In the Middle Ages, Macrobius' commentary was one of the most frequently used sources for elementary astronomical and cosmographic knowledge, and it is to this kind of knowledge that the present visualization of the dream of Scipio—who lies asleep below—refers. The spheres of the fixed stars and the planets are surrounded by the outermost *primum mobile,* the ancestral souls met by Scipio in his dream standing in and about a depiction of the Milky Way. The earth is portrayed at the center of these heavenly orbs with edifices of Rome and Carthage adorning its face. Other pictures and diagrams from, or related to, Macrobius' commentary can be seen elsewhere (Illustrations 248, 250, 251, and 289), as well as some from the same fourteenth-century manuscript reproduced here (Illustrations 15 and 181).

Library of Congress Cataloging in Publication Data
Murdoch, John Emery, 1927–
Antiquity and the Middle Ages.
(Album of science)
Bibliography: p. 369
Includes index
1. Scientific illustration—History. 2. Science, Ancient. 3. Science, Medieval. I. Title. II. Series.
Q222.M87 1984 509'.01 84-1400
ISBN 0-684-15496-X

This book published simultaneously
in the United States of America and in Canada—
Copyright under the Berne Covention

1 3 5 7 9 11 13 15 17 19 Q/C 20 18 16 14 12 10 8 6 4 2

Printed in the United States of America

Contents

Acknowledgments

THE IMPORTANCE of scientific illustrations and diagrams within the development and dissemination of early science was an aspect of this particular segment of scientific history of which my knowledge and appreciation were at best marginal before I. Bernard Cohen asked if I would be willing to write the present *Album,* covering antiquity and the Middle Ages. It is to him, then, that I owe the motivation and the occasion for learning whatever I have about the significance of pictorial material for ancient and medieval science. This learning required that I peruse manuscripts not for their texts, but for their figures and miniatures, a new and most pleasant experience the indulgence of which was also set in motion by Bernard Cohen's invitation to take up the task of writing this volume—a task for which, once again, he provided constant encouragement and more helpful suggestions than I can acknowledge.

From the very beginning, I knew that in attempting to write and carry out research in an area that, for me, harbored so many unknowns, I would require advice and critical comment from more than a handful of colleagues. Indeed, my need for counsel and the length of time over which I have sought it have rendered these willing colleagues so numerous that I am certain not to remember and thus be able to thank them all. This confessed, I should begin my list of those to whom I am indebted with A. I. Sabra, to whom all-too-frequent appeal was made for advice on things Arabic. Valuable help relative to diagrams in Arabic science was also received from Edward Kennedy, David King, and Jamil Ragep. John North and Owen Gingerich have considerately looked over what I have said about many of the astronomical illustrations in this volume as well as suggested or furnished some of these illustrations for inclusion. Karen Reeds and Linda Voigts have made similar suggestions for the areas of materia medica and botany. The fact that Bruce Eastwood was engaged in his own research in the history of early medieval astronomical and cosmographic diagrams during much of the time I was at work on the present volume conveniently made his advice exceedingly valuable, to say nothing of the generosity with which he consistently offered it. Special mention should also be made of Michael McVaugh, who placed at my disposal the slides of medical illustrations collected by the late Loren McKinney.

Others whose aid and counsel should be recorded are: Marie-Thérèse d'Alverny, Guy Beaujouan, Joan Cadden, Marshall Clagett, Edward Grant, Bert Hansen, David Hughes, Christina Kasica, David Lindbergh, William Newman, Charles Schmitt, Edith Scylla, Dirk Struik, and Stephen Victor.

Reasons of time and space have prevented me from taking into account all of the suggestions made by all of these colleagues, but those I have been able to incorporate have surely made the end product the better for them.

Finally, I must express my gratitude for the many hours Ruth Bartholomew and Valerie Lester consumed in the typing of the text of this volume as well as for the effort, advice, and above all patience given by my editors at Scribners: James Maurer, H. Abigail Bok, Janet Hornberger, Marshall De Bruhl, and also, for the interest he has expressed, Charles Scribner, Jr.

Foreword

THIS IS THE THIRD volume in the series to appear, although it is the first in chronological order. *The Album of Science* has been conceived as a pictorial record of the scientific enterprise, an attempt to show in images what science was like in the distant and recent past and to convey a sense of the perception of science by men and women living in different ages, both scientists and non-scientists. It covers the sciences from Antiquity through the Middle Ages and early Renaissance, that is, the whole panorama of development up to the time of the Scientific Revolution. It thus completes the early part of the sequence, since the two previously published volumes present the subject "From Leonardo to Lavoisier" (in a volume by I. B. Cohen) and the panorama of "The Nineteenth Century" (by L. Pearce Williams). Two other volumes currently in preparation will be concerned with the physical sciences and the biological sciences in the twentieth century.

Like the other volumes in the series, this one does not attempt to be a record of great discoveries. To have done so would have been inconsistent with the general aim of the series, and it would also have been an impossibility—since much of the scientific activity in Antiquity and in the Middle Ages did not consist of making great discoveries. Furthermore, this volume does not even attempt to show how science was done in all the different periods, since the lack of adequate pictorial material would have rendered this impossible.

John Murdoch, the editor of this volume, has had a particularly difficult task for two reasons. First, there is the above-mentioned lack of the wealth and variety of pictorial materials and objects available for the later periods. Furthermore, there is the additional problem that so much of the scientific activity of the Middle Ages consisted of innovative alterations, emendations, and extensions of older ideas rather than the charting of revolutionary new paths such as occurred in the seventeenth century and afterwards, with the establishment of "modern" science as we know it. In addition, there have not been preserved the large numbers of scientific instruments that exist for the later periods, although there are a number of drawings which enable us to know their construction, their mode of use, and even their size.

In many ways, therefore, John Murdoch's volume may very well be the most original in the series, since he has had not only to find illustrative materials in obscure collections in libraries from all over the world, but has had the additional task of trying to reconstruct the way in which scientific knowledge was transmitted and to do so by means of graphics, a task that has never before been attempted.

One of the most impressive results of John Murdoch's book is the way he has pictorially documented the variety and scope of interest in scientific matters over several millennia. In particular, he has shown some of the perils that were inherent in the days before printing, when every hand-written copy of a manuscript text had to have the illustrations added separately by an artist. Fortunately for the reader, he has included the ways in which knowledge was organized for the purposes of teaching and memorization and has indicated the ways in which mathematics was developed for its own sake as well as for its use in the other sciences. Many readers will be agreeably surprised to see the materials that he has assembled and interpreted on the practice of science—the workshops and institutions devoted to science and the way in which science developed under the patronage that made the enterprise financially possible.

There are also materials concerning planetary theories, the development of alchemy, and the theory of medicine. Finally, there is a presentation of ideas concerning the structure and organization of the universe and its component parts.

Because so much of this material is strange, even to historians of science who have worked in the more recent periods, John Murdoch's volume differs from the others in this series in that the explanatory texts are fuller. Charting new ways in the discipline of history of science, in which accounts of scientific thought and activity in Antiquity and the Middle Ages have been confined to texts and words rather than images, he has produced a work which will, without doubt, be an eye-opener for even specialists in early science, to say nothing of those who cultivate the history of science at large. Above all, John Murdoch has kept in mind the importance of having his subject matter be available for a larger reading public than the specialists for whom carefully edited literary texts are the normal fare. Indeed, this mode of combining explanations and images may be the best way of introducing this subject to the general reader. Here is an important visual demonstration of the importance of the scientific enterprise as a part of the main history of intellectual culture, going back to the earliest records of civilization.

I. Bernard Cohen

Introduction

THIS VOLUME IS INTENDED primarily as an anthology of the kinds of pictorial and diagrammatic materials to be found in the scientific literature of antiquity and the Middle Ages. It attempts to address itself, as it were, to the question of what types of "visual aids" accompanied the written works of early science. Furthermore, the emphasis will fall upon the *kinds* of pictures and diagrams themselves and not upon the scientific entities or doctrines to which they relate. It will be, for example, far more important to illustrate different *types* of geometrical diagrams, even if they be of the same theorem or problem, than to attempt to include diagrams relating to as many different theorems or doctrines as possible.

None of the *Albums of Science* has been intended as an illustrated *history* of science, and in the present volume the absence of this intention is even more apparent. For not only is it true that this volume has not been designed as an illustrated history of science, but it could not have been so designed, at least if one were to provide an illustrated history which in any adequate way represented the substance and development of science in the period in question.

That it is impossible to compile such an illustrated history of ancient and medieval science can easily be seen if one realizes in the most general way just what the science of that period was. One summary manner of calculating its content is to begin with antiquity and realize that Greek mathematics plus the works of three men—Aristotle, Ptolemy, and Galen—provide one with the lion's share of what counts. Add to this the medieval Arabic and Latin commentaries on, and developments of, these four Greek elements, and one obtains a similar nucleus of the scientific enterprise during the whole Middle Ages. However, apart from the diagrams populating mathematical treatises and the works of Ptolemy and his successors (many of which will find their place in the present volume), the illustrations are notoriously few in the manuscript copies of Aristotle and Galen and of their medieval translations and seemingly endless commentaries. In attempting to compile an illustrated history of science for this earlier period, one would, then, have to leave out a very major portion of that science. Greek and medieval science without proper attention to Aristotelian natural philosophy and Galenic medicine—both of which dominated the Western tradition for almost two millenia—would be a misrepresentation of the grossest sort. Thus the present volume is not, nor could it be, an illustrated history of ancient and medieval science. Nevertheless, the explanations given of the illustrations it contains will of necessity reveal fragments of that history. Although without proper historical coverage or balance, these fragments will say at least something of the kinds of concerns that made the enterprise of science in antiquity and the Middle Ages what it was.

The pictorial and diagrammatic materials selected for this *Album* can be conveniently divided into distinct types, a division that is at least partly reflected in the distribution of the chapters themselves. To begin with, there is the "illustrative material" that is necessary to the understanding of the science in question—diagrams and figures accompanying mathematical and astronomical texts being the most notable cases of such material. Although it is perhaps possible that one might understand a rather complicated geometrical proof without the assistance of appropriate figures, proper comprehension of the proof is greatly facilitated by the presence of figures and in some instances would be virtually impossible without them.

A quite different kind of "required illustration" can be found in medieval herbals and in works dealing with astronomical constellations. These two types of works often functioned as "field guides" enabling one to identify the scientific object in question. Although, again like a modern field guide, some text was also often helpful in effecting the identification, the accompanying illustration was of even greater assistance. The pictorial material in herbals and constellation treatises differed from the required diagrams in mathematical works, since in the former observable objects were being represented. Yet this should naturally not be taken to mean that these objects necessarily had been observed by their illustrators; clearly, in most instances they were not. Instead illustrations were copied from manuscript to manuscript. When no copy was available from which an illustrator could derive his own representation, in more instances than not, he worked, not from any observed object, but from a *text* which provided him with the information he needed. To be sure, there were cases—such as the depiction of an instrument or of the locus of some scientific endeavor—where familiarity through observation may well have been involved in the resultant illustration (provided the depiction was not simply taken from some earlier manuscript); but even then it seems clear that "textual guidance" frequently played a role in making the illustration what it was.

A third kind of illustration was even more definitively shaped by the text being illustrated: the representation of a conception, doctrine, or theory set forth or argued for. At times the text itself may have called for an illustration with a phrase such as *sicut haec figura docet* ("just as this figure shows"), whereupon the required picture or diagram followed (unless the scribe merely left a space for the figure and it was never filled). In other instances, the reader provided his own illustration, feeling that some kind of pictorial material was necessary to explain the theory given by the text. Yet in most cases such figures or schemata became part of tradition and simply passed from manuscript to manuscript, whether announced by the text itself or created by a later scribe or reader.

It is the occurrence of these different kinds of illustrations (required mathematical ones, those representing objects, and those representing theories), together with the different functions that they served relative to the scientific texts to which they were attached, that has provided the framework upon which chapter divisions in the present volume have been based. Other considerations with respect to type and function of illustrations have also been taken into account in establishing these divisions. Distinctions based upon differences between the scientific disciplines themselves have been kept to an appropriate minimum. For it will be far more instructive to see, side by side, the same kind of illustration applied to material from two or more disciplines than to have such illustrations separated by a "disciplinary gap" which, from the point of view of the history of scientific illustration, would be an adventitious one. Thus, I have gone beyond the categorization of pictorial materials into the kinds already mentioned and have included a section treating of distinct illustration types at an even more specific level: *rotae* or circular schemata of almost any sort, dichotomies and "tree diagrams," squares of opposition, and mnemotechnic and symbolic material.

In all of the categories of pictorial material, but especially in the case of the last mentioned more specific types, illustrations have been selected from different areas of learning, particularly from non-scientific ones. This seemed appropriate since many of the types of illustrations in scientific works often found even more extensive application in other kinds of works which, like scientific ones, had a didactic intent. If, as in the present volume, an attempt is made to provide a history—however partial—of the use of illustrations in works of learning, then examples of these other applications will help to fill out this history. It is with this in view that several instances of the occurrence of *rotae* or wheel-diagrams, for example, have been chosen from theological and religious works as well as from scientific and philosophical ones.

Similarly, some idea of the variation suffered by a single kind of illustration can also be provided by contrasting instances of the more artistic and careful rendering of a professional scribe or illustrator with the efforts of a non-professional, in some cases, a student. Although only a few such instances have been included in this volume, it is hoped that the resultant repetition will again prove informative with respect to the history of the utilization of illustrative material in scientific works.

Another, even more important, kind of repetition will be found in this volume: examples of different attempts to illustrate the very same object, theory, idea, or text. Although these examples will often illustrate stylistic variations that occurred over time, they will also draw attention to the fact that in some instances an illustration was not produced simply by copying it from another manuscript but by a scholar trying to reproduce his own understanding of the theory or text that was being illustrated. At times there is good evidence in the variations found in a set of illustrations that the illustrators' understanding of the text or idea on which they were based must also have varied. What is more, even when a given illustration seems to have been copied from manuscript to manuscript, in some instances notable differences can be seen in the sequence of reproductions. Sometimes such differences were due simply to differences in style; but at other times they seem to have arisen from deliberate attempts to alter that being copied. Again, an attempt has been made to include in the illustrations selected for this volume at least samples of all of these kinds of variations.

A final note about the anthology of material constituting the present volume has to do with the rather obvious imbalance in favor of the medieval Latin West. This was inevitable. For if one sets aside the relatively few instances of illustrative material to be found in Egyptian and Mesopotamian sources and focuses on Greek antiquity, then the number of illustrations and diagrams—particularly the number of different *kinds* of such pictorial material—is extremely small. More than that, Greek manuscripts of ancient works are themselves medieval products (the earliest Greek manuscript of Euclid, for example, is of the ninth century). Thus any diagram or figure in such a manuscript is not an ancient Greek production, but a Byzantine one, often a thousand or more years removed from the time of the work being illustrated. Secondly, although Arabic manuscripts do not have the same problem of remoteness as those in Greek, the frequency with which they are accompanied by illustrative material (especially if one does not consider mathematical diagrams) is no greater. This is due in part to the fact that the illumination of Arabic manuscripts, discouraged as it was by Islam, was not the outlet for artistic enterprise that it was in the Latin West (although compensation is had in the care and elegance of much Arabic calligraphy). This means that medieval Latin manuscripts of scientific texts provide a much richer source from which to draw illustrations. Even if special effort is made to have something Greek or Arabic when it is appropriate and available, the outcome is bound to be a spectrum of illustrations that is decidedly Latin, especially if one hopes to remain faithful to the criteria of selection that have been outlined above.

Other limitations and asymmetries in the material in this volume are deliberate. Little attention has been paid to illustrations in works belonging to the early history of technology. Finally, there is an intentional prejudice in favor of illustrations in works describing and explaining the use of scientific instruments and against photographs of extant instruments themselves. Although naturally a matter of choice, the exercise of this prejudice has seemed to be a corollary to the attempt to have the central concern of this volume be that of the occurrence of pictorial and diagrammatic material in ancient and medieval *written* scientific works, the only sources available from which we can learn of the role of visual facilitation in early scientific activity.

Part One

SCIENCE AS LIVRESQUE

1. A miniaturist and his apprentice. A medieval illuminator, Hildebertus, suffers an unwanted interruption of his work. A twelfth-century Bohemian manuscript of St. Augustine's *City of God* has preserved a humorous episode in the trials of this profession. Seated in front of a desk, whose stand appears to be a carving of a very medieval-looking lion, Hildebertus is about to throw his sponge at a mouse that has unceremoniously already tumbled a chicken and a bowl from "Hildebertus' table," and is in the process of devouring the remainder of the illuminator's meal, perhaps bread or cheese. In addition to the implements of his trade appropriately fixed on Hildebertus' desk, there is an open book which reads "Wretched mouse, you enrage me often enough to deserve God's damnation." Below, unperturbed, Hildebertus' apprentice Everwinus patiently designs a rinceau.

THE CONTEMPORARY NOTION THAT observation and experiment are necessary ingredients of a discipline called science which is carried out in laboratories or other specialized, professional locales is not a conception properly applicable to antiquity and the Middle Ages. Then, science was *livresque.* It was not just set down in books; it was largely carried out in books.

It is true, of course, that appeal to experience and observation did occur in the early stages of astronomy, medicine, and natural history; and the results of these observations have been preserved in a great number of astronomical tables, medical case histories, and works such as Aristotle's *Historia animalium.* But once such results and the theories that were derived from them had become ensconced in the written word, there was little return to the empirical—in many cases, none at all.

Ancient and, especially, medieval science was an intellectual activity that concerned itself almost totally with the analysis of written texts or, in the case of university lectures or disputations, of what would become written texts. It was an activity that only infrequently required paying attention to nature itself. In fact, in some segments of later medieval science, one has something akin to a "natural philosophy without nature" or a kind of analysis that concerned itself not with situations that might obtain in this world, but with those that could conceivably obtain in any possible world. Yet even in those ancient and medieval works where nature did occur as the proper referent, it was largely nature as spoken of or written about, not nature as then and there observed. And that, of course, was something for which books alone would suffice.

From another perspective, one might say that coming to know about the spread of scientific ideas in antiquity and the Middle Ages requires that we know about the dissemination of books. Unfortunately, our sources for this kind of knowledge are meager. We may know, for example, that twelfth-century travelers to Spain returned to France or England armed with veritable libraries, but we are mostly ignorant of what books they had collected. In other instances we may know that a particular manuscript was at an earlier date in one locale and at a later date in another, but we know next to nothing about what may have happened to the manuscript in moving from the one place to the other.

Still, if we are uninformed about such details regarding the dissemination of books, we do know that such dissemination occurred, and often that it did so in a remarkably short period of time. An idea "published" at Paris, for example, might be present, and even under discussion, at Oxford barely six weeks later. Similarly, if interest in a given work or author was exceptionally keen, multiple copies could be, and were, produced and distributed in relatively short order. Thus, to his great dismay and disappointment, St. Bernard of Clairvaux complained that the books of his bitter enemy, Peter Abelard, were flying about (*volant*) everywhere, traveling from nation to nation, creating darkness instead of light with their virulent pages.

As the medieval university came to be the focal point of learning, the dissemination of books became even more efficient. Students needed them, and at times were even required to bring to class the text being "expounded," so that they could better follow what was being said, and even gloss their copies with the substance of the comments to which they were privy. The number of booksellers or stationers, and even of book producers, increased markedly. Although they were often severely regulated to guard against their saddling students with an inferior product, they also were sometimes freed of taxes precisely because their wares were so sorely needed. When Roger Bacon complained about such *stationarii* because they foisted corrupt copies of required books upon their clients, the vehemence of his complaint was, to an extent, a function of his recognition that these books were necessary for science and for learning in general. Some hundred years later, Chaucer was an accurate commentator on the scene: "And out of olde bokes, in good feith, cometh al this newe science that men lere."

CHAPTER ONE

The Production of Books and Texts

THE BOOKS OF SCIENCE in antiquity and the Middle Ages were of various forms and materials. With the exception of those produced in the few fifteenth-century years belonging to this very long period of time, all of them were handwritten. The scientific texts belonging to Babylonian astronomy and mathematics were "books of clay," the tablets constituting them being produced by using a stylus to impress wedge-shaped signs into clay that was later dried and baked. The earliest Egyptian and Greek texts were written on papyrus, a material derived from the stems of marsh plants. Although some Egyptian texts were inscribed on other substances—on leather, for example—papyrus was clearly the standard material. Since papyrus is rather fragile and does not easily submit to folding, the form of these early books was that of scrolls or *rotuli*.

At a later date, parchment made from the skins of animals (such as calves or sheep) replaced papyrus as the standard material for writing and the form of books became that of the codex, flat leaves of parchment fastened together at their edges like the pages of a modern book. At a still later date paper replaced parchment as a more easily available and economical substance on which to write, although parchment continued in use even into the earlier years of printing. As the most cursory examination of the following pages will reveal, almost all of the figures have been taken from codices, a factor that accurately reflects the fact that far and away the greatest share of the documents of ancient and medieval science have come down to use in this particular "book form," especially if one is considering pictorial or diagrammatic material.

Yet whatever form or substance a book may have had, its production during antiquity and the Middle Ages was a laborious and time-consuming task. Professional scribes who could carry out this task were an absolute necessity. Although one could copy a needed book oneself, it seems that such an alternative was not adopted with any frequency before the later Middle Ages.

Our knowledge of precisely how scribes worked and how books were produced in general is far fuller for the Latin Middle Ages than it is for earlier periods. In some instances we know that an author himself produced at least a first draft of a work, and in others we know that he dictated to a scribe. But the predominant way of producing a book was through copying. This was, at best, an extraordinarily slow process, even when professional scribes were at one's disposal.

When the rise of the medieval university created a need for a greater number of books, which had to be produced more rapidly and at less cost, another manner of producing copies was developed. Although richer students might be able to hire others to copy needed books (indeed, in some instances we know that unusually well-to-do students even hired help to carry their weighty volumes to and from class), those who were not so financially endowed had to copy what they needed themselves.

Stationers went into the business of providing such students with the primary materials to be copied, and in order to maximize the number of copies that could be simultaneously produced of a given work, they rented out gatherings of four sheets. The student could borrow these gatherings, called *peciae,* one by one, until a copy of the whole book had been made. If he were lucky, he might even be able to rent the relevant *peciae* in their proper order and not have to leave "open space" in his copybook while awaiting the arrival of a *pecia* which was "out to another student."

Since both parchment and paper were costly commodities and, especially, since one wished to be as expeditious as possible in copying books, every effort was made to put as much text as possible on a page and to do so in the shortest possible time. One method of accomplishing this was to write in an abbreviated form, a kind of medieval shorthand. Extremely rare in Arabic manuscripts, such abbreviation was present in Greek codices, but its greatest frequency is in the Latin manuscripts of the later Middle Ages. One could, for example, write ꝑb° for 'probatio', ꝯcūqȝ for 'cuivscumque' or ꝯ° for 'conclusio'. Since abbreviations varied from one period to another and from locale to locale, at times a fair amount of context would be needed to determine just which word was meant.

In spite of such disadvantages, abbreviated script did facilitate the production of needed books. Nevertheless, however much facility this or other techniques of saving time or effort may have introduced, the copying of manuscripts remained a tedious affair. Rather obvious testimony to this fact can be found in the "laments" which scribes (professional or not, no matter) occasionally appended to the results of their efforts. Most frequent are the expressions of thanks to the Almighty that the wearisome task is finished, but we also find more colorful scribal comments, often in verse, at the ends of texts: "At this pay, never again" (*pro tali precio, nunquam plus scribere volo*); "The book finished, this scribe is going to dance" (*libro completo, saltat scriptor pede leto*). Often the scribe suggested that he be rewarded for his accomplishment: perhaps with a fat goose or a pretty girl, but most frequently with drink: "Let wine be given to the better scribe," was one wish, "so that he who has written may continue to write and drink of good vintage" (*detur scriptori, vinum meliori; qui scripsit scribat, et bona vina bibat*).

For all the toil the copying of books entailed, the end products were the real rewards. Indeed, long before the appreciation of books as being absolutely essential to university learning, we are witness to someone like Alcuin of York (eighth century) telling his peers that it is far better to write books than to plant one's vineyard, since the latter would only serve one's stomach, whereas the former would serve one's soul.

2. The writing of papyri. A Fourth Dynasty relief (2680–2565 B.C.) on a mastaba at Giza depicts four Egyptian scribes properly implemented with rush pens and palettes. Spare pens are at the ready behind each scribe's ear. One might imagine that the scribes were in the course of producing a document such as the mathematical Rhind Papyrus (see Illustration 106). The status of scribes in ancient Egypt was one of considerable importance. They were part of the educated populace, a factor that carried great esteem among a largely illiterate people. "Decide to become a scribe," we are told, "it is a fine profession suited to you. You will summon one, and thousands will answer your call. You will be free to come and go, and not be yoked like an ox which is bartered. You will take precedence over others."

3. (Right) Authors and their entourage. An Iraqi manuscript of the thirteenth century depicts two authors, possibly as they dictate to their scribe. The codex contains the text of the *Epistles* (*Rasāʾil*) of the Ikhwān al-Ṣafāʾ—the so-called Brethren of Purity—and the authors of this work are correctly identified in the inscription at the top of the miniature. The *Epistles* was an important tenth-century encyclopedic work covering mathematics, dialectic, natural science, philosophy, mysticism, astrology, and magic. Only two of the five known authors of the *Epistles* appear in this miniature (the other three are depicted in another miniature of the same manuscript). Students or other auditors appear in the balcony above, while servants of one sort or another, in proper Arabic fashion appropriately pictured in a smaller size and with coarser features befitting their lower status, are to the left and right of the main group of three.

نقل من تتمة صوان الحكمة لظهير الدين أبي القسم البيهقي ان خمسة من الحكماء اجتمعوا
وصنفوا رسايل اخوان الصفا وهم ابو سليمن محمد بن مسعر البستي ويعرف بالمقدسي وابو الحسن
علي بن هرون الزنجاني وابو احمد النهرجوري والعوفي وزيد بن رفاعة

4. The monk as scribe. This thirteenth-century English manuscript, perhaps written at St. Albans, contains fine miniatures done by the school of Matthew of Paris (*fl. ca.* 1236–1249), who was himself a scribe, illuminator, goldsmith, and historian of St. Albans. In this miniature a monk, seated at his desk, performs his required scribal duties. Presumably some kind of lamp hangs over his desk, on which the codex he is preparing is appropriately stationed. He holds the two most important implements of his trade: in his right hand, a quill pen—which began to replace the reed pen about the sixth century—and, in his left, a knife, a tool he could put to the triple purposes of sharpening his pen, smoothing out rough places on his parchment, and making erasures. Two other necessities do not seem to be present: his inkpot and the manuscript from which he was copying. We know a fair amount about *scriptoria,* or the writing rooms in which manuscripts were produced. An interesting morsel of information concerns the location of the *scriptorium* in a monastery: near the kitchen, the better to prevent frozen fingers that might not properly function in their assigned task. Another picture from this manuscript is given in Illustration 206.

5. Medieval dictation. Most medieval books were produced by being reproduced; they were, that is, copied from previously existing manuscripts of the same work. But what of the "original"? It could have been written by the author himself—and we do possess autograph copies of a fair number of scientific and philosophical works—or it could have been dictated. This illustration, from a twelfth-century manuscript written at the Abbey of Saint-Martin in Metz, shows such an activity in progress: the historian and hagiographer Sulpicius Severus (fourth–fifth century) dictating (perhaps his *Life of St. Martin*) to Abbot Richer. Although this miniature is not from a scientific or philosophical work, we do know that dictation was frequently employed in those disciplines as well. Some accounts of dictation in these areas have come down to us, most probably because of the unusual prowess of the dictator in question. Al-Tawḥīdī tells us, for example, that the tenth-century Nestorian Christian philosopher and logician Abū Bishr Mattā ibn Yūnus (sometime teacher of al-Fārābī) would dictate, often while drunk, to his students, and would charge a fee for such a service. After all, they were being taught! A bit less colorful is what we know of St. Thomas Aquinas and his secretaries. His fourteenth-century biographer Bernard Gui relates that he would dictate to three, or even four, secretaries at the same time, and on different subjects. He even reports that one of his secretaries claimed Thomas would sometimes fall asleep while dictating, but would go on dictating, the secretaries continuing to write none the less.

CHAPTER TWO

Resultant Books and the Formats of Texts

PART OF THE VISUAL facilitation of learning was by the format or "layout" of the relevant learned texts themselves. Many scientific texts were not simply read and absorbed, but were returned to time and again for information, so it was quite important that ways be developed to make such repeated reference relatively simple. Divisions of the text could be "highlighted," perhaps by the insertion of large initials or by writing the first few words of each new section in a decidedly larger hand or in ink of a different color. Running heads might be inscribed at the top of each page to announce just what *quaestio, distinctio,* or *articulus* was under discussion. Rubricators might introduce paragraph signs as "flags" within the text, or place locator numbers or similar indicators in the margins. Often the readers created their own marginal "signposts" to mark notable divisions or points in a text. All of these provided visual clues enabling one to refer with greater ease to a given text and to determine its structure.

If the text in question was a commentary, it was important to have the comment set off in an easily identifiable way from that being commented upon. In some instances the text occupied the center of the page, while the commentary filled all available marginal space. More frequently the text was divided into small, manageable sections, the relevant commentary then directly following each such section. As a further aid to ready reference, one could number each of these sections, thereby providing later authors with a simpler way to refer to a given passage in a text or commentary than by quoting its first few words.

6. Aristotle in triplicate. This unique manuscript of the thirteenth century presents a rarity: three different Latin translations, elegantly written in parallel columns, of a single work: Aristotle's *Physics.* One of the translations was made by James of Venice (that on the left) from a Greek text of Aristotle; the other two were made from the Arabic, one (on the right) by the indefatigable translator Gerard of Cremona, the other (which was the translation accompanying the commentary of Averroës) possibly by Michael Scot. The page represented here is the beginning of Book II of the *Physics,* as shown by 'capitulum I' in the right margin. But the medievals divided their Aristotle not only into book and chapter, but also into separate passages or *textus,* each bearing its own number. This allowed quick and ready reference to "the Philosopher," as Aristotle was known. Here, in the upper right margin *textus* 83 (lxxxiii), constituting the final passage of Book I, is indicated.

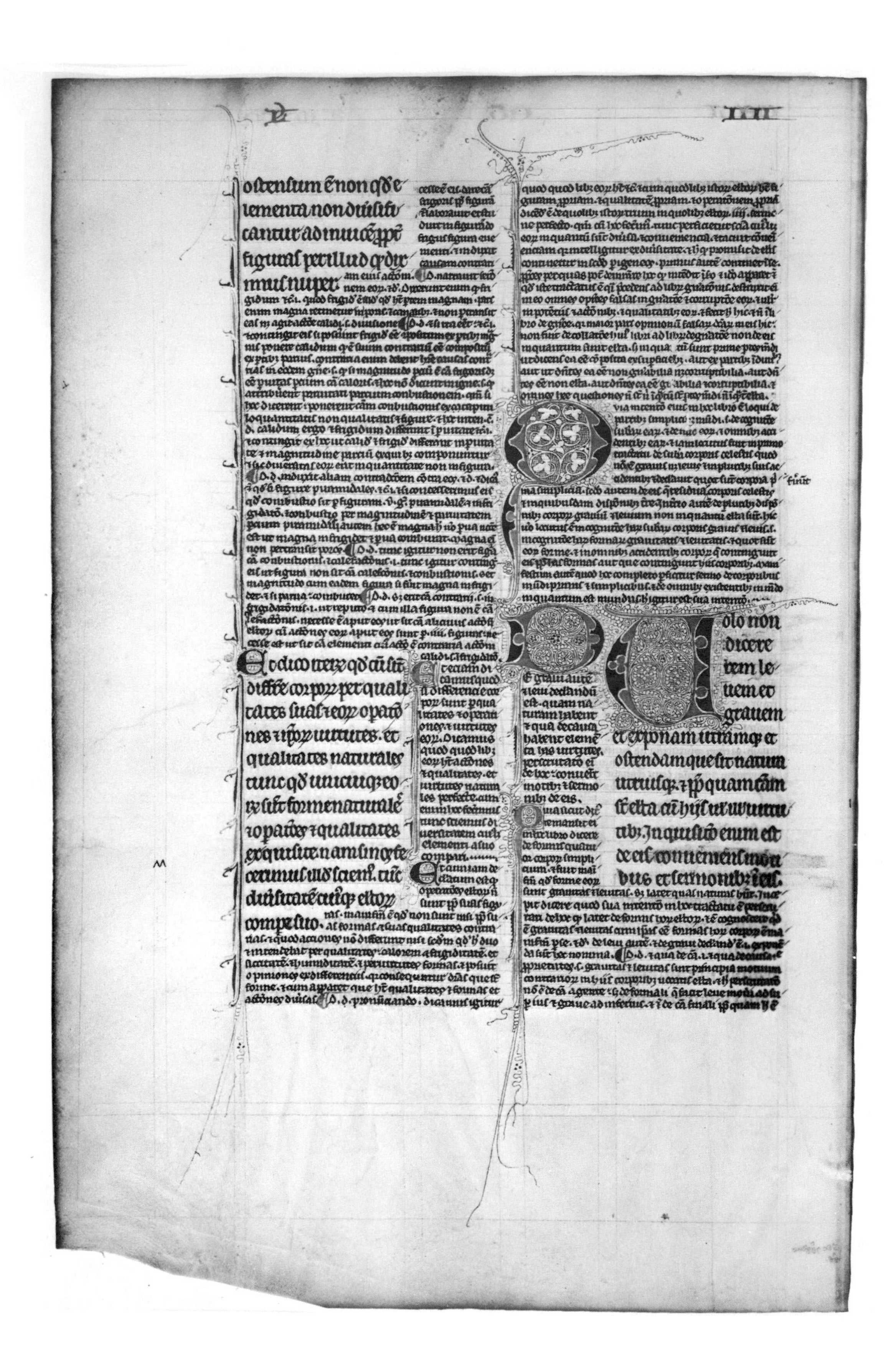

7. (Left) The commentary format. Again a text of Aristotle, his *De caelo,* but this time with the commentary of the twelfth-century Arab philosopher Averroës (Ibn Rushd). The text itself is set off from the commentary by larger writing, a practice that was followed in many medieval commentaries. As was often the case in medieval copies of Averroës' commentaries, this thirteenth-century manuscript presents two translations of Aristotle's text: one, made from the Greek, appears in the largest hand; the other, made from the Arabic that was used by Averroës, next to that from the Greek, here appears in a hand slightly larger than that reserved for Averroës' comments. The medievals themselves almost always referred to the translation made from the Greek, even when they were simultaneously utilizing the commentary of Averroës, who naturally quoted a different text of Aristotle in his comments. But this presented little trouble for Latin medieval scholars. Averroës was constantly utilized by these scholars, so much so that he was labeled "Commentator," just as Aristotle was "Philosophus." The illustration is of the end of Book III and the beginning of Book IV of the *De caelo.*

8. A well-used text. Perhaps the most important work in all of medieval medicine was the *Kitāb al-Qānūn* of the eleventh-century philosopher-physician Avicenna (Ibn Sīnā). (Translated as the *Liber canonis,* it became equally significant in the Latin West.) It seems that the first complete commentary on this magnum opus of Avicenna was the *Sharḥ al-Qānūn,* written by the thirteenth-century physician Ibn al-Nafīs (much mentioned in histories of science for his realization of the pulmonary circulation of the blood). In turn, an epitome of that commentary was written, the *Al-Mūjiz al-Qānūn,* and that is the work whose opening page is depicted. (There were commentaries on the *Al-Mūjiz,* but they need not concern us.) After a brief proem the larger writing announces the beginning of the first chapter. There is also, however, an interlinear Persian translation. All of this then is completely surrounded (in a manner and to an extent that Arabic manuscripts often seem to present in the extreme) by marginal annotations. Clearly this was a work that was read and studied (at least through its opening pages). The original format of the text—initially an elegant and easily read *naskh* script of twelve lines per page—has become quite obscured.

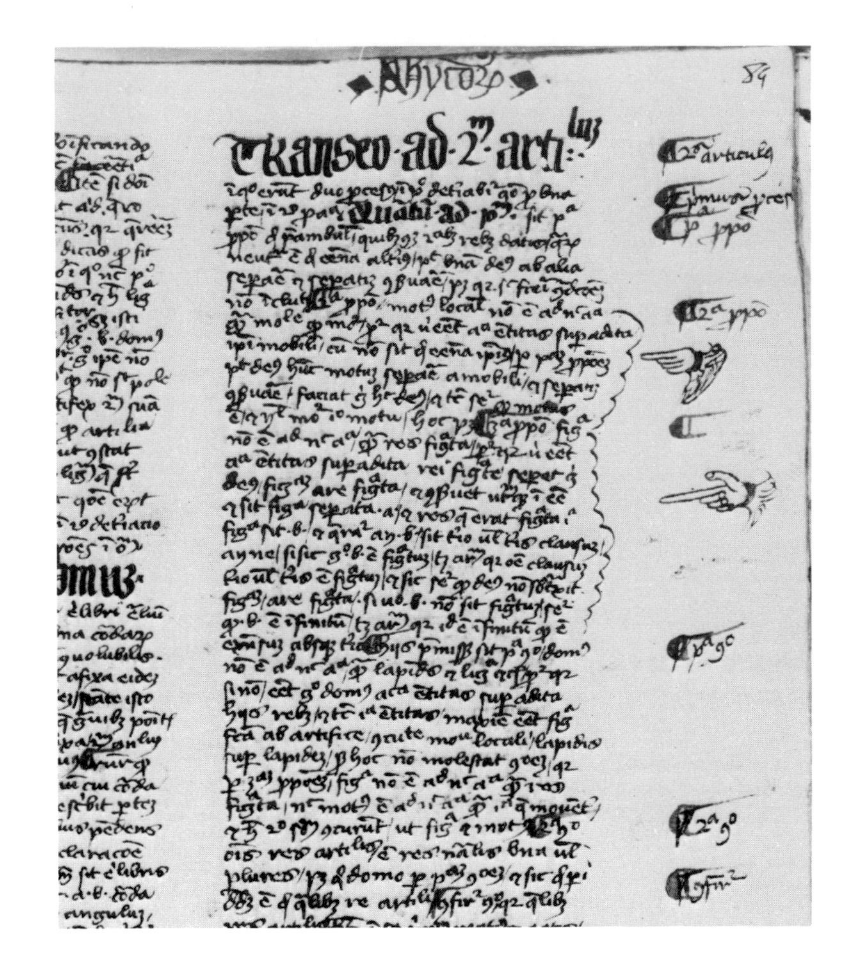

9. Finding one's way around a medieval text. Medieval scientific and philosophical works often were extraordinarily prolix and complicated, and without the help of an index—although some texts did come to carry indices—locating just what was where could frequently prove to be a problematic and time-consuming task. At times one had the luxury of "text numbers" such as those in Illustration 6. Another kind of "locator" was provided by the marginal articulation illustrated here. The present manuscript, written in 1397, contains the *Exposito* and *Quaestiones* on Aristotle's *Physics* of Blasius of Parma, a physician and natural philosopher who flourished in the latter part of the fourteenth century. The particular *quaestio* of which a segment is shown here is "whether, in constructing a house, a builder makes something that is distinct from natural things," a question that related to Aristotle's discussion in Book II of the *Physics* of the distinction between the artificial and the natural. Blasius has divided his *quaestio* into four "articles," and it is the beginning of the second of these, which contains his own solution or *determinatio* of the question, that is pictured here. The entire article will be divided, Blasius tells us, into two sections (*processus*); the first of these sections begins with three *propositiones* and then goes on to draw a number of *conclusiones*. The marginal articulations on the right point to exactly where each of these elements is to be found. Beginning at the top of the margin, in properly abbreviated Latin, with *secundus articulus,* there are "indicators" for *primus processus, prima propositio, secunda propositio* (the third *propositio* is missing from the empty "paragraph sign") and, next, *prima conclusio, secunda conclusio,* and so on. The second conclusion possesses a confirming proof that is also marginally flagged by 'confirmatur'. The hands with pointing index fingers are intended to draw attention to especially important moves in the argument. Here we seem to be directed to the fact that, if one maintains a distinction between two "things"—between, for example, a moving body and its motion or a figured thing and its figure—then God should be able to cause these things to exist separately (which, incidentally, He cannot do in the present case). This is an instance of the frequent fourteenth-century appeal to God's absolute power: because anything can be or can occur *de potentia Dei absoluta* which does not involve a contradiction, this was only another way of determining what was logically, in addition to physically, possible.

CHAPTER THREE

The Production of Illustrations and Diagrams

OUR KNOWLEDGE OF HOW manuscript illustrations were produced is infinitely more complete in the case of the well-known artistic miniatures of illuminated medieval manuscripts than in that of the scientific illustration or diagram. One of the reasons why this is so is due to the fact that we possess medieval treatises *De arte illuminandi* telling us in detail of some of the techniques involved. To be sure, some scientific illustrations are artistic miniatures themselves (those in herbals, bestiaries, and collections of constellation figures, for example) and there our knowledge of manuscript illumination in general is quite relevant. On the whole, though, scientific illustrations belong more to what we would call drawing than to painting, and in that area our knowledge of medieval procedures leaves much to be desired.

We naturally do know that, for all types of scientific illustration, the text was almost always written first, appropriate spaces being left for the subsequent insertion of illustrations, should they be placed in the text itself and not assigned a marginal location. We also know that at some point in the thirteenth century, illuminators became detached from their hitherto monastic or ecclesiastical places of employ and set up workshops of their own. Presumably a similar change occurred with respect to the illustrators of scientific manuscripts, especially those of greater elegance and artistic merit.

On the other hand, most scientific manuscripts bearing illustrations did not receive the artistic care and skill given (say) an illuminated psalter or book of hours. The scribes, or even readers, of scientific manuscripts often provided the illustrations and diagrams themselves. Some clearly seem to have been executed by a student who presumably felt that a "picture" would be of help in understanding the text at hand. In some cases the earliest copies we have of a given text bear no illustrations at all. Subsequent copies, though, may carry figures in the margins, and still later ones have the same figures incorporated into the text, a procedure that reflects various stages in the dissemination and understanding of the work in question.

In most cases an illustration was copied from its occurrence in an earlier manuscript. This did not necessarily mean that it was a direct copy; the reproduction might instead have taken place through some kind of imitation. Yet whatever the degree of accuracy of reproduction involved in copying earlier illustrations or diagrams, the copying itself was very much part of an accepted tradition, just as it was in the case of reproducing texts in general. Instances of charges of plagiarism were almost nonexistent.

10. The production of illustration and text. Perhaps the most important illustrated scientific work of late antiquity and the Middle Ages was the *Materia medica* of the first-century physician Dioscorides, numerous copies of which were available in Greek, in Arabic, and in Latin—a fact that will be reflected in the number of illustrations in the following pages drawn from some of these copies. The medieval Dioscorides is, however, not only an invaluable source for the history of the illustration of plants; the earliest Greek manuscript we have of the *Materia medica* also includes a number of frontispieces depicting its author and other medical scholars. One of these is represented here in a fifteenth-century Greek descendant of the sixth-century codex. The central goddess figure holds a mandrake, presenting it for inspection to both illustrator and author. Although the goddess is not named in this codex, she is identified as a personification of Epinoia (the power of thought or inventiveness) in the original manuscript from which the present one derives. While Dioscorides is presumably writing a description of the plant, the illustrator, with his paints on a table at his side, is intently constructing his picture "from nature," his attention presently fixed on the mandrake itself and not on his creation. For examples of resulting illustrations of the mandrake see Illustration 194; for illustrations from the sixth-century Greek codex which was the ancestor of that used here, see Illustrations 188, 193, and 202.

11. An illustration from sketch to final version. We know that a variety of kinds of preliminary sketching were employed in preparing miniatures for illuminated manuscripts, but most of this knowledge relates to codices in fields other than that of science. In that area our examples of such preparatory sketches are very few indeed. Hence the present instance of two stages in the creation of a portrait of an important personage in the history of medieval science is something of a rarity. This fourteenth-century manuscript contains the *Liber astrologiae* of Georgius Zothorus Zaparus Fendulus, a fourteenth-century compiler of things astrological drawn from earlier treatises in the field, particularly from the Latin translations of the works of the ninth-century Arabic astrologer Abū Ma^{c}shar. Indeed, it is a portrait of that Arabic scholar that is depicted here, the words 'Albumasar Philosophus' being inscribed on the hand-held book in the finished version of the portrait. The "preliminary sketch" is a rather finished product, so much so that one wonders why such a careful preparatory drawing was needed in order to "process" the finished one some ten pages later in the manuscript. For other illustrations taken from this codex—which is, as a whole, a kind of "picture book" for astrology—see Illustrations 230 and 232.

12. The unfinished illustration. It was standard medieval practice for a scribe to leave spaces for the illustrations or diagrams that were to be inserted later—sometimes much later—by an illuminator or, perhaps, by the scribe himself. In some cases this task was carried through only "halfway," as in the present instance: a fifteenth-century manuscript of the logic, here entitled *Expositio summularum,* of the fourteenth-century Master of Arts, Jean Buridan. The text immediately preceding that shown here has a battery of rules governing the opposition, equipollence, and conversion of modal propositions, and concludes by telling us that everything just said will be crystal-clear from the following diagrams (*ut clare patent in figuris sequentibus*). Two diagrams do follow, but the first is incomplete. Only the "skeleton" is present; but that alone tells us that the diagram was to be a "square of opposition," a type of figure employed extensively in logical texts. We shall have occasion later to examine squares of opposition and will then, so to say, "turn the page" to reveal the second *completed* diagram announced in the present manuscript (see Illustration 61). For yet another diagram from this manuscript, see Illustration 55.

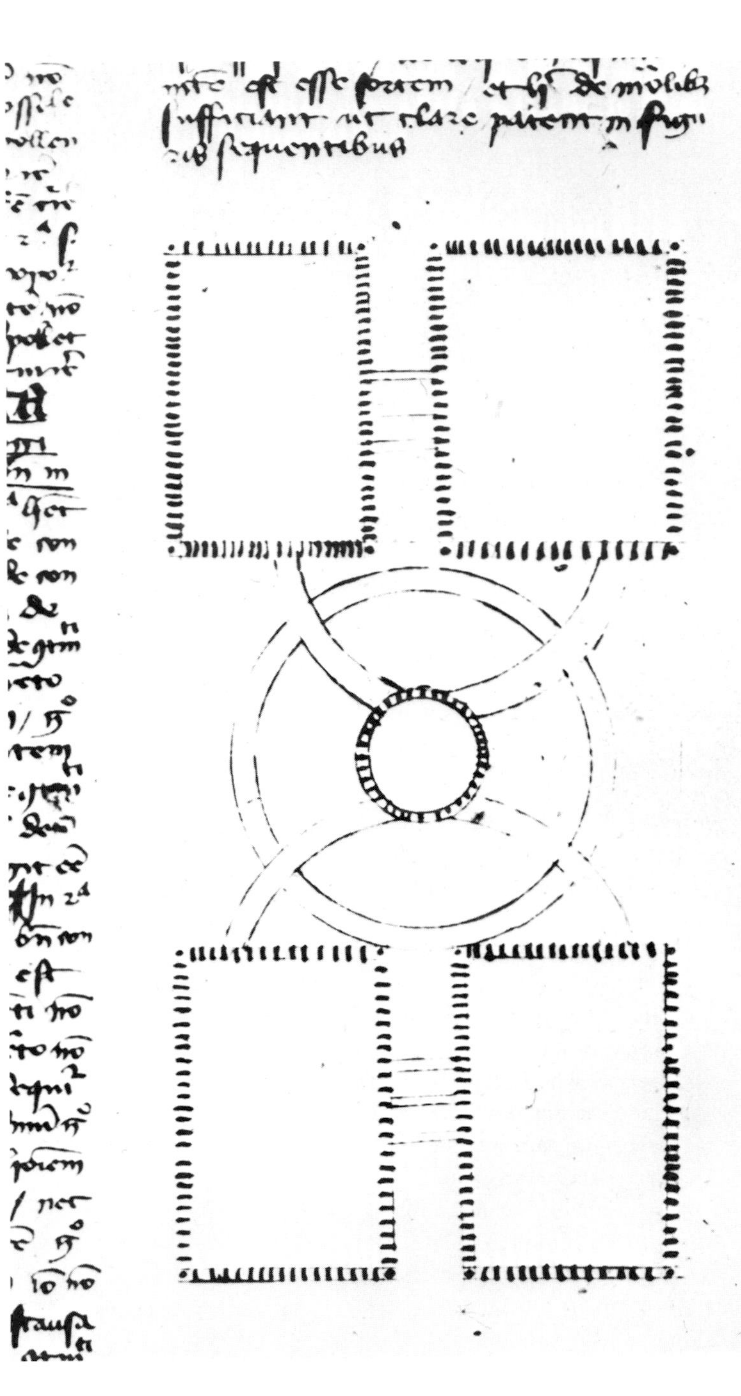

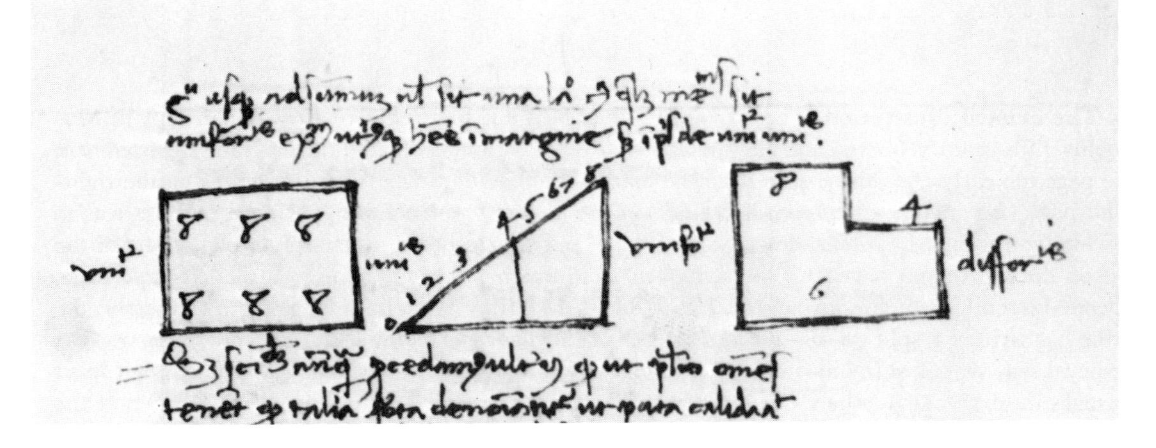

13. (Above) From margin into text. The diagrams in this sixteenth-century manuscript apparently were drawn in the margin of an earlier manuscript, since they are here announced by the words "You have an example of each [of the conceptions in question] in the margin (*exemplum utriusque habes in margine*)." It seems, therefore, that some scribe had at some time brought the diagrams from their marginal home into the body of the text. The present text is a commentary on parts of the *Liber calculationum* of the fourteenth-century scholar Richard Swineshead. The diagrams represent three different ways that a quality of constant or varying intensity may be distributed over a given subject. In the first the quality is uniformly distributed in degree 8 over the whole subject (here labeled *uniformiter uniformis*); in the second diagram (labeled together with the third as *uniformiter difformis*) the quality varies uniformly from degree 0 to degree 8, while in the third the first half of the subject is uniform in degree 8, the second half in degree 4, giving the whole a "denomination," as the diagram indicates, of 6. For other illustrations of the doctrine or theory in question see Illustrations 142–145; and for other illustrations from the present manuscript see Illustrations 143, 144, and 255.

14. (Below) A test diagram. This circular diagram or *rota* was an extremely frequent visitor on the pages of medieval manuscripts and we shall later have occasion to say much more about it and about others like it. For present purposes, however, it is not the diagram itself—which gives, according to the seventh-century Bishop Isidore of Seville, Aristotle's theory of the four elements—but rather the single word 'probatio' in its center that is of interest. Note first of all that this diagram appears in a manuscript of the tenth-eleventh century that contains a text that bears in no way at all upon the diagram. What, then, explains its presence in this codex? The word 'probatio' gives the likely explanation: in some sense the diagram was a "test." Perhaps it was executed as a kind of "test run" for the drawing of the diagram. Or perhaps it was drawn as a "test of the quill pen," a *probatio pennae*. In any event, the presence of the term 'probatio' implies that some kind of trial was involved. It seems clear, though, that the scribe who first drew the diagram did not label his work as such, since it is a later hand that has written the key word, and hence, it appears, a later scholar who saw fit to draw attention to the fact that the diagram was, in some sense, but a "test."

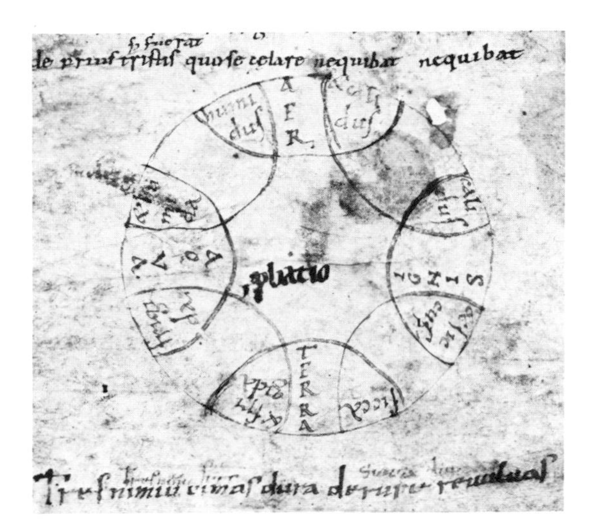

53

linea recta

linea curva

linea tortuosa [illegible]

superficies plana

superficies curva

[illegible] recti

circumferencia

dyameter

centrum

semidiameter

semicirculus

[illegible] equilaterus seu ysopleurus

orthogonius

ambligonius

oxigonius

quadratum equilaterum

tetragonus longus seu quadrangulus

pentagonus

exagonus

eptagonus

octagonus

nonagonus, decagonus [illegible]

due linee [illegible] paralellae

17. (Left) Model geometric figures. Although not a model or pattern book in the same sense as those in Illustrations 16 or 18, this fourteenth-century manuscript contains one page that may very well have functioned as a "model" for the proper identification and drawing of geometrical figures. The figures included are not at all keyed to some specific geometric work—for an example of that, see Illustration 111—they cover, rather, all manner of figures that one might meet in any (plane) geometrical work or in any work applying or speaking of geometry. The standard plane geometrical figures are almost all present and inscribed appropriately with their names. However, in the series of (presumably regular) polygons illustrated at the extreme right, the scribe has stopped short with *octagonum,* but has indicated the succeeding figures he has not drawn with the concluding words: *nonagonum, decagonum, et cetera.* Most of the names are, as one would expect, either Latin or transliterated Greek. An exception is the set of parallelograms (the first a rhombus) and trapezoids at the top of the right-hand column: they bear the standard Arabic names for those figures, if in a somewhat erroneous form: *elmahym* (for *elmuain*), the rhombus; *elmifarifa* (for *elmuharifa*), the trapezoid. Some of the figures are meant not simply to illustrate, but also to show or prove some "geometrical fact." Thus, the third figure from the top on the extreme left is meant to make it clear that the shortest line between two points is a straight one. Similarly, the first two circles in the center are intended not only to show what the center, circumference, diameter, and radius of a circle are, but also to prove that a diameter divides a circle into two equal parts. All the figures deal with plane geometry, although it seems that the sixth figure from the top on the left labeled "superficies curva" is to represent a curved surface in contrast to the plane surface depicted above it, and not simply a plane surface with curved sides. The present page in this manuscript directly follows a series of philosophical questions on geometry and directly precedes a listing of the enunciations of propositions from Books I–III of Euclid.

18. A model book with a built-in possibility for reproduction. This thirteenth-century manuscript is, in terms of subject matter, a bestiary. Secondly, it is, like the manuscript in Illustration 16, a model book in which one could find the proper form for a drawing one might wish to make of some given animal. But this particular manuscript was furnished with an "extra"—perhaps at the time of its composition or perhaps later—that would provide even more. As the picture on the right shows, each "model" was pinpricked so that it could serve as a kind of stencil that would automatically reproduce the form of the animal in question. One could presumably utilize the form provided by the resulting pin holes by some method of pouncing, for example, by pouncing with some substance like charcoal dust wrapped in a small rag. The result would be a dotted outline of an illustration in which one was interested. Although we know of reproducing illustrations or patterns by pouncing from other areas of the history of art—in textile design and painting, for example—this seems to be the earliest known instance of a model book that includes a means of reproducing illustrations. No copy made by such a technique from the present manuscript has yet been uncovered. The illustration reproduced is that of a stork as the words 'De cyconia' on the round "frame" indicate.

19. Tradition in copying illustrations. Although it is unlikely that one of the two manuscripts represented here was copied from the other, they are undoubtedly "relatives," though whether "first cousins" or something more or less distant it is difficult to say. Both manuscripts are of Dutch provenance, the one whose figures constitute the upper row dating from the mid-fourteenth century, that constituting the lower row from 1370–1380. Comparison of their representations of the same three animals clearly shows that they both derive in one way or another from some common ancestor. The stance of the animals shows that illustrations such as these were not designed through an interpretation of the text they were to serve, but copied from earlier illustrations. Just as, therefore, one can speak of a textual tradition and attempt to establish a stemma or "genealogy" reflecting that tradition, so one can often speak of an "illustration tradition" and establish its stemma as well (although, admittedly, modern editors of texts have not, until recently, paid much attention to this latter kind of tradition; they should, for it might well assist them in constructing the relations of the relevant textual tradition itself). The work represented by both manuscripts depicted here is Jacob of Maerlant's *Der naturen bloeme* (*The Flowers of Nature*), a thirteenth-century free translation or paraphrase into Flemish verse of the *De natura rerum* of the Dominican Thomas of Cantimpré, who was also a thirteenth-century Fleming. Jacob had previously written poems and had translated French romances into Flemish, but when he turned his hand to translating the *De natura rerum* of his fellow countryman, he claimed that he was doing so for those "who are annoyed by romances and tired of lies." As a translation, howsoever free, Jacob's source was, of course, the work of Thomas of Cantimpré; but Thomas' encyclopedic work in turn went back to the likes of Aristotle, Pliny, and Solinus (to cite only its more ancient sources). The popularity of Thomas' compilation can also be measured by the fact that it was translated into German in the fourteenth century by Conrad of Megenberg. The three animals whose representations have been selected above from our two manuscripts are, from left to right: the *catapleba* or (in the lower row) *cathapelaba* (*catoblepas* in Pliny), a mythical beast rumored to inhabit the headwaters of the Nile; the *damma,* or twisted-horned antelope; and the *duram,* reputed to be a savage beast capable of great speed, whose deliberately discharged droppings would prevent hunters and their dogs from following its trail.

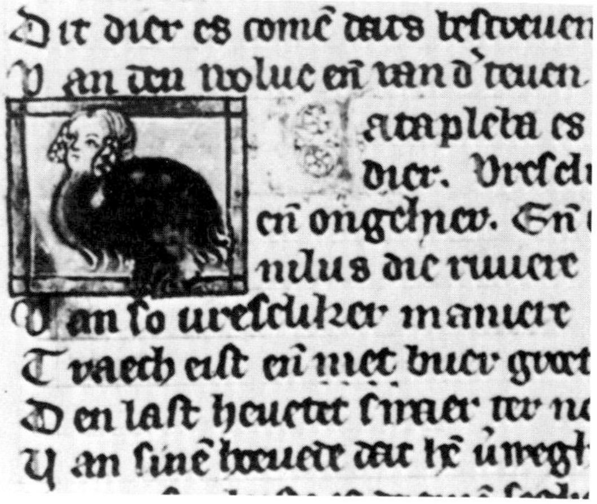

Amma dats [illegible]
dier dat also [illegible]
Es van der groete [illegible]
een gheet.
En staert na sierre groete waert
Et levet sere starc die hoeren
Navuse riectet ende es snel
En can sijn leuen ouden wel
Sine hoeren sijn .ij. voete lanc
Op waert recht es haer ganc

Waer so si hem hene keren
Alle die hen op hem gapen
Bede ridders ende papen
[illegible]uvam sprect[illegible]
Dat een [illegible]
beeste es. Vermaten
staerc ende snel. Als [illegible]
die iaghers sijn te fel.
Ende sijt moete hebben gheiaghet
So dat hem sijns selues mishaghet
Aenstet dus tongaen haer strec

Vie dier es comen dats [illegible]
Van den wolf ende vander [illegible]
[illegible]athapelabe
es een dier
vreeselic ende [illegible]
ghebner. Ende [illegible]
op Iulius die [illegible]
te. Vander vreese[illegible]
liker maniere
[illegible]vaech eist ende met bueie groot
[illegible]en laet heeftet swaer ter noot
Van sinen hoefde dat hem verwegh[illegible]
Van deser beeste es datmen seghet

[illegible]r so si hen hene keren
[illegible] die heren op hem gapen
[illegible]e ridders ende papen.
[illegible]uvam spreect a[illegible]
dat .i. wrede
[illegible]e es. Vermaten
[illegible] ende snel. Als [illegible]
[illegible]aghers sijn te
Ende sijt moete
[illegible]en gheiaghet
dat hem sijns selues wanhaghet
[illegible]nstet dus tongaen haer strec
[illegible]worpet vte sinen drec

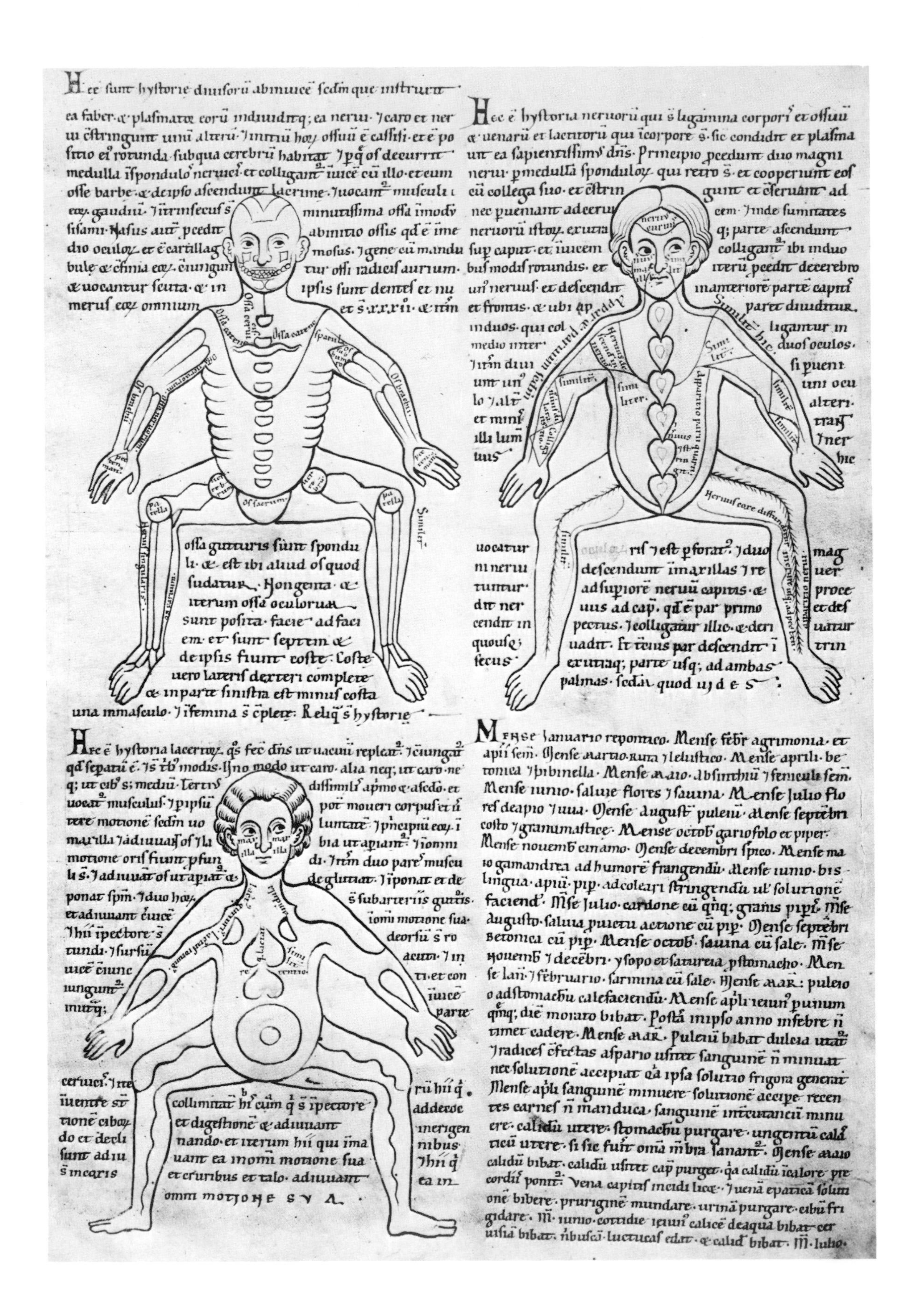

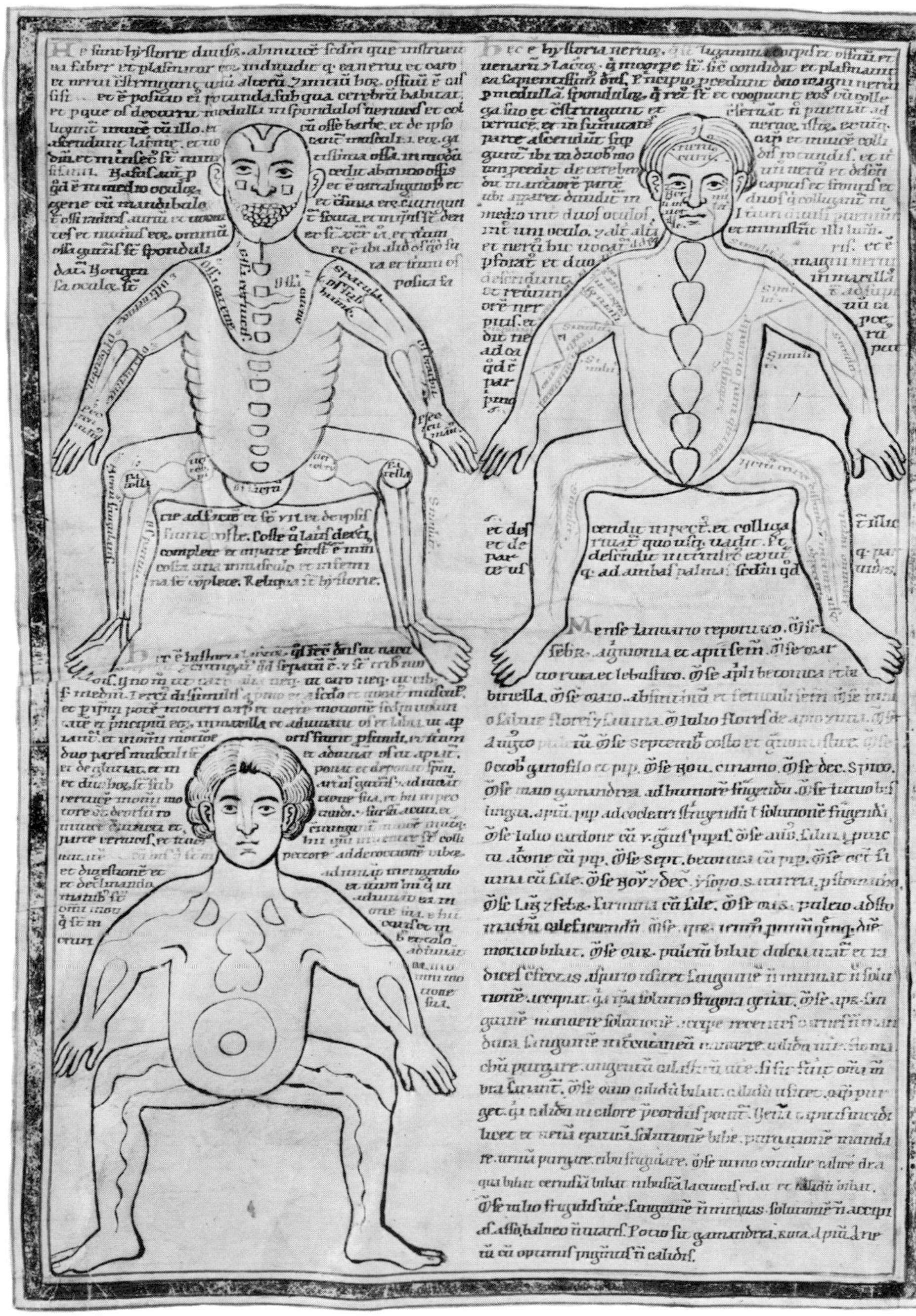

20. Illustrations with immediate offspring. At times the relation between two manuscripts is such that we are certain that one was copied from the other. This is surely the case here. The codex on the left was written in the mid-twelfth century at the monastery of Prüfening in Bavaria; that on the right was copied from it in 1247 by a monk named Conrad at the monastery in Scheyern. There is a page-for-page matching in this section of the two manuscripts, the only significant differences being in the arrangement of the text, the absence in the thirteenth-century copy of the text within the figure at the bottom left, and the fact that the copy contains three more lines of text at the bottom of the right column. The illustrations in question are three of five "anatomy men" who, as a whole, had such an established tradition in the Latin Middle Ages that they have become collectively known among historians as the *Fünfbilderserie*. In the squat position that was so often characteristic of anatomy illustrations (see also Illustration 209), the figure at the top left in both manuscripts is a "bone man"; at the bottom left, a "muscle man"; and at the top right, a "nerve man," the relevant anatomical features and parts being described in the text surrounding the figures and within the figures themselves as well. The two figures of the five in these manuscripts that are not reproduced here are the "artery man" and the "vein man" (although one should note that some codices related to this "series" contain either more or less than five figures). For another illustration from the first manuscript, see Illustration 288.

Part Two

STANDARD SCHEMATA AND TECHNIQUES FOR THE VISUAL FACILITATION OF LEARNING

21. Arbor virtutum. The medieval scholar was all too often faced with problems of classifying things, of drawing distinctions, or of analyzing wholes into their constituent parts. One of the most popular manners of succinctly expressing the results of such classifying and dichotomizing was the construction of *arbores* or trees, their branches graphically representing each stage of the process of conceptual division. Among the *arbores* most frequently found were trees of the virtues, as depicted here, and their sisters, trees of the vices. This fourteenth-century manuscript gives an especially elegant example of an *arbor virtutum.* The root of the tree, and hence of the virtues, is *humilitas,* its trunk the way of life, and its fruit that of the spirit. The tree itself branches into both the four cardinal virtues (prudence, fortitude, justice, and temperance) and, appropriately loftier in the tree, the three theological virtues (faith, hope, and charity). Each of these seven virtues is in turn divided into seven more specific virtues, the whole giving a visual epitome of the basic conceptions within Christian moral theology—even though the more fundamental of these conceptions were taken from Greek antiquity. In addition to their appearance in the tree itself, the four cardinal virtues are personified by the four maidens standing below it. Other pictures from this manuscript can be seen in Illustrations 45, 58, and 276.

AS ONE LEAFS THROUGH ancient and medieval nonmathematical science texts, there appears to be no limit to the kinds of diagrams, illustrations, and visual devices used to explain these texts and render them more easily comprehensible. Among these, several forms and techniques were notably recurrent not only within a given segment of science, but within the literature of most scientific disciplines and in nonscientific writings as well. It is appropriate then to begin our anthology with a sampling of standard forms that are found, for the most part, accompanying the texts of nonmathematical science in general. Diagrams of mathematics and the exact sciences express a need for the visual so different from that of the rest of science that they will be considered separately.

In nonmathematical science the first thing to note is that diagrammatic and pictorial material is very rarely found in Aristotle and Galen, whose works formed the backbone of nonmathematical science in antiquity and the Middle Ages. Most is found in manuscripts of encyclopedic works or "handbooks." Our "basic library" would consist of manuscripts (from almost any century) of the following: Chalcidius' (fourth century) *Commentary on Plato's Timaeus,* Macrobius' (fourth-fifth century) *Commentary on Cicero's Dream of Scipio* (a handbook on cosmography), the *De nuptiis Philologiae et Mercurii* (*On the Marriage of Philology and Mercury*—a handbook on the seven liberal arts) of Martianus Capella (fifth century), the *Institutiones humanarum litterarum* of Cassiodorus (sixth century), the *Liber etymologiarum* and the *De natura rerum* of Isidore of Seville (seventh century), the *De natura rerum* of the Venerable Bede (eighth century) as well as the supplementary material in many manuscripts of his computistic work the *De temporum ratione,* and the *Dragmaticon* and *De philosophia mundi* of William of Conches (twelfth century). If we move beyond the domain of the standard forms to cover pictorial material of a more strictly representational sort, we would have to include herbals, bestiaries, and constellation treatises.

Apart from this latter segment, the encyclopaedic or handbook works that frequently carried visual material in their manuscript copies had much in common. To begin with, "encyclopaedic" did not mean something of massive proportions. Each of the works listed could be comfortably situated between the covers of a single volume, some of them slim volumes at that. Not that they did not have coverage; they merely had it compendiously. Hundreds of pages in Pliny's (first century) *Natural History,* for example, became only twenty-five in Macrobius, and in turn a bare two paragraphs in Isidore of Seville. Secondly, as compilations of one sort or another, these late antique and early medieval encyclopaedic handbooks were derivative, often at third or fourth remove, from earlier sources, which were almost always ultimately Greek. Moreover, the derivative knowledge they compiled was in many instances simply just compiled and did not represent a systematic investigation of conceptions, doctrines, or problems. Finally, and most importantly, these handbooks were didactic in tone and purpose and the object was to teach fundamentals. What more fertile ground could there be, then, for the creation of diagrams and figures? One had in them a pedagogic tool of real value. This was surely realized by the authors of these handbooks, and even by the later scribes and readers who sometimes confirmed this realization by adding even more diagrams and figures, some of which were taken from one handbook and put to use in another.

CHAPTER FOUR

Tabulated Information

ONCE AGAIN WE MUST set aside mathematical and astronomical works, since there tables served a purpose quite unlike that of those which must occupy our attention now. For these latter tables were not concerned with the presentation of data or results (like rising and setting or multiplication tables, for instance), but rather with presenting the substance of some text or doctrine in a more compendious, visually communicable, form, even if such a form were not much more than a "table of contents" of some book or text. Most of the following Illustrations are examples of different ways and formats of exhibiting what is said in some text, or of presenting the very text itself, in tabular or chart form. One Illustration, however, is more comprehensive in its coverage, giving the chapter headings of a standard scientific work. There were also means other than such a table of headings by which the medieval scholar was able to set forth in handy form the contents of a given scientific text. There were, for example, straightforward lists of the major points made by an author in some work. Aristotle was, as one would expect, frequently favored with such treatment. The lists might be called *conclusiones* or *auctoritates Aristotelis* or *Compendium librorum naturalium,* but the matter presented was much the same: a kind of *florilegium* of the more important conceptions and doctrines held and argued for by Aristotle in a given work or group of works. There were also elaborate alphabetically arranged lexica of his works, each of them, like each *florilegium,* often covering a separate division of Aristotelian philosophy: one lexicon or *florilegium* for logic, one for natural philosophy, one for moral philosophy. Although naturally much longer than *tabulae* presenting the points made in a page or paragraph or two of text, these lexica and *florilegia* were directed toward the same goal of distilling the substance of something into a more manageable form, and of doing so in such a way that the format of the "distillate" was one that had visual indicators that facilitated its use.

22. A Medieval Table of Contents. The very heart of medieval Latin scholastic philosophy and science was the *quaestio* and any number of questions devoted to a given work—Aristotle's *Physics,* for example—carried lists of the *quaestiones* they contained, an addition that made finding one's way through the work a less arduous task. Less frequent were "tables of contents" for a specific work, even for standard texts. The present illustration gives the beginning of one of these less prevalent tables: that for Boethius' (sixth-century) *Arithmetica.* The left-hand column gives the rubric or heading for each chapter of this work, while the right column gives what often served in place of page numbers in such a medieval table—the initial words of each chapter plus the beginning words of each major paragraph within the chapter (these groups of words distinguished here by an appropriate superscript *a, b, c, d,* etc.). Curiously, this table occurs in a thirteenth–fourteenth-century manuscript that does not contain the Boethian work to which it belongs, a factor that seems to testify to the importance assigned to the table alone. Boethius' *Arithmetica,* which was a free redoing of the Greek *Introduction to Arithmetic* of Nicomachus (first century), remained the standard text on the subject throughout the Middle Ages and was never replaced by translations of the arithmetical Books VII-IX of Euclid's *Elements* or the more advanced *Arithmetica* of Jordanus of Nemore (thirteenth century). Diagrams from other medieval copies of Boethius' *Arithmetica* can be seen in Illustrations 85, 94, 95, 97, and 172.

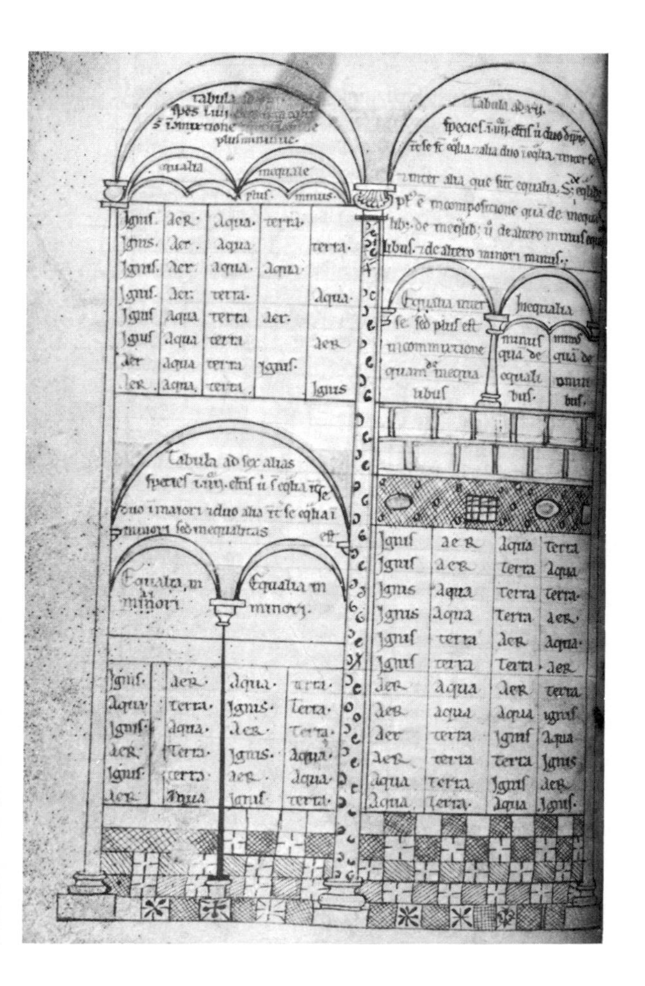

23. Combinations of the Elements. One substantial section of this anonymous twelfth-century work on natural philosophy and medicine is devoted to the four elements—how they were created, what they are, the etymology of the term 'elementum', what their motions are, and, especially, what possible combinations *(commixtiones)* they can suffer. The present table is part of the author's effort to address himself to this last named subject. He considers combinations of fire, air, earth, and water in groups of two, three, and four, specifying whether the constituents of these groups are equal or unequal to one another. In the tables depicted here, that at the top left sets forth eight possible combinations of the four elements in which three are equal, but the fourth is either greater than or less than these three. The table below gives the "six species" of combining the four elements into unequal groups of equal pairs, while the table on the right considers the twelve combinations or species resulting from two of the elements being equal in amount while the remaining two are respectively less and greater than these equal pairs. If one adds to all of these possible combinations those which are presented in tables on the preceding and following pages in this manuscript, one arrives at a grand total of 142 possible cases taken into account. The colonnade setting in which the author or scribe has presented these tables reminds one of the Canon Tables for the Gospels (see the following Illustration) which were so popular in the earlier Middle Ages. Perhaps they were consciously modeled on them.

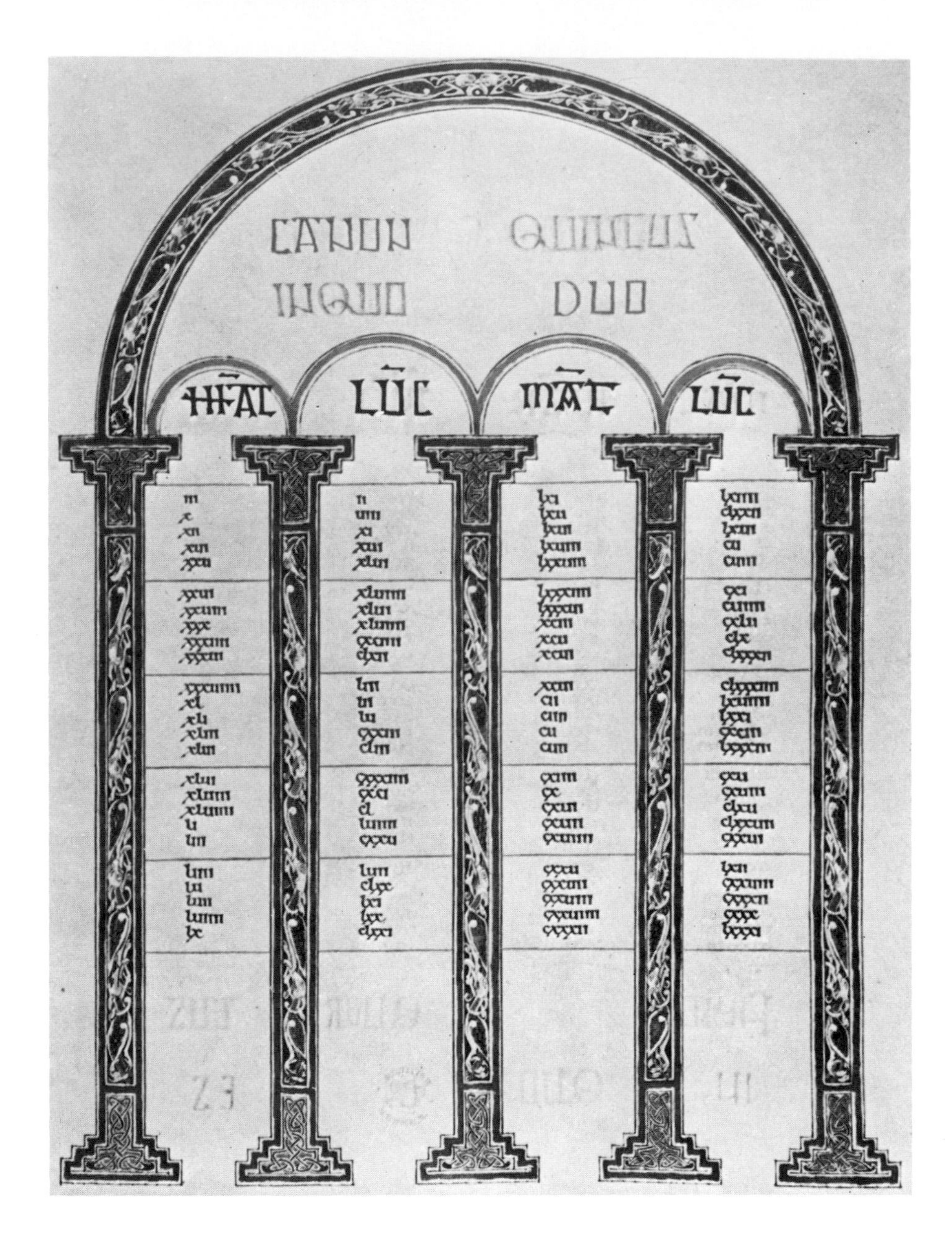

24. A Canon Table for Matthew and Luke. Eusebius of Caesarea (fourth century) devised a system of numerical tables (*κανόνες*) which established concordances for the four Gospels, thus enabling the reader to locate parallel passages with ease. Since there was then no division of the Gospels into chapter and verse, Eusebius drew his numbers from the divisions known as the Ammonian Sections (possibly due to Ammonius Saccas [third century], but now usually thought to be an invention of Eusebius himself). Undoubtedly the most artistic and decorative of medieval *tabulae,* these Canon Tables quite frequently served as prefaces or appendices to both Greek and Latin manuscripts of the Gospels, especially before the thirteenth century. The present example is from the seventh-century Book of Lindisfarne, the earliest codex of Hiberno-Saxon Gospels to present the Tables in arcaded form. There are 16 such Canons in this manuscript. The fifth, represented here, gives a concordance for Matthew and Luke.

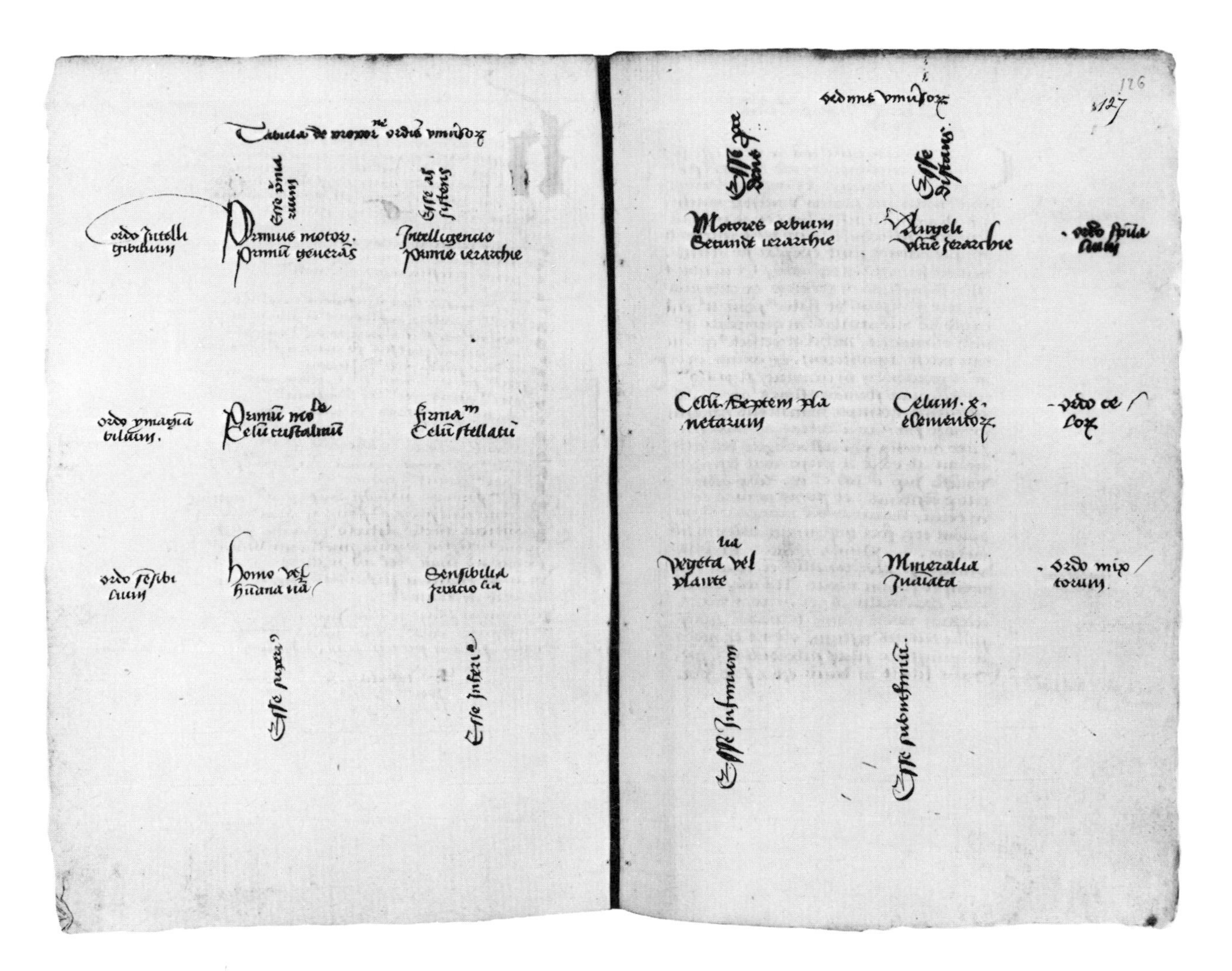

25. On Order in the Universe. This fifteenth-century manuscript contains a *Compendium of Natural Theology Taken from Astrological Truth* by an otherwise unknown Curatus de Ziessele iuxta Brugas (most likely the present-day Belgian town of Sijsele, barely ten kilometers from Bruges). It seems that this sole extant manuscript of this work contains only a part of the whole, one which was originally presented as a disputation at the University of Louvain. Its subject was (as the heading of the left-hand page of the present Illustration tells us) "the proportion of the order of the universe." Its basic concern was, in other terms, the "scale of being" of the universe, the sum and substance of which is given in the two-page table represented here. The "scale" runs from God *(primus motor, primum generans)* at the upper left down to inanimate minerals on the lower right. The top row on the left gives the order of intelligible beings: God as primary being and, as *esse assistens,* the Intelligences of the first hierarchy, while the same row on the right (there specified as the order of spiritual beings) contains, as *esse procedens,* the movers of the celestial orbs belonging to a second hierarchy and, as *esse distans,* angels of the ultimate hierarchy. The scale continues in the middle row, labeled the order of imaginable beings, moving from the *Primum mobile* or crystalline heaven through the firmament or heaven of the fixed stars to the order of heavens on the right containing the heavens of the seven planets and that of the four elements. The scale comes to an end in the lowest row moving through the order of sensible being on the left of man and animals to the order of mixed being on the right consisting of plants and minerals. Each element in this lowest row is correspondingly characterized as *esse superius, inferius, infimum,* and *subinfimum.*

26. Ready Reference Astrology. In Dante's eyes, the thirteenth-century scholar Michael Scot was one "who verily knew every trick of the art magical" (*Inferno* XX, 116). Although Dante names no source for his judgment, it was surely the likes of Michael's *Liber introductorius* that he had in mind. For this work was one of the most comprehensive original medieval Latin compilations dealing with astrology, one which contained considerable information on natural philosophy, meteorology, and music as well as on astronomical and astrological matters. The present fourteenth-century manuscript of the *Liber introductorius* contains numerous "visual aids" for understanding the text, that pictured here being a chart of the "aspects of the moon" (as is duly announced in the left margin). The aspects were various symmetric arrangements of points on the ecliptic which the planets could occupy relative to one another, four such aspects being traditionally distinguished: opposition (when the planets in question would be at opposite ends of a diameter or 180° from one another), quartile (planets at 90° from one another), trine (at 120°) and sextile (at 60°). The vertical columns in Michael's chart give each of these aspects (in reverse order), plus conjunction (which was also sometimes considered an aspect if not classified as such). However, since he was here interested only in the aspects of the moon, his chart supplies these aspects relative to each of the six remaining planets, each planet covering a horizontal row. The resulting thirty squares in the chart (each related to the successive "days of the moon") enabled Michael to summarize the astrological information he wished to impart to his readers: whether the moon in this or that aspect relative to this or that planet was a good or bad sign and, hence, whether this or that action would be advisable at such a time. Other pictorial material from this manuscript of Michael Scot is given in Illustrations 80, 101, 231, 251, and 277.

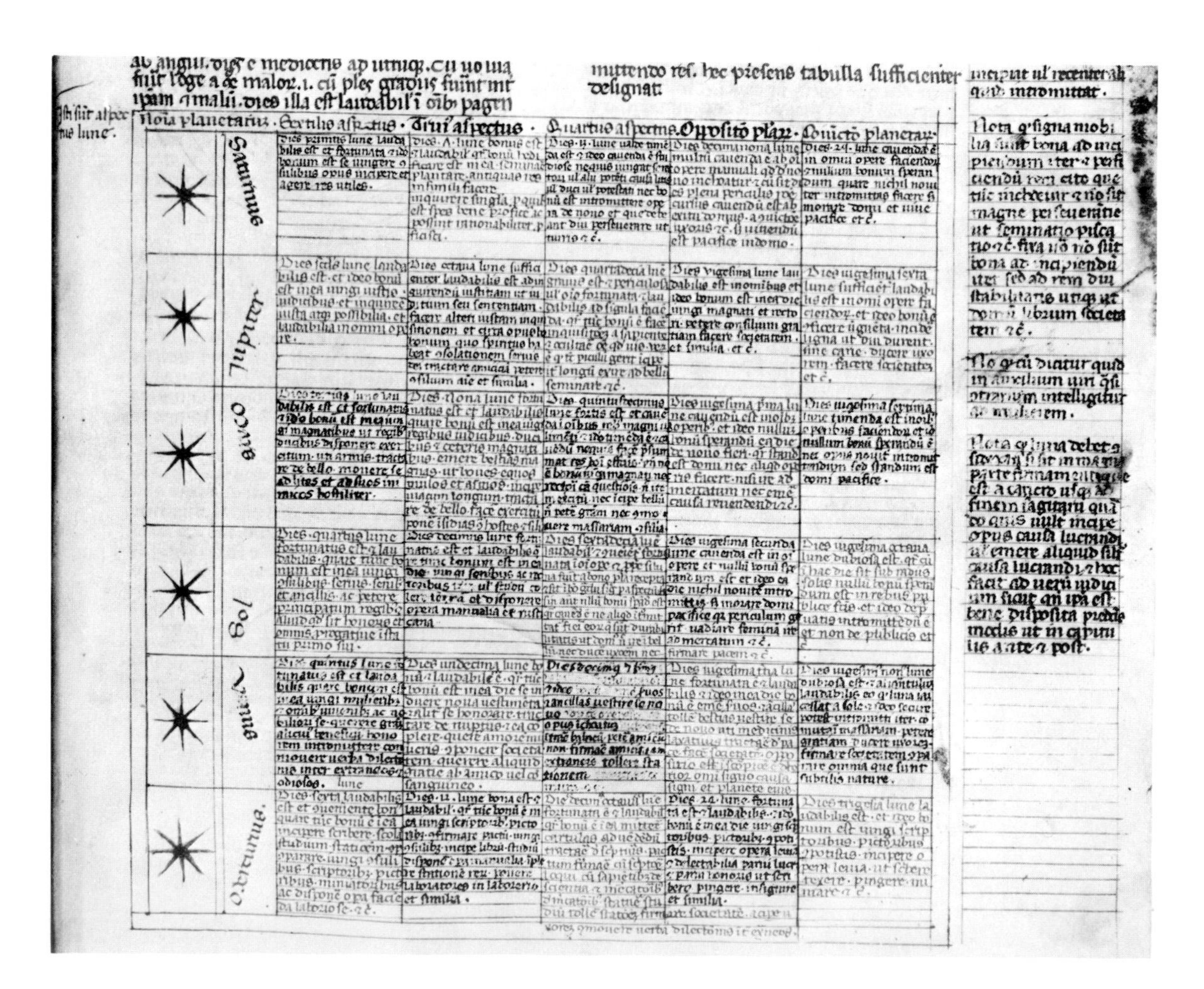

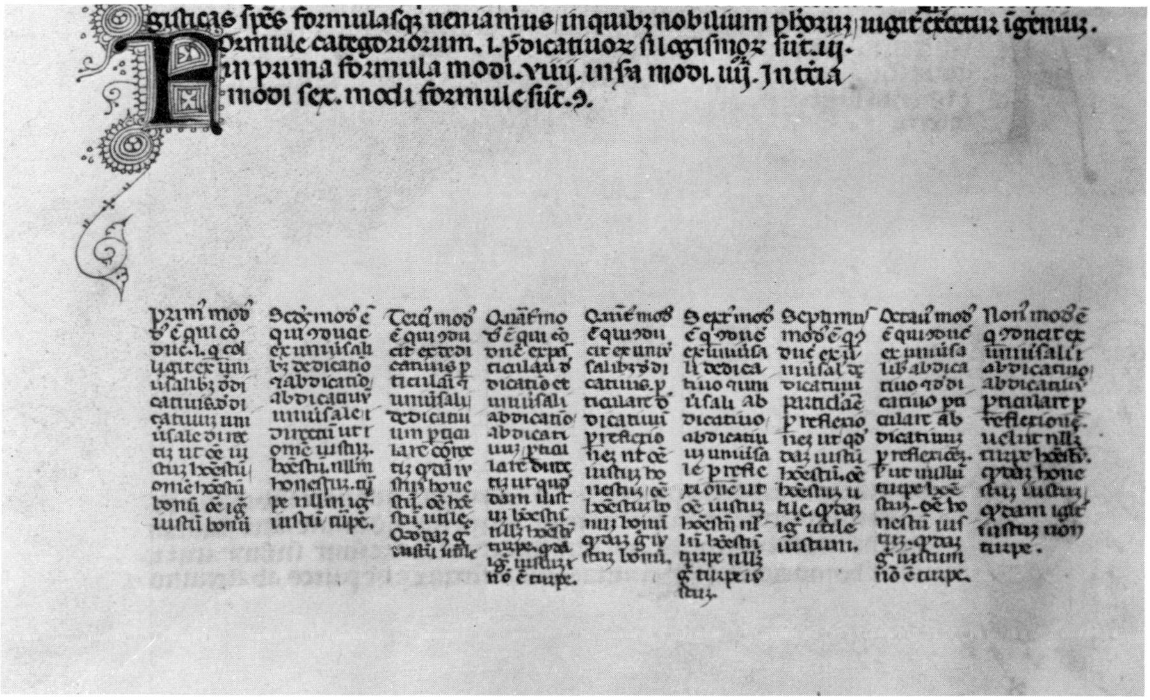

27. Turning Text into Table. In some instances a *tabula* occurs in a copy of a text without any announcement by the text itself. Instead an enterprising scribe has turned part of the very text in question into a *tabula.* One elegant instance of such a procedure can be seen in this fourteenth-century manuscript of Cassiodorus' *On Secular Letters.* Constituting Book II of the *Institutiones* of this sixth-century Ostrogothic statesman and scholar, this work was a concise encyclopedia of the seven liberal arts: grammar, dialectic, rhetoric, arithmetic, geometry, astronomy, and music. Its elementary review of these disciplines made this an especially popular "handbook" throughout the Middle Ages. The section depicted here is from Cassiodorus' treatment of dialectic, in particular his exposition of Aristotle's doctrine of the categorical syllogism. Thus, he tells us that the *formulae* (we would say "figures") of the categorical syllogism are three in number and that in the first figure there are nine valid "moods" *(modi),* in the second, four, and in the third, six. Cassiodorus then proceeds to explain all of these valid moods, giving appropriate examples for each, and it is just that which our fourteenth-century scribe has turned into a *tabula.* The nine valid moods of the first figure are those "tabulated" here. The first column of the *tabula* reads: "The first mood is that which concludes, that is, deduces, a universal affirmative directly from [two] universal affirmative [premises]; for example: 'Everything just is honorable; everything honorable is good; therefore, everything good is just'." This is an example of the first valid mood of the first figure of the categorical syllogism—one later given the mnemonic name *Barbara.* Note that Cassiodorus reverses the normal order of the major and minor premises in his examples, giving the latter first. The nine valid moods of the first figure given by Cassiodorus are the four standard ones plus five "indirect" valid moods (which are equivalent to the normal valid moods of the fourth figure, a figure not explicitly recognized by Aristotle nor, hence, by Cassiodorus). For more on the syllogism, its valid moods, and *Barbara,* see Illustrations 55, 60, and 70. For other material from this manuscript, see Illustrations 30, 119, 170, and 177.

CHAPTER FIVE

Dichotomies and Arbores

ONE OF THE MOST noticeable features of scholastic literature—that of most science included—was the ever prevalent tendency to draw *distinctiones;* "seldom affirm, seldom deny, always distinguish" was a refrain often uttered by the later Humanist critics of scholasticism. Yet this very characteristic of scholastic thought often meant that, when it was a question of illustrating some aspect of that thought, a diagram that dichotomized was most natural and fitting. Such a technique of diagraming also fit well with the medieval logic of definition through the dichotomizing effected by the successive application of *differentiae* to a sequence of *genera* (as depicted in Illustration 44).

Similar in some respects to a modern outline in dividing and classifying a subject, these dichotomies occur in medieval manuscripts in many forms and many degrees of care in execution. They might be structured simply by employing a device similar to the modern bracket or brace, they might be divisions applied to a text itself, or they might appear as chains of circles arranged in various positions. Further, at one end of the scale, there are "professional" dichotomies artistically drawn by scribes whose business it was to rubricate and illustrate texts, while at the other extreme there are those that were hastily sketched by a student, at times in the margins of a book in order to render more apparent what was being said.

A much more elegant kind of dichotomy is found in the *arbores* or trees that populated so many illustrated medieval manuscripts. Applied to all manner of subject, these *arbores* had a solid authoritative base in the Bible. The *Book of Genesis* speaks (II, 9, 17) of the Tree of Life *(lignum vitae)* planted by God in the Garden of Eden (to which Illustration 40 is related) and the Tree of Knowledge of Good and Evil *(lignum scientiae boni et mali)*, the eating of whose fruit led to expulsion from the Garden. Alternatively, the genealogy of Christ from David in the opening verses of the Gospel of Matthew was taken to refer to the Tree of Jesse found in the prophetic words of Isaiah (XI, 1, 10). These symbols appeared frequently in religious art and literature, the *lignum vitae* receiving a special canonization by the importance assigned to it by St. Bonaventure (thirteenth century), and it is very likely that these biblical occurrences of *arbores* favorably affected the popularity of their being applied to many disciplines.

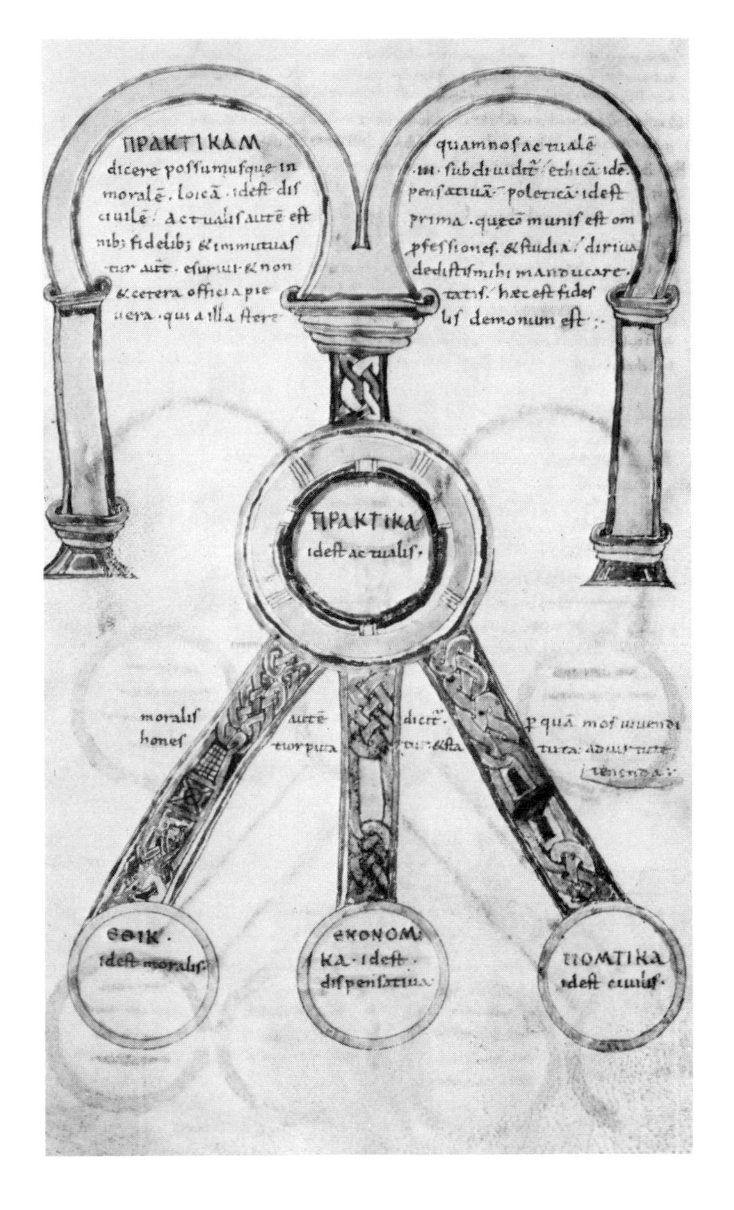

28. Divisio scientiarum I. From the early Middle Ages through the thirteenth century, it was not an infrequent concern of medieval scholars to address themselves to the problem of what we would today call the "classification of the sciences." In particular, it was a special preoccupation in the twelfth century in the Latin West when whole treatises were devoted to the problem. These treatises utilized not only earlier Latin sources (such as Boethius) in discussing the problem, but also employed material that was newly translated from the Arabic. Classifications of this sort provided natural occasions for pictorial representation through dichotomies. That given here is from a North Italian manuscript of the eleventh century and thus antedates the extensive Latin concern with such classification in the twelfth century. In diagramming the division of one "genus" of the sciences, that called "practical," the author of this early dichotomy has tried to make his points in both Greek and Latin, but not without some stumbling when dealing with the former. The word 'practica' was evidently not well known to Latin readers in the opinion of our author (even though it was already used by Boethius in the sixth century), for he glosses this Greek term by translating it as 'actualis'. Similarly, the division of practical science into *ethika, ekonomika,* and *politika* is Latinized as *moralis, dispensativa,* and *civilis.* Moreover, this author or scribe was concerned not only with turning Greek into Latin but also with turning the topic to Christian purposes. He notes that practical science is common to the faithful in the exercise of their religious duties, and even quotes the Gospel of Matthew (XXV, 42: *esurivi et non dedistis mihi manducare*). The particular division of the practical sciences given here is Aristotelian if not of Aristotle himself: later Greek scholars maintained it was the division intended by Aristotle and it is from such sources that it came to the Latin West. One finds it, for example, in the sixth century in Cassiodorus and in the seventh in Isidore of Seville.

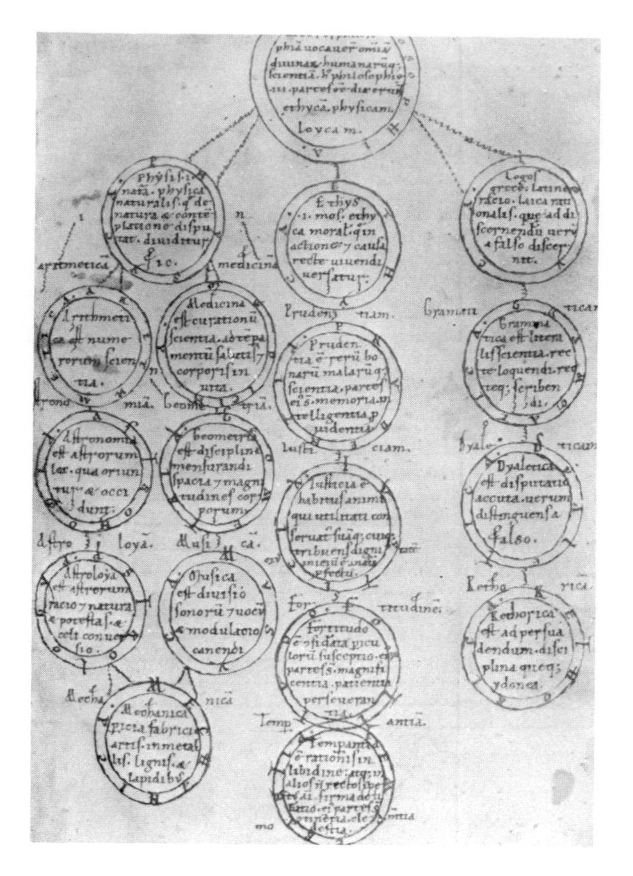

29. Divisio scientiarum II. Beginning from the assertion that the ancients called wisdom "philosophy," this twelfth-century dichotomy goes on to divide this *summa disciplina* into *physica, ethica,* and *loyca* (this last understood, not as logic proper, but generically as rational science, that being a closer translation of the Greek term λόγος). This threefold division was ascribed to Plato by St. Augustine (*De civ. Dei* VIII, ch. 4) and was considered Platonic throughout the Middle Ages. On the right, the chain of circles divides *loyca* into the traditional trivium of grammar, dialectic, and rhetoric, the interior of the circles giving the function of each of these disciplines. In the center, *ethica* is divided into the four cardinal virtues of prudence, justice, fortitude, and temperance, but perhaps the most interesting part of the dichotomy is the division of *physica* given on the left. As was traditional in classifying the sciences according to the Platonic triad, *physica* or, in Latin, *scientia naturalis* here contains the quadrivium of arithmetic, geometry, astronomy, and music. But our dichotomy adds three other links to the *physica* chain: medicine, astrology (which concerns the nature and power of the heavenly bodies, as distinguished from astronomy, which deals with their rising and setting), and mechanics (in the sense of the art of fabricating things in metal, wood, or stone).

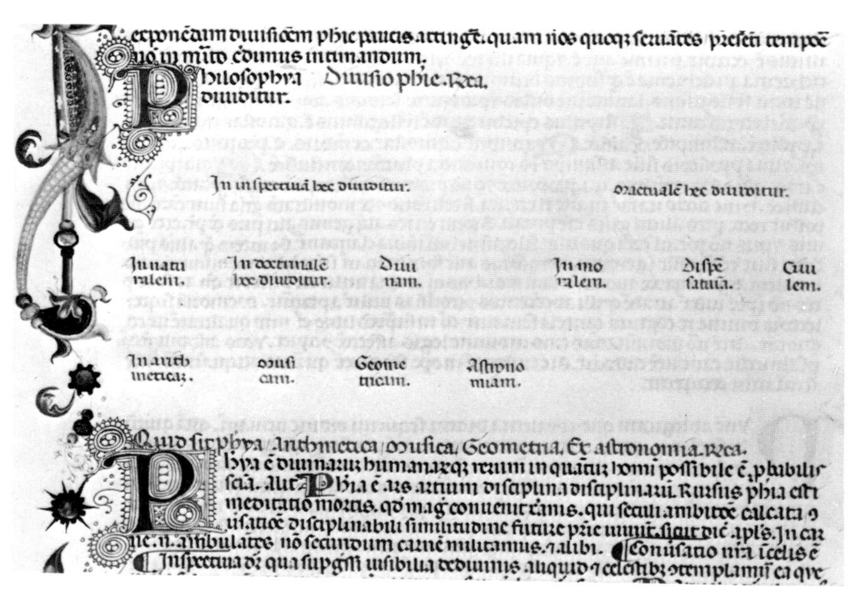

30. Divisio scientiarum III. The same fourteenth-century manuscript of Cassiodorus as in Illustration 27 here gives the two-part "Aristotelian" classification that appears in the following Illustration, but it omits the Greek-based primary terms 'theorica' and 'practica,' using only, respectively, the Latin 'inspectiva' and 'actualis'. [Note that in Illustration 31 the later scribe had inserted 'inspectiva' as a Latin variant for the 'contemplativa' of the original diagram.] A more important difference, however, is the fact that the present case gives the text of Cassiodorus' *On Secular Letters* itself, turning it into a dichotomy, just as in Illustration 27 the scribe had turned part of the text into a table. Other pictures from this manuscript are given in Illustrations 119, 170, and 177.

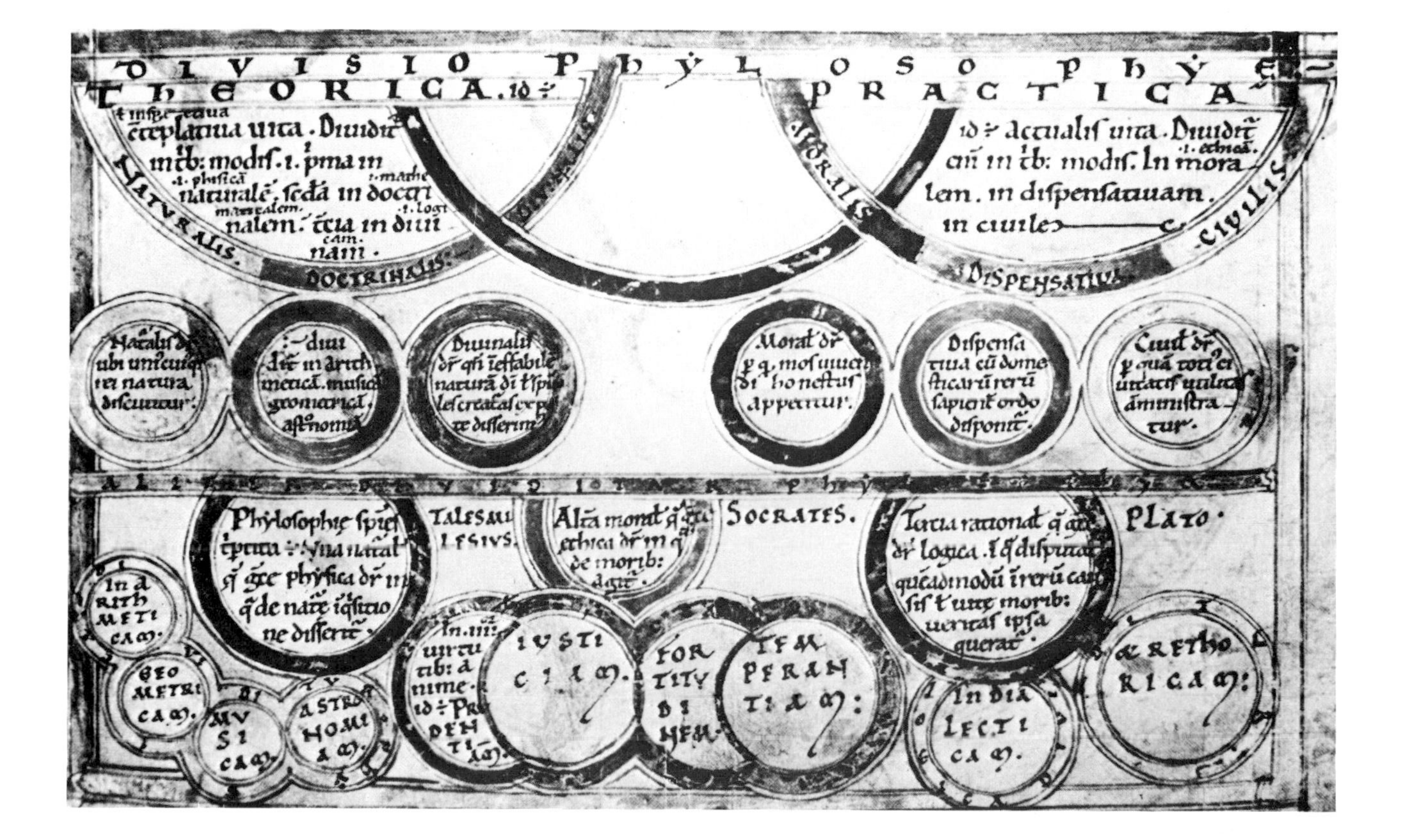

31. Divisio scientiarum IV. Another twelfth-century manuscript reveals a more elaborate dichotomy of different classifications of the sciences. The bottom half of the diagram gives the tripartite Platonic division we have already noted in Illustration 29, omitting the addition of medicine, astrology, and mechanics in this earlier dichotomy (and also inadvertently omitting grammar from the division of the rational sciences), but adding the names of three ancients who were practitioners of philosophy: Thales of Miletus, Socrates, and Plato. The top half of the diagram is "Aristotelian," combining what Aristotle himself maintained in various texts with what later Greek scholars added to him. These later scholars held that Aristotle recognized only two main branches of knowledge: theoretical and practical. We have already seen the breakdown of the latter member of this pair in Illustration 28. The ancestry of the division of the theoretical sciences here given at the top left is more problematic. The first chapter of Book V of the *Metaphysics* contains what is most likely Aristotle's own view of the division of the sciences; it is threefold: practical, productive, and theoretical. He then goes on to divide the theoretical into *mathematica, physica,* and *theologica* (which for Aristotle means metaphysics). This latter subdivision is here grafted on to the putative twofold Aristotelian division. *Physica* is appropriately given as *naturalis* (and described as being involved when the nature of any *thing* is investigated); mathematical science is termed *doctrinalis* (and further subdivided into the quadrivium); and *theologica* is Latinized as *divinalis* (a science said to concern the ineffable nature of God or created spiritual beings [that is, angels], hardly the core of Aristotle's *Metaphysics*). A later scribe has tried to be helpful by adding the appropriate transliterated Greek to some of the Latin terms: there is a proper superscript 'physicam' for 'naturalem' as well as 'mathematicalem' for 'doctrinalem'. But what seems to be a scribal slip has turned 'divinam' into 'id est, logicam' (change 'id est'—here abbreviated as '·i·'—to 'theo' and all would be well!). For other diagrams from the present manuscript, see Illustrations 51, 53, 69, 249, 279, 285, and 290.

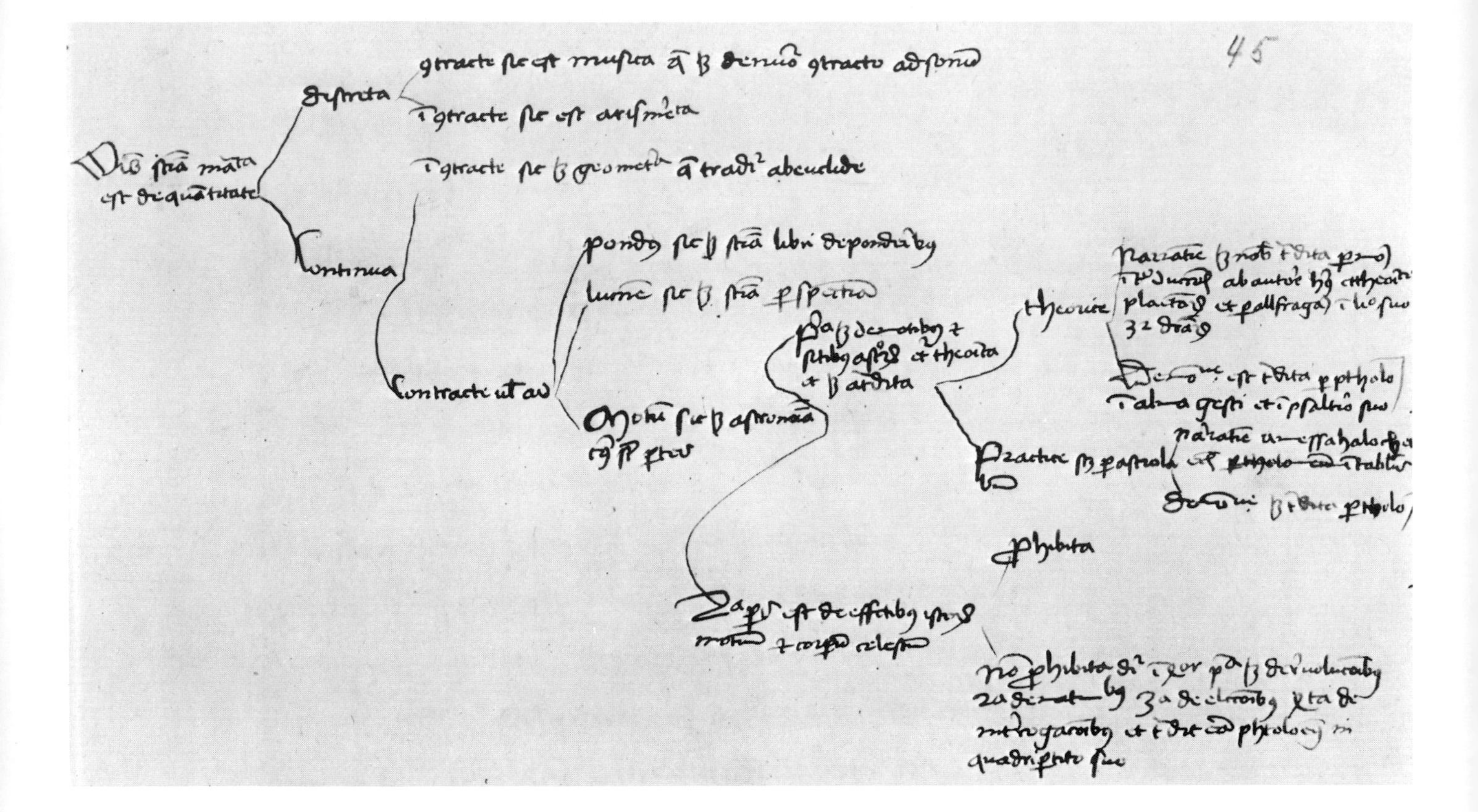

32. Divisio scientiarum V. This dichotomy, from a fifteenth-century manuscript containing a variety of works in mathematics and natural philosophy, was very likely drawn by a student. The classification is based not so much on theory as on the scientific works available at that time. It begins from the standard view that all mathematical science is concerned with quantity, of which there are two sorts, discrete and continuous (a division that goes back to Aristotle's *Categories*). Discrete quantity in general *(incontracta)* is number, which naturally gives us the science of arithmetic; but if discrete quantity is related or limited *(contracta)* to sound, then we have the science of music. The major portion of our dichotomy, however, is devoted to the subdivision of continuous quantity. If considered generally, we have geometry and the work transmitting that science is, as our dichotomy says, that of Euclid. On the other hand, continuous quantity can be considered as limited to weight (in which case the basic text is the *Liber de ponderibus,* probably the work usually—but erroneously—ascribed to the thirteenth-century Jordanus de Nemore), to light (giving us the science of optics, or *perspectiva*), or to motion (in which case astronomy is involved). Astronomy itself has a number of parts and it is in the display of these that our student dichotomy is most detailed and interesting. The first major part of astronomy, it tells us, deals with the motions and positions of the heavenly bodies, and this can be either theoretical or practical. If theoretical, it can be merely expository *(narrativa)*, in which case the relevant works are the *Theorica planetarum* (probably the work of that name written by—again probably—the thirteenth-century Italian Gerard of Sabbioneta) or the *Liber differentiarum* of the ninth-century Arabic astronomer Al-Farghānī. Alternatively, the theoretical science can be demonstrative; then Ptolemy's *Almagest* is the major work to be studied. In its practical aspect, astronomy is carried out *per astrolabium,* in which case we again have both expository works (the Latin of the eighth-century Arab Māshāʾallāh) and demonstrative works (again Ptolemy, presumably his *Planesphaerium*). The second major part of astronomy is astrology, which deals with the effects of the heavenly bodies and their motions. Here, one branch of our dichotomy comes to a dead end, for one part of this science is *prohibita.* The permissible part is fourfold: on revolutions of the heavenly bodies, on nativities, on elections, and on interrogations—all of which, our dichotomy notes, can be studied in Ptolemy's *Quadripartitum* (or *Tetrabiblos*), although, in fact, some of the topics would not be found there.

33. What Types of Fallacies and Why? This circular dichotomy is found in a fourteenth-century manuscript that is predominantly logical in content. It approaches its subject by quoting—in its center—Aristotle's definition of *elenchus* or refutation as it appears in Chapter 5 (167a 23–28) of his *Sophistici elenchi,* the reason for this apparently being Aristotle's claim that some fallacies arise because one does not have in hand a proper definition of this term. The main business of the dichotomy is, however, not refutation, but fallacies. All thirteen of the fallacies treated by Aristotle are duly represented: six that arise from an ambiguity of language *(in dictione)*, namely, those due to equivocation, amphiboly, composition, division, accent, and figure of speech; and seven that arise apart from any ambiguity of language *(extra dictionem)*, namely, those connected with accident, a confusion of relative and absolute *(secundum quid et simpliciter)*, *ignoratio elenchi,* begging the question or *petitio principii,* false cause *(non causa ut causa)*, consequent, and many questions. For some reason or other our dichotomy neither mentions Aristotle's basic division of fallacies into *in dictione* and *extra dictionem* nor arranges the dichotomized circle in accordance with it. Helter-skelter as the arrangement may be, the "spokes" of the dichotomy attempt to give, in as few words as possible, the reasons each fallacy may occur. Those of equivocation or amphiboly, for example, arise because one is not dealing with one and the same *thing* in the argument in question, while those of composition and division arise because some *expression* involved is not the same.

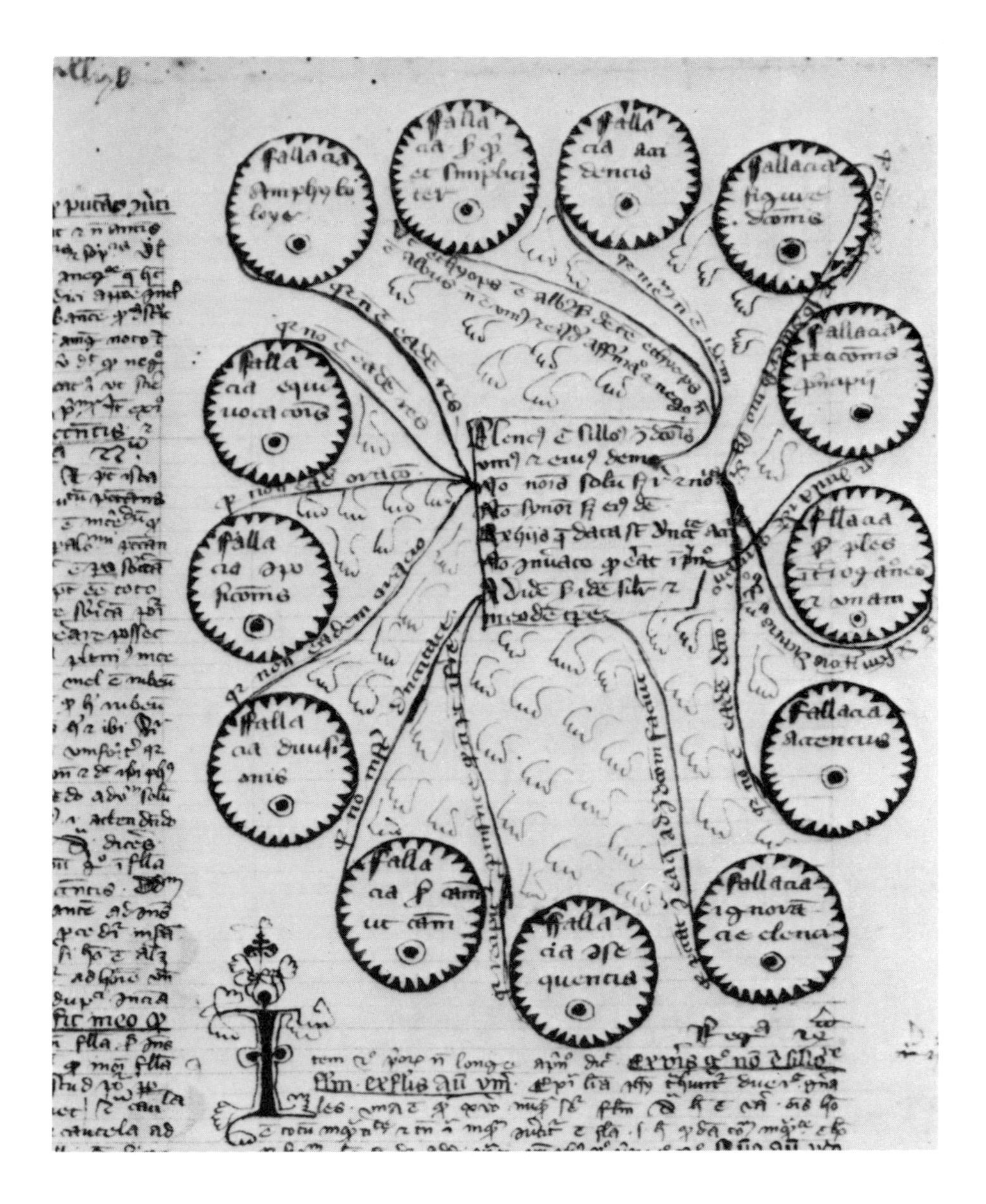

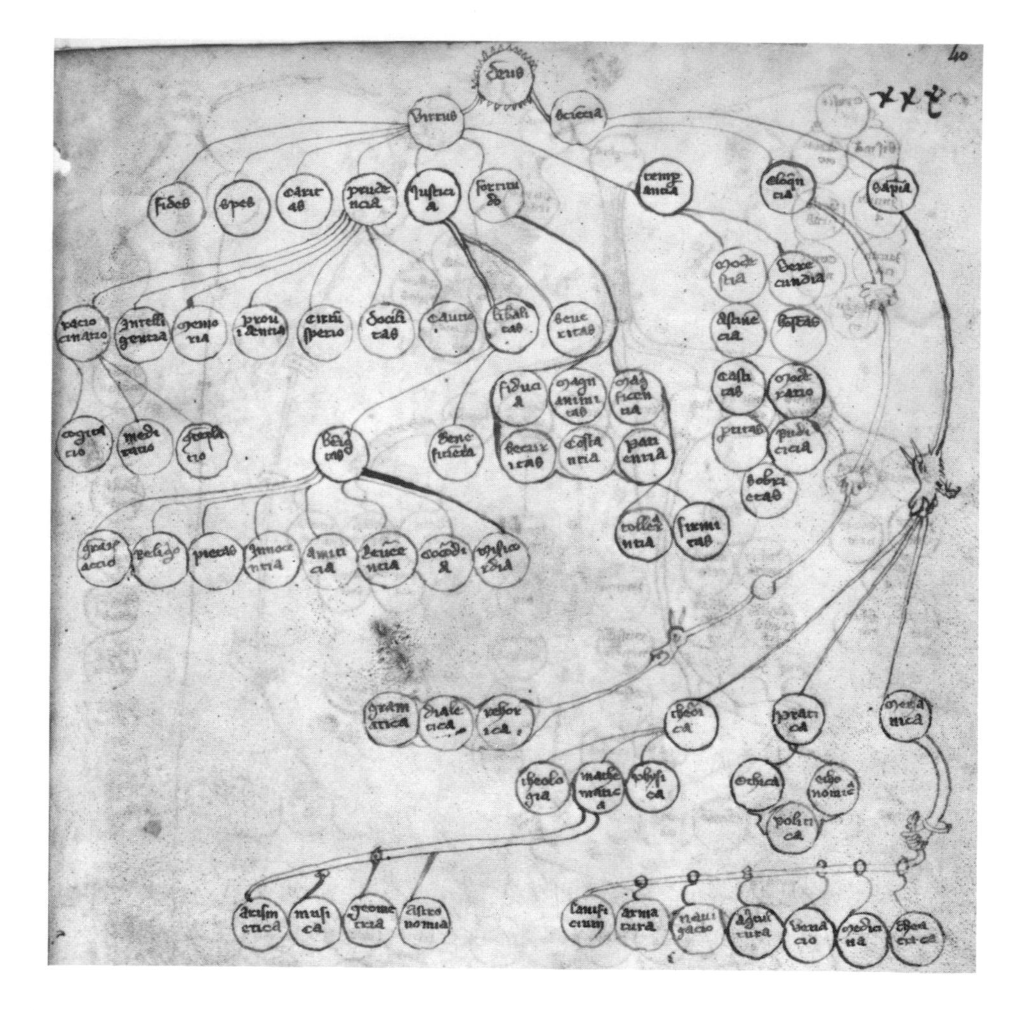

34. Virtues, Vices, and the Sciences. These two fourteenth-century dichotomies display an interesting connection between sciences on the one hand and virtues and vices on the other. Starting with God in the uppermost position, one first reaches virtue *(virtus)* on the left, this element then being divided into the same cardinal and theological virtues we have already seen in Illustration 21. But the subsequent subdivision is quite different and at times surprising. For example, *prudentia* (the fourth circle from the left in the third row from the top) has as the first three of its seven species *ratiocinatio, intelligentia,* and *memoria,* the first of these in turn having *cogitatio, meditatio,* and *contemplatio* as its subspecies (a triad that may have been derived from the twelfth-century scholar Hugh of St. Victor). Of greater interest, however, is the treatment of the second main element proceeding from God: i.e., *scientia.* This is immediately split into *eloquentia* and *sapientia.* The division of *sapientia* in turn yields the Aristotelian classification of the sciences we have already seen in other pictorial forms in Illustrations 30 and 31 with the important additional branch (again from Hugh of St. Victor) of the mechanical sciences, of which there are seven species (the circles on the lower right): textile making, armaments, commerce, agriculture, hunting, medicine, and theatrics. If we return to *eloquentia*, the other main branch of *scientia*, we see that this in turn yields the trivium, another reflection of the influence of twelfth-century scholars such as Hugh of St. Victor and John of Salisbury while at the same time a nod to Aristotle, who felt that, as a nonsubstantive science, logic stood outside of, and was propaedeutic to, philosophy as a whole. The central "lesson" of this whole dichotomy is, however, the explicit filiation of "good" science or knowledge with things virtuous and even with God. If we turn the folio, we find a second dichotomy, here given on the opposite page. In place of God, all begins from the Devil, the initial element corresponding to virtue in the first dichotomy now being vice (*vitium*). The subdivision is again nonstandard, although the seven deadly sins do find their place relatively high in the sequence of de-

(*Continued*)

scending species. However, it is the other initial offspring of the Devil which is intriguing: *ignorantia* or *sophystica.* The first branch of this species (descending on the left) leads to fallacies *(paralogisme)*, whereupon the subdivision yields the thirteen types of fallacy we have seen in the preceding illustration. The other branch of ignorance and sophistic knowledge leads to the strange label 'parosophyca' at the upper right, which in turn is split into *magica* and *abusio.* All manner of *abusiones* descend vertically at the extreme right, including such things as disobedient students, obstinate old men, and Epicurean meals. The dividing of magic for the most part once more follows Hugh of St. Victor, the first division consisting of the four circles at the bottom right giving the species *maleficium, sortilegium, augurium,* and *prestigium*, with a much less visible circle under the first of these carrying the inscription *mantica.* It is presumably the division of these mantic arts that the six unconnected circles at the bottom left are to effect: *geumantia, hidromantia, aeromantia, pyromantia, creomantia,* and *nigromantia.* Again, the most intriguing message of the whole dichotomy is the connection it makes with the vices and things diabolical, on the one hand, and "evil" kinds of knowledge, including errors in logic, on the other.

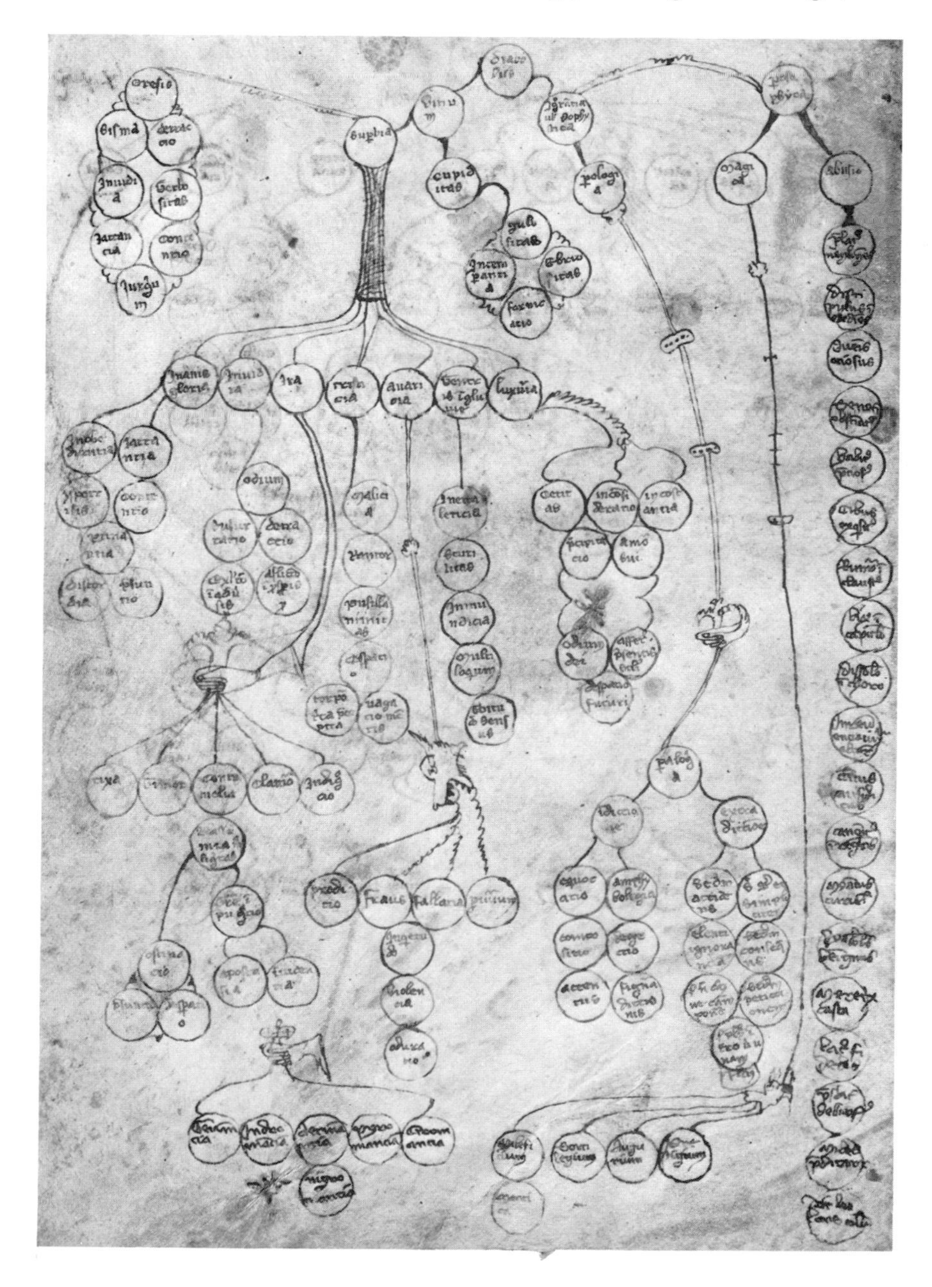

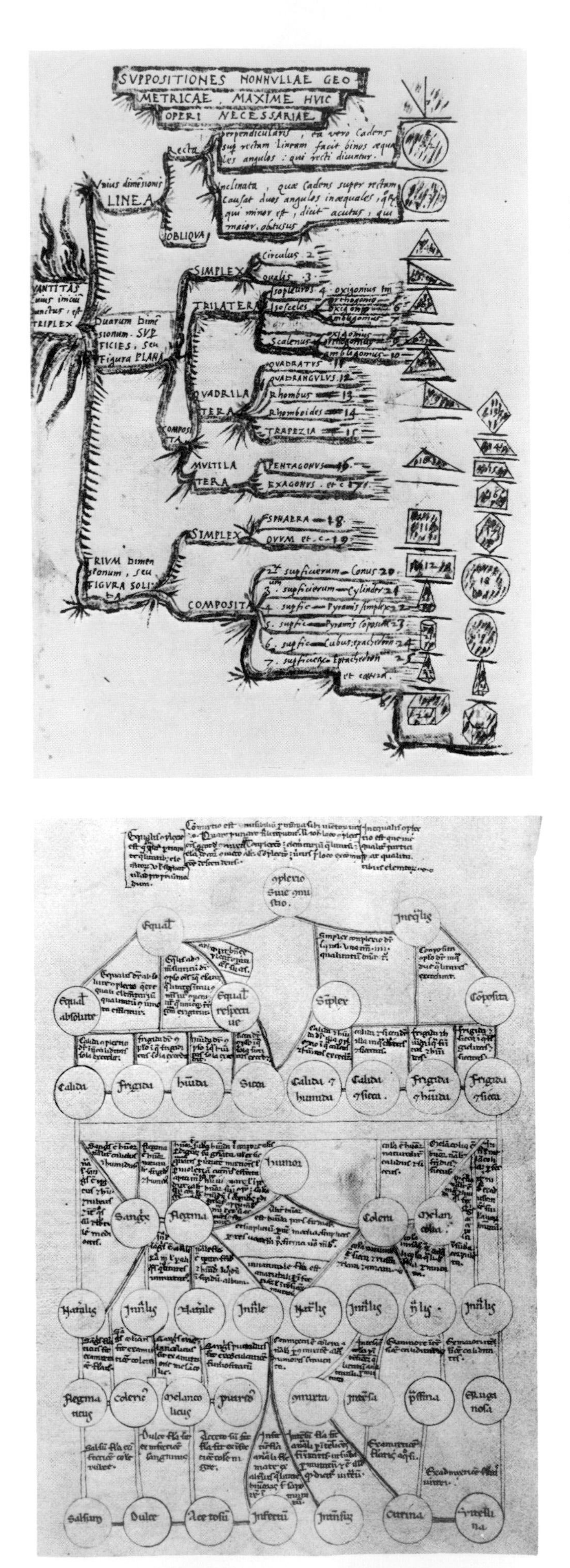

35. A Dichotomy with Figures. The mathematical proemium to a 1554 manuscript of an anonymous work, *On the Parts and Causes of the Astronomical Sphere*, contains the present dichotomy of geometry. In addition to giving and classifying the names (with a few lines here and there of explanation) of different sorts of geometrical lines, figures, and solids, it also provides drawings of each entity in question, keying each drawing to the relevant name by numbers. For other figures from this manuscript see Illustrations 169, 254, and 281.

36. The Parts and Practice of Medicine. This fourteenth-century manuscript contains eleven full-page dichotomies of a complexity seldom witnessed in diagrams of this sort. The complexity derives not only from the number of "elements" that are "dichotomized" (there are over 550 in all), but also from the attempt to cram explanatory text into almost all imaginable space. The dichotomies begin with one devoted to the classification of the sciences, where medicine holds a central place, and run on through all manner of medical theory, anatomy, physiology, pathology, and so on, following Galen and Avicenna's *Canon,* in idea if not word. The dichotomy pictured here occurs on "page 2" of the series. The top half sets forth the doctrine of complexions, that is, those overall qualities in a whole body or in some member of a body resulting from the mixing and interaction of the four primary qualities (hot, cold, dry, and moist) within that body or member. As the dichotomy points out in an easily comprehensible fashion, such mixing of the primary qualities, and hence the resulting complexion, can be either equal or unequal. Equal and unequal mixing are in turn divided into absolute or merely relative equality on the one hand, and simple or composite inequality on the other (simple obtaining when there is an excess or dominance of only one primary quality, composite when such occurs with respect to two qualities). The lower half of the dichotomy treats of the doctrine of the four humors: blood, phlegm, red bile, and black bile, and begins by setting forth the pairs of primary qualities that are characteristic of each. Each of the humors can, however, occur in a natural or unnatural—we might say normal or abnormal—form, and this distinction occupies the next level of the dichotomy, while its last two ranks draw attention to other properties the humors might exhibit.

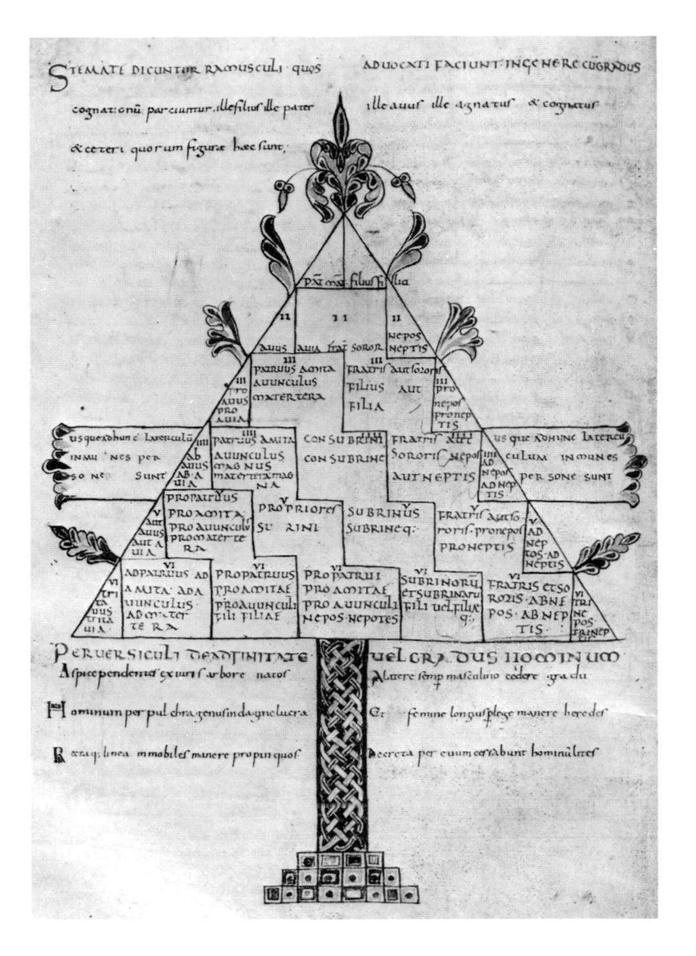

37–38. Consanguinity Trees I and II. One of the earliest medieval uses of the *arbor* or tree to illustrate some point in a written text was genealogical. And of such a use one of the very earliest instances must be that found in manuscripts of the *Liber etymologiarum sive originum* of the seventh-century bishop, Isidore of Seville. The *Etymologies* contains two consanguinity trees (plus a circular diagram addressed to the same subject, for which see Illustration 50). Illustration 37 is an example of Isidore's first type of tree as found in a ninth-century manuscript. Although the appropriate label is missing from this example, the very apex of the triangular tree is to represent the person *(ipse)* whose genealogy is being diagrammed. Going down the right-hand side of the triangle are one's offspring, from son and daughter *(filius, filia)*, grandson and granddaughter *(nepos, neptis)*, to the sixth generation great-great-great-great-grandson and granddaughter *(trinepos, trineptis)*. Isidore has also indicated the generations by inserting Roman numerals at the relevant levels in his tree. The left side of the triangle yields direct ancestors, again through six generations (from *pater* and *mater* to *tritavus* and *tritavia)*. This leaves the interior parts of the tree: four stair-step figures and one lone rectangle at the lower left. The longest of the stair-step figures begins at Level II with brother and sister and runs through their offspring, while that beginning at Level III covers paternal and maternal aunts and uncles and their progeny, that beginning at Level IV great-aunts and uncles, and so on, the final element, in the lone rectangle, being the sisters and brothers of one's great-grandparents. The illustration of Isidore's second type of genealogical tree (Illustration 38) is taken from another ninth-century manuscript of his *Etymologies*. When compared to the first type of tree, its advantages seem obvious. Thus, taking the "zero point" just below the rather stern-looking father and mother, one can proceed through seven generations or levels in any direction: straight down through one's offspring, straight up through direct ancestors, and up and horizontally out the tree's "branches" through the successive offspring of aunts and uncles, great-aunts and great-uncles, and so forth, simply following the numerals as one "turns the corner." Equally helpful, each horizontal level covers a single generation, unlike Isidore's first tree. What is more, maternal relatives are neatly separated on the right half of the tree from paternal ones on the left. Clearly, this type of consanguinity tree was much easier to put to effective use.

39. Consanguinity Tree III. This mid-thirteenth-century French manuscript gives a much later version of the second Isidorean tree. It is, however, rather drastically altered in such a way as to make it almost useless. There is no problem in moving straight up and down through direct ancestors and descendants, but tracing one's way through aunts, uncles, and cousins of any level becomes a mind-boggling task. One must, for example, vertically ascend to one's maternal grandmother *(avia)*, next move one step horizontally to the right to one's maternal great-uncle and aunt *(avunculus magnus, matertera magna)*, then back down on a 45° slope through the descendants of one's great-uncle and aunt. As the briefest examination of such a path will show, one cannot even be guided by an appropriate succession of numerals. Finally, note that in this version Isidore's second tree has been further complicated by the addition in its lower branches of information concerning one's brother and sister or brothers and sisters and their descendants, this graft perhaps deriving from the information given in Isidore's first tree, shown in Illustration 37. In any case, if it was true that Isidorean trees of this sort were used to reach a determination with respect to various legal questions having to do with kinship, this thirteenth-century copy would have been of little use in this regard. Such was often the case, however, when more artistic and stylized versions were made of many kinds of diagrams and figures; being decorative took the place of being functional. In the present example, the person who apparently made things more artistic but less useful is identified at the base of the tree as one Gautier Lebaube.

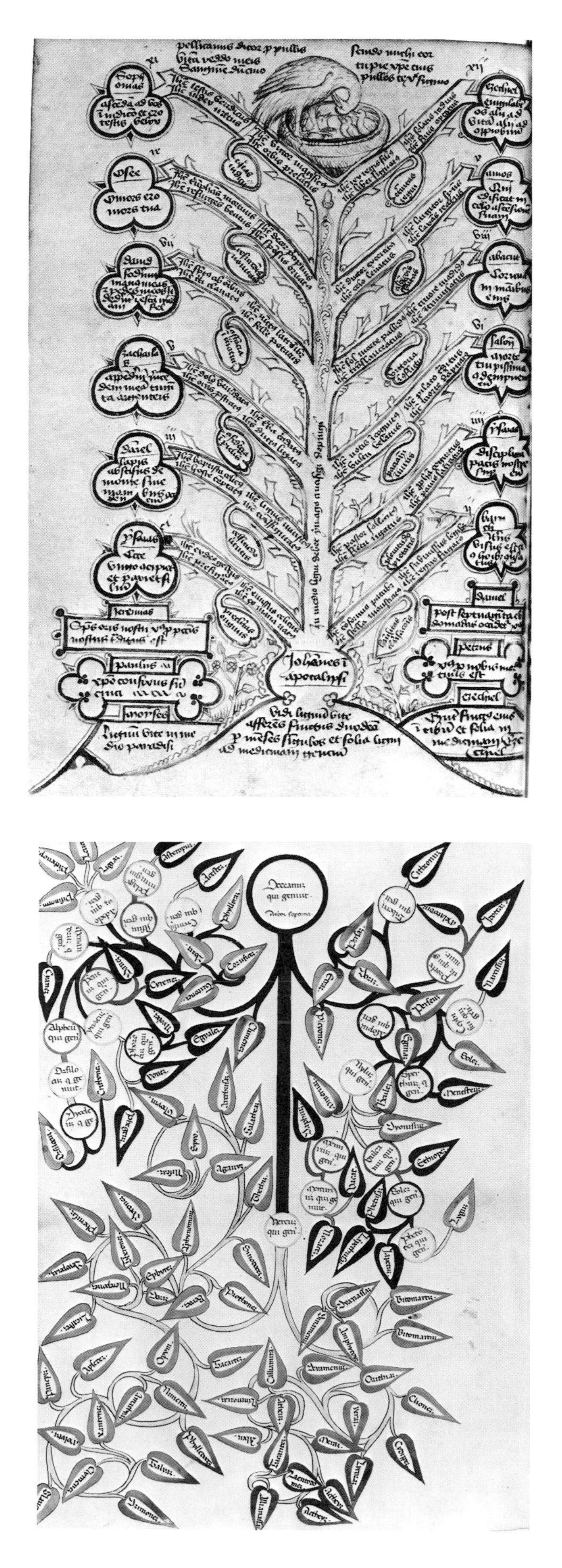

40. Arbor Prophetarum. This fifteenth-century manuscript tree of prophets is modeled after a tree of life *(lignum vitae)*, and is in effect an interpretation of that tree as described in *Revelation* XXII, 2. Two other biblical references to a *lignum vitae* appear at the base of the tree (*Genesis* II, 9; *Ezechiel* XLVII, 12). The twelve fruits that the tree will bear according to *Revelation* have here become twelve prophets, their messages apparently viewed as *medicina gentium* afforded by the leaves of the tree (again according to the passage from *Revelation*). The names of the prophets (Sophonias, Hosea, David, Zachariah, Daniel, Isaiah, Ezechiel, Amos, Habakkuk, Salomon, Isaiah [for a second time], and Baruch) are inscribed in scalloplike figures at the end of each branch, with biblical quotations—though not always from the Book of the prophet in question—placed beneath each name. The names of six other biblical authors appear below the branches (Jeremiah, Paul, Moses, Daniel, Peter, and Ezechiel), again with appropriate quotations. Each branch carries four different characterizations of Jesus, together with pendant "pods" enclosing various inspirational phrases. The pelican and its young at the very top of the tree symbolize the sacrifice of the Crucifixion and the sacrament of the Eucharist. Medieval bestiaries recount the legend of pelicans, who were believed to kill their young when these offspring opposed them, but who later would pierce their own breasts, and with the blood that poured from their bodies nourish their dead young and restore them to life. This was likened to man's rebellion against his Creator, the denial of God's grace which followed, and the sacrifice of Christ, who by pouring forth his own blood restored man to eternal life. In medieval illustrations the nest of a pelican is often shown on the top of a crucifix, and the inscription here on the trunk of the tree even tells us that "there should be an image of a crucifix depicted in the center of the tree."

41. The Kinship of Pagan Gods. In the later years of his life, Giovanni Boccaccio wrote more and more in Latin. One of his efforts in this was his *Genologia deorum gentilium* (in 15 books), which was a kind of mythological encyclopedia, and was often plundered by other medieval scholars for the mass of information it contained. Each of the first thirteen books of the *Genologia* contained an elaborate genealogical tree that presented in summary form the substance of what Boccaccio had said. The example given here is from a late fourteenth-century manuscript. Beginning in the circle at the top of the page with Occeanus as the begetter of all things (a reference to Homer's *Iliad* XIV, 202, 246), this copy of one of Boccaccio's trees is an especially good instance of how comprehensive and complex a didactic tree could be in medieval sources.

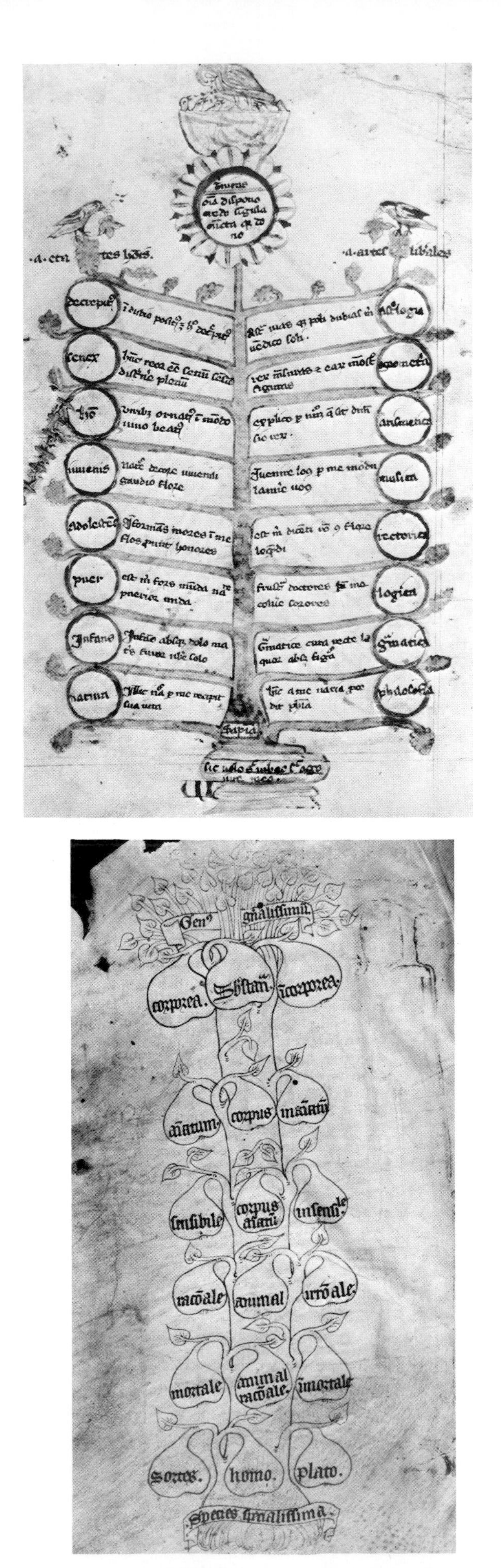

42. Knowledge and the Ages of Man. A fourteenth-century manuscript of astronomical and mathematical works includes this tree setting forth a correspondence between the seven liberal arts on the right and the seven ages of man *(infans, puer, adolescens, iuvenis, homo, senex, decrepitus)* on the left. The lowest branches represent philosophy and nature as the "origins" of the arts and the ages; the base of the tree is wisdom *(sapientia)*, while the tree's pinnacle represents the Trinity surmounted by the nesting pelican already discussed in Illustration 40. The verses written on each branch are perhaps the most interesting features of the present tree—each verse speaking directly to the art or age of man carried by its limb. For example, in the case of geometry, we learn that by it one reveals the measures of things and their figures *(rerum mensuras, et earum monstro figuras)*. For arithmetic we are told that by it one explains the division of things according to number *(explico per numerum, que sit divisio rerum)*. That these verses were not of the present author's or scribe's own making can be seen from the fact that some of those relating to the ages of man are found—albeit often in slightly altered form—in the figure given in Illustration 58.

43. The Tree of Porphyry. The present illustration, in a fourteenth-century manuscript, is of a didactic tree that has long gone under Porphyry's name: the *Arbor Porphyriana.* It displays how one is to proceed from the *summum genus* (here called *genus generalissimum*), substance, to a definition of man as the *species specialissima* by moving through the series of descending subaltern genera, each subaltern genus distinguished from the genus directly above it by an appropriate differentia. Thus, substance as the *summum genus* at the very top of the tree is divided by the two differentiae corporeal and incorporeal, corporeal yielding the subaltern genus body (since a corporeal substance is a body). Body in turn is divided by the differentiae animate and inanimate, the former yielding the subaltern genus animate body, which, in turn, by the differentiae sensible and insensible, gives us animal. But animal can be divided by rational and irrational, the former giving us the definition of man as a rational animal. The tree adds Socrates and Plato at its base as sample individuals of the species man. Note that the series of differentiae on the right side of the tree are not relevant since they are not constitutive of the subaltern genera forming the tree's center. The tree most likely received its name from the fact that in his *Isagoge* Porphyry (see next illustration) noted that any differentia which is divisive of some genus is constitutive of the subaltern genus immediately below it.

44. Aristotelian Logic. This decorative fourteenth-century manuscript is a rather massive collection (over 1,000 pages) of Latin translations of Greek works and original Latin texts treating all seven liberal arts. The page represented here gives the opening lines of the *Introduction (Isagoge)* to Aristotle's *Categories* written by the third-century Neo-Platonist Porphyry. Under the title *Praedicabilia,* this work became the first logical text to be studied in the curriculum during the Middle Ages. The illuminated 'C' that begins this text shows a teacher presumably trying to inculcate the fundamentals of logic into two students. They stand at the base of a tree which gives, in essence, all of these fundamentals. The first leaf growing from the trunk is 'word' *(sermo)*. But words can be either noncomplex or complex (that is, combined into phrases or sentences). The *incomplexus* main branch leans to the left and first leads to Porphyry's five predicables (already in Aristotle's *Topics,* although in a slightly different form), shown as the five leaves: *genus, species, differentia, proprium*, and *accidens.* Yet since the study of Porphyry's predicables was to serve as an introduction to Aristotle's *Categories,* it is these which we find—under their Latin name *Predicamenta*—capping this left-hand branch. These ten "predicaments" were considered the ten most general kinds of being: *substantia, qualitas, quantitas, relatio* (often also simply *ad aliquid* in Latin), *actio, passio, quando, ubi, situs,* and *habitus* (the last six in English usually being action, passivity or being acted on, time, place, situation or posture, and possession or state). Turning to the right main branch labeled *complexus,* the first two leaves represent the main parts of any proposition: subject and predicate. Moving up this branch we first come to this sentence or proposition itself. Seven additional leaves give the seven kinds of propositions: either categorical or hypothetical, universal or particular, affirmative or negative, and 'simplex' (although one would have expected 'singularis'). The display of the different possible "modes" or modalities of a proposition forms the termination of this right-hand branch: a proposition may be either necessary, impossible, possible, or contingent. For other diagrams from this fourteenth-century manuscript see Illustrations 59, 60, 62, 104, 178, 241, and 252.

45. A Dichotomy Bearing Angel. Many-feathered wings of angels were at times employed to serve a function similar to that of *arbores* or of more straightforward dichotomies. In the present example, taken from a fourteenth-century manuscript, the six wings of a cherubim are used to represent the six kinds of meritorious acts that will, by "incrementa virtutum," enable one to draw close to God: love of God, love of one's neighbor, confession, satisfaction (that is, reparation for an injury committed), purity of mind, and purity of body. Each wing in turn reveals the more specific actions that will lead to the more generic merit in question. Purity of body, for example, may be increased through the free giving of alms, the holding of vigils, the obtaining of education, the exercise of prayers, and fasting. The cherubim stands on a wheel that gives a further dichotomy: that of the seven works of corporeal mercy (drawn largely from Matthew XXV, 35–36). For other pictures from this manuscript, see Illustrations 21, 58, and 276.

CHAPTER SIX

Rotae and Circular Diagrams

THE CIRCULAR DIAGRAMS KNOWN as *rotae* or wheels were constant companions of many early medieval handbooks of learning, the most notable instance of this phenomenon undoubtedly being the *De natura rerum* of Isidore of Seville. Indeed, these *rotae* were viewed as being so central to this work that, already in Pre-Carolingian times, in its manuscript copies it is often called the *Liber rotarum* instead of the *De natura rerum.* What is more, these particular circular diagrams were in no sense the invention of some scribe who deemed them to be useful aids for understanding the text; Isidore himself explicitly called for their presence, a request that was of influence well beyond the boundaries of his own work. For in many instances it is clear that *rotae* migrated from Isidore's *De natura rerum* to manuscripts of the handbooks of Macrobius, the Venerable Bede, William of Conches, and others, and at times even occurred alone without any accompanying text, both of these phenomena being sure indications of the importance and informativeness medieval scholars believed them to have. Of course, many circular diagrams were brought to bear in the explanation of the medieval universe where they fit well with its spherical form. The *rotae* of present concern, however, were meant to explain doctrines and ideas that did not have (save in a few instances) such a neatly matching "circular form." Like dichotomies, their central purpose was to divide and classify, or to set in an easily grasped sequence, elements within these doctrines and ideas. Some *rotae,* however, were designed to do more than classify things and set them in order; they were meant to reveal contrariety between some of the elements they displayed, the opposing spokes of a wheel or the ends of a diameter furnishing a convenient means of representing that contrariety (the second Illustration below gives an example of such a use of a *rota*).

There was an extremely wide variety in the kinds of doctrines and ideas that were classified by these wheeled diagrams, but their most popular use—especially in manuscripts of works deriving from the earlier Middle Ages—seems to have been in the visual explanation they provided of calendaric information on the one hand (the cycles of months or seasons, the factors involved in the proper dating of Easter, and so on) and of elementary astronomical and cosmographic information on the other (such things as the order and orbits of the planets, the signs of the zodiac, the various terrestrial zones, winds, and elements). Since "opposites" are to be found among such things as the seasons, the elements, and the winds, the relevant diagramming *rotae* could bring to bear their ability to exhibit contrariety as well as to classify. Later medieval works and manuscripts utilized wheels to represent more complicated mathematical doctrines and the technical aspects of such disciplines as logic, but their preponderant use in scientific literature remained in the calendaric and astronomical-cosmographic domains. Outside of science, *rotae* are found, like dichotomies and *arbores,* in the manuscripts of writings belonging to all corners of medieval thought, and in religious and moral works in particular. Yet unlike dichotomies and *arbores,* they only infrequently appear in unprofessional, student-produced forms.

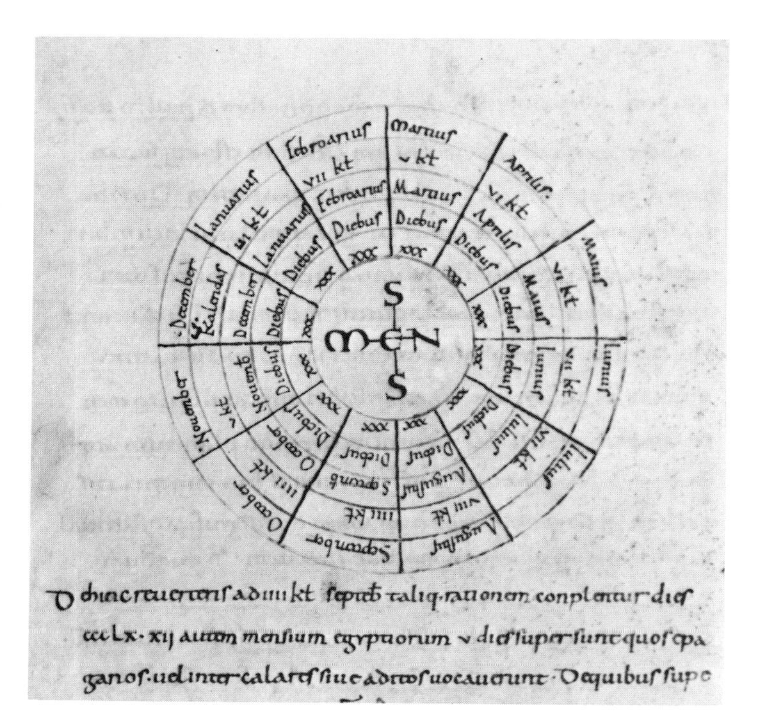

46. (Above) A Wheel of the Month. The *De natura rerum* of the seventh-century bishop Isidore of Seville explicitly calls for seven *figurae,* all but one of them *rotae* or wheels. The first of these is from a tenth-century manuscript (two others appear in the three Illustrations which follow, the remaining three in Illustrations 279, 280, 286, and the seventh, non-*rota* figure, in Illustration 247). As is obvious from its name, the present *Rota mensium* was to provide in pictorial form a complete roster of the months, the number of days in each, and the date of the first of each month in terms of the Roman kalends. Since Isidore is here concerned with Egyptian months, each month contains 30 days in his wheel. The Egyptian calendar consisted of 12 months of 30 days each plus 5 additional days at the end of each year (an addition which is not reflected in Isidore's *rota,* although it is duly noted in his text). His concern with Egyptian months also explains the information he gives regarding the kalends of each month. The kalends was the first day of each month in the Roman calendar; but, as Isidore explains, the first day of each Egyptian month precedes the kalends by a varying number of days, thus requiring the kind of concordance represented in his *rota.* Other figures from this manuscript are given in the two Illustrations which follow.

47. (Below) A Wheel of the Seasons. The second *rota* occurring in Isidore's *De natura rerum* was that of the year *(annus)*. Here each season occupies a position on the wheel directly opposite that of the contrary season and is at the same time characterized by its primary qualities. Since these primary qualities are shared by the seasons taken two by two, they serve as natural bonds *(vinculi naturales)* or common elements *(communiones)* uniting them. Thus, spring *(ver)* occurs at the top of the wheel and is wet and hot, taking part of the circular segment *(humidus)* to the left and part of the segment *(calidus)* to the right. But summer *(aestas)* on the right of the wheel is hot and dry, being united to spring by sharing part of the circular segment *(calidus)*. Proceeding clockwise around the remainder of the wheel, we have autumn *(autumnus)* at the bottom, dry and cold, and winter *(hiemps)* at the left, cold and wet. Isidore's *rota* also associates each season with a cardinal direction, something that is not explained in the text of the *De natura rerum,* but that is mentioned in his *Liber etymologiarum* (V, 35). In it he tells us that spring is to be associated with the east *(oriens)* because it is then that all things arise *(oriuntur)* from the earth. Summer is related to the south because of its greater heat, winter to the north because of its cold, and autumn to the west because then serious illnesses occur and leaves fall from the trees (where the obvious linguistic connection is between setting or falling [*occiduus*] and west [*occidens*]). One can also note in this copy of the *Rota anni* that, as in the case of all of his *rotae,* Isidore clearly intended their presence in his text since the final words above the *rota* explicitly tell us so: *haec figura est* (with which one can compare the similar texts immediately preceding the *rotae* in Illustrations 48 and 279).

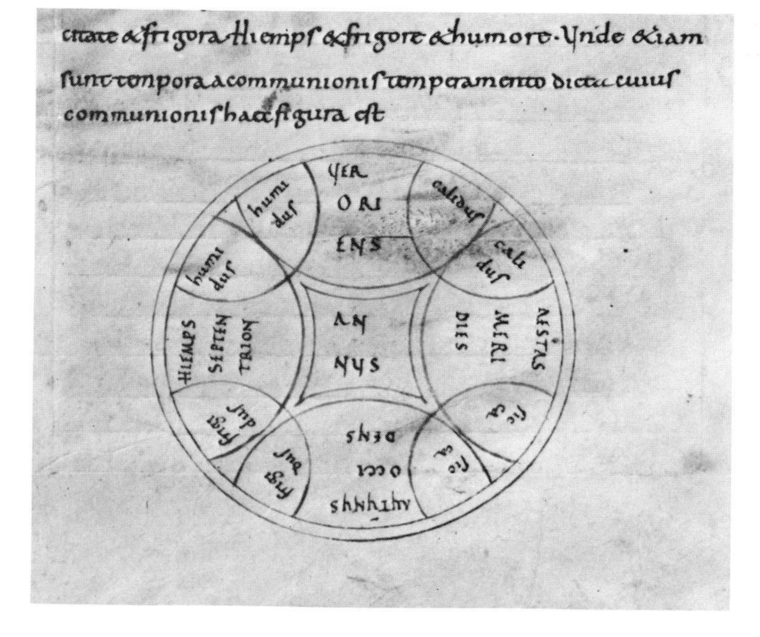

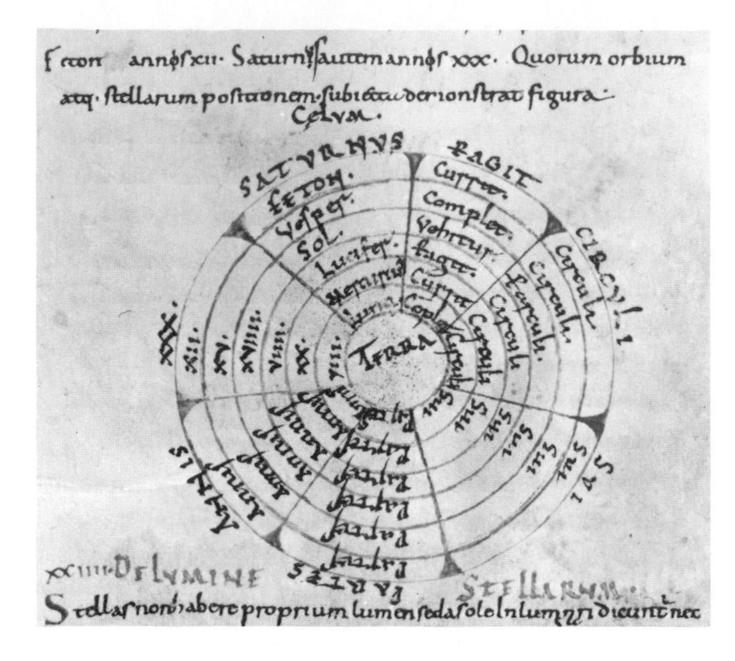

48. A Wheel of the Planets I. These are two examples of the fifth *rota* in Isidore's *De natura rerum,* both from tenth-century manuscript copies of the work. The wheel reveals the times of the orbits of each of the planets. The values are clearly visible in the *rota* on the **left.** Beginning at the top of the *rota* with Saturn, which completes its orbit in 30 years, the next outermost circle gives the corresponding period for Jupiter (here called Feton = Phaeton) as 12 years. Then, moving toward the center through successive circles, the period of Mars (here called Vesper, a name usually reserved for Venus as the Evening Star) is given as 15, that of the Sun as 19, that of Venus (here Lucifer) as 9, that of Mercury as 20, and finally that of the Moon as 8. The numbers given by Isidore are common ones which he could have derived from any number of sources.

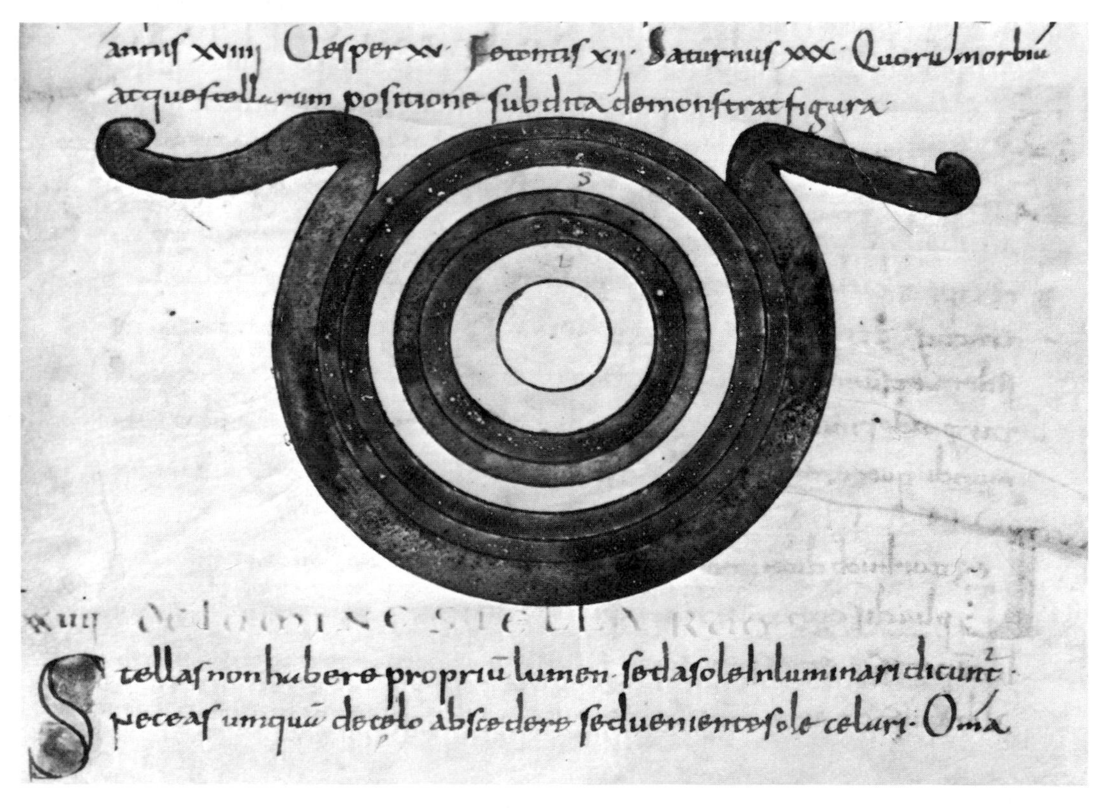

The *rota* **above** from the other tenth-century Isidorean manuscript is quite different. In fact, were it not for the text immediately above it assuring us that it too is a *rota planetarum,* it would be difficult to identify it. If there originally was some text written within this *rota,* a later overlay of pigment has totally obscured it. Although perhaps decorative, it is a *rota* with no didactic value at all. For another *rota* from this manuscript, see Illustration 279.

49. (Below left) A Wheel of the Planets II. The kind of information about planetary periods that was given by Isidore in his *rota planetarum* was so common in "handbooks" of the earlier Middle Ages that we cannot claim that the present *rota* is necessarily a development of the Isidorean model itself. Found in a late twelfth-century manuscript containing *computus* (that is, Easter-dating) material, the wheel pictured here contains the data found in Isidore (albeit with different values for some of the planets); but it also contains a fair amount of supplementary information. Most clearly distinguishable are the names of the zodiacal signs inscribed in the third outermost circle. Between this circle and the circumference we are told the date on which the Sun enters each zodiacal sign, while the outermost circle itself specifies the yearly orbit of the Sun and the Moon in days as well as the time they take to traverse each sign (which naturally amounts to a simple division by 12). However, the more interior circles not only carry the relevant information about planetary periods, but also tell us of the qualities of the outer planets: Saturn, for example, is cold, Mars fiery, so that Jupiter's position between them renders it temperate. We are also informed that the two inferior planets are sometimes *ante,* sometimes *post,* the Sun. For another circular image from this manuscript, see Illustration 234.

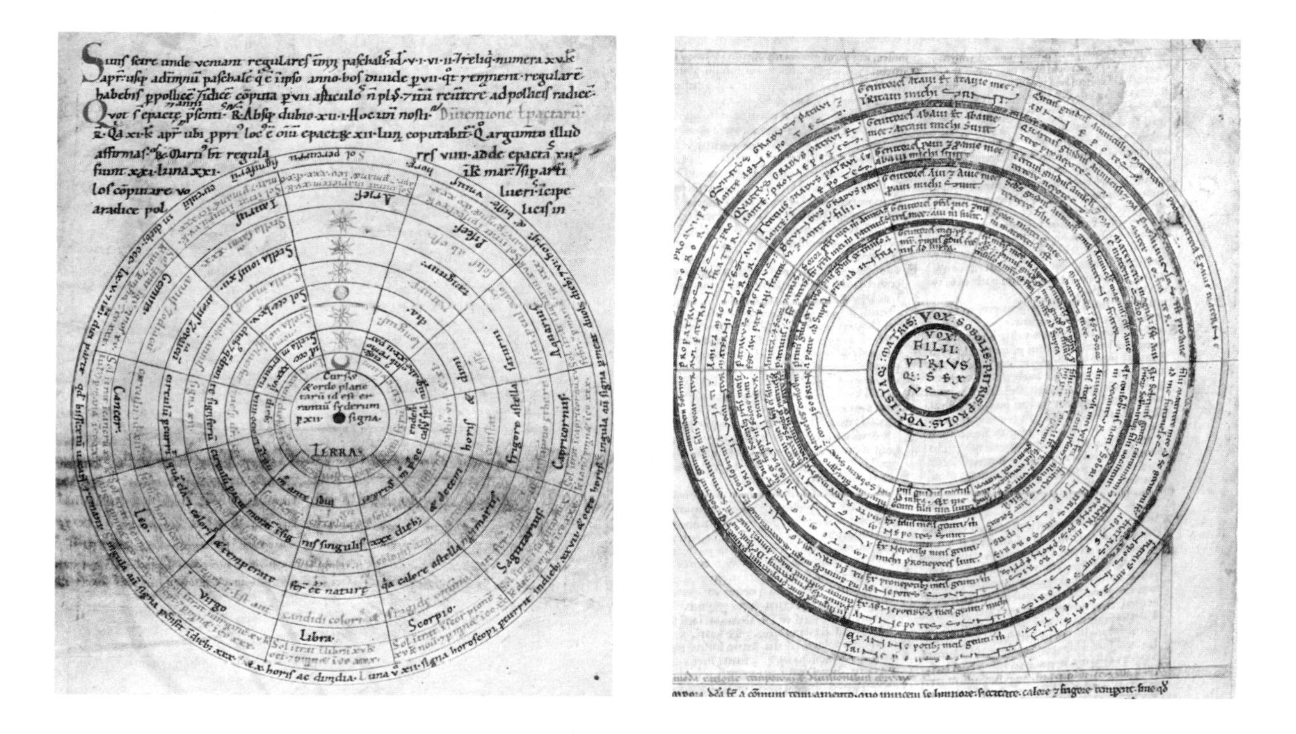

50. (Above right) A Wheel of Consanguinity. Illustrations 37–39 provide examples of the most common medieval way of diagramming consanguinity: by *arbores,* the progenitor being Isidore of Seville. Isidore's *Etymologies* also contain another way of visualizing kinship: by a *rota.* A late thirteenth-century example is given here. It does not appear in a copy of Isidore's *Etymologies,* but reproduces his *rota* as faithfully as do most earlier manuscripts of the text itself. One major difference between this figure and Isidore's *arbor consanguinitatis* (in either version) is that it exhibits more of that said in Isidore's text, giving, for example, not simply lines of relation, but definitions of the things so related. We are told that an *amita* is the sister of one's father and not merely presented with *pater* and *amita* side by side in the same genealogical "row." However, to add such information could cause "spatial problems." The present *rota* alleviates such problems by announcing (in its center) that the word *filius* ("son") will at times be used to cover both sexes and hence to apply to both paternal and maternal offspring. We have here one of the earliest uses of altering the meaning of a term so that it might function better within the confines of a particular diagram. In terms of the content of the *rota,* one's ancestors *(genitores)* are successively reached by proceeding directly upward from the center, while descendants *(geniti)* are covered by moving in the opposite direction. The left half of the upper semicircle treats ancestors and offspring relating to paternal uncles and aunts; that on the right covers the same on the maternal side. In both cases ancestors are distinguished from offspring by the terms *ad supra* and *ad infra.* The remainder of the *rota* addresses, among other things, the problem of variety in cousins and the offspring of one's siblings, giving relevant definitions as well as lines of filiation. For other diagrams from this manuscript, see Illustrations 131, 251, and 280.

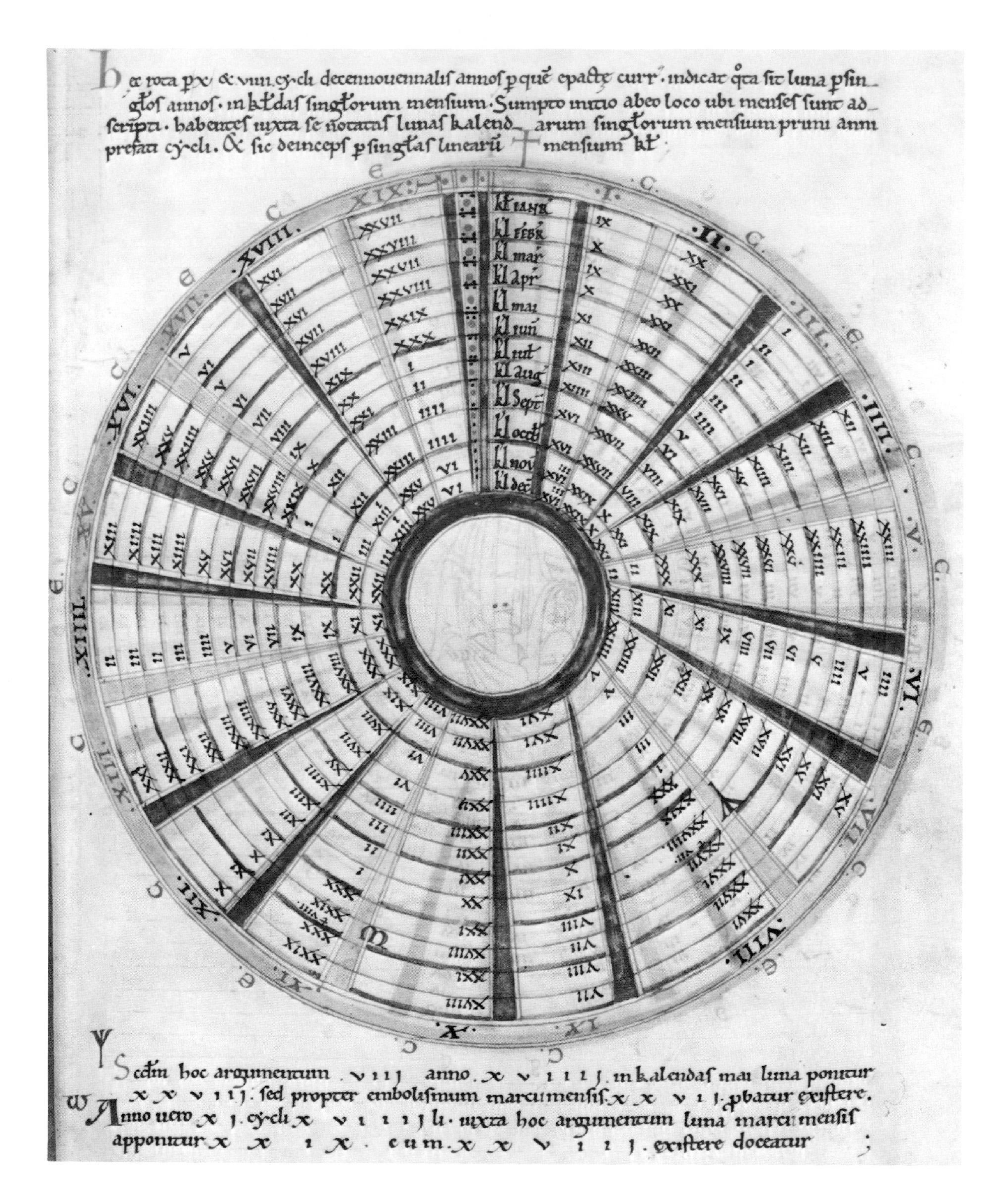

51. Rota computistica. The medieval term 'computus' referred to the discipline for calculating the date of Easter and, as a function of that, the dates of other movable ecclesiastical feasts. The calculation was problematic because Easter is fixed in the lunar calendar but not in the solar calendar. The crux of the problem is that 12 solar months contain more than 12 lunar months, an excess of roughly 11¼ days. The problem was finding a way "to take care of" that excess—finding some total number of solar years that will contain a whole number of lunar months. This was accomplished by "intercalating" an added lunar month for some of the years in a cycle corresponding to a selected number of solar years. Any year in a given cycle that had an added lunar month was called an "embolismic" lunar year; those in the cycle without an addition were called "common." In attempting to solve the "Easter problem" the

(*Continued*)

most successful cycle proposed was that of 19 solar years and 7 embolismic years. This cycle, known as the "Metonic Cycle," which yielded a correspondence of lunar and solar cycles that was off only a single day in 300 years, is the subject of the computistic wheel pictured here, as is evidenced from the words *rota per x et viiii cycli decennovalis* in the first line at the top of the page. The purpose of the *rota* is to tell one the age of the Moon (*quota sit luna*) on the first day or kalends of each month for each year of the cycle. Each of the 19 years forms a "spoke" of the wheel. Those which are embolismic are noted by an 'E' at the circumference; those which are common by a 'C'. The present rota is from a twelfth-century manuscript and is related to the major computistic treatise of the eighth-century Venerable Bede, the *De Temporum ratione* (Chapter 20). The *rota* was not part of Bede's text—he scorned those too ignorant or too lazy to calculate. Nevertheless, a great number of manuscripts of Bede's work do contain some kind of labor-saving table. The letters ω and Ψ appearing in the bottom half of the *rota* mark corrections necessary in the eighth and eleventh years of the cycle, corrections (from Chapter 45 of Bede) that are spelled out in the text at the bottom of the page, where the same letters appear again. This *rota* alone would not suffice for finding the date of Easter for a given year. In addition, one would need some source or table telling just where the year in question falls within the 19-year cycle, plus some table relating the relevant date to the days of the week. But such additional tables were available in the Middle Ages. For other material from the present manuscript, see Illustrations 31, 53, 69, 249, 279, 285, and 290.

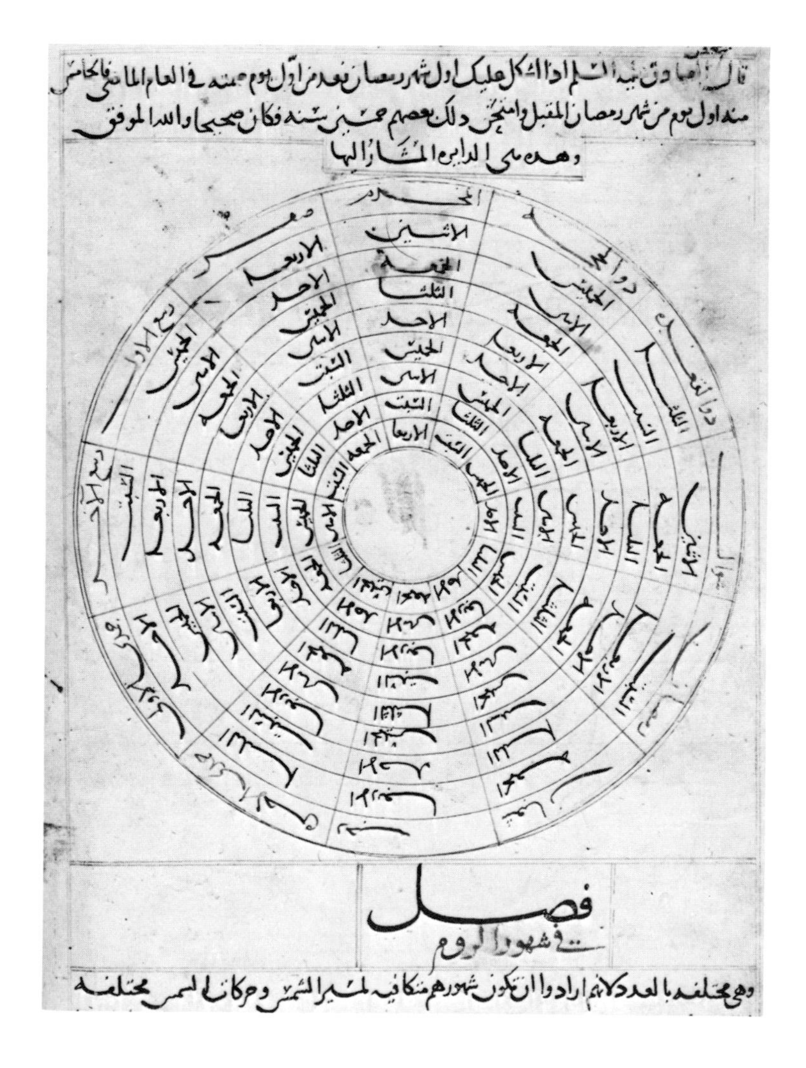

52. A Muslim Calendaric Wheel. The purpose of this *rota*, part of the *ʿAjāʾib al-Makhlūqāt (The Wonders of Creation)* of the thirteenth-century encyclopaedist al-Qazwīnī, is to facilitate the identification of the day on which any month in the Muslim calendar begins. We are to begin with the present Hijra year, or any Hijra year in which we might be interested, and to cast out, or subtract, eights from it until we reach a remainder less than eight. Then, selecting the month in which we are interested on the circumference of the *rota,* we count down the relevant sector toward the center a number of steps equal to that remainder and there find the day on which that month begins in the Hijra year in question. However, if the remainder is zero, we must count down eight steps. In introducing this *rota,* al-Qazwīnī tells us that its use will save laborious calculations, an attitude quite the contrary of that of the Venerable Bede. For other pictorial material from the present manuscript (which was written in 1280, three years before al-Qazwīnī's death), see Illustrations 126, 203, 208, and 253.

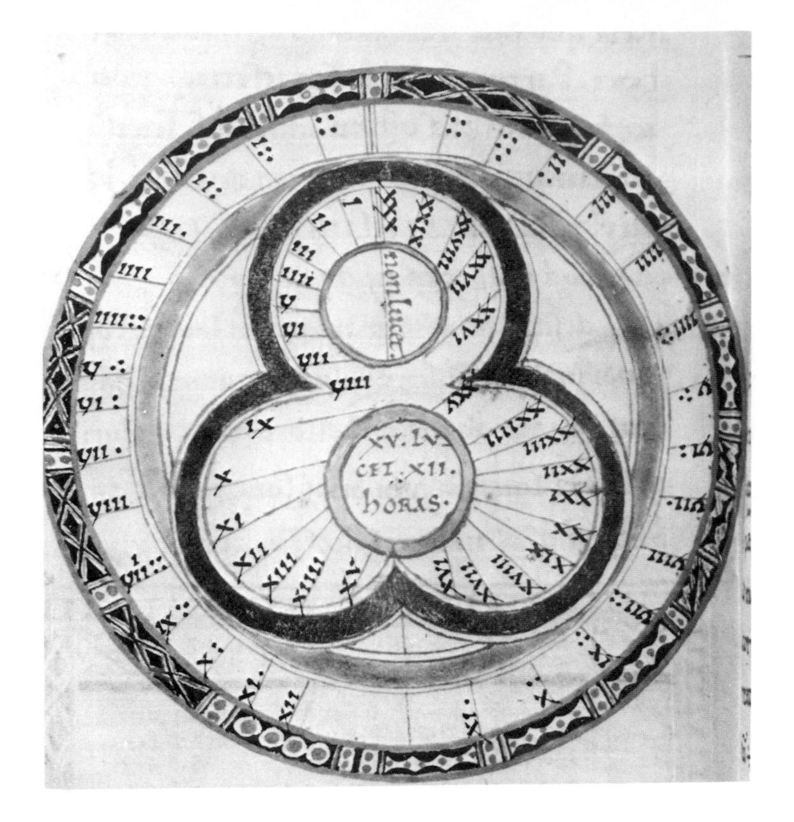

53. (Above) Measuring Moonshine. Found in the same twelfth-century manuscript as Illustration 51, this wheel illustrates Chapter 24 of Bede's *De temporum ratione.* The concern of this chapter is to determine how many hours any given moon will shine *(quot horis luna luceat)*. The present wheel gives a direct one-to-one correspondence between the age or days of the moon (increasing, counterclockwise, in the scalloped figure in the center) and the duration of moonlight (given on the wheel's circumference). The duration is given in hours and *puncti* (which, as Bede makes clear elsewhere, equal ⅕ of an hour when lunar calculations are involved), the latter conveniently represented on the wheel by dots. Thus, a one-day-old moon will shine for ⅘ of an hour, one nine days old for 7⅕ hours, one twenty-three days old for 5⅗ hours, and so on. Maximum moonlight occurs on the fifteenth day of a moon, as the inscription in the small central circle explicitly tells us. The small circle above it reveals that a moon does not shine (*non lucet*) on its thirtieth day. Once again, the figure is not part of Bede's text itself, but rather another "visual aid" added for the sake of easier comprehension. Bede himself gives a rule for calculating moonlight duration (a rule that seems to be derived from Pliny, *Hist. nat.*, XVIII, 70): from day 1 of the moon through day 15 multiply the preceding time (assuming the moonlight for day 1 as 4 *punctos*) by 4 and divide by 5; from day 16 on subtract the number of the day from 30 and multiply by 4 and divide by 5. For other figures from the present manuscript, see Illustrations 31, 51, 69, 249, 279, 285, and 290.

54. (Below) Rising and Setting Points. As the most important astronomical work of antiquity, the *Almagest* of Ptolemy (second century) was translated into Latin a number of times in the Middle Ages, both from the Arabic and directly from the Greek. The present *rota* is taken from a thirteenth–fourteenth-century manuscript of the Greek–Latin translation. Occurring at the end of Book VI of the *Almagest,* the *rota* gives a diagrammatic representation of the distances in the horizon (hence the heading *orizonton descriptio*) of rising and setting points of the ecliptic calculated relative to East and West of the horizon (that is, the rising and setting amplitudes of these points). The points in question are the end points of all of the zodiacal signs, their rising and setting amplitudes being given for the latitudes of each of the seven geographical climata that were standard in antiquity. In the present copy of this *rota,* south *(meridies)* is at the top, east *(oriens)* at the left. The seven climata are inscribed in order of decreasing latitude on the concentric circles at the top center of the *rota,* from *Boristhenes* through *Rhodi* in middle position to *Merois.* The bottom half of this vertical diameter lists the hours of the longest day for each of these climata as well as the heights of the pole for the same. The zodiacal signs appear on the circumference of the *rota*; the values for the rising and setting of each then are given in the relevant "columns" extending toward the center of the wheel. The values are given in Roman numerals in degrees and minutes, with minutes written to the left of the degrees. An added feature of this *rota* is the listing of the 16 directional winds in its center. For more on winds and climata, see Illustrations 279–280. The ultimate purpose of this Ptolemaic *rota* was the application of its information to a classification of the "inclinations" of solar and lunar eclipses; these "inclinations" in turn were of significance for weather prognostication.

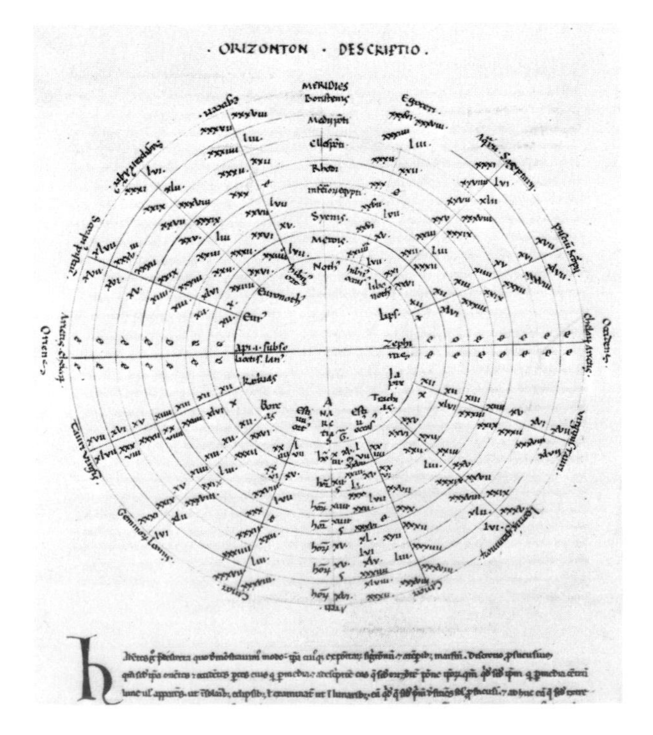

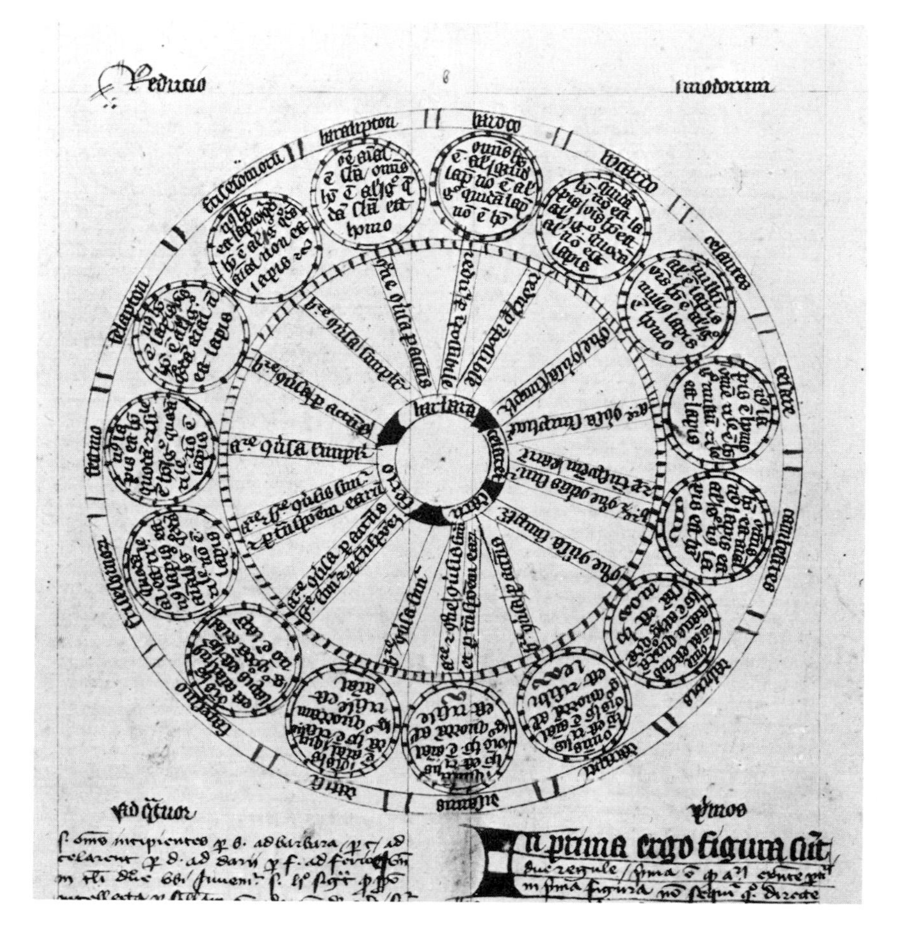

55. All Must Be Perfect: The Reduction of Syllogisms. For Aristotle, some syllogisms were merely valid, while others were that plus perfect. Perfect syllogisms were those of the first figure; imperfect, but nevertheless valid, ones were those of the second and third figures. The validity of the latter figures could be established by reducing their valid moods to a valid mood of the perfect first figure. It is this technique of reduction with which the *rota* pictured here is concerned. It is drawn from a fifteenth-century manuscript of a logical writing by the fourteenth-century Parisian scholar Jean Buridan. The mnemonic names (*Barbara, Celarent, Darii,* and *Ferio*) of the four valid moods of the perfect first-figure syllogism are engraved on the small central circle, while mnemonic names of the valid moods of the other figures (plus those of the indirect moods of the first figure) populate the wheel's circumference; examples of each of these latter valid moods are appropriately ensconced in circles directly below their mnemonic names. The "spokes" specify how to reduce the imperfect valid moods to the perfect ones of the first figure. The mechanics of reduction is tolerably simple if one has a few fundamentals of syllogistic in hand: The defining characteristic of a first-figure syllogism is that the "middle term" is the subject of the major premise and the predicate of the minor premise. For example: "No man is a stone; every risible being is a man; therefore, no risible being is a stone" is an instance of the valid mood of the first-figure syllogism mnemonically called *Celarent;* the middle term involved is "man." In second-figure syllogisms the middle term is the predicate of both the major and minor premises. For example, a syllogism in *Cesare* (on the right of the *rota* slightly above center) might be: "No stone is a man; every risible being is a man; therefore, no risible being is a stone." Using our *rota* we can learn how to establish the validity of this syllogism by reducing it to one of the valid moods of the perfect first figure. The instructions on *Cesare's* "spoke" read: "By simple conversion of the major premise" (*maiore conversa simpliciter*), that is to say, convert "No stone is a man" directly into "No man is a stone." With this new premise, we have a syllogism of the first figure: namely, that in *Celarent.* In some cases, however, the required reduction is not so straightforward. Nevertheless, if we know something of the mechanics of reducing, not directly, but *per impossibile,* the instructions inscribed on the spheres of our wheel enable us to effect the appropriate transformations. The *rota* does contain some error (notably listing the valid mood *ferison* as *frisesomoron*—which occurs rightly elsewhere in the *rota*—and incorrectly connecting it to *Barbara* rather than *Ferio* for reduction), but otherwise gives the relevant information. For more on the construction and function of mnemonic names for the syllogism, see Illustration 70. For other figures from the present manuscript, see Illustrations 12 and 61.

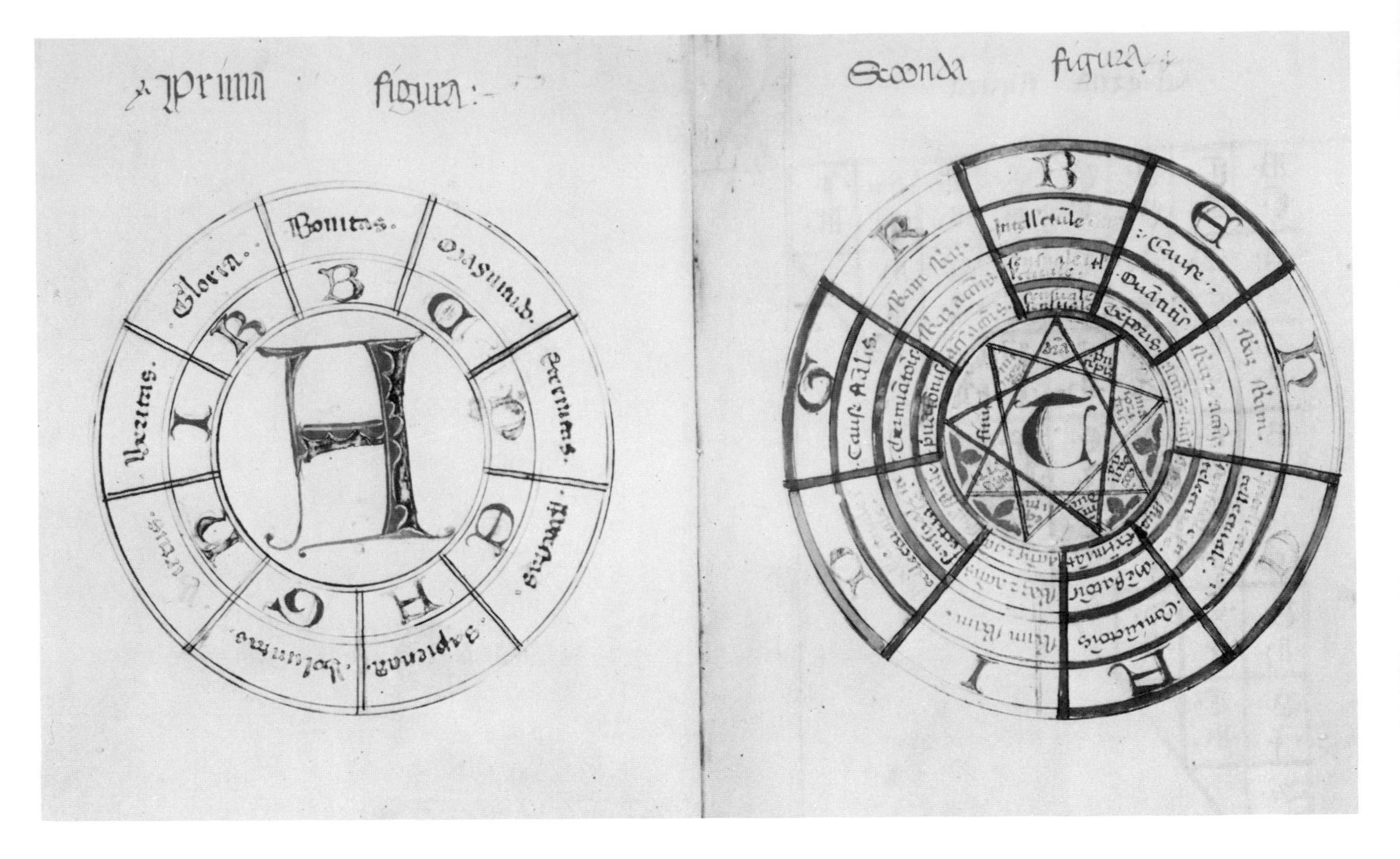

56. The World by Combinations. The thirteenth-century Catalan scholar Raymond Lull devised a system which, he believed, would enable one to know all natural things and would serve as a panacea for law, medicine, science, and theology as well. Called the *Ars generalis,* the heart of the system was the combination of concepts in such a way that all possible alternatives would be exhausted. The concepts in question were divided into six groups of nine each, the first two of which are illustrated here in two *rotae.* Taken from a fifteenth-century manuscript where they precede the commentary on Lull's *Ars generalis* written in 1433 by Juan Bulons of Barcelona, these *rotae* reveal the "Alphabetum" which enabled Lull to effect his multitudinous combinations with greater ease. In the *prima figura* at the left, the letters B through K represent, respectively, nine absolute principles: Goodness, Magnitude, Eternity (but in most other diagrams, Duration), Power, Wisdom (elsewhere, Knowledge), Will, Virtue, Truth, and Glory. Here, as combinations, one can see, Lull claims, "subject change into predicate and vice versa, as when one says 'Bonitas magna,' 'Magnitudo bona,' etc." Further, this first *rota* covers whatever exists (*quidquid est*), since whatever exists is either good or great, etc. The second *rota* gives a set of nine relative principles, grouped by threes in the vertices of three equilateral triangles. Thus, the triangle with a horizontal base and vertices directed toward B, C, and D carries Difference, Concordance, and Contrariety, which govern the relations between the intellectual and the sensible (here inscribed in the concentric circles surrounding the triangle). The HIK triangle yields Excess, Equality, and Defect, covering the relations between substance and accident, while the remaining EFG triangle contains Principle, Middle, and End, the relevant concentric circles now specifying the different types of each member of this final triad. In succeeding *rotae* and *tabulae,* the Alphabetum B through K is related to nine basic kinds of questions, nine types of subject, nine virtues, and nine vices. Given all these added groups of basic concepts, one scholar has calculated that, given Lull's manner of combining these concepts and his intention to do so exhaustively, he was faced with 531,441 combinations. One would think that that would have exhausted Lull, if not, as he hoped, all knowledge.

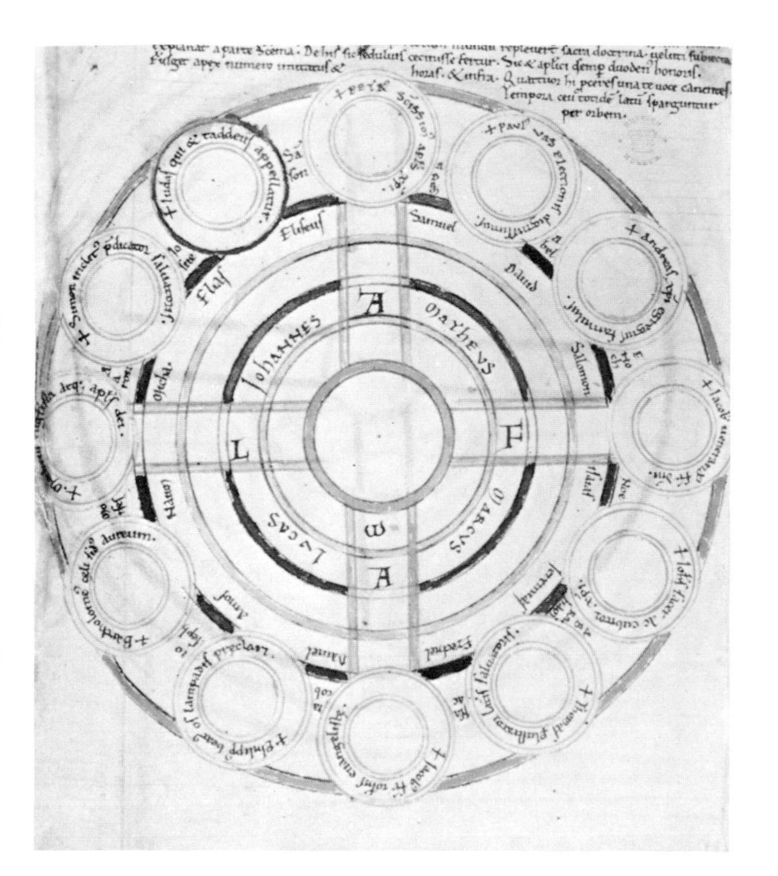

57. (Left) Circles and the Bible. The *rota* of this twelfth-century manuscript serves to present biblical personalities of import. The four Evangelists occupy the central position, and the twelve Apostles fill an outer rank of circles where, in addition to their names, one is reminded of their qualities: Peter (in the topmost circle) as *sanctissimus,* Paul (immediately to the right of Peter) as the most honorable chosen vessel (*vas electionis dignissimus,* a reference to *Acts* IX, 15), and so on. The names of 24 Old Testament patriarchs and prophets occupy positions both between these Apostolic circles and in a circle of their own immediately beneath them. In the center, the names of the four Evangelists are separated from one another by the letters A-L-A-F which, when seen in conjunction with the ω directly above the second A, reveal God to us as Aleph, or Alpha, and Omega, Beginning and End (*Revelation* I, 8). Other diagrams from this twelfth-century manuscript can be seen in Illustrations 247 and 263.

58. (Right) A Circle of Life. Entitled "Twelve Properties of the Human State," this fourteenth-century *rota* presents the Ages of Man together with a poetic explanation of each. The names and symbols of the four Evangelists fill the four corners, while the face of Christ serves as the center, around which is a verse telling us that the Holy Trinity creates, sets in order, and gives to us each and every thing *(Sancta Trinitas omnia dispono, creo singula cunctaque dono)*. The sector at the very top of the wheel tells us its purpose: a visual presentation of the nature of man, what he is, what he will be. The spokes of the wheel define each Age of Man and set forth the reason or nature *(ratio)* of these Ages, all in verse, some of the couplets being different versions of those we have already seen in Illustration 42. The Ages of Man of this *rota* add five to the seven that were paired with the seven liberal arts in the *arbor* or that earlier Illustration: birth *(nascens)* at the very beginning, and then, at the end, feeble, sick, dying, and dead *(imbecillis, infirmus, moriens, mortuus)*. For more on the Ages of Man, see Illustrations 289–90; and for other figures from the present fourteenth-century manuscript, see Illustrations 21, 45, and 276.

CHAPTER SEVEN

Squares of Opposition

THE REPRESENTATION OF CONTRARIETY that we have seen to be an occasional feature of *rotae* has become the central function of this particular type of visual aid. Originating within logic, where the relations between specific kinds of propositions called for definite kinds of opposition and connection, these square diagrams afforded a convenient means of representing the increased number of factors in need of explanation. Not only contrariety, but contradiction and subalternation, could be unambiguously pictured for easy comprehension, thus neatly covering the basic relations ensconced in the logical pairs affirmative–negative and universal–particular. As will be made evident by some of the following Illustrations, the relations displayed by these "logical squares" could become extraordinarily numerous and complex; more numerous when there was an increase in the number of elements being related, more complex when the nature of the relations involved needed more explanation to show their fit with simple contrariety or contradiction. The more "advanced" squares of opposition that attempted to account for these more numerous and complex relations almost always functioned as mere aids in comprehending their subject, since they were unable to present in a manageable visual form all of the necessary ingredients and features contained in that subject. The square might reveal, for example, that the kinds of propositions at hand were contrary or contradictory to one another in some special sense—naming that special sense, but not explaining what it was; that the consumer of the square was expected to know through other means.

Although the most frequent use of squares of opposition, especially in their more complex, advanced form, was decidedly the province of logic, they were also applied to other sciences that dealt with elements having relations to one another that were similar to those obtaining between affirmative and negative, or universal and particular, propositions. At times the contrariety, contradiction, and subalternation of these nonlogical elements did not exactly correspond to those relations as existing between propositions, but the match was always close enough to make the square of opposition in question quite effective as an explanatory visual device. In most instances—complicated or simple, logical or extra-logical, no matter—squares of opposition were applied in the explanation of a doctrine or conception which could be found in any number of medieval scientific texts. Less numerous are cases in which (as in Illustration 64) we find them employed to facilitate the understanding of a single argument that appeared only within a specific text.

59. The Fundamental Relations of Propositions. It was traditional in Aristotelian logic that categorical propositions (that is, nonhypothetical or nonmodal ones) could be either affirmative or negative, universal or particular. We witnessed that division as part of Illustration 44. However, that division alone leaves untreated relations between the propositions in question. Historically, the visual presentation of these relations was accomplished by a "square of opposition," a particular schema that became so classical that it is often still employed. In the square before us, the upper left corner is occupied by a universal affirmative proposition: "Every man is just." The upper right carries the universal negative: "No man is just." The lower left gives the particular affirmative: "Some man is just," and lower right the particular negative: "Some man is not just." At the top we are told that universal affirmative and universal negative propositions are contraries; at the bottom, that particular affirmative and particular negative propositions are subcontraries. We are also informed that the "difference" between any such contrary propositions is one of "quality," which is to say affirmative differs from negative in quality. The sides of the square inform us that universal and particular propositions (whether affirmative or negative) are subaltern, the latter of the former, the difference now being one of quantity (since universal and particular do differ in quantity). Finally, the diagonals of the square tell us that the universal affirmative is the contradictory of the particular negative and the particular affirmative the contradictory of the universal negative. The present square of opposition is from a fourteenth-century copy of Boethius' work *On the Categorical Syllogism* (see also Illustration 62).

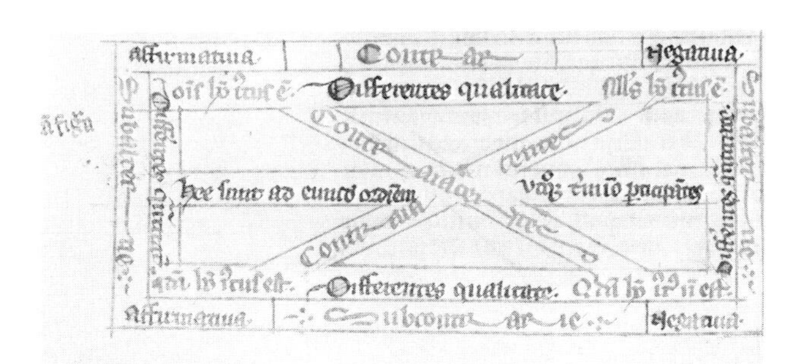

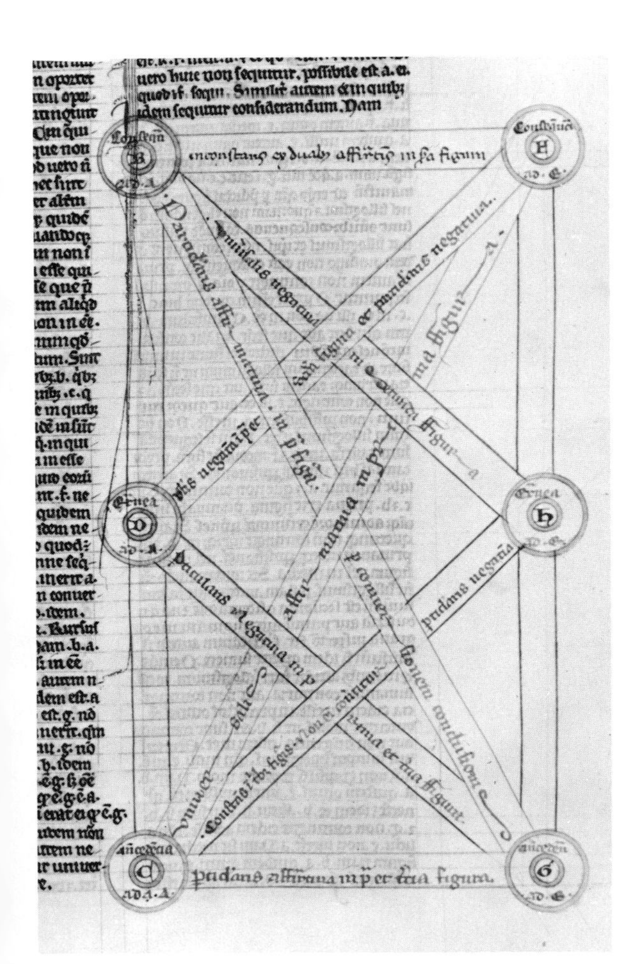

60. The Construction of Syllogisms. Aristotle addressed himself to problems concerning the discovery or construction of categorical syllogisms, the problems that became known in the Middle Ages as relating to *inventio medii,* that is, the discovery of the middle terms that made syllogisms "work." This fourteenth-century six-element square of opposition puts the substance of Aristotle's complex text into a more visually comprehensible form. One must, however, turn to the text itself to find out, to begin with, just what the six elements of the diagram are. The variable A is any predicate and the variable E any subject. Then B (in the upper left corner of the square) stands for the consequents of the predicate A (in the sense that, whenever one has A, one has B), F (in the upper right corner) for the consequents of the subject E. At the bottom of the square, C represents the antecedents of A (in the sense that, whenever one has C, one has A), G the antecedents of E. Finally, in the middle of both sides we have D as attributes which cannot possibly belong to A (here termed *extranea*) and H as those which cannot possibly belong to E. With this in hand, the diagram now reveals in shorter order the "rules" spelled out at greater length in Aristotle's text. If, for example, some C is the same as some F (which is to say that some antecedent of the predicate A is identical with some consequent of the subject E), then, as the inscription on the diagonal connecting C with F tells us, we have a syllogism in the first figure (the middle term being the subject of the major premise and the predicate of the minor premise) with a universal affirmative conclusion. For, following the meanings of consequent and antecedent as used here, C as the antecedent of A means that any time we have C we have A or (1) All C is A; similarly, F as consequent of E means that any time we have E we have F or (2) All E is F. But when, as in the present instance, the C in question is identical to the F in question, then (1) and (2) yield a syllogism whose conclusion is (3) All E is A. Thus, we have "discovered" the middle term C = F which makes the syllogism "work." The base and the other diagonals of the square furnish similar information for other of its six elements which, taken two by two as being identical, provide the middle terms for other syllogisms. Two such combinations (C = H and B = F) will not, however, result in a syllogism; but that, too, is noted in our square. The present figure is from the same manuscript as the preceding illustration; so, too, are Illustrations 44, 62, 104, 178, 241, and 252.

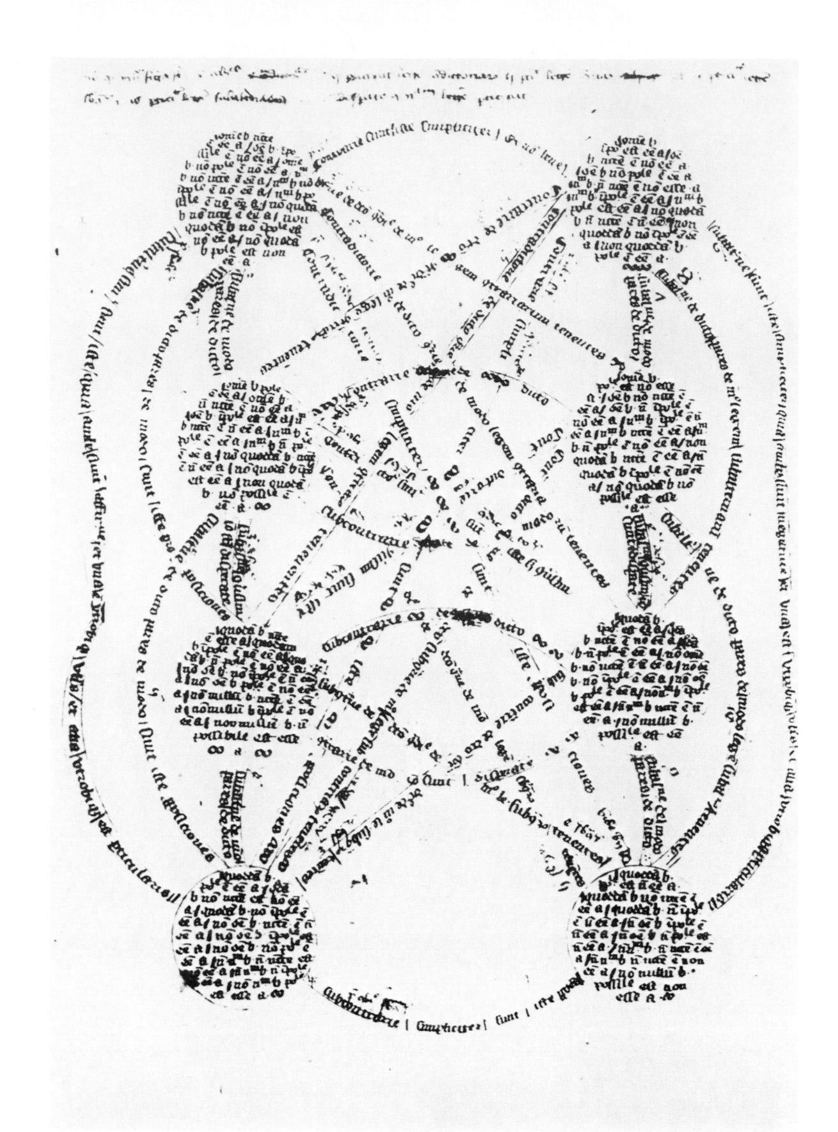

61. The Relation of Propositions: Advanced Course. The opposition of propositions that is diagramed in the square of opposition in Illustration 59 was basic, but elementary, in medieval logic. If one wished to treat the kinds of relations between noncategorical propositions, that is, between the likes of modal, hypothetical, and composite propositions, then the task of diagramming, let alone setting forth in a text, all of the twists and turns of the relevant rules was a more ambitious undertaking—as the present two squares of opposition clearly reveal. That above, taken from the same fifteenth-century manuscript of Jean Buridan we saw in Illustrations 12 and 55, is not simply an eight-element square; it attempts to include in each of these circular elements examples of all of the types of quantified modal propositions whose relations with others it seeks to explain. For example, the first three of the nine entries in the circle at the upper left are: 'For every B it is necessary that it be A,' 'For every B it is impossible for it not to be A,' 'For every B it is not possible for it not to be A.' Since there are also nine sample propositions in each of the other circles of the square, Buridan was in effect diagramming the relations between 72 different kinds of modal propositions. Some of these are, as his "connecting lines" read, simply contrary or simply contradictory, but in most cases things were more complex. Unlike the relations of quantified categorical propositions where one had simply to deal with the relations effected by the presence of the terms 'all,' 'none,' 'some,' and 'not,' one now had all of these variables *plus* the complications effected by the presence of varying modes like necessary vs. possible or necessary vs. impossible. In many cases the result was propositions that were not *simply* contrary, contradictory, subcontrary, or subaltern, but (for example) contrary in one way, contradictory in another, *secundum legem.* That is to say, in some instances one could not determine the overall opposition of two propositions merely on the basis of their form, but one had to appeal to different elements within the proposition that suffered opposition in different ways and also to the laws *(leges)* governing propositions whose opposition was determinable by form alone. Thus, the proposition (at the upper left in the square) 'For every B it is necessary that it be A' and the proposition (in the second circle from the top

(Continued)

at the right) 'For every B it is possible that it not be A' are related as contraries in terms of what they assert *(de dicto)* and contradictories in terms of their modes (necessary vs. possible). However, both of these oppositions are determined in accordance with the law governing ordinary contraries *(legem contrariarum tenentes)*, a stipulation that means that, although both propositions cannot be simultaneously true, they can both be simultaneously false. The second square of opposition shown below appears in a sixteenth-century printed work: the *Expositio ... in primum tractatum Summularum Magistri Petri Hispani* by the sixteenth-century Spanish scholar Juan de Celaya. The original treatise which is here expounded was written by the thirteenth-century philosopher and physician Peter of Spain (later Pope John XXI), and became a standard logical handbook in the Middle Ages, commented upon by numerous later medievals. Celaya's commentary contains the present twelve-element square. Although he inscribes only a single proposition in each circle of the square, like Buridan he specifies relations of opposition that are *secundum legem* as well as *simpliciter*. He has diagrammed hypothetical and composite propositions of various sorts (the six on the right being simple contradictories of the six on the left). This renders the kind of relations he must deal with less complex than those treated by Buridan, in spite of the fact that his square is much more difficult to comprehend. Indeed, in this case the attempt to visualize things seems to have gone berserk; one wonders how it could have served as an effective shortcut for any kind of understanding.

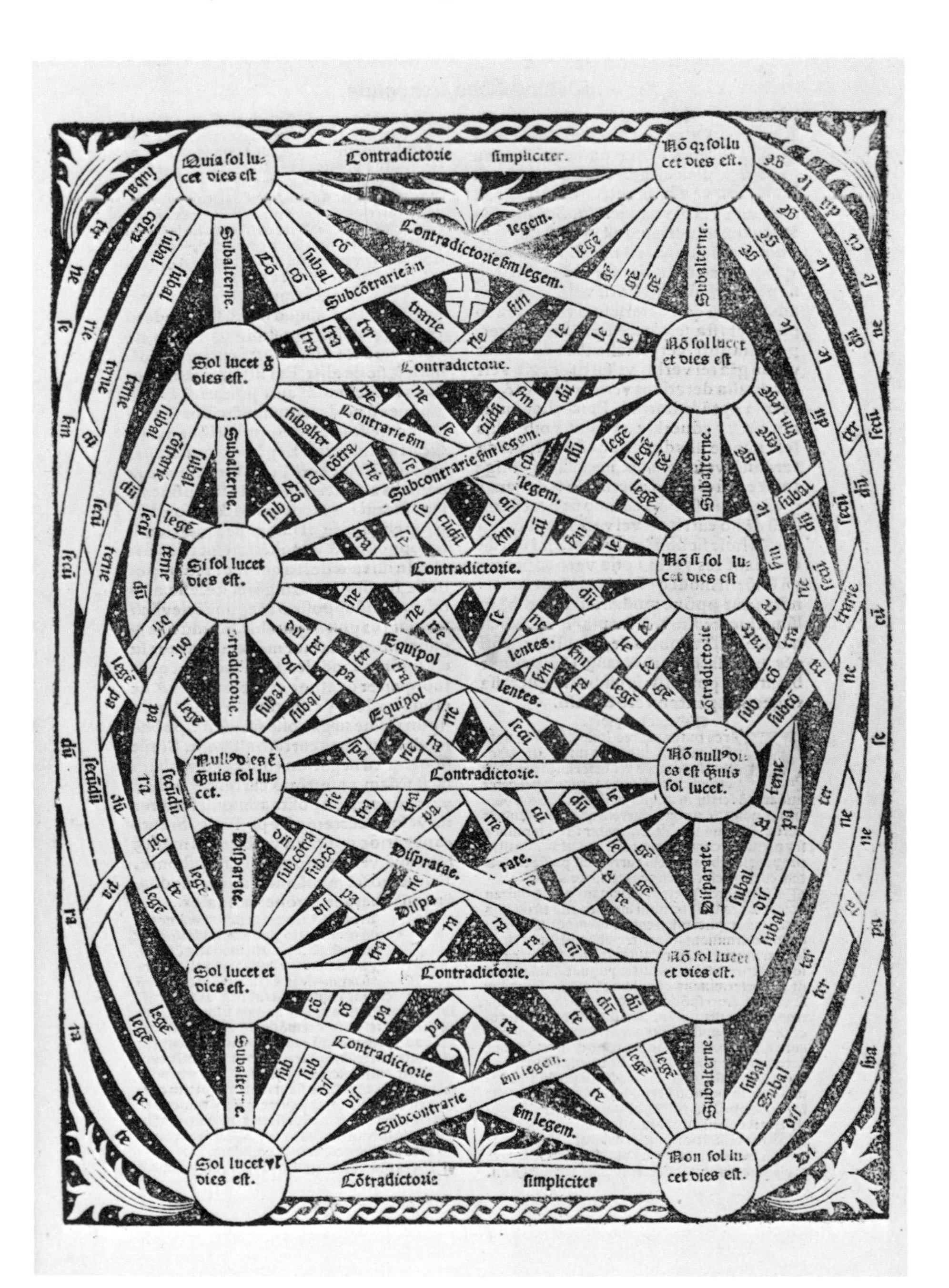

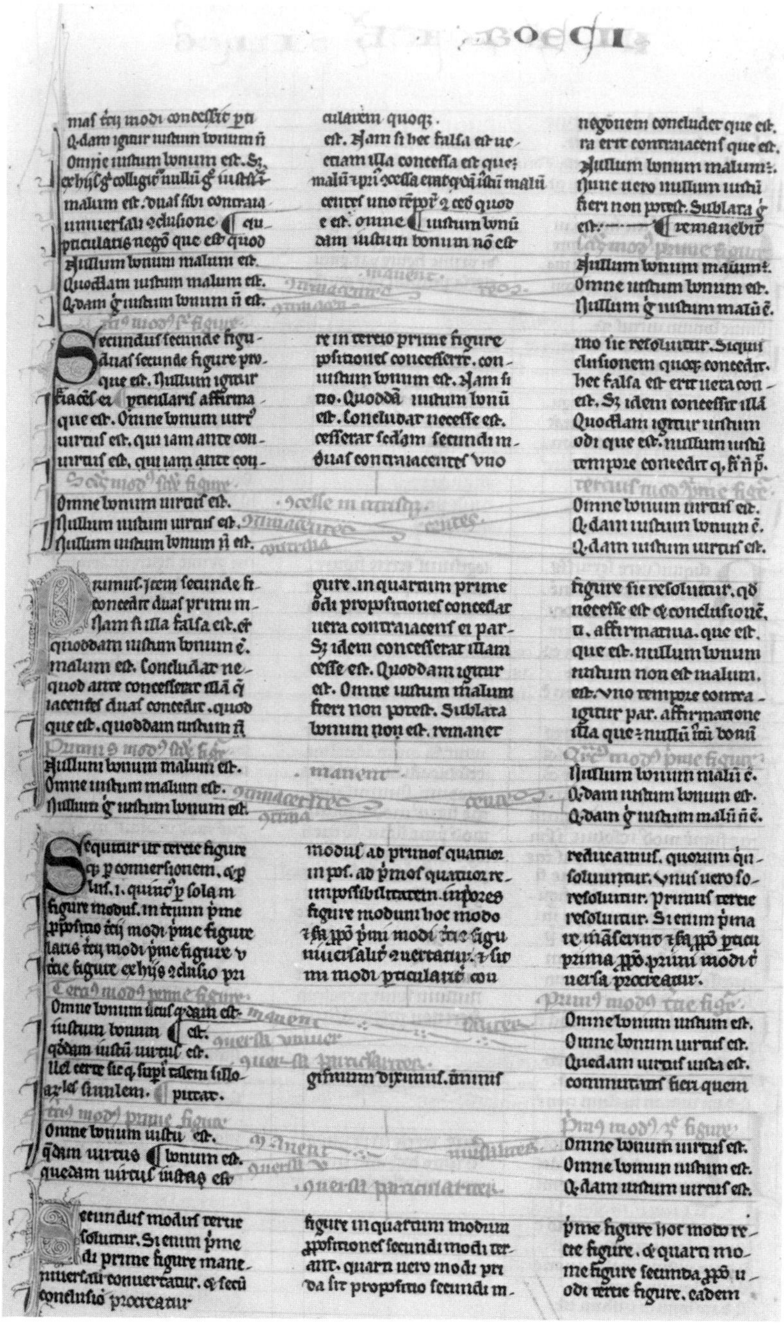

62. Opposition Within the Text. Illustration 55 presented us with a *rota* that set forth the doctrine of the reduction of moods for the categorical syllogism. These intertextual squares of opposition address themselves to at least part of the same doctrine. Taken from a fourteenth-century copy of Boethius' *On the Categorical Syllogism,* the moods of each figure are taken one by one and resolved (*resolvitur,* in place of *reducitur*) into the relevant moods of the first figure. Unlike the *rota* of Illustration 55, the mnemonic names for the valid moods do not appear here, since they were not yet invented in Boethius' time. Instead, Boethius numbers each mood relative to the figure to which it belongs. His procedure can be illustrated if we take as a sample the central paragraph beginning with the initial P. The first mood of the second figure, Boethius tells us, is resolved into the fourth mood of the first figure. Using his example, the first mood of the second figure is: 'No good thing is evil; every just thing is evil; therefore, no just thing is good.' The "resolution" establishes the validity of this syllogism. For take the particular affirmative which is the contradictory of its conclusion: (a) 'Some just thing is good'; combine this with something already conceded by the second figure syllogism whose resolution is at stake, namely: (b) 'No good thing is evil', and one must conclude: (c) 'Some just thing is not evil.' But (b), (a), and (c) form a syllogism in the fourth mood of the first figure, so our resolution is complete. However, it remains to show how this validates the initial second-figure syllogism. We do so by noting that (c) contradicts the already conceded second premise of the second-figure syllogism: 'Every just thing is evil'; since one cannot concede two contradictories at the same time, we must "remove" the premise that gave rise to the contradiction, namely, (a). But this leaves the contradictory of (a), namely, 'No just thing is good,' which is the conclusion of the second-figure syllogism we were "testing." One must conclude, then, that this original syllogism is valid. At the conclusion of this explanatory text, Boethius tabulates the two syllogisms involved in the resolutions, that of the second figure *(primus modus secunde figure)* to be resolved on the left and that of the first figure *(quartus modus prime figure)* to which it is resolved on the right. This sets the stage for the scribe of our copy to construct a partial square of opposition between the two syllogisms. The diagonal labeled *contrajacentes,* which runs from the conclusion of the second-figure syllogism on the left to the second premise of the first-figure syllogism on the right, indicates that these are contradictory, and corresponds to the first step given by Boethius. The second step—namely, the mutual concession of a common premise—is put into our square as the horizontal "manent," which says, in effect, that the two propositions thus connected remain unchanged. The final move in Boethius' procedure is reported by the remaining diagonal (again labeled *contrajacentes*), pointing out which proposition will have to be "removed" and which will, therefore, remain. The other intertextual squares of opposition function in a similar way in illustrating Boethius' resolution of other imperfect valid moods. That the introduction of these squares into the text was planned in advance is evident from the fact that the text itself is in three neatly separated columns, even though one still reads the text continuously across the whole page. The separation into columns provided for the subsequent introduction of the explanatory squares of opposition. For other figures from this present manuscript, see Illustrations 44, 59, 60, 104, 178, 241, and 252.

63. The Mathematics of Musical Intervals. Although the primary and most effective use of squares of opposition in the Middle Ages was in logic, they were occasionally employed as "visual aids" in other disciplines as well. This eleventh-century manuscript of Boethius' *De institutione musica* shows an example of its use in that liberal art. By inscribing the numerals XII, IX, VIII, and VI in the corners of a square (proceeding clockwise from the upper left), one could display in an immediately comprehensible way the four basic intervals of musical theory. For the numbers 12, 9, 8, and 6 were the basic "building blocks" for the ratios determining these intervals. Thus, the ratio of 12 to 6 (which is 2 to 1) is that of the octave or *diapason* (as indicated on the left side of the square). Similarly, the ratio of 12 to 9 or 8 to 6 (both equal to 4 to 3) gives the interval of a fourth or *diatessaron* (as written on the top and bottom of the square), while 12 to 8 and 9 to 6 (equal to 3 to 2) yield the fifth or the *diapente* (as on both diagonals). That leaves the ratio of a whole tone *(tonus)*, appropriately given in the ratio of 9 to 8 on the right side of the square. For other diagrams from this manuscript, see Illustrations 102–103 and 282.

64. Aristotelian Natural Philosophy in a Square of Opposition. The fourteenth-century Parisian mathematician and philosopher Nicole Oresme translated a number of Aristotle's works into French, commenting on them at the same time. The present figure is from a late fourteenth-century manuscript of Oresme's translation and commentary on Aristotle's *De caelo,* the *Traité du ciel et du monde,* written in 1377. In attempting to establish the eternity of the world (*De caelo* I, ch. 12), Aristotle had argued that being generable and being corruptible or destructible implied one another (that is, if something is generated or capable of being generated, then we can infer that it will pass away or is capable of passing away, and vice versa). On the basis of this mutual implication of generable and corruptible, he went on to establish the mutual implication of the *un*generable and the *in*corruptible. For, Aristotle argued, if one assumes that being incorruptible does *not* imply being ungenerable, then, since any given thing whatever is either ungenerable or generable, being incorruptible must imply being generable. However, by hypothesis, being generable and being corruptible imply one another, which means that being incorruptible implies being corruptible, which is impossible. Therefore, being incorruptible must imply being ungenerable. Q.E.D. It is this argument that Oresme has tried to elucidate with his square of opposition. He has translated Aristotle's "ungenerable" and "generable" as, respectively, "*sans commencement*" ("without beginning," on the upper left) and "*avoir commencement*" ("having a beginning," lower right) and his "incorruptible" and "corruptible" as, respectively, "*sans fin*" ("without end," upper right) and "*avoir fin*" ("having an end," lower left). In order to facilitate his exposition, he adds, again following Aristotle, the variables E, I, T, and Z next to the circles containing the foregoing labels. To represent Aristotle's claim of mutual implication between E and G on the one hand and T and Z on the other, he has labeled the two horizontals in his square "convertibles," while the diagonals carry the notice "contradictoires" to cover Aristotle's claim of the impossibility of implication obtaining between the elements at their ends. Although the square of opposition alone would surely not have allowed a reader to construct Aristotle's argument, having it at one's side while reading Aristotle's text and Oresme's commentary would undoubtedly have facilitated the reader's understanding of the material. One should note by way of conclusion that, given the Christian doctrine of a temporal creation, Oresme was loath to admit the absolute validity of arguments supporting the eternity of the world. He therefore closed his account of this particular segment of Aristotle by saying that the premise on which Aristotle had built his argument (namely, the mutual implication between generable and corruptible) was false. For other "visual aids" from this manuscript, see Illustrations 254–255 and 278.

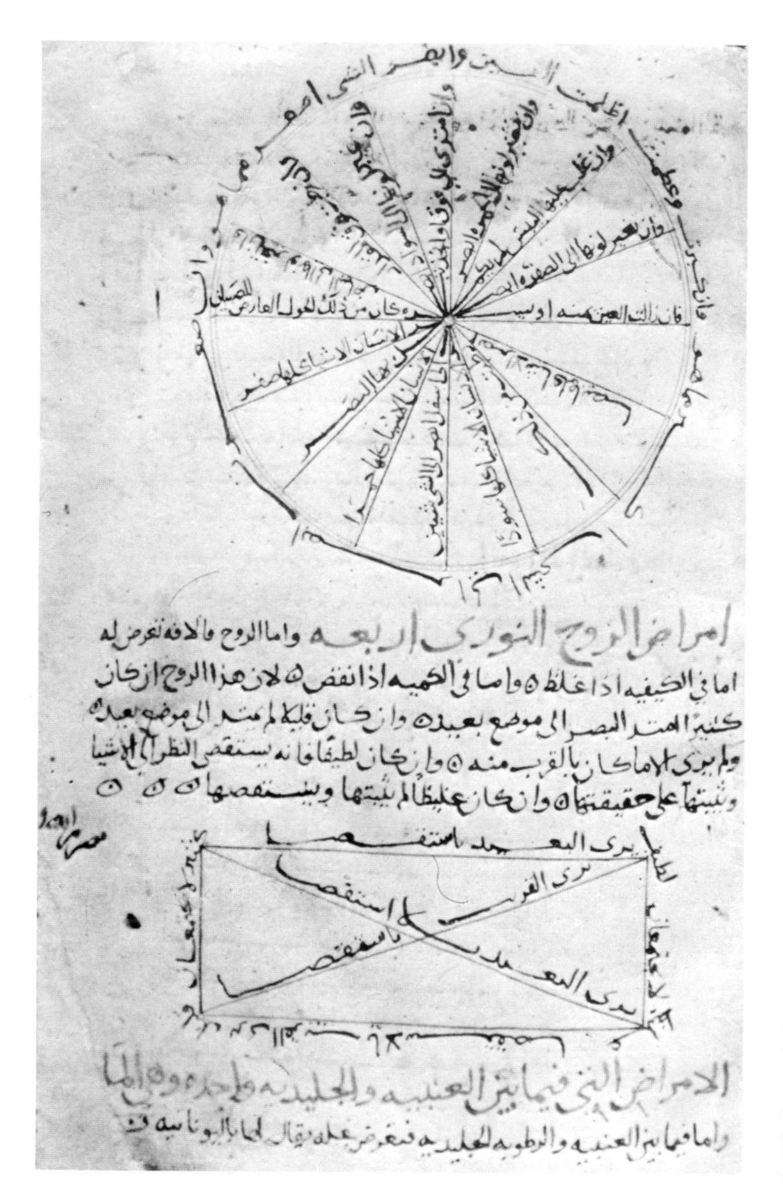

65. Eye diseases in Circle and Square. Treatises on the eye were common in medieval Islam, and an especially popular "handbook" of this genre was the *Tadhkirat al-kaḥḥālīn (Memorandum for Oculists)* of the eleventh-century oculist ᶜAlī ibn ᶜĪsā of Mawṣil, known as Jesu Haly in the Latin West. Consisting of three books, the *Tadhkira* covered all aspects of ophthalmology, the *rota* and square of opposition pictured here being taken from a twelfth-century manuscript of the third book. The subject is the variety of diseases of the crystalline or aqueous humor. Twelve such diseases are displayed in the *rota,* the first two being described on its circumference: if the humor increases in amount, then it darkens the eye and things will be seen smaller than they really are, while its diminution will have the opposite effect. Furthermore, beginning with the horizontal diameter of the *rota* and proceeding clockwise, the spokes tell us that if the crystalline humor is displaced to the left or the right, the squinting that we find in children occurs; if the humor turns yellow, then things appear yellow (a notion that most likely in some way refers to jaundice); if the humor dries up, then vision will be less perfect (other manuscripts say that the eye will then become bluer, a phenomenon historians have taken to refer to glaucoma); if the humor reddens, then things will be seen as red; if the humor is displaced up or down, double vision will occur; if it blackens, then things will appear black; if the humor liquefies, then the eye itself will become wetter; and, finally, if the crystalline humor becomes whiter, then objects will appear white. The square of opposition below this *rota* approaches disorders of the humor in a different fashion. The humor can be either greater or smaller in quantity (respectively, the upper and lower corners on the left) or thin or thick in quality (respectively, upper and lower corners on the right). Greater quantity and thinness results in seeing near things clearly (and the upper horizontal connecting these two variables carries a label to that effect), while smaller quantity and thickness entails an unclear vision of near things (so specified on the lower horizontal line). On the other hand, greater quantity combined with thickness (see the diagonal going from the upper left to the lower right) indicates vision at a distance, while the other diagonal, representing the combination of smaller quantity and thinness, indicates the clear vision of near objects. Finally, the two sides of our square draw attention to the rather obvious fact that great and small, thin and thick, cannot coexist in the humor. Although much of what one finds in these two diagrams of ᶜAlī was derived from Galen, his *Tadhkira* was still a convenient source for such information, both in Islam and, upon its translation, in the Latin West. Other diagrams from this manuscript can be seen in Illustration 213.

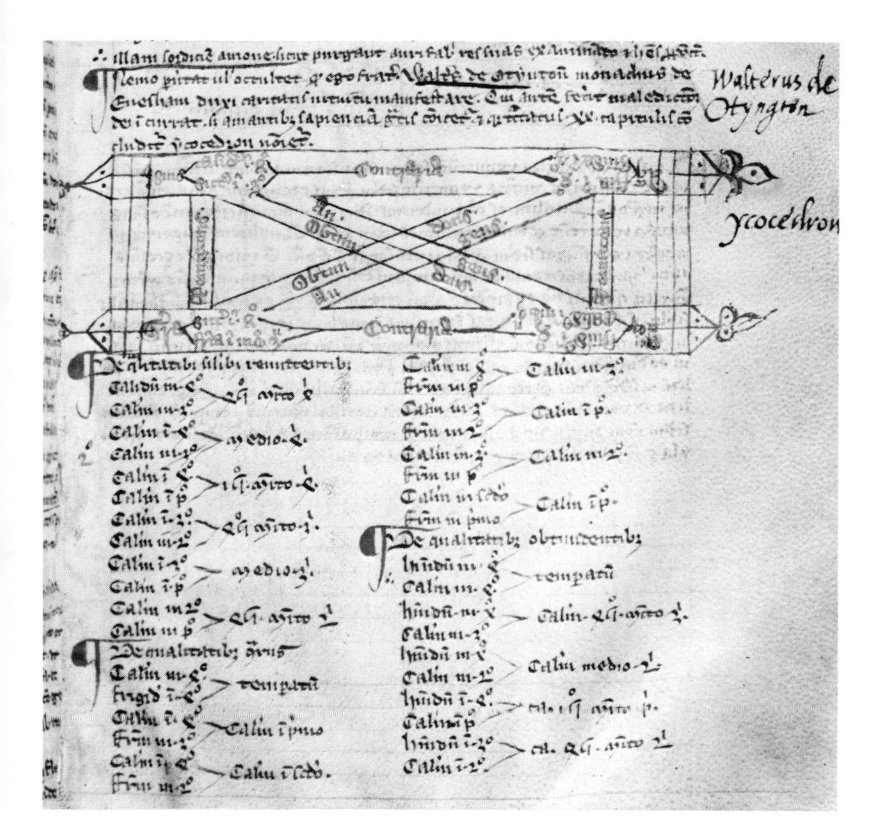

66. The Opposition of the Elements. This square is taken from a fourteenth-century manuscript of a brief but curious alchemical treatise called the *Icocedron,* a name deriving from its twenty chapters. Written by Walter Odington, who was apparently a Benedictine monk at the Abbey of Evesham in Worcestershire in the early years of the fourteenth century, the final chapters of this work speak of the mixing and contrariety of the elements and it is this matter to which the square of opposition pictured here relates. Fire *(Ignis)* and Earth *(Terra)* occupy the corners on the left, Water *(Aqua)* and Air *(Aer)* those on the right (the inscriptions for the last two to be read by turning the square upside-down). As was standard, Fire and Water, Earth and Air, are specified as contraries and hence have positions at the ends of the sides of the square. The text of the *Icocedron* treats not only the contrariety of the elements interpreted in terms of their primary qualities, but also addresses itself to the degrees *(gradus)* of these qualities as possessed by the elements. Clearly the elements represented are elements as they occur in mixtures *(prout in mixto)*, since it is in that state that they possess their primary qualities in the degrees indicated in our square: Fire, hot in 4°, dry in 3°; Water, cold in 4°, wet in 3°; Earth, dry in 4°, cold in the middle of 3°; Air, wet in 4°, hot in the middle of 3°. The vertical sides of the square are both labeled *remittens,* most likely a reference to the fact that the common qualities on these sides (dry on the left, wet on the right) are remitted, or decreased, in degree as one moves from bottom to top, an interpretation that receives support from the table or list of similar remitting qualities *(de qualitatibus similibus remittentibus)* directly below the square on the left. This list and those following it, both on the present and the next page of the work, give the result in degrees and minutes of combining certain pairs of qualities, where the combination would presumably occur through the mixing of elements possessing those qualities (such as hot with hot, hot with wet, and so on). Two of these lists tell us more about our square of opposition: that on the lower right dealing with weakening qualities *(de qualitatibus obtundentibus)* and that on the next page treating of strengthening qualities *(de qualitatibus augentibus)*. Taken together, they explain the diagonals of our square, since the words 'strengthening' and 'weakening' *(augens, obtundens)* are written on both of them. Thus, the diagonal joining the upper left with the lower right connects hot with wet and hot with dry, precisely those pairs of weakening and strengthening qualities listed in the tables. The other diagonal is more problematic, since Odington nowhere tells us which qualities other than the pairs hot–wet and hot–dry are strengthening and weakening. Yet if we use the remaining diagonal as evidence, presumably the pairs it connects, cold–wet and cold–dry, would qualify.

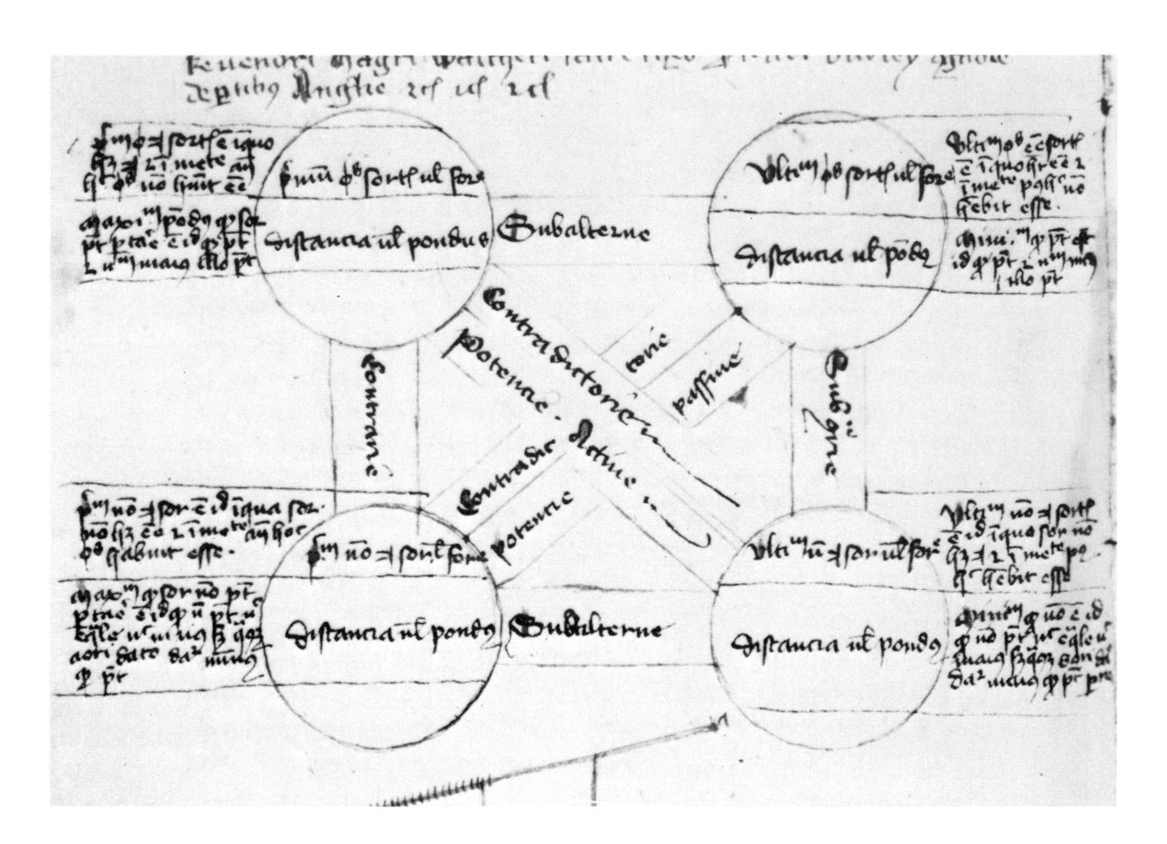

67. Setting Limits to What Exists and Can Be. In the later book of his *Physics,* Aristotle was much concerned with problems surrounding the continuity of motion and change and devoted considerable effort to establishing the inadmissibility of the atomistic or indivisibilist composition of magnitudes, motion, and time. One part of his effort in this regard was his attempt (*Physics* VI, chapter 5) to show convincingly that there is no "primary when" or first moment of continuous change or motion, although there is such a "first" moment at the end of such a change signifying that the change has been completed. In a second passage in his *Physics* (Book VIII, chapter 8), he set forth the related doctrine that if one is concerned with instantaneous change, say becoming white after being nonwhite, then there is a first moment of the existence of that into which the subject changes, in this case the state of being white. In developing Aristotle, the medieval Latins combined what he had said in these two places and formulated a comprehensive doctrine of "first and last instants of being." Thus, it was decided that if one were dealing with a "successive thing," such as a given continuous change or motion, then that change or motion was limited *extrinsically* at both of its extremities: there was a last instant of the change *not yet* existing limiting its beginning and a first instant of it *no longer* existing limiting its end. On the other hand, given a "permanent thing" all of whose "parts" exist simultaneously, such as Socrates' existence or some object being white, then there *was* a first instant of that existence or being *intrinsically* limiting its beginning in time, although, like motion or change, it was extrinsically limited at its end by a first instant of nonbeing. The most popular exposition of this whole doctrine was the treatise *On First and Last Instants (De primo et ultimo instanti)* of the fourteenth-century Oxonian, Walter Burley. It is at the end of an early fifteenth-century copy of this treatise that the present square

(*Continued*)

of opposition is found (see Burley's name in the explicit directly above the square). As its first function, the square treats the first and last instants of being that limit a permanent thing. Hence, the top inscription in the upper left circle reads: "first instant of Socrates or a form," the lower left: "first of nonbeing of Socrates or a form," the upper right: "last instant of Socrates or a form," and the lower right: "last of nonbeing of Socrates or a form." Since these labels taken alone are not quite adequate, further elucidation is given by definitions just outside each circle. For example, on the upper left we are told that the first instant of Socrates' existence is that in which he has being and immediately before which he did not have being. Similarly, on the lower left we are informed that the first instant of Socrates' nonexistence is that in which he does not have being and immediately before which he did have being. Since Socrates' existence is a "permanent thing," these two kinds of instants are those that serve to limit his existence, although our square does not provide us with this information. Correspondingly, the last instants of existence and nonexistence on the right (both furnished with similar explanatory texts outside their circles) are not proper limits for Socrates' existence (since there is no last intrinsic limit at the end of his existence and no last extrinsic limit at the beginning of his existence; things are, as we have seen, just the other way round: intrinsic at the beginning, extrinsic at the end, which are exactly those limits specified on the "true," left-hand side of our square). Although the square does not furnish one with such information about which kinds of limits do obtain and which do not, it does present us with something which is not, curiously, spelled out in Burley's text. The square is rotated 90° counterclockwise from its usual position so that its left side tells us that the first instant of Socrates' existence and the first instant of his nonexistence are contraries (just as the proposition "Socrates now exists" and "Socrates does not now exist" are contraries). Accordingly, on the right, a last instant of Socrates' existence and a last instant of his nonexistence are described as subcontraries (even though there are no such instants limiting his existence). The diagonals stipulate that (1) a first instant at the beginning of Socrates' existence and a last instant of nonbeing at the beginning of his existence are contradictory (which is true since one must opt for either an intrinsic or an extrinsic initial limit, the former always being the choice made) and that (2) a first extrinsic instant of nonbeing at the end of his existence is contradictory to a last intrinsic instant of his existence. Lastly, the top and bottom of the square claim that first instants of existence and nonexistence on the one hand and last instants of existence and nonexistence on the other are subaltern to one another, the former to the latter. The contention of our square of opposition asserting that such and such kinds of instants as limits are contraries and such and such others are contradictories is well taken, since there is a clear parallel between what these relations stood for in a traditional logical square of opposition and what we have here. However, it would seem that, given this much fit, the scholar who constructed our square went somewhat overboard and tried to relate to the elements before him according to the other logical relations expressed in the more traditional squares. Yet it is hard to see just how the kinds of instants he specifies as subaltern are really subaltern in any parallel sense, or why last instants of existence and nonexistence are *sub*contraries rather than mere contraries. We here have a case, it appears, of the application of a traditional diagram to a doctrine "taking over" and implying things that are not really there. (All of this, however, is only the first installment presented by our square of opposition. The lower entries in each circle, as well as the external explanatory phrases at their sides, treat of yet another medieval doctrine of ascribing limits: that usually termed *de maximo et minimo*. Precisely what our square tells us about this second doctrine will be explained below, in the picture credit to the present Illustration.) As a whole, the square is tolerably informative, even if one needs to know something in advance about the theories it is designed to explain. Yet this does not mean that the square did not serve its purpose. For that purpose, as in the case of so many other medieval diagrams and schemata, was not to present one with a whole, readymade theory, but rather to facilitate one's learning of a theory from some text and, after that, perhaps to remind one of the theory's salient points. For other diagrams from the present manuscript, see Illustrations 142 and 275.

CHAPTER EIGHT

Mnemotechnic Devices and Symbolic Notation

SCIENTIFIC WORKS IN VERSE form were not an infrequent medieval occurrence. There was, for example, the *Macer Floridus de virtutibus herbarum* (possibly composed by the eleventh-century scholar Odo of Meun), in 2269 verses describing the properties of some 70-odd herbs; the eleventh-century *Liber lapidum,* a medical work on gems in over seven hundred hexameters written by Marbod of Rennes; and the *Regimen sanitatis,* originally connected with the medical school of Salerno in the twelfth century, but later to receive numerous additions and to become one of the most popular writings of the whole Middle Ages. Finally, there was the thirteenth-century Franciscan Alexander de Villedieu who specialized, it seems, in metrical epitomes of various branches of science: the *Doctrinale puerorum* on grammar, the *Carmen de algorismo* on arithmetic, and the *Massa compoti* dealing with the computistic art of the dating of Easter. All of these works were naturally designed to facilitate the elementary comprehension of their subjects. But verses were also applied in another way in learning science: as brief metrical texts to be memorized which, once in the mind, could be applied like rules to recall the scientific fact in question or to make some inference or carry out some operation. The brief mnemonic poems shown in Illustrations 69 and 70 are from areas of medieval science in which such verses were most needed and useful: logic and computus. Easily learned quatrains and the like were not, however, the only mnemonic devices applied in medieval science. The human hand often served as a kind of memory bank for a theory or as a mechanism for calculating results from given data. And as one moves toward the Renaissance, mnemonic pictures are found whose purpose it was to bring to mind the content or contentions of some given doctrine or work. Finally, having a mnemonic function of their own, symbols were devised in order to be able to present the substance of certain areas of knowledge in a more compendious fashion. There was the medieval technique of abbreviation used in the writing of scientific texts, but we should not mistake these abbreviations for symbols in the same sense as those used, for example, in alchemy and astrology. The occurrence of '·a·' and '·b·' in logical texts, for instance, should not be viewed as symbols, or even variables, standing for the propositions to which they refer; they are simply abbreviations for 'maior' and 'minor' as standing for the major and minor premises of a syllogism. Although one may not be able to draw an absolutely unambiguous line between ancient and medieval abbreviations and symbols, it seems clear that almost all abbreviations were not viewed, like the medieval idea of a symbol, as being signs or "characters" of some entity, relation, or event.

68. (Above) An Astronomical Acrostic. This square of text consists of twelve verses, the initial letters of each line vertically spelling out **ΕΥΔΟΞΟΥ ΤΕΧΝΗ**, that is, "The Art of Eudoxus." The document from which these verses are reproduced has become known as the "Eudoxus Papyrus." Eudoxus of Cnidus (fourth century B.C.) played a very important role in the early history of Greek mathematics (chiefly for the development of a generalized theory of proportion and a rigorous method for calculating the area of curvilinear figures and the volume of curviplanar solids) and was the author of a theory of homocentric spheres designed to account for planetary motions. However, the contents of the present papyrus have nothing to do with this astronomical accomplishment. The ascription to Eudoxus rests solely on his name in the acrostic (although a very late source does say that he wrote a poem entitled "Astronomy"). As a whole, the papyrus contains an elementary compilation of astronomical facts (the orbits of the planets, the circles of the celestial sphere, the lengths of the seasons, the visibility of fixed stars, etc.). Written about 190 B.C., the present text (largely in prose) goes back to an original didactic poem composed, perhaps, about 300 B.C. The papyrus contains a number of rather crudely drawn figures (see Illustration 220), a feature that makes it the earliest of all illustrated Greek papyri. In addition to being an acrostic, of the twelve verses given here, the first eleven contain 30 letters, the twelfth 35, giving a total of 365, most likely a rather unusual reference to the 365 days of the Egyptian year. The content of the acrostic can best be given by its translation: "Herewith I will reveal to you all the subtle composition of the heavens, and give you certain knowledge of our science in a few words. There is nobody so wanting in intelligence that it will seem strange to him, if he understands these verses well. The line stands for a month, the letter for a day; the letters provide you with a number equal to the days which a Great Year brings. Time brings to men a yearly circle, as it governs the starry signs: of which none outrivals another, but always all come to the same point, when the time comes round."

69. (Below) A Medieval "Thirty Days Hath September." Mnemonic verses are found in many areas of medieval learning, but they were especially popular in computus literature. Methods of remembering what numbers corresponded to what or how to calculate this or that were of premium value to scholars in many fields, and verses—of which the present is a simpler example—often provided the answer. The subject of these verses is the allocation of *regulares feriales* (solar regulars), that is, numbers assigned to each month which, when used in conjunction with other information, enabled one to determine the day of the week on which the first of the month would fall. With a copy of Bede's *De temporum ratione* one could page through the volume and easily discover the proper assignment of weekdays and months, but having the present verse in one's head could save the trouble of such a search. Either way, the following correspondence would result: January–1, February–5, March–5, April–1, May–3, June–6, July–1, August–4, September–7, October–2, November–5, December–7. In turn, each of these numbers stood for a given day of the week, beginning with Sunday = 1 and following their usual order. Yet one needed other information in order to apply these *regulares feriales* in determining the day of the week for the first of any month. This was supplied by knowing the concurrent or solar epact for any given year (also a number from 1 to 7), but tables giving the *concurrentes* for various sequences of years were plentiful. Barring that, one could appeal to Bede, who had furnished a method of calculating the concurrent for a given year. To return to our mnemonic verse: although the *regularis* for the months mentioned in each line are written to the right (except for the fourth line where May has the number 3 and August 4, the latter being omitted), this information would naturally not be present in the memorized verse itself. The relevant numbers are written out in the verse (a helpful later scribe or reader has written *regularibus* over most such "numeral-words"): *binis (Januarius, Octimber), quinis (Februarius, Mars, November), unis (Aprilis, Julius), ternis (Maius), quaternis (Augustus), senis (Junius), septemis (September, December)*. Knowing *concurrentes* and *regulares feriales* would, incidentally, provide the kind of supplementary information needed to permit a computist *rota* such as that in Illustration 51 to allow one to calculate the date of Easter. For other figures from the twelfth-century manuscript from which the present verse is taken, see Illustrations 31, 51, 53, 249, 285, and 290.

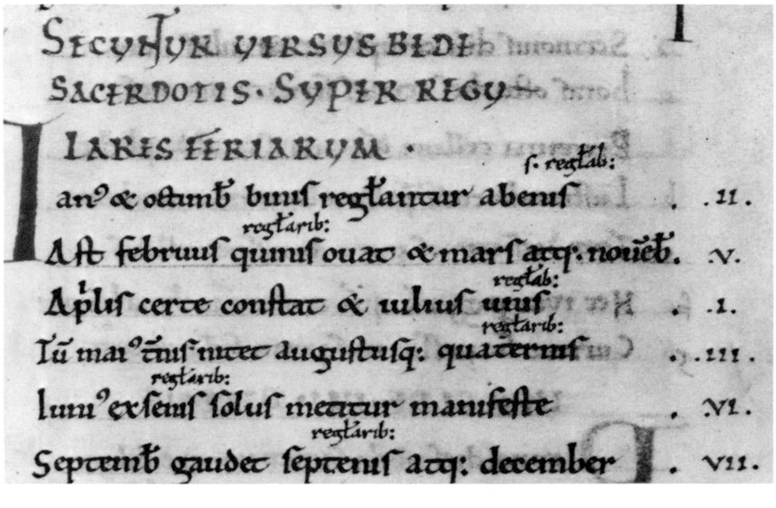

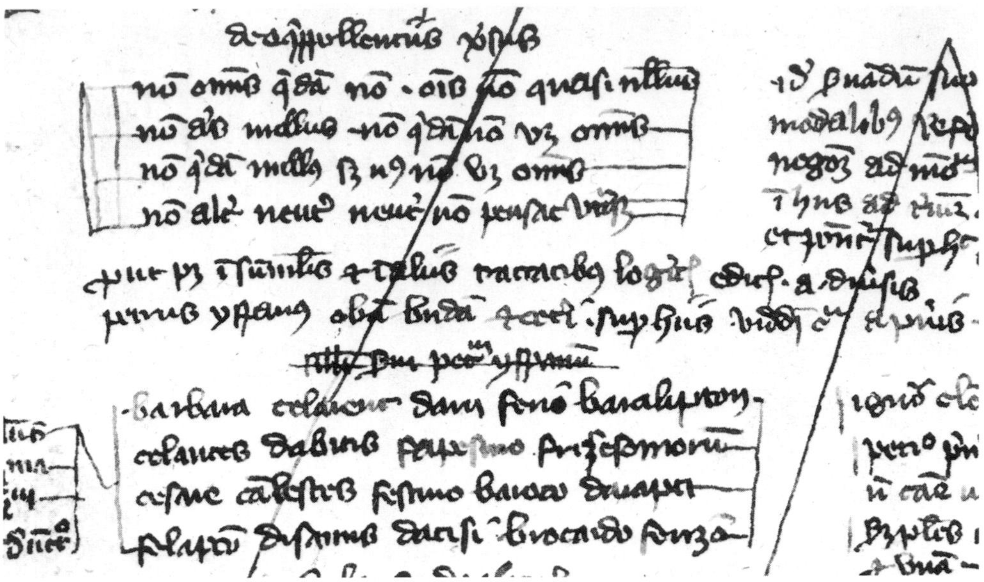

70. Barbara and Her Friends. The only medieval rival to computist literature in the employment of mnemonic verses was logic. The two such verses pictured here are taken from a notebook of a student in Paris from the third quarter of the fourteenth century. In the text between the two verses we are told that they relate to material "in *Summule* and other treatises in logic written by different scholars, such as Peter of Spain, Ockham, Buridan, and others." Furthermore, the crossed-out heading for the second verse gives the ascription: *Sillogismi secundum Petrum Hispanum* ("Syllogisms according to Peter of Spain"), Peter being the earliest (thirteenth century) of all of the authors named on this page. Since by the fourteenth century his *Summulae logicales* had become *the* logical handbook (see the figures from commentaries on it in Illustrations 12, 55, and 61), it is one of the most likely sources for any student's jottings about logic. The notebook containing the metrical jottings pictured here is primarily a theological one and its student owner or a later reader has had the present "logical page" deleted by a diagonal line, part of which is visible here. The above verse relates to equipollent propositions *(de equippolentiis versus)* and specifies which logical "quantifiers," to use a modern term, are equivalent: 'not every' = 'some not,' 'every not' = 'none,' 'not some' = 'none,' 'not some not' = 'every,' etc. The lower verse on syllogisms, however, contains much more information. Certainly the most popular piece of logical mnemonics, it reads (correcting a few scribal errors in spelling): *Barbara, Celarent, Darii, Ferio, Baralipton, Celantes, Dabitis, Fapesmo, Frisesomoron Cesare, Cambestres, Festino, Baroco, Darapti Felapton, Disamis, Datisi, Bocardo, Ferison.* These are the mnemonic names for all of the valid moods of the categorical syllogism, some of which we have had occasion to mention earlier (Illustrations 27 and 55). The names give in sequence the valid, direct, and indirect moods of the three figures of the syllogism. Recall (Illustration 55) that the different figures were distinguished by the variant positions occupied by the middle term: for the first, subject of the major premise, predicate of the minor; for the second, predicate in both; for the third, subject in both (there was even a mnemonic to help one learn this: *Sub pre prima, bis pre secunda, tertia bis sub*). However, in order to appreciate just how much syllogistic was crammed into our mnemonic quatrain, one other piece of information is necessary: that the four basic types of (quantified) categorical propositions were traditionally represented by the vowels A, E, I, and O, indicating, in sequence: *(A) All ducks are aquatic - universal affirmative; (E) No mountains are edible - universal negative; (I) Some philosopher is tired - particular affirmative; (O) Some scientist is not asleep - particular negative.* These four vowels as they occur in the first three syllables of any of the names in our verse tell us what kind of proposition functions as, respectively, the major premise, the minor premise, and the conclusion of the valid mood in question (for example, all three are universal affirmative, A-propositions for the first figure *Barbara,* while in the second figure *Festino,* the major premise is a universal negative, E-proposition, the minor a particular affirmative, I-proposition, and the conclusion a particular negative, O-proposition). But our verses contain much more hidden in their consonants. The initial consonants of the names for all imperfect valid moods (that is, all but the first four direct moods of the first figure) announce the valid mood of

(*Continued*)

the first figure to which the mood in question is to be reduced for validation (thus, *Cesare* reduces to *Celarent, Datisi* to *Darii, Festino* to *Ferio,* etc., all of which is neatly displayed in Illustration 55). There remains, however, the problem of *how* to effect these reductions; once again the consonants of the mnemonic names come to the rescue. 'M' *(muta)* indicates that the premises must be transposed; 'S' *(simpliciter)* indicates that the premise, or conclusion, that is represented by the preceding vowel is to be converted simply; 'P' *(per accidens)* indicates that the same is to be converted by "limitation": and finally, when it appears other than as the initial letter of the name, 'C' *(conversio syllogismi)* indicates that the reduction must proceed *per impossibile* (see again the example cited in the caption to Illustration 55). This is all; other vowels are nonsignificant. As an illustration of how our verse works, let us look at an example of *Cambestres* or *Camestres* as it usually is in later texts (listing initially the vowels representing its constituents): *(A) All ruminants are even-hoofed; (E) No horses are even-hoofed; (E) Therefore, no horses are ruminants.* The 'M' in our mnemonic name tells us to transpose the premises, while the first 'S' following an 'E' informs us that the original minor premise is to be converted simply. These two moves yield: *(E) No even-hoofed animals are horses; (A) All ruminants are even-hoofed; (E) Therefore, no ruminants are horses.* But this is a first-figure syllogism in *Celarent,* so our validating reduction has been accomplished. However, the final 'S' of *Cambestres* calls for the simple conversion of the conclusion, thus yielding the conclusion of our original.

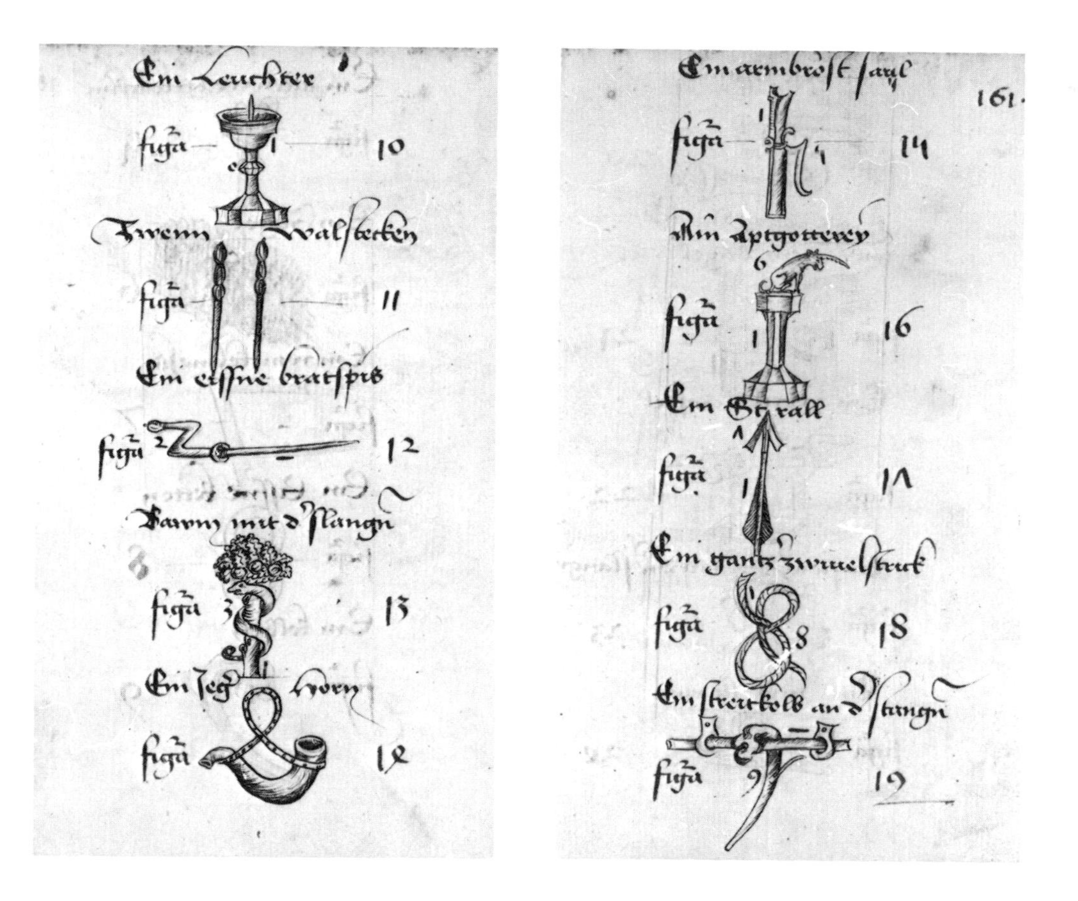

71. (Above) Mnemonics for Numerals. Written *ca.* 1470–75 by a scribe named Bernhard Hirschvelder, this Middle High German manuscript contains rather outlandish mnemonic pictures of the most unusual, and in some cases "portmanteau," objects designed to facilitate the learning of Arabic numerals from 1 to 99. Those for 10 through 19 are given here. Throughout this mnemonic work, it is often rather difficult to see just how the picture resembles the numeral whose learning it is to facilitate, so the frequent repetition of the ciphers of the numeral next to the part of the picture to which they presumably belong is an added aid. The objects depicted are also named: a candlestick, two wooden clubs, an iron roasting spit, a snake on a tree trunk, a hunting horn, a crossbow jamb, an idol, an arrow, a twisted knot, and a hurlbat on a rod.

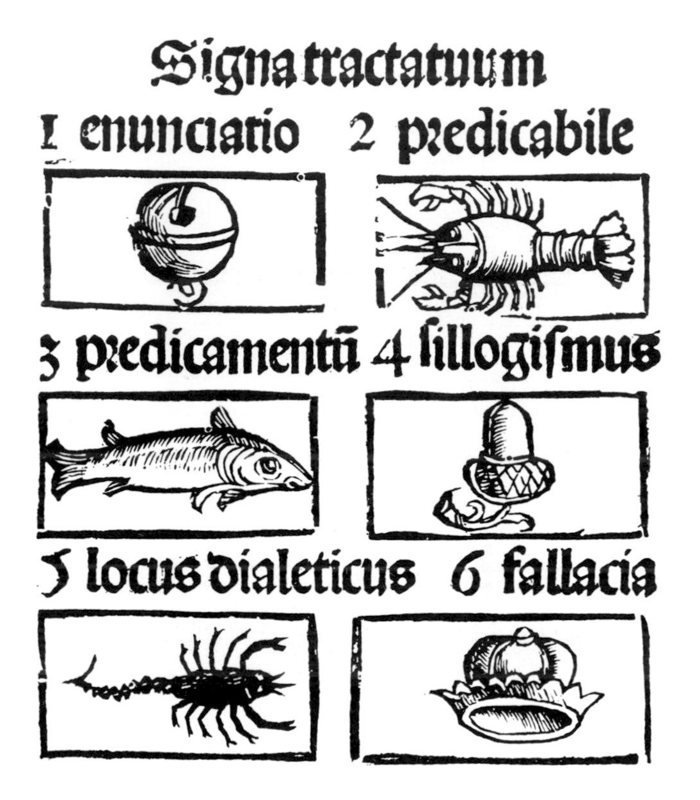

72. A Logical Card Game or the Whole of Dialectic by Memory. Such is the subtitle of an introductory mnemonic logic written by the Franciscan satirist and anti-Lutheran, Thomas Murner (1475–1537): *Logica memorativa: Chartiludium logice sive totius dialectice memoria,* published in Strasbourg in 1509. Apparently composed during his humanist wanderings in Cracow, Murner's little treatise had for its source the traditional late medieval handbook of logic: Peter of Spain's *Summulae logicales.* The division of Peter's work dictated that of Murner's. The first six miniature pictures *(signa tractatuum)* at the top left represent the first six tracts of Murner's work, which themselves follow Tractatus I–V and VII of Peter's *Summulae.* The *signa tractatuum* on this page do no more than announce to the reader which areas of logic will be covered. Their real importance occurs when the objects they depict occur in later pictures or, as Murner calls them, *imagines* (such as that at the right above) throughout the remainder of the book, informing us just what logical realm is being treated by the image in question. Thus, the fish in the *imago* at the top right (p. 77) tell us that we are in the realm of *predicamenta.* We are, however, at the very end of that realm since the specific topic our image is intended to explain visually is that of the *postpredicamenta* which were treated in the final chapters of Aristotle's *Categories* and came to be regarded as consequences of the *predicamenta* or categories properly so called. They were five in number: opposition, priority, simultaneity, motion, and state or condition *(habitus).* Murner purports to explain these five elements by means of his mnemonic images, keying his explanation to the picture by inscribing numbers next to the relevant objects within it. (1) Here the tonsure at the rear of the man's head reminds one of the "definition" of the *postpredicamenta,* that is, they came *after* the *predicamenta* proper. (2) Arrows in a target, by which one is to understand—though it is not made clear how—relative opposition, such as that between father and son. (3) A burning oil pot, which is to draw the opposition of contraries to mind, as the opposition of water and fire in oil. (4) A headband which is to stand for privative opposition,

(*Continued*)

presumably because one can use it to cover the eyes, thereby obtaining the privative opposition: sight vs. blindness. (5) The cross of St. Andrew on the thigh of the figure at the left to represent contradictory opposition. (6a) An hourglass for priority in time. (6b) A crowing cock for priority by nature since its "song" is given earlier in the day than that of other animals. (6c) A cape to remind one of priority of order; presumably the cape worn by those in religious orders was the connection in Murner's mind. (6d) A beret to indicate priority in honor. (6e) A cross on a rosary, which, as the starting point of prayer, precedes the actual recitation of the prayer and therefore represents the paradigm case of causal priority: that of the thing being prior to the spoken word standing for the thing. (7a) A joining of heads to remind one of simultaneity in duration. (7b) A turning backward of the feet to stand for the simultaneous conversion of a logical consequence. (7c) The separation of heads to recall simultaneity in logical division, such as the division of animal by rational and irrational. (8a–f) The six species of motion: a womb to represent the motion of generation; a worm for motion of corruption; the greater size of one shin to remind one of the motion of augmentation and the smaller size of another shin for motion of diminution; white and black on the target to recall qualitative motion or motion of alteration; the swaying of a balance bar of a clock to stand for local motion. (9) Seven beads on the rosary to bring to mind the seven types of a state or *habitus*. Murner's *Logica memorativa* contains 51 pictures similar to that just analyzed. Intended as a mnemonic device—or even a card game (although the relevant playing cards were never, it seems, produced)—to make learning the fundamentals of logic, of "all of dialectic," an easier task, it is doubtful how effective Murner's work may have been in realizing that goal. If one were to understand, and not simply memorize, the multitudinous correspondences he gives between visual object and idea, one would wonder whether the "visual velocity" associated with such learning could have been very great. For another picture from Murner's *Logica*, see Illustration 180.

73. Picturing the Divisions of Speculative Theology. At the mid-point of the twelfth century, Peter Lombard composed a work known as the *Libri Sententiarum* or *Books of the Sentences,* its content being the whole of doctrinal theology, presented in the form of a canon of authorities (Scripture, Church Fathers, Councils, etc.) for each doctrinal thesis or question, together with a brief judgment of his own relative to that being discussed. Other *Sententiae* were also written in the twelfth century, but Peter's completely captured the field and became the standard theological text upon which all aspiring Bachelors had to lecture and comment in obtaining their degrees in theology. Each of Peter's four books covered separate ground: the first, God and the Trinity; the second, The Creation and Created Beings Before the Coming of Christ; the third, Christ Relative to the Incarnation and Redemption, Sins, Commandments, and the Virtues, both Cardinal and Theological; the fourth, the Sacraments and the Novissima, that is, Death, the Day of Judgment, and heaven and hell. These books were further divided into *distinctiones* covering the separate topics under their charge. The subject of each of the first twenty distinctions of Book II are those whose mnemonic pictures are given here. Their "fit" with the actual subject of each distinction is often not obvious. Some notion of the "fit" these pictures have, or do not have, with the actual subject of each distinction can be seen from the pictures for the following *selected distinctiones* (reading each page in order from top left to bottom right): (1) the creation of heaven and earth: the hand of God over the sphere of the universe; (2) the creation of angels: the hand of God touching a circle-enclosed angel; (5) angels with a confirmed loyalty to God: an angel holding a rosary faced by the Holy Ghost as a white dove; (6) fallen angels, especially Lucifer: a falling angel separated by flames from a devil holding a torch; (12) introduction to the six days of creation, discussing the matter and the four elements of the world, and the "earth without form and void" (*Genesis* I, 1): the hand of God over an amorphous form representing prime matter (to be compared with Illustrations 273–274); (13) the first day of creation—*fiat lux:* the hand of God over a burning torch or candle; (14) the second, third, and fourth days of creation—*fiat firmamentum in medio aquarum . . . fiat luminaria in firmamento celi* ("let there be a firmament in the midst of the waters . . . let there be lights in the firmament of the heavens": God's hand, the sun, moon, and a star, with the earth and water, or perhaps the firmament and the so-called celestial waters, below; (16) the creation of man: God's hand and a naked human; (18) the creation of woman: the hand of God drawing Eve from Adam's side; (19) the fall: Eve plucking the apple from a tree, with a skull at her feet.

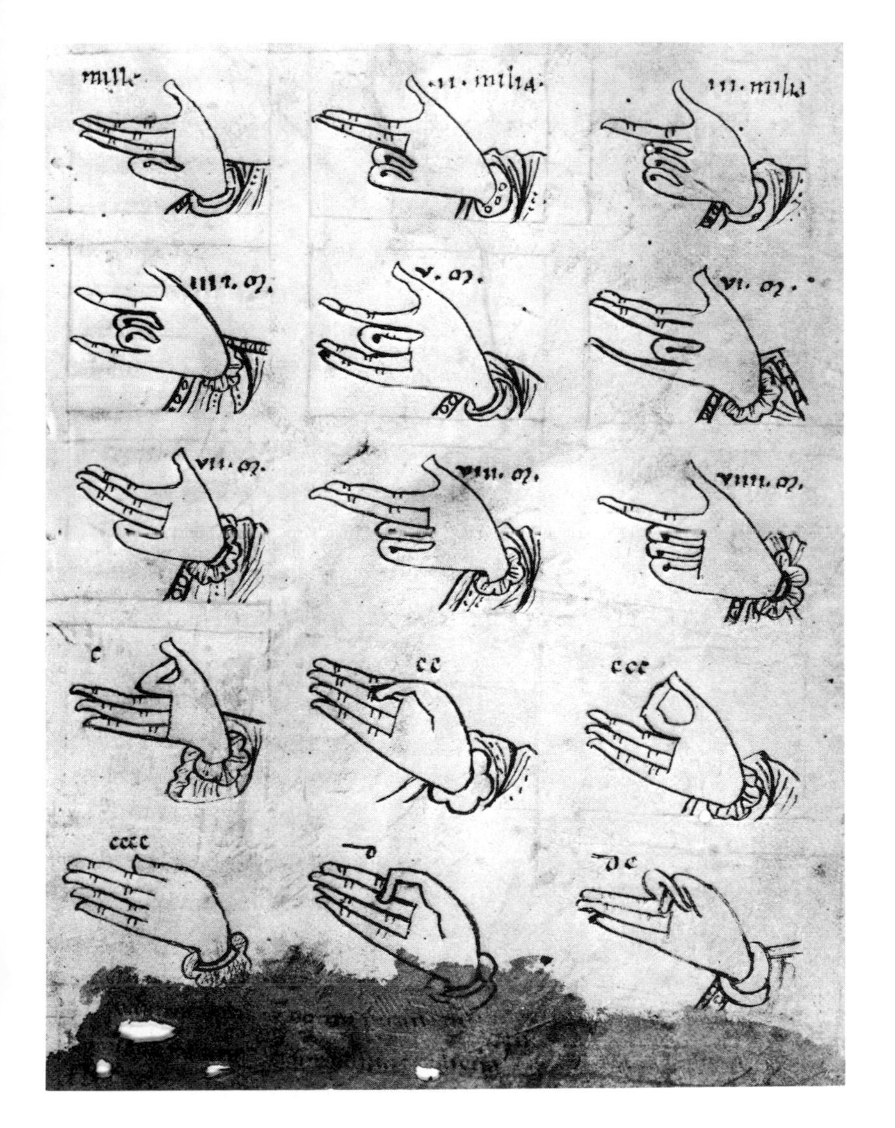

74. Literally Manual Arithmetic. This twelfth-century manuscript gives an example of what is perhaps the most elementary and straightforward use of the human hand to calculate: finger reckoning. There is appreciable evidence of advanced reckoning of this sort in antiquity as well as of its extensive use in the Middle Ages, both Islamic and Latin. Fundamental to any such reckoning is, of course, the representation of numbers by the bones and joints of the fingers and the various distinct relative positions they can assume. It is this kind of representation which is pictured here: the top three rows give the configurations of the hand for each thousand from 1000 to 9000, while the bottom two rows do the same for each hundred from 100 to 600. Note that all of these configurations are of the right hand. This is so because one always began with the left hand, usually with the little finger, which allowed one to reserve the right for higher numbers, a procedure that was behind remarks like that of Juvenal (*Satira* X, 248–249), who deemed the man happy who could live so long as to number his years on his right hand. The whole process was well described by Bede in an introductory chapter of his *De temporum ratione* (from which one can compare the following lines relevant to the present picture): "The reckoning is made first on the left hand, as follows: when you say One, bend the little finger of the left hand, and place it on the middle of the palm. When you say Two, bend the fourth finger and place it likewise. When you say Three, do the same with the third finger. . . . When you say Ten, put the tip of the first finger on the middle joint of the thumb. . . . So far for the left hand; to represent a Hundred, make the same sign on the right hand as you did for Ten on the left. . . . A Thousand on the right hand is as One on the left, Two Thousand on the right as Two on the left, Three Thousand on the right is Three on the left, and so on with the rest up to Nine Thousand." Of course, such finger notation is not yet finger computation. But that too was developed: simple cases of multiplication and division as well as of addition and subtraction.

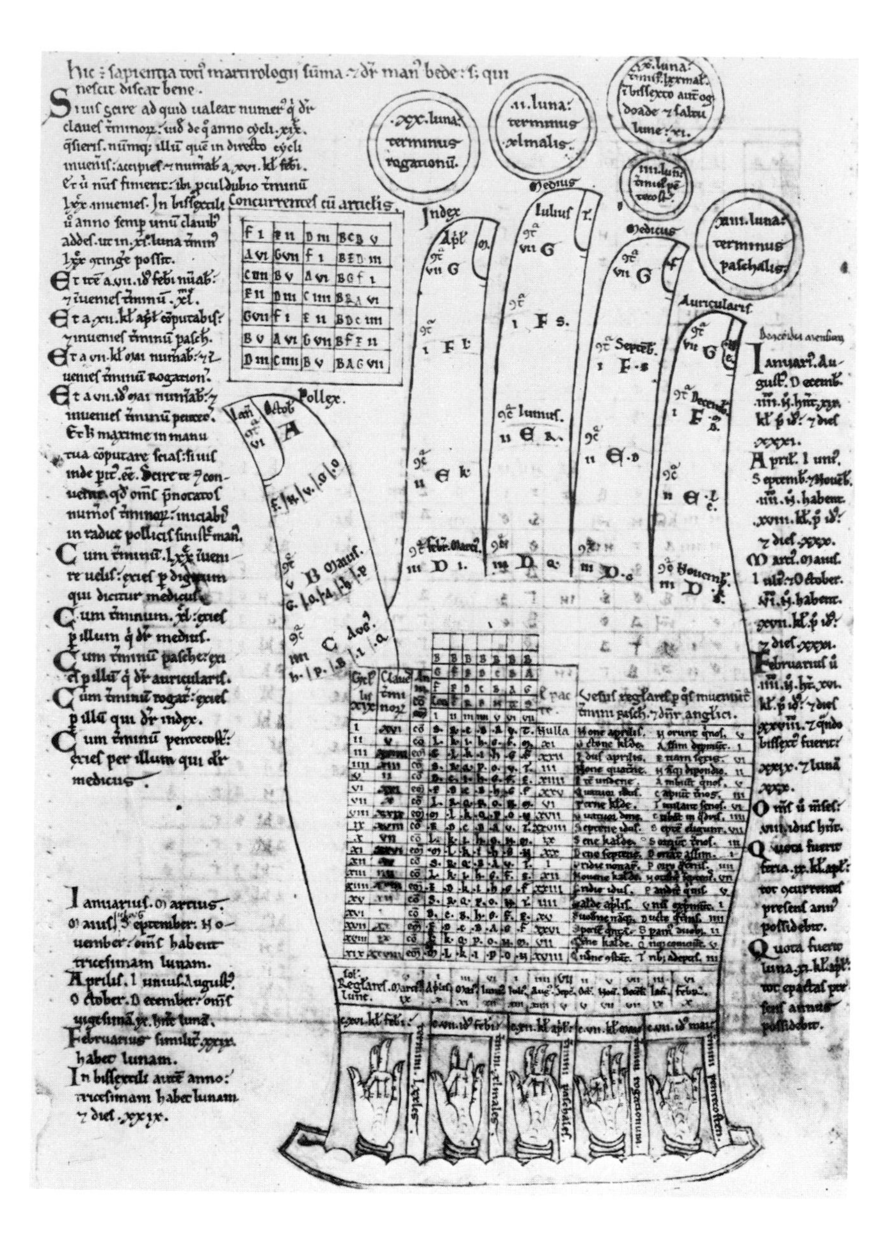

75. The Hand of Bede. In addition to providing a fairly extensive account of finger reckoning, the Venerable Bede also devoted part of Chapter 55 of his *De temporum ratione* to the computistic use of the hand. Although Bede himself did not include or call for drawings of hands, they soon came to accompany his work or, more frequently, to be drawn independently, in which case they were often explicitly labeled, as here, "the hand of Bede"—*et dicitur manus Bede* is part of the heading in the present example of a computistic hand, taken from a twelfth-century manuscript. The hand here depicted contains a great deal of information relevant to the computus genre, much of it independent of the application of Bede's hand as such. The standard names for the digits are given above them (from thumb through little finger): *pollex, index, medius, medicus* (so-called because of the doctrine that a nerve ran between that finger and the heart), *auricularis* (because it was that most easily inserted into the ear). Circles above the four fingers specify what "days of the moon" serve as limits for the religious feasts of Rogation, Quadragesima, Septuagesima, Pentecost, and Easter, while, at the very bottom of the hand, these limits are illustrated by various digital configurations and associated with their relevant monthly dates. The square between the thumb and the index finger relates the *concurrentes cum articulis,* that is, with the bones of the hand. The table in the palm covers such other fundamentals of computus literature as the nineteen-year cycle, the limits for the date of Easter, and the so-called *claves terminorum* (numbers giving the distance in days obtaining between fixed days and the Easter limits on the basis of which one can determine the limits of other feast days), while directly below there appears a listing of the solar regulars (Illustration 69) and lunar regulars for each month (the latter giving the age of the moon on the first of each month for each year of the *cyclus decennovalis*). The columns flanking the hand contain other data and instructions pertinent to the computist art: a concordance of months, the number of days in the various lunar months, and from what value, or what part of the hand, one should begin in calculating the limit of various feasts.

76. The Hand of Guido. The utilization of mnemonic hands was also part of medieval musical theory where they may well have developed under the influence of their use in computus literature. One such musical hand totally eclipsed all others in terms of popularity: the so-called Guidonian Hand. The name derived from the early eleventh-century musical scholar, Guido of Arezzo, even though the hand itself does not receive treatment in his influential writings. Basic to these Guidonian Hands was their illustration of the sequence of 20 notes in the scale, representing this sequence by the letters Γ A B C D E F G a b c d e f g aa bb cc dd ee (corresponding to the modern sequence G A B c d e f . . . c′ d′ e′) beginning with Γ at the end of the thumb, running down the thumb, across the bottoms of the proximal phalanges up the little finger and then in a counterclockwise spiral over the tips of the fingers, ending with dd at the outermost joint of the middle finger, ee having a special position over the tip of that finger. Many copies of Guidonian Hands contained a good deal more information. The one pictured here is taken from a manuscript written in 1274 which is unique in its inclusion of a human, one of whose magnified hands provides the "diagram." In addition to the foregoing letter notation, it represents the 20-note sequence (although ee = e′ is omitted) in two other manners: first, numerically as *puncti* (often abbreviated as a mere 'p') with accompanying Roman numerals: second, by solmization, that is, by the syllables ut–re–mi–fa–sol–la, a system invented by Guido himself, the symbols themselves being the initial syllables of a familiar hymn to St. John (note that both 'fa' and 'mi' are written at the tip of the little finger and the first joint of the ring finger, 'mi' representing b-flat, although the corresponding letter notation is missing). The solmization sequence appears twice on the circumference of the circle enclosing our picture as well as on the hand itself. A final piece of information in our copy of the Guidonian Hand is its indication of the seven *claves* (at times abbreviated by 'c') over the 20-note sequence. These *claves* relate to the hexachord, a sequence of six diatonic tones, having a semitone interval in the middle (for example, in modern notation, the sequence G A B c d e). There are seven of these hexachords "overlapping" within the 20-note sequence of our hand, and a *clavis* indicates the note on which each hexachord begins. Further, since Guido of Arezzo used the solmization syllables ut–re–mi–fa–sol–la to designate the hexachord, seven of these sequences also begin at the same positions within the hand (which explains the occurrence of two or three of these syllables to represent some of the notes). The popularity of the Guidonian Hand in medieval musical theory is perhaps best expressed by the fact that, as late as the sixteenth century, nondiatonic music was objected to because it was "not in the hand" *(non est in manu)*.

77. Symbols for the Musical Scale. Systems of abbreviating or representing the succession of musical tones have varied greatly throughout history. The letter notation and solmization depicted in the previous Illustration were part of that history; so too is the present Illustration from a fifteenth-century manuscript of a chart giving the Greek symbolization that was transmitted to the Latin West by Boethius in his *De institutione musica,* Book IV, Chapter 3. Setting aside for a moment the symbols themselves, something should be said first of the chart. The column at the right is headed *nomina cordarum,* that is, the names of the tones or, better, the names of the tones associated with the strings of an instrument, the kithara. The names that follow are those of the two-octave Greek scale, here in ascending order (*Proslambenomenos* to *Nete hyperboleon*) rather than the usual descending one. They give what has become known as the Greater Perfect System and, at one and the same time, the Lesser Perfect System. The names across the top of the chart are the so-called "Greek modes," from (the Latinized) *Hypodorius* on the right to *Hypermisolydyus* on the left. The "crossing" of the modes and the tones represents what occurs when, moving through the modes from right to left, we step up one tone or semitone for the initial tone of our scale. However, the symbols within the chart are of greater present interest. To begin with, there are two symbols or *notulae* for each tone: the one above for voice (*dictionis* or *verborum*), that below for an instrument *(percussionis)*. The symbols themselves are letters of the Greek alphabet in their normal form as well as in varying forms and positions: ⊢ *(tau iacens)*, ɔ *(sigma conversum)*, π *(pi graecum deductum)*, V *(lambda supinum)*, etc. Whether, as Boethius claimed, a given melody would, by means of these symbols, endure in one's memory *(in memoriam posteritatemque duraret)* is far from certain.

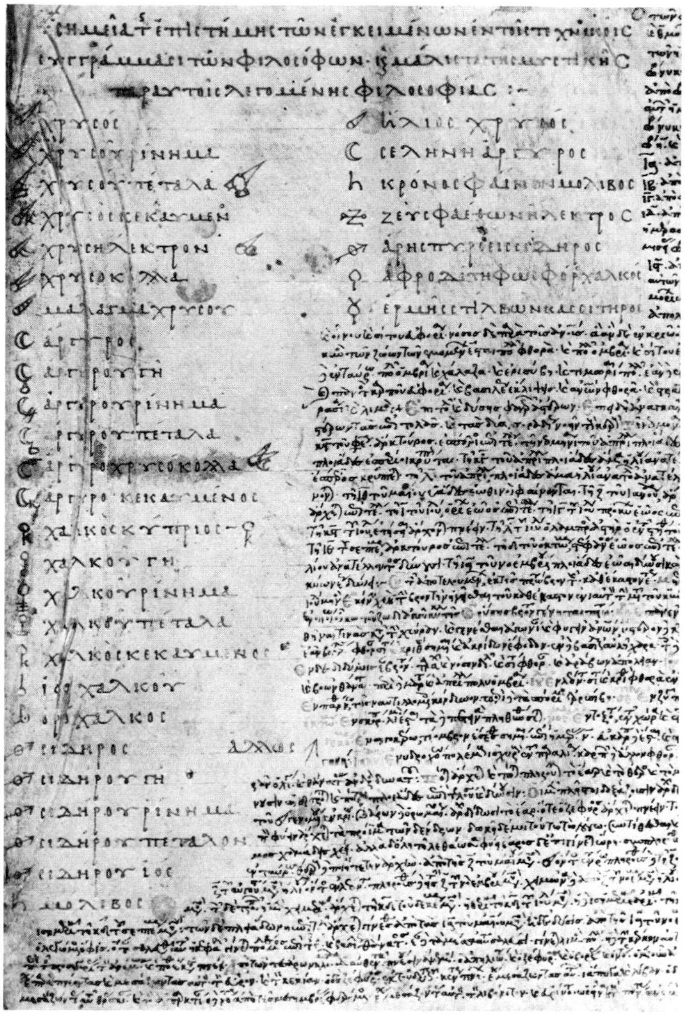

78. (Above) Signs of Knowledge. Contained in an eleventh-century manuscript, this list of alchemical symbols carries the title: "Signs of the knowledge found in the technical writings of the philosophers, particularly those in the so-called mystical philosophy." The shorter column on the right relates each of the planets to the seven metals, preceding each coupling with the relevant symbol or sign, often giving two Greek names (compare Illustration 48) for the planet in question: (1) the sun, gold *(Hélios, chrysos)*; (2) the moon, silver *(Seléné, argyros)*; (3) Saturn, lead *(Kronos* or *Phainōn, molibos)*; (4) Jupiter, electrum—an alloy of gold and silver *(Zeus* or *Phaethōn, elektros)*; Mars, iron *(Ares* or *Pyroeis, sidéros)*; (6) Venus, copper *(Aphrodité* or *Phōsphoros, chalkos)*; (7) Mercury, tin *(Hermés* or *Stilbōn, Kassitéros)*. The longer column on the left lists different types of these metals, in each instance the symbol being some variant of the symbol held in common by a planet and its matching metal. At the top of this list gold as such *(chrysos)* appears together with the standard symbol for the sun: (a variant symbol for the same was often ☉). Then follow gold filings *(chrysou rhinéma)*, gold leaf *(chrysou petala)*, and calcined gold *(chrysos kekaumenos)* together with their symbols, all constituted of relevant additions made to the common symbol of gold

(Continued)

and the sun. The next entry, electrum, has a more instructive symbol; since electrum is an alloy of gold and silver, and silver is the metal allied with the moon, the resulting symbol is a combination of the symbol for the sun and that for the moon (☾). The list continues with two more types of gold, then moves to silver for six entries, copper for six more, iron for five, and a last entry for lead (although entries for lead and other metals continue on the next page). Occasionally, alternative symbols are given by a later hand to the right of the name, one of them even being announced as an alternative *(allōs,* sixth entry from the bottom). The smaller writing (of the fourteenth century) on the page is a fragment of the so-called *Liber Geoponicus,* an anonymous Byzantine work on agriculture.

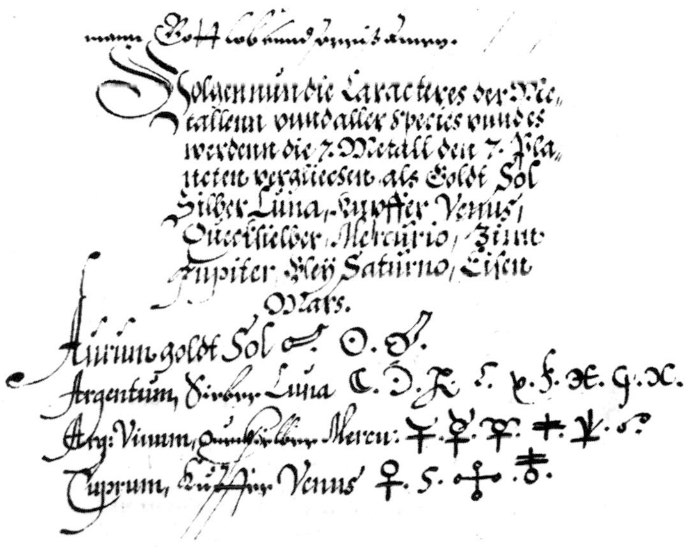

79. (Above) Variant Alchemical Symbols. This sixteenth-century, Middle High German alchemical manuscript gives many of the same symbols found in the eleventh-century Greek manuscript of the preceding Illustration, but adds a fair number of variants which the author apparently found to be employed in other alchemical literature. An opening paragraph refers to these symbols as the "characters" of the metals *(die caracteres der Metallenn)* and singles out the correspondence of the metals to the planets *(werdenn die 7 Metallenn den 7 Planeten verglieesen)*, although now mercury or quicksilver is related to Mercury and zinc to Jupiter (yet in the list on the following page Jupiter is matched with tin). The four metals, with their corresponding planets, at the beginning of the list pictured here are: *Aurum, goldt, Sol; Argentum, silber, Luna; Argentum vivum, quicksielber, Mercurius; Cuprum, kupffer, Venus.*

Aries	Taurus	Gemini	Cancer
Leo	Virgo	Libra	Scorpio.
Sagitari'	Capicōi	Aquarius	Pisces
Capud draconis		Cauda draconis	

80. (Left) An Exposition of the Zodiacal Signs. Probably the most familiar of medieval signs or symbols are those relating to the twelve signs of the zodiac. They are displayed here in a table from a fourteenth-century manuscript of Michael Scot's *Liber introductorius.* Michael tells us that the twelve signs in his table are prefigured as inscribed characters *(prefiguratis in modum karacterarum)*. Alternative forms are given of some of the symbols or characters. At the bottom of this table, Michael presents two other symbols: the head and tail of the dragon *(capud draconis; cauda draconis)*, that is to say, the ascending and descending nodes of the lunar orbit. Like the planets and other celestial bodies, they too came to have considerable astrological significance. For other figures from the present manuscript, see Illustrations 26, 101, 231, 251, and 277.

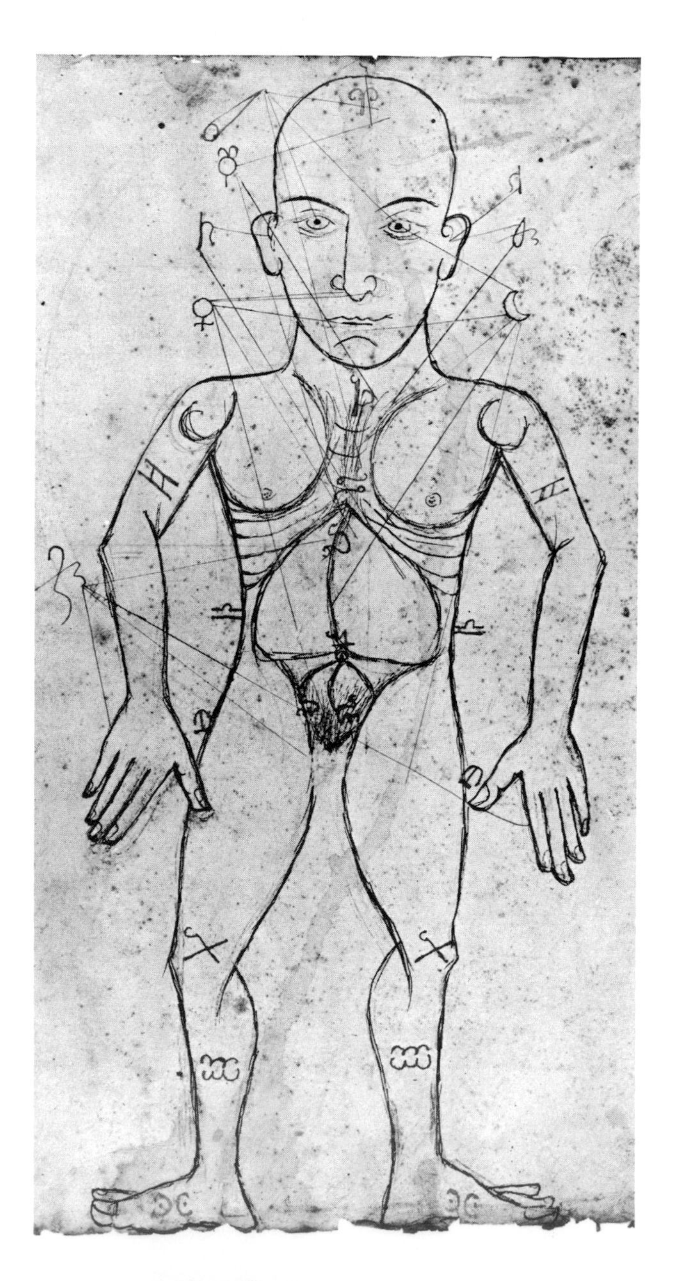

81. (Right) Symbols Applied. If not one of the most frequent, certainly often the most artistic, application of symbols occurred in the medieval figures known as "zodiac men." Each zodiacal sign was related to a part of the body, the first sign Aries assigned to the head, the last, Pisces, to the feet, with the others following, top to bottom, in the standard zodiacal order. The planets were also paired with body parts, and such a correspondence is given along with the zodiacal one in this sixteenth-century Greek manuscript. Both correspondences played an important role, of course, in astrological medicine. In many zodiac men, the names of the signs and their pictorial representations were also related to their proper locations within the "microcosmic" human body. In this instance, however, only the symbols for the zodiacal signs and the planets are given, the symbols for the former often differing in lesser or greater degree from those seen in Michael Scot's table in the preceding illustration. For more on zodiac men, see Illustrations 266 and 267.

Part Three

EXACT SCIENCE AND THE MATHEMATICAL "ILLUSTRATION": *Tables, Figures, and Diagrams Required by the Substance of the Discipline*

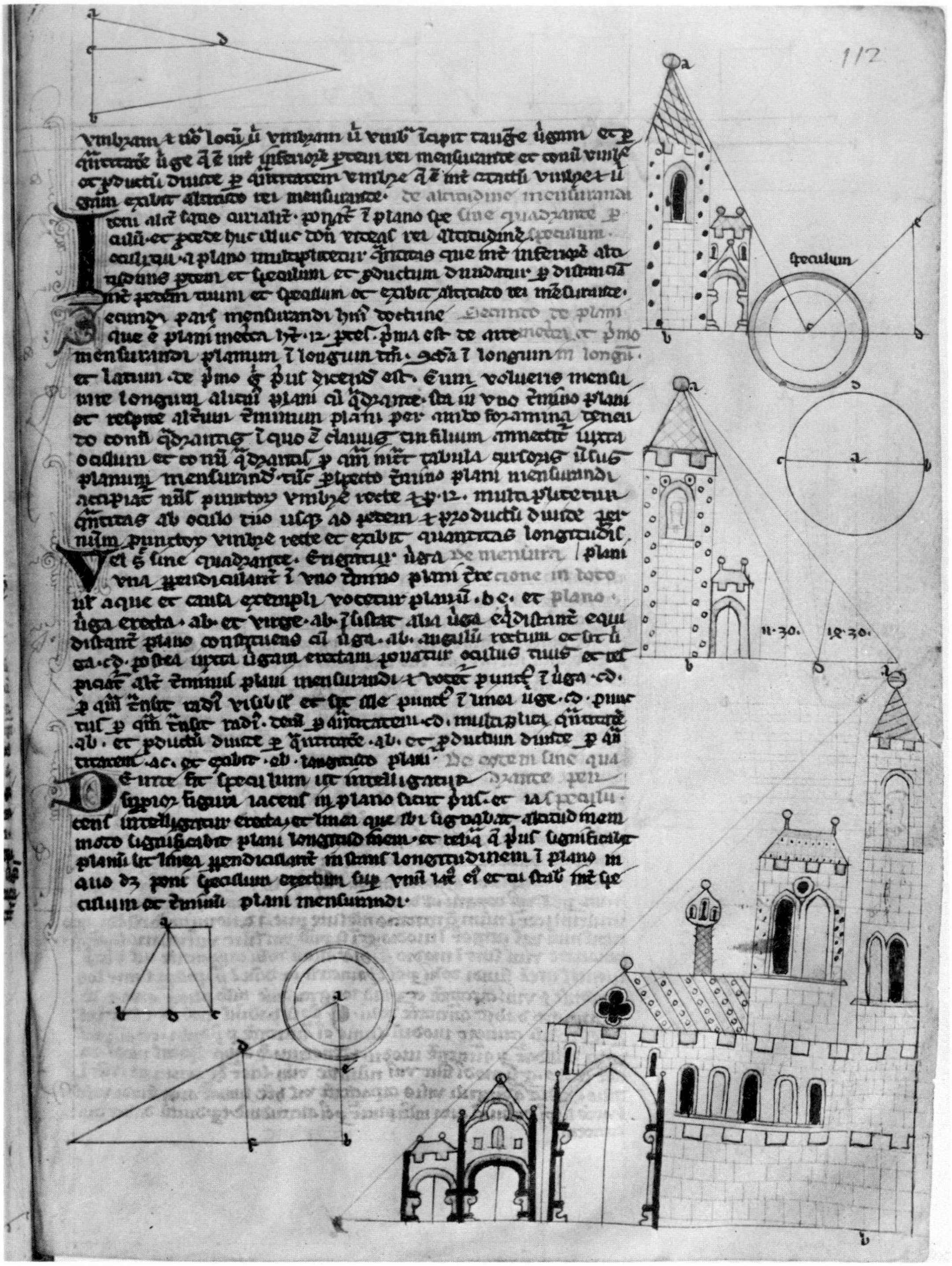

82. The Medieval Geometry of Measuring Towers. This thirteenth-century manuscript of an anonymous treatise on the geometrical and astronomical quadrant provides three illustrations of different ways of measuring the heights of towers. The first makes use of a mirror (*speculum*) and appeals to the properties of similar triangles. The second depicts a method based on two different sightings of the summit of a tower, the sightings presumably accomplished by means of a quadrant or some other instrument not shown here. The third and largest figure measures the tower in question by means of an isosceles right triangle. Tower measurement was standard material in every medieval practical geometry, and more of it can be seen in Illustration 152.

MATHEMATICAL SYMBOLS, DIAGRAMS, and tables are unique in the early history of "scientific illustrations." For, although the primary function of most of the pictorial material reproduced in the present volume was to facilitate the understanding of some scientific text, or of some small part of such a text, this was not the case with mathematical "illustrations." Their function was instead almost always not to facilitate understanding, but to make it possible. How could one possibly follow, for example, a complicated geometrical proof without the appropriate geometrical figures accompanying the text of the proof or, at least, without drawing one's own figures while puzzling one's way through the text? Or how could one easily understand complicated relations between numbers or kinds of numbers with nothing more than the natural language of some arithmetical text? That one could not follow a proof or understand complicated relations under such circumstances seems an almost inevitable reply to these rhetorical questions, but there is confirming evidence in more than a few mathematical manuscripts themselves. It is simply that when we do come upon a copy of some mathematical text without appropriate diagrams, it shows all the signs of a clean, unused copy: in particular, the occasional marginal jottings, or even marginal "flags" to indicate a point of importance, are absent. Thus, in most instances, the kinds of symbols, diagrams, and tables we regard as natural concomitants of doing mathematics were a necessary part of the ancient and medieval mathematical texts and not something *ad commodem lectoris.*

More than that, early mathematical diagrammatic material that we would not regard as a natural part of the enterprise of mathematics was also often a necessity, and not just a luxury, for the ancient and medieval scholar. One of the reasons for this necessity of the not strictly necessary was an occasional lack of adequate symbolization. To be sure, there was no want when it came to the symbolization of particular numbers, even if the results of some of the efforts in this regard were cumbersome and not especially effective in facilitating calculation. However, in some instances there was a less than adequate symbolization of mathematical relations or operations and that was what at times gave rise to the (to us) rather curious diagrams or tables that accompany some mathematical texts.

Yet it would be wrong to go beyond this to claim, as some historians have done, that a lack of adequate symbolization (particularly in medieval Latin mathematics) was a hindrance to mathematical development. For there were scholars such as Nicole Oresme and Richard Swineshead (see Illustrations 142–144) who, without adequate symbolization, achieved substantial accomplishments in the application of mathematics to natural philosophy, even if these accomplishments did not lead (again *pace* some historians) in the direction of analytic geometry or calculus. Here, then, was at least one area in which the "necessary" diagram or symbolization was absent and the mathematics was done nonetheless, although admittedly some readers of the likes of Swineshead felt the necessity of visual aids for a better comprehension of it all.

CHAPTER NINE

Tabulae: Calculational and Stored Information

ALTHOUGH THE LINE of division is far from rigorous, unlike the tabular illustrations that have been sampled in Chapter Four, the tables discussed in this chapter did not function as outlines or tables of contents or as visual epitomes of a doctrine particular to some specific text. The role was, for the most part, to present results that could be applied in understanding other, more advanced, aspects of the science in question and in developing one's expertise in that discipline. Like the table (Illustration 85) drawn from Boethius, the purpose might have been to present in a more visually comprehensible form a doctrine discussed in the accompanying text, but even then the doctrine was one that would be needed in a great part of the remainder of the mathematics at hand. Alternatively, the table might present a particular worked-out example of a mathematical procedure, whether the procedure in the accompanying text used the same example (Illustration 86) or was explained in general terms (Illustration 87). In such cases, the tables were very much like a modern worked-out homework problem, although often in a more elegant form, with all steps of the solution given an appropriate label.

Tables were most important, however, in astronomy. The information they provided was always directly relevant to carrying on the work in the field of astronomy itself. Such is the case with the three astronomical tables given in Illustrations 88–90, which represent but a very small sample. Thus, in addition to trigonometric and mean motion tables like those appearing here, there were tables dealing with planetary latitudes, planetary stations and retrogradations, solar and lunar parallax, and eclipses, to cite but a few of those found in Greek, Arabic, and Latin manuscripts. There were also visibility tables, star tables, geographical tables containing latitude and longitude information for numerous cities, astrological tables (one instance is given in Illustration 90), and chronological tables treating of all manner of calendrical issues.

The amount of "numerical information" contained in these tables varied from case to case, although the efforts of Ptolemy in this regard frequently served as a guide as to just how complete the entries of a table should be. In some instances, tables contained information that would provide a scholar with the wherewithal to calculate values in addition to those appearing in the table itself (see Illustration 89). At other times, an effort was made to give as many entries as possible, saving the prospective user the labor of calculating those additional or "in between" values that may have been absent in earlier tables. Attempts at such exhaustiveness were particularly frequent in later Arabic astronomical works.

Related, if at times only remotely, to astronomical tables were tables dealing with computus problems surrounding the dating of Easter and other movable religious feasts (of which Illustration 51 is an example in *rota* form), tables found in almanacs (Illustration 91), and even tables in many copies of traditional medieval calendars. In these cases, of course, the application was usually concerned with more practical matters than the theoretical aspects of astronomy as such.

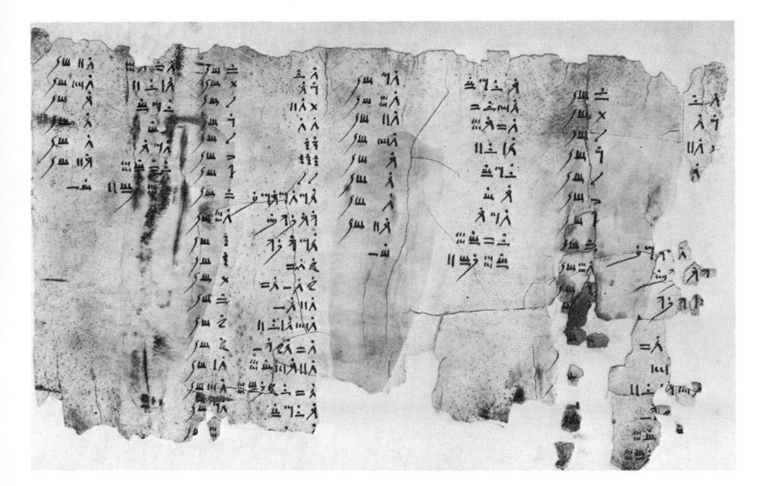

83. (Above) Egyptian Mathematics on Leather. In 1858 a Scottish lawyer named A. H. Rhind purchased this leather roll in Luxor, along with the more famous mathematical papyrus that bears his name (see Illustration 106). Although the roll bears no date, the writing on it strongly suggests that it derives from about the same time as the Rhind Papyrus, thus from some point in the seventeenth century B.C. Written in hieratic, a cursive form of hieroglyphics, the text gives, in duplicate, 26 sums of unit fractions. Thus, the top entry in the two columns at the extreme right (repeated in the first line of the two columns just before that on the extreme left) tell us that $\frac{1}{8} = \frac{1}{10} + \frac{1}{40}$. Other entries specify $\frac{1}{2} = \frac{1}{6} + \frac{1}{6} + \frac{1}{6}$; $\frac{1}{14} = \frac{1}{21} + \frac{1}{42}$; $\frac{1}{8} = \frac{1}{25} + \frac{1}{15} + \frac{1}{75} + \frac{1}{200}$. Yet these (possibly schoolboy) sums are quite integral to a particular, and most important, feature of Egyptian mathematics. Excluding $\frac{2}{3}$, for which there was a special sign (and possibly also fractions consisting of unity less an aliquot part, such as $\frac{7}{8}$, $\frac{99}{100}$, etc.), the Egyptian scholar had no way of writing fractions other than unit fractions, that is, other than those whose numerator was 1. Indeed, the Rhind Papyrus opens with the so-called $2/n$ table giving the double of all unit fractions with odd denominators from $\frac{1}{3}$ to $\frac{1}{101}$ in terms of other unit fractions. Our leather roll is not concerned (save for the single instance of $\frac{2}{3} = \frac{1}{3} + \frac{1}{3}$) with sums of unit fractions yielding the double of other unit fractions, but rather with such sums as result from the breaking up of a single unit fraction. Yet it appears that such a breaking up or dividing of unit fractions may well have been involved in determining just how to obtain the unit fraction sums needed for the Rhind $2/n$ table. Since, in turn, the multiplication and division of fractions in general involved the Egyptian scholar in the doubling of his unit fractions, both the Rhind table and the present leather roll were more central to his mathematics than we might initially expect.

84. (Below) Pythagorean Triples. One of the most familiar pieces of elementary geometry is the so-called Pythagorean theorem: If z is the hypotenuse of any right triangle and x and y the two sides about the right angle, then $z^2 = x^2 + y^2$. When one determines whole numbers that satisfy this relation, these numbers are said to be Pythagorean triples (for example, 3, 4, and 5; 5, 12, and 13; 9, 12, and 15). This Babylonian tablet, of unknown provenance but datable through its script to the period 1900–1600 B.C., relates to just such triplets. Again taking z as the hypotenuse, x as the shorter side about the right angle, and y as the longer side, the column in the center of our tablet (carrying a heading "solving-number of the width") tabulates 15 whole number values for x, and the column immediately to its right (headed "solving-number of the diagonal," that is, of the hypotenuse) gives 15 corresponding whole number values for z. The larger column at the extreme left lists 15 corresponding values for x^2/y^2 (or, in another interpretation, assuming that a sexagesimal unit of 1 preceding each entry was broken away at the left of the tablet, z^2/y^2). The column at the extreme right merely gives the ordinal number of each line: first, second, third, etc. All of the numbers given are, as in Babylonian mathematics in general, sexagesimal. It is still a matter of some controversy just how these fifteen instances of Pythagorean triples were constructed. Most of the relevant numbers are too large to allow of their discovery through trial and error (for example, the fourth line gives $z = 5, 9, 1$; $x = 3, 31, 49$; and $y = 3, 45, 0$; decimally, this is $z = 18541$, $x = 12709$, and $y = 13500$).

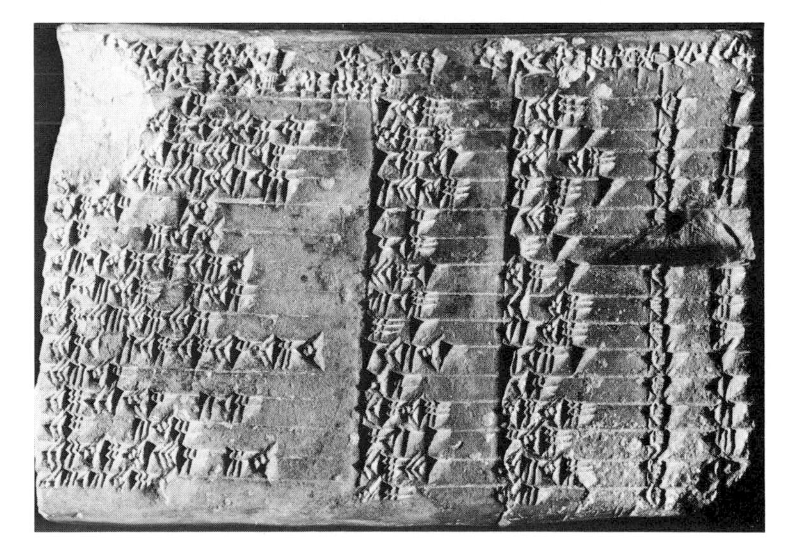

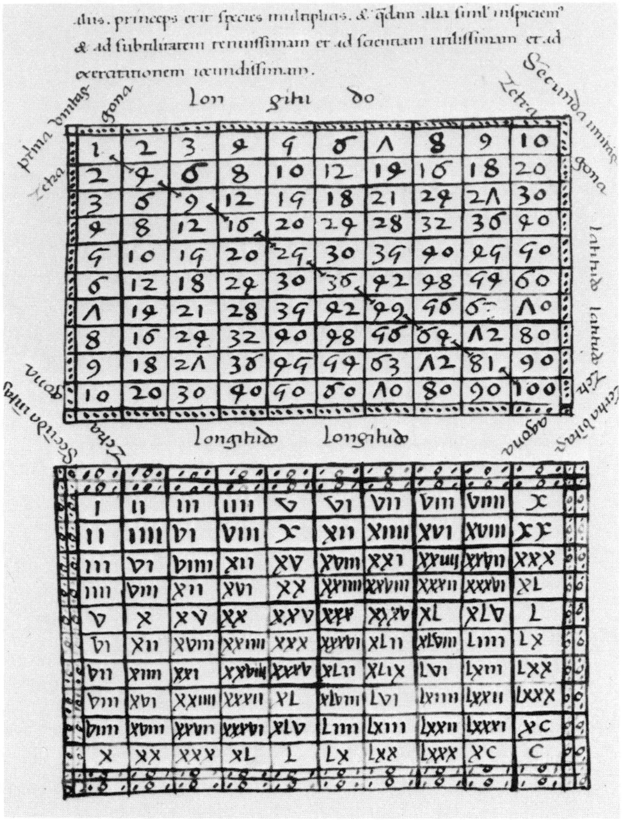

85. Multiplication and More. This fifteenth-century humanist manuscript of Boethius' *Arithmetica* gives both an Arabic and a Roman numeral version of what clearly looks like an ordinary multiplication table for 1 through 10. It is, indeed, that, but Boethius (*Arith.* I, 26–27) considered the primary function of this table—or *descriptio,* as he calls it—to be a visual aid in coming to know some of the various kinds of relations or ratios that can obtain between numbers. Thus, if we compare the entries in any horizontal or vertical line with, respectively, the corresponding entries in the horizontal line at the top or the vertical line at the extreme left, we will always obtain multiple ratios (that is, in our terms, ratios of the form $n:1$). Alternatively, if we compare entries in the third horizontal line or the third vertical line with corresponding entries in the second horizontal line or second vertical line, we will have sesquialtern ratios (that is, ratios equal to 3:2), while effecting a similar comparison between the fourth horizontal or vertical lines with the third horizontal or vertical lines will yield sesquitertian ratios (that is, ratios equal to 4:3). Both 3:2 and 4:3 are species of superparticular ratios (that is, of the form $n + 1:n$) and Boethius tells us that if we continue to compare the entries in lines in the fashion just indicated, we will obtain yet other species of the superparticular. He also notes something "divine" about this table: that all entries on the diagonal from top left to bottom right are square numbers. The words *longitudo* and *latitudo* on the bottom and side of the upper table are Boethius' way of referring to horizontal and vertical. The inscriptions *prima unitas tetragona, secunda unitas tetragona,* and *tertia unitas tetragona* (first, second, and third square unities) written at the corners of the table refer to older Pythagorean terms for 1, 10, and 100. Almost all earlier manuscript copies of Boethius' *Arithmetica* utilize Roman numerals in the text and accompanying illustrations, but the greatly increased use of Arabic numerals throughout the Middle Ages has apparently given rise to the "double table" in this late manuscript. For other figures from this manuscript, see Illustration 94.

86. Abacist Division. This twelfth-century manuscript presents the text of an anonymous tenth-century treatise on calculation with fractions. As is evident from the present figure, the methods of calculation relate directly to the abacus, that is, to a columnar table on which the numerals 1 to 9—zero being absent—assume differing values according to the columns in which they are inscribed. There are three columns in each of the two sample calculations represented by the abacist table depicted here: hundreds (C), tens (X), and units (I); this final column is also used for the entry of fractions. One could calculate either by arranging in the appropriate column counters on which some indication of the numerals 1 to 9 was inscribed or, as in the present instance, by writing some representation of the relevant numerals in the columns. The table below has, for the most part, used Roman numerals, with occasional abacist forms of Arabic numerals (for example, = 2, = 6, = 8), as well as the standard signs for fractions (for example, = $^{11}/_{12}$ = *deunx*, = *semis* = $^{1}/_{2}$, = $^{1}/_{288}$ = *scipulus*). Thus, I and in the third line of the C and X columns represent 120; II and IIII in the eighth line of the X and I columns, 24; etc. (the XL in the X column is an error for IIII = 40). The table gives two examples of division by fractions: that on the left, 120 by $11\,^{11}/_{12}$; that on the right, 120 by $11\,^{1}/_{288}$. The division proceeds "by difference"; that is, in place of utilizing the given divisor, one increases it to the next highest multiple of 10, of 100, etc. (in the left-hand example, the $11\,^{11}/_{12}$ on the first line is increased by a difference of $8\,^{1}/_{12}$ on the second line to 20) and then employs this augmented divisor throughout on successive steps of the division in question. Thus, instead of following the modern procedure of subtracting the product of the divisor and each partial quotient from the relevant partial dividend, one subtracts the product of the augmented divisor and each partial quotient (here called *prima, secunda,* and *tertia medietas*) and then adds the product of the difference and this partial quotient to obtain the next dividend. The answer to the division on the left is given by the four Roman numerals in the last line of the units (I) column (I, I, II, VI) together with information to the right of these numerals specifying a remainder of $^{11}/_{12}$ (*remanente* —here wrongly written). This is, in modern terms, 10 plus a remainder of $^{11}/_{12}$ths of $11\,^{11}/_{12}$ths. Curiously, the present table is not viewed as an explanation of the accompanying text; instead, the text is said to explain the table *(explanatio presentis tabelle)*, an unusual bit of evidence of the central importance assigned to figures such as the present one.

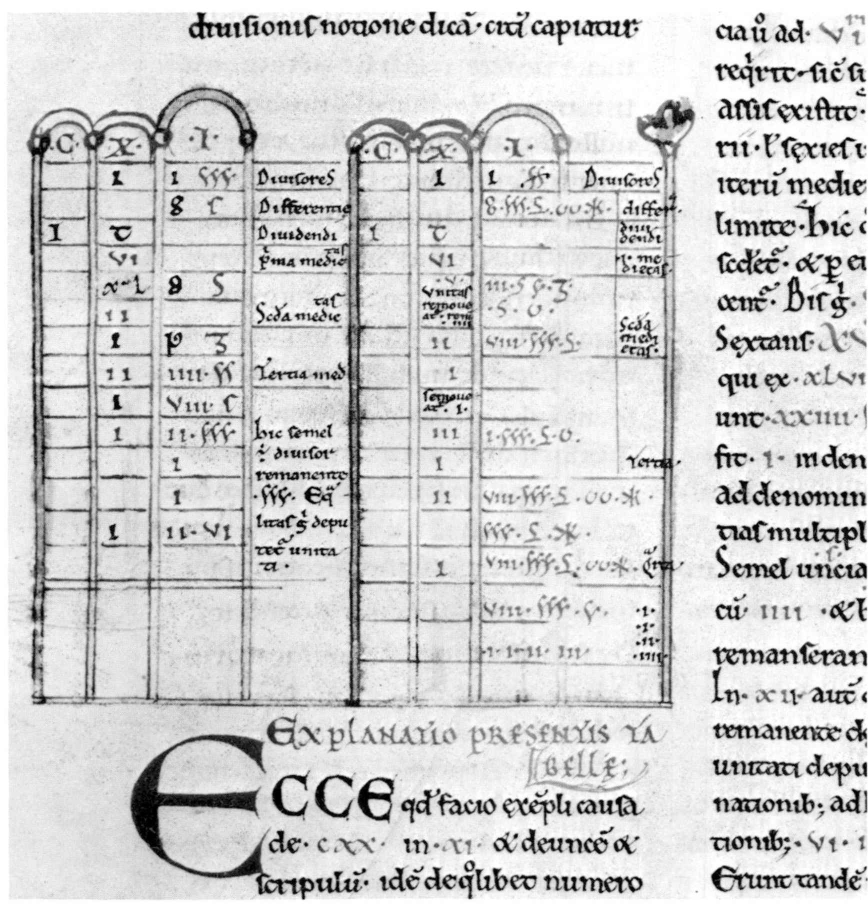

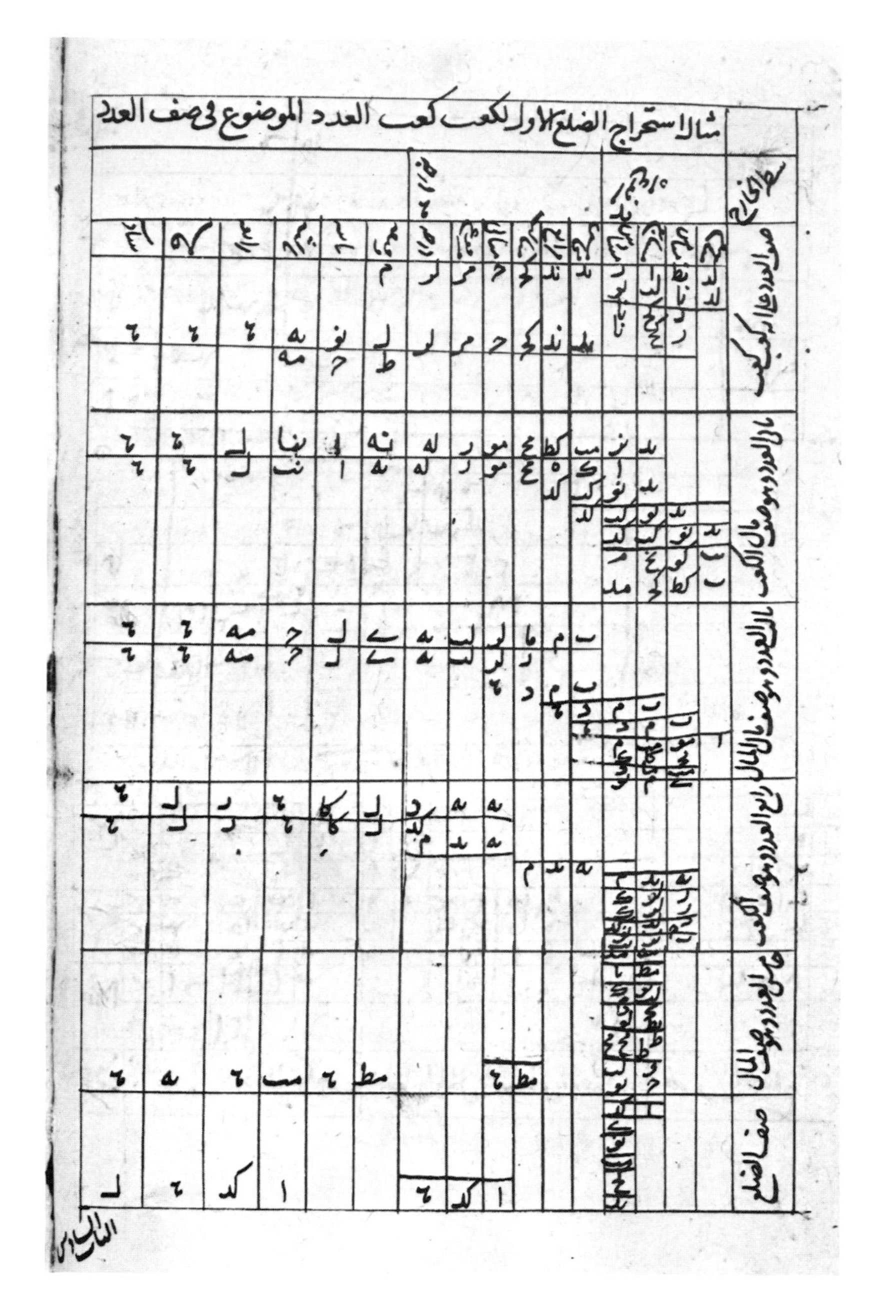

87. Tabulation of a Root Extraction Procedure. One of the most impressive encyclopedic mathematical works of late medieval Arabic science was *The Key of Arithmetic (Miftāḥ al-ḥisāb)* of the fifteenth-century Jamshīd Ghiyāth al-Dīn al-Kāshī (*d.* 1429), perhaps the most important scholar among those who flourished in Samarkand under the patronage of the Sultan Ulugh Bēg. Al-Kāshī's *Miftāḥ al-ḥisāb* consists of five books, the third of them devoted to the "calculation of the astronomers"—that is, to sexagesimal arithmetic—and it is the final page of one of the chapters of this third book which is reproduced here from a sixteenth-century manuscript. The chapter itself is concerned with the extraction of roots and sets forth a procedure for deriving the *n*th root of any sexagesimal, concluding by giving several examples of the procedure in question, the last of which is pictured here: to find the sixth root of 34, 59, 1, 7, 14, 54, 23, 3, 47, 37; 40 (decimally, of 352554535943221657 ⅔). Al-Kāshī does not work out the example in his text, but instead presents it as already worked out in the present table. Beginning with an inscription at the top informing us that we have here "an example of extracting the sixth root of the number put in the row of the number," the body of the table consists of six horizontal rows corresponding to steps outlined in the general procedure of root extraction by al-Kāshī in his text. Each of these steps is in effect "named" in the vertical column at the right, beginning at the top with "the row of the number" (in which, accordingly, there appears the number whose sixth root we are to find), proceeding through rows of the fifth power, fourth power, cube, and square, to the "row of the root" at the bottom. Finally, the second line from the top in our table is "the line of the result" and thus gives the sought-for sixth root: 14, 0; 30, given to one fractional sexagesimal place (although the figure for 30 minutes is missing in the table); the result decimally corresponds to 840.5. Taken as a whole, therefore, the table can be viewed as a convenient visual means by which to present a sample calculation, each step of the calculation receiving an appropriate label. In this, the present table functioned much like the abacist table in the preceding Illustration.

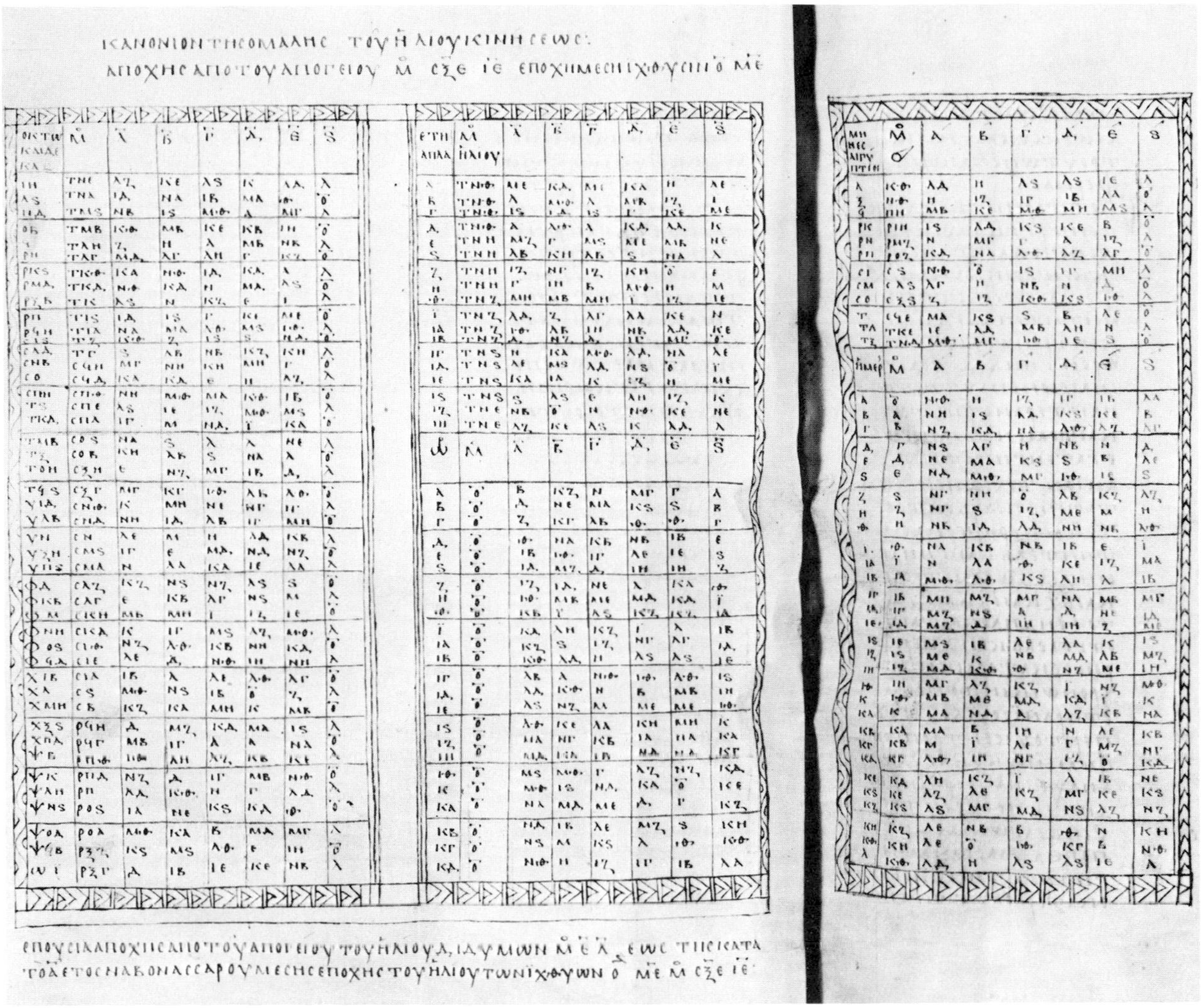

88. The Mean Motion of the Sun. Appearing in a ninth-century manuscript of Ptolemy's *Almagest,* these tables constitute the whole of the second chapter of Book III of that work. In the preceding chapter Ptolemy had investigated the length of the tropical year and had determined that the value given by Hipparchus (*ca.* 150 B.C.) was indeed the correct one: 365; 14, 48 days (equivalent to 365 days, 5 hours, 55 minutes, 12 seconds). He concluded this account by dividing this value into the 360° of the sun's orbit in a tropical year, thus obtaining a sexagesimal figure of 0; 59, 8, 17, 13, 12, 31 as the sun's mean motion for a single day, at the same time calculating similar values for the solar mean motion for a single hour, a single (30-day) month, a single 365-day Egyptian year, and a period of 18 years. Using these parameters, Ptolemy then set up the five tables pictured here, employing as points of reference (as the legends to the tables tell us) the longitude of the solar apogee and the mean longitude of the sun relative to the epoch of Nabonassar (= noon, 26 February 747 B.C.). The entries in all the tables are given in degrees and 6 sexagesimal places for fractions of degrees. The whole column at the left gives 45 entries for the mean solar motion over 18-year intervals; the top of the center column, entries for 18 yearly intervals; the bottom center, 24 entries for hourly intervals; the top right, 12 entries for monthly intervals; and the bottom right, 30 entries covering the days of a (30-day) month. From the standpoint of the history of scientific illustrations, it is interesting to note that considerations of "layout" dictated the 18-year intervals constituting the first table. For, as Ptolemy himself makes clear, the 45 lines required for this particular table if it is to cover an adequate number of years neatly balance the 42 lines (plus headings) needed for the four tables making up the other two columns. Indeed, the results of an identical "Ptolemaic" consideration in the construction of another table can be seen in Illustration 89. For another figure from the present manuscript (which is one of the two earliest extant Greek copies of Ptolemy) see Illustration 125.

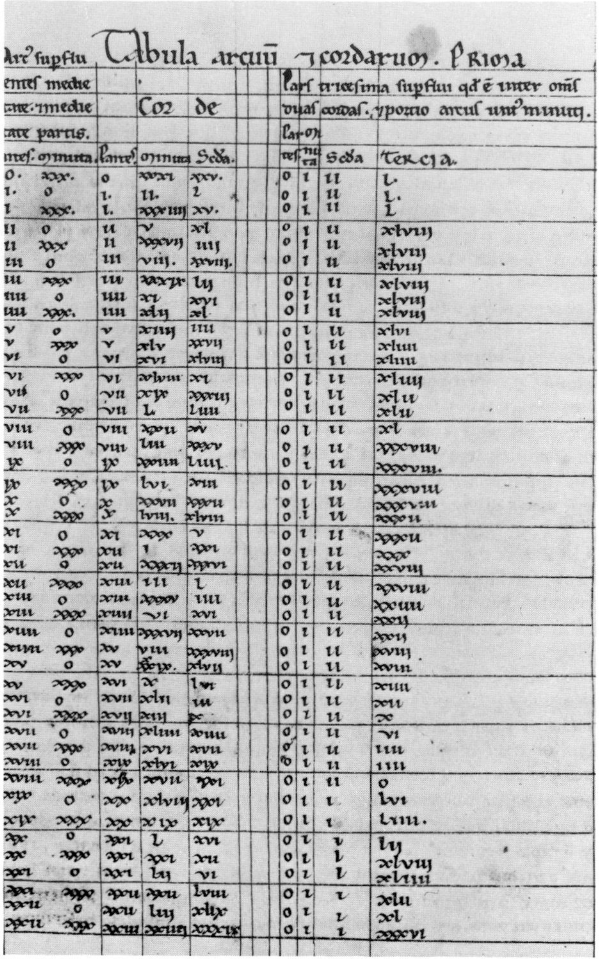

89. Ancient Trigonometry: The Calculation of Chords. The basic trigonometric functions as we know them were not a part of ancient mathematical astronomy; chords subtended by given angles or arcs of a unit circle served in their stead (a "substitution" that was quite effective, given the fact that the chord subtended by any angle is equivalent to twice the sine of half that angle). Central as calculations by chords were to astronomy, Ptolemy began the mathematics of his *Almagest* (Book I, chapters 10–11) by constructing a table of chords. The beginning of this table is pictured here from a twelfth-century manuscript of Gerard of Cremona's Latin translation (from the Arabic) of Ptolemy's work. Values are given for chords subtended by all arcs from ½° to 180° in ½° intervals. The extreme left column of the present portion of the table lists in Roman numerals arcs from ½° (O XXX) to 22°30′ (XXII XXX). The next three columns (carrying the common heading 'corde') give the lengths of the corresponding chords expressed in terms of units *(partes)* and sexagesimal fractions (*minuta* and *secunda*), the units referring to the radius of the circle in question with this radius arbitrarily set at 60 units. The remaining columns give values carried to three sexagesimal places for 1/30 of the difference between the corresponding chord length and that of the next largest chord (values which can be utilized to calculate the lengths of chords of arcs not listed in the initial column at the left). There are in all eight tables like the present one, each of them containing 45 entries. Once again, Ptolemy notes that "for the sake of symmetry" this is the best arrangement. Considerations of "design" were, one might say, already being applied to the papyri of Ptolemy's original text.

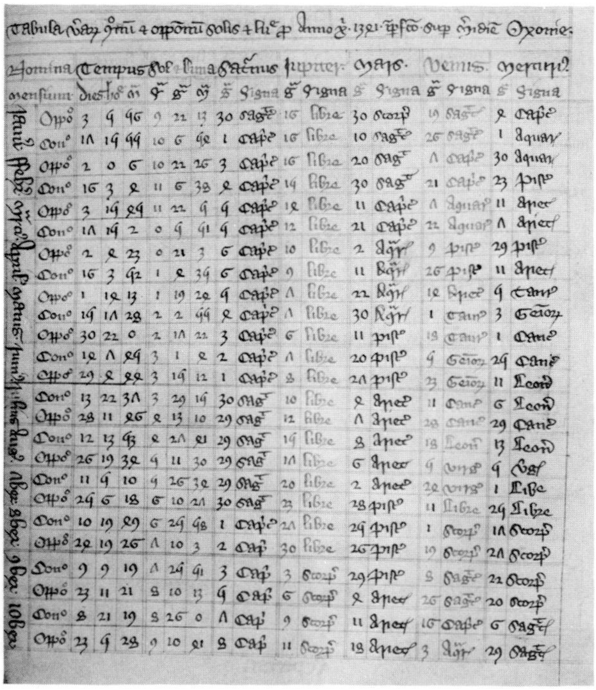

90. An Almanac for Medieval Scholars. This fourteenth-century manuscript containing some 238 pages of assorted astronomical and meteorological information served as a kind of almanac for students and masters at the Oxford colleges of Merton and Exeter. The page here reproduced gives a table of the times of conjunction and opposition of the sun and the moon for the meridian of Oxford for the year 1341 together with an indication of the positions of the planets at these times. The column at the extreme left lists the names of the months, the next seven columns specifying the times of opposition and conjunction by day, hour, and minute (with the hours calculated beginning at noon), and the position of these oppositions and conjunctions by zodiacal sign (given by number), degree, and minute. The remaining columns list the positions of the five planets at the times in question by zodiacal sign (now by name and not number) and degree. For example, the fourth line of the table reveals that on 16 February 1341 the sun and moon were in conjunction in 6°38′ of Aquarius at 3:04 P.M., at which time Saturn was in 4° of Capricorn, Jupiter in 14° of Libra, Mars in 30° of Sagittarius, Venus in 21° of Capricorn, and Mercury in 23° of Pisces. From the times and positions given here, it is clear that they were derived from some version or other of the thirteenth-century Alfonsine Tables, perhaps that compiled by the fourteenth-century astronomer and mathematician William Rede, since he at one time owned, and had even written part of, the Oxonian almanac from which our table is drawn.

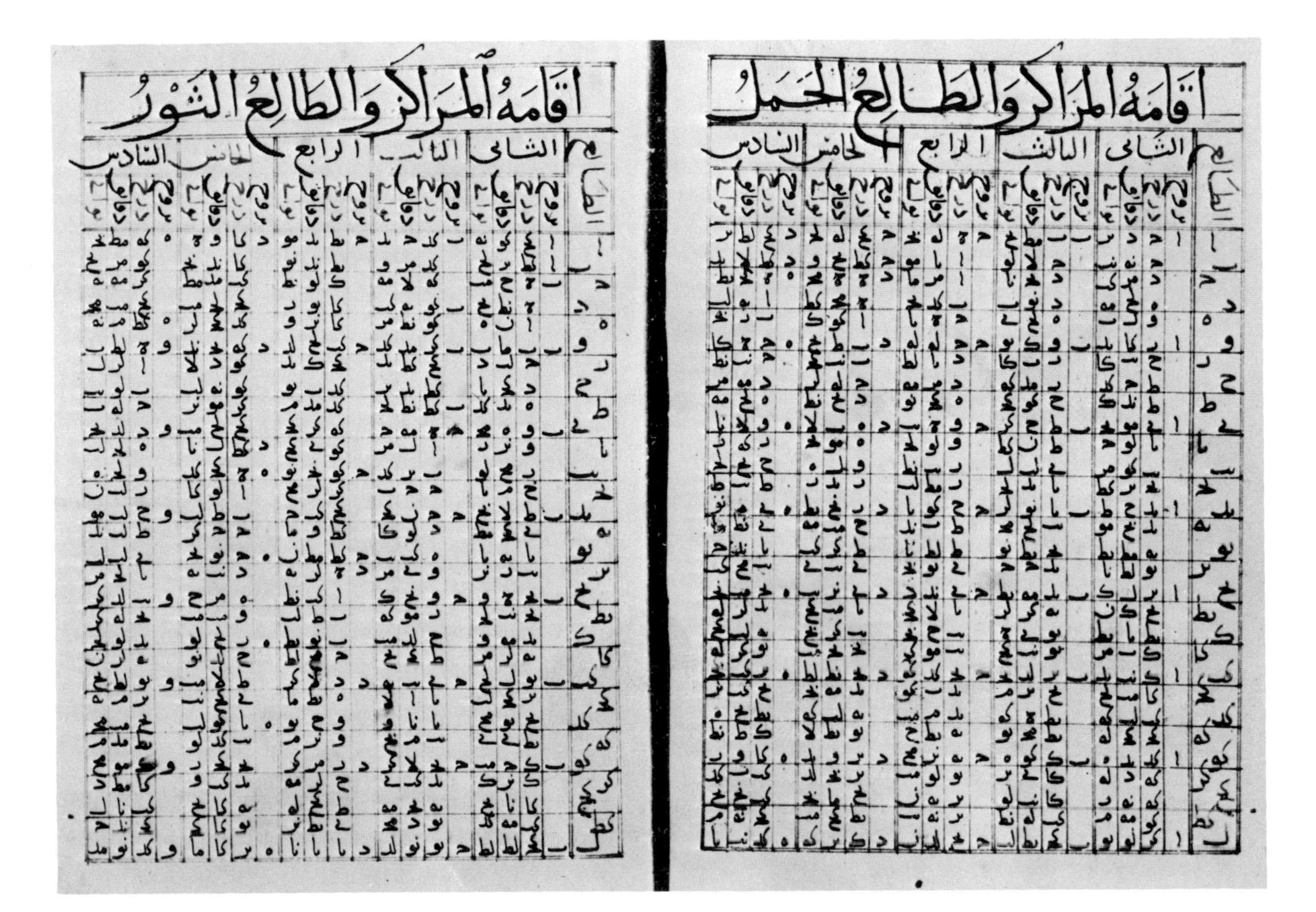

91. An Islamic Zij. If tables were an essential part of Greek mathematical astronomy, they were that plus a quantitatively major part of what Islamic scholars did with this astronomy. In Arabic, any astronomical table was called a *zīj*. The two pages pictured here are taken from a thirteenth-century manuscript of the *Zīj* of Jamāl al-Dīn al-Baghdādi (ca. thirteenth century) and give the first two tables of twelve concerned with the equalization of the astrological houses calculated for the latitude of Baghdad. These astrological houses (in Arabic *buyūt*) are twelve segments of the ecliptic, beginning with some ascendant point on the ecliptic (that is, some point on the ecliptic rising relative to the horizon) calculated or "equalized" in the direction of increasing longitudes and having equal right ascensions. As was customary, the present *zīj* gives a table of equalization for the ascendant in each of the twelve zodiacal signs. The two reproduced here are for Aries (on the right) and Taurus. Thus, one begins with the columns at the extreme right which give each of the thirty different degrees of Aries or Taurus for the ascendant in question and then reads to the left to obtain the beginning of the second house, the third house, and so on, up to the sixth house in the last column. The seventh through twelfth houses did not appear in tables such as these since they were diametrically opposite to the corresponding six houses already given. The values in the present table have been calculated to three sexagesimal places, which means to a precision of better than one part in 200,000. More than that, the values given in this particular *zīj* have been recalculated by means of a computer and have been found, almost without exception, to be precise to the last digit. Considering the extreme complexity of many of the calculations involved, this was an impressive medieval accomplishment.

92. Numbers Plus Pictures. This sixteenth-century German manuscript gives an example of the quite frequent late medieval and Renaissance practice of coupling dates and numerical data for eclipses with little pictures of how they would appear. The legends on these tables *cum figuris* tell us that the lunar and solar eclipses given are calculated for the meridian and (for the solar eclipses) the latitude of Ulm. The information given covers 1516–1563 for lunar eclipses and 1516–1570 for solar eclipses (the later years appearing on the next page not reproduced here). Each entry begins with the year and month of the eclipse in question, then lists the half duration in hours and minutes, and concludes with the magnitude of the eclipse given in *puncta ecliptica et minuta.* The date and time of each eclipse is not given, but the final lines on the left-hand page inform us that this can be determined from a previous table (which occurs a few pages earlier in the present manuscript) giving the times of conjunction and opposition of the sun and the moon (here symbolized by the usual ☌ and ☍) for the relevant year and month. It is evident that the numerical data given in this illustration were taken from some printed edition or editions of the *Almanac* or *Ephemerides* prepared by the fifteenth–sixteenth-century scholar Johann Stoeffler and his continuators. Preparing a manuscript copy of a printed work or, especially, of some part of a printed work was a frequent occurrence in sixteenth-century science. When we turn to the miniature pictures of eclipses (which are also found in Stoeffler), all is not as proper as one might like. The little circles are accurate in depicting total vs. partial eclipses, but in the case of the latter the difference in shading does not always correspond to what part would be darkened. Similarly, the abbreviations *sep (septentrionalis)* and *mer (meridionalis)* written on the "face" of the circles is meant to indicate that the moon will be, respectively, either north or south of the ecliptic during the eclipse in question, but that information is also occasionally incorrect.

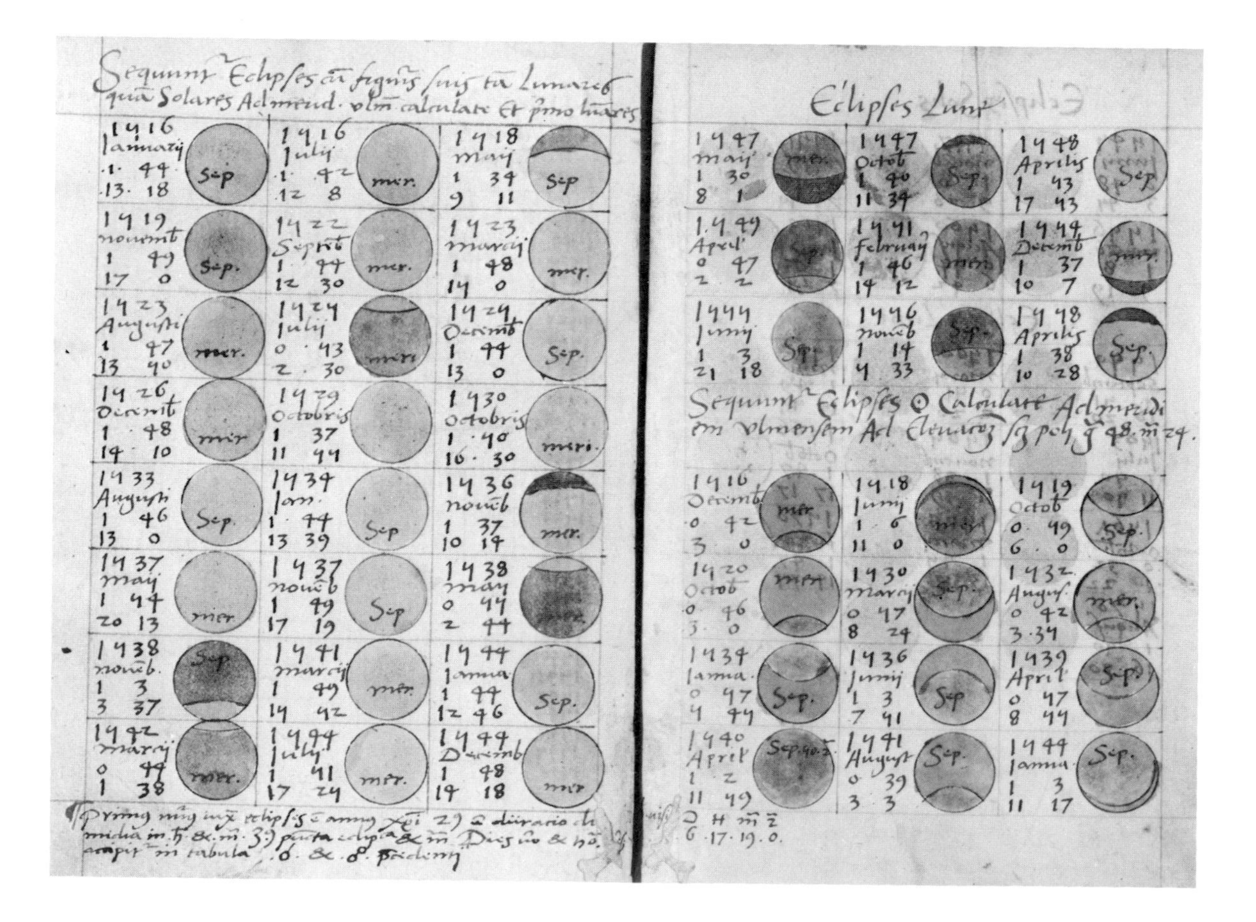

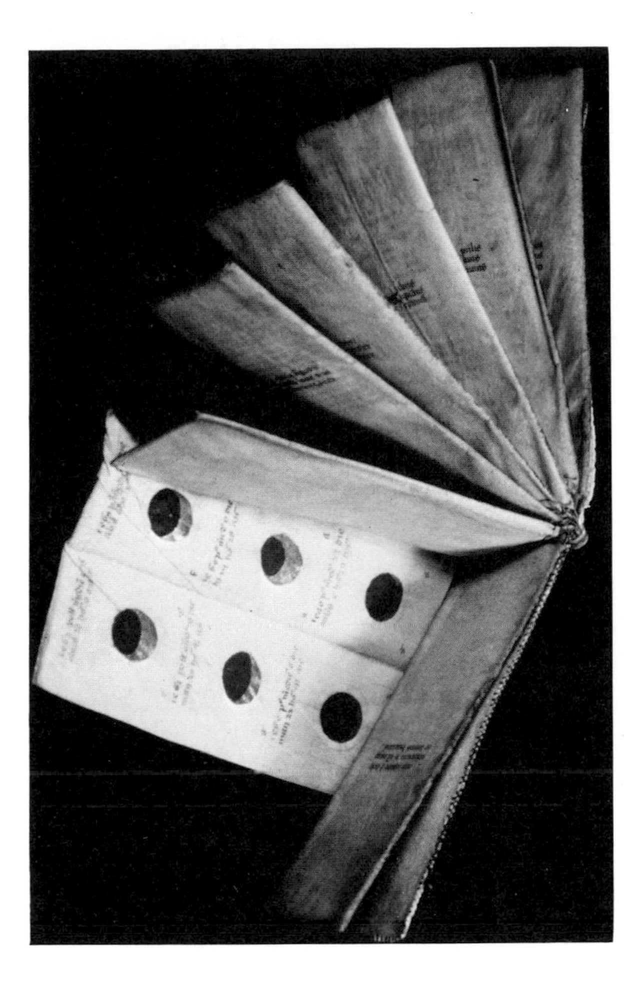

93. Data at the Ready. The illustration at the left is of a fifteenth-century girdle almanac, the sheet that is here unfolded containing eclipse information for the years 1462–1470 quite similar to that depicted in the preceding Illustration, although with less detail than the latter since data for only the time of the eclipse in question is given. The pictures of each eclipse are strikingly reproduced in blue and gold with contrasting red and black script for their legends. The almanac contains calendaric information and tabular data concerning planetary positions for the later years of the fifteenth century and the beginning of the sixteenth as well as the eclipse material shown here. This additional information was arranged in gatherings covering three months each, the slightest glance at the "label" on the outside of each gathering revealing which months were its concern. Several of these "labels" are visible on the unfolded gatherings at the bottom. Girdle books such as that depicted here were attached to one's belt, conveniently positioned for ready reference. They were written and put together in such a way that opening or unfolding them while still fastened at waist level would display their contents "right side up" in a properly readable orientation. The woodcut at the right is taken from a fifteenth-century edition of the *Pronosticatio in latino* of Johann Lichtenberger, an extremely popular astrological work—appearing in numerous German and Italian as well as Latin editions—that related its predictions especially to notable planetary conjunctions in the years 1484–1485 and to the solar eclipse of March 16 of the latter year. Our interest in the present picture is in the girdle book, most likely an almanac similar to the one in the illustration at the left, hanging from the belt of the monk on the right who, pointing toward the heavens, is presumably explaining a prognostication of some sort to his fellow monks. For another girdle book and its contents, see Illustration 267.

CHAPTER TEN

The Representation of the Properties of Numbers

THE ILLUSTRATIONS OF TABLES in the preceding chapter were all, in one fashion or another, concerned with numbers. This concern was always computational: The table either depicted how to calculate or represented the results of calculations which were to be applied in subsequent scientific pursuits (although Illustration 85 is something of an exception in this regard).

The concern of the illustrations in the present chapter also is numbers, but it is a concern of a quite different sort. In present-day terms, it would fall under number theory. In ancient and medieval terms, however, this was simply arithmetic, to be distinguished quite sharply from the very different discipline of computation which we now cover by the word "arithmetic." Indeed, this ancient-medieval distinction was even reflected in the use of different terms to designate the disciplines in question: The art of calculation was for a Greek *ἡ λογιστικὴ τεχνή*, for an Arab *ᶜilm al-ḥisāb,* and for a medieval Latin *calculatio;* whereas the science of arithmetic was, correspondingly, *ἡ ἀριθμητικὴ ἐπιστήμη*, *ᶜilm al-ᶜadad,* and *arithmetica.*

One of the charges of ancient and medieval arithmetic was the investigation of the properties of particular kinds of numbers: odd versus even numbers; prime versus composite numbers; linear, plane, and solid numbers (of which square and cube numbers are species of the last two); or such things as perfect or friendly numbers. (The former is any whole number which is the sum of its proper divisors—for example, $28 = 1 + 2 + 4 + 7 + 14$—and the latter are any two whole numbers such that each is the sum of the proper divisors of the other—for example, 220 and 284.)

On the other hand, arithmetic was also concerned with the different kinds of relations that could obtain between numbers: They could be mutually prime or have a common measure; and, most important, they could stand to one another in different kinds of ratios, various ratios in turn standing in different kinds of proportion. What is more, that part of arithmetic concerned with ratios, means, and proportions was the very backbone of the mathematics of music theory.

The basis for all this was, as is to be expected, Greek. Books VII through IX of the *Elements* of Euclid (third century B.C.) were arithmetical and contained material relevant to most of the concerns just mentioned. What was not to be found there could be had from the *Introduction to Arithmetic* of Nicomachus (first century), although there was some overlap of this work with that of Euclid.

When one turns to the Latin Middle Ages, perhaps the most significant fact to note is that although scholars then possessed Euclid in a variety of different translations, arithmetic was more frequently learned from Boethius' version of Nicomachus. In the thirteenth century, Jordanus de Nemore produced a more elaborate and sophisticated arithmetic that derived from the Boethian tradition, but Boethius himself more than held his own when it came to just what was studied in the medieval curriculum. Alternative arithmetics like that of Thomas Bradwardine in the fourteenth century were little more than epitomes of straight Boethius. The terrain of the mathematics of music theory was also basically Boethian, since it was the *De institutione musica* of that sixth-century scholar which dominated the discipline in medieval faculties of arts.

In order to present the broadest possible spectrum in illustrations dealing with the properties of numbers, we have taken all of those constituting the present chapter from medieval Latin manuscripts, the greater share of them from copies of the works of Boethius and Jordanus de Nemore. They vary from the most elementary to the most complex and, in some instances, have been selected to show the unhelpful, even the unusable, diagram or figure, an entity that seems to have been much more common when it came to the arithmetic involved in music.

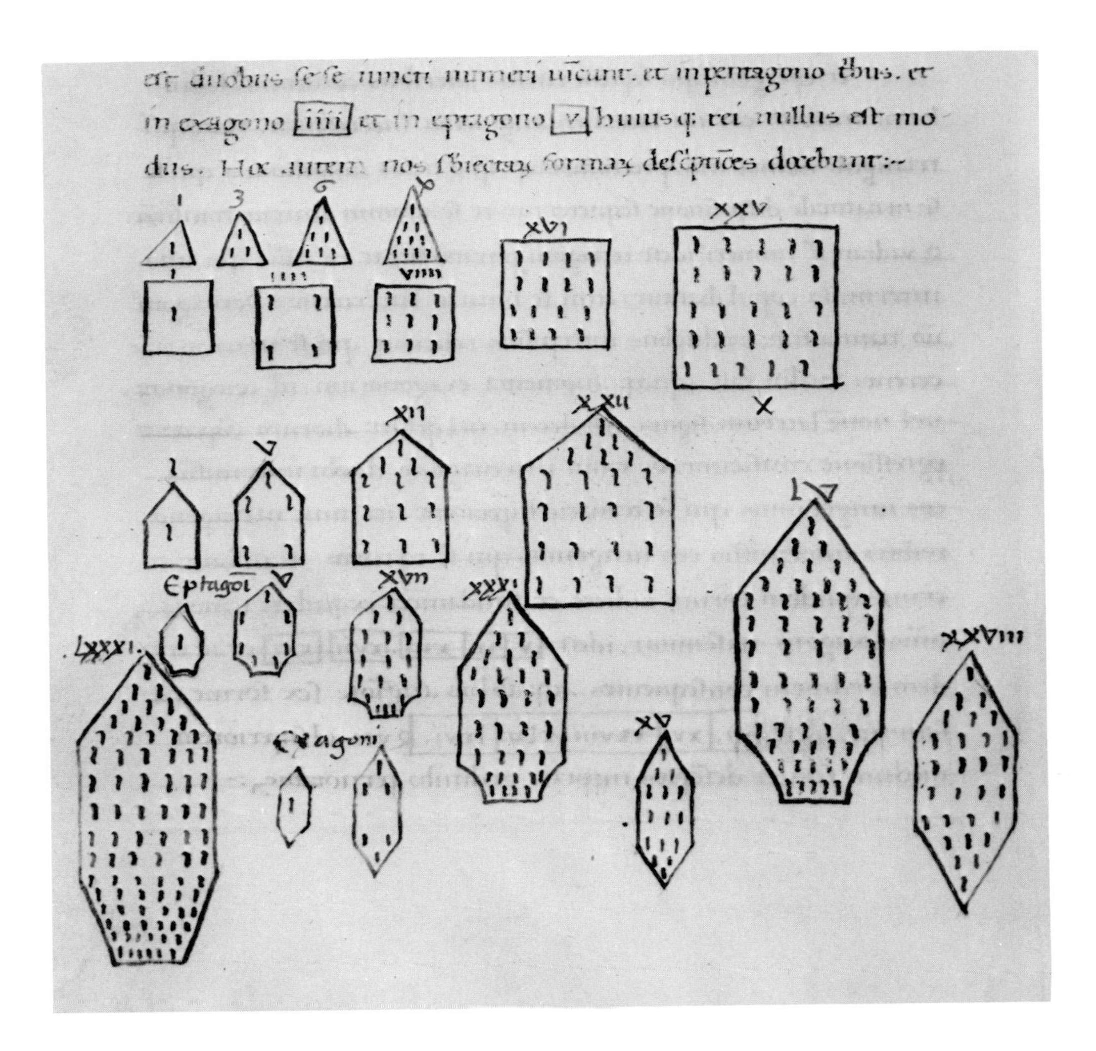

94. Geometric Arithmetic. The notion of figurate or polygonal numbers appears to have been part of the so-called Pythagorean arithmetic of the fifth century B.C., but a reasonably full treatment of the theory of such numbers is preserved for us only in the much later works of Nicomachus of Gerasa (fl. ca. 100 A.D.) and Theon of Smyrna (fl. early second century). In turn, Boethius' (sixth century) *Arithmetica,* being but a Latin paraphrase of the *Introduction to Arithmetic* of the first of these two Greek scholars, made the doctrine of figurate numbers a permanent part of medieval mathematics. The present illustration, taken from a fifteenth-century manuscript of Boethius' work, gives the figures that were typical for triangular, square, pentagonal, hexagonal, and heptagonal numbers. The "doctrine" behind these numbers was simply that the sum of any number of successive terms of a series of natural numbers in arithmetic progression beginning from 1 would yield the figurate number in question. Thus: (1) $1 + 2 + 3 + 4 + \ldots + n = [n(n + 1)]/2$ gives the triangular numbers 1, 3, 6, 10, . . . presented in our illustration; (2) $1 + 3 + 5 + 7 + 9 + \ldots + (2n - 1) = n^2$ gives the square numbers 1, 4, 9, 16, 25, . . .; (3) $1 + 4 + 7 + 10 + \ldots + (3n - 2) = [n(3n - 1)]/2$ gives the pentagonal numbers 1, 5, 12, 22, . . .; (4) $1 + 5 + 9 + 13 + \ldots + (4n - 3) = 2n^2 - n$ gives the hexagonal numbers 1, 6, 15, 28, . . .; (5) $1 + 6 + 11 + 16 + 21 + 26 + \ldots + (5n - 4) = [n(5n - 3)]/2$ gives the heptagonal numbers 1, 7, 17, 34, 55, 81, . . . (although 34 is mistakenly given as 31 in our illustration). Each 1 is considered as being the polygonal number in question only in potency *(vi et potestate)*, while all others are such numbers actually *(actu)*. In the text (belonging to Book II, ch. 16) directly above our illustration, the numbers iiii and v are part of Boethius' explanation of the constant differences involved in the series, generating, respectively, hexagonal and heptagonal numbers. Note that any such constant difference is always 2 less than the "side" of the polygonal number associated with it. There is little doubt that the pictorial representations such as that given here of figurate or polygonal numbers were every bit as important to the medieval scholar in coming to understand the theory of such numbers as was the text of Boethius (or any other similar arithmetical tract). For other illustrations from Boethius' *Arithmetica* or from the present manuscript, see Illustrations 22, 85, 95, 97, and 172.

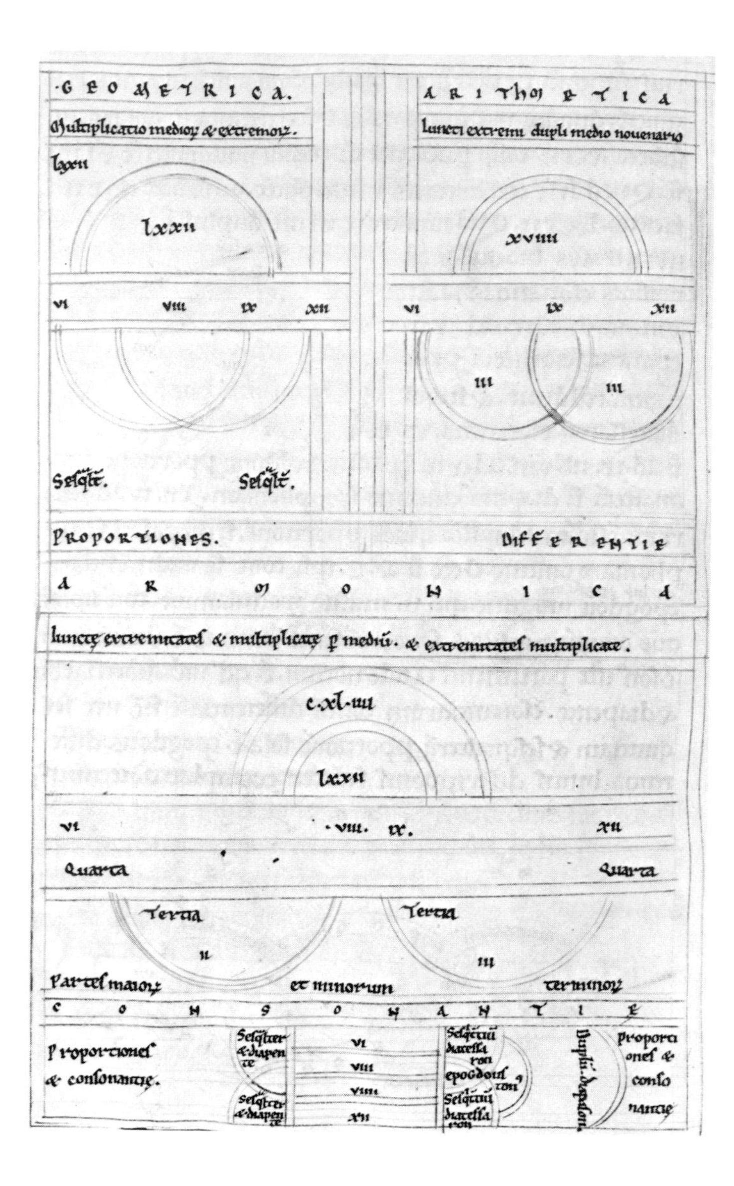

95. From Arithmetic to Music. The present full-page diagram, here taken from an eleventh-century manuscript, falls at the very end of the second and final book of Boethius' *Arithmetica.* Boethius has been expounding the doctrine of arithmetic, geometric, and harmonic means and proportions (or, as he would call them, proportionalities), and in his final chapter he cites the numbers 6, 8, 9, and 12 (here in Roman numerals) as an "exemplar" containing all of these means and proportions and, as a result, being the repository of *omnes musicas consonantias.* The basic mathematics of the kinds of proportions behind Boethius' concern at this point is, of course, the following (where b is the mean in question): arithmetic: $a - b = b - c$; geometric: $a{:}b = b{:}c$; harmonic: $a - b{:}b - c = a{:}c$. Using his "exemplar" of 6, 8, 9, 12, geometric proportion occupies the upper-left corner of the diagram, where 8 and 9 are two means between 6 and 12 as extremes. In such a case, we are told (in the second line from the top) that the product of the means equals the product of the extremes (thus: $6 \times 12 = 72 = 8 \times 9$). Further, the bottom of this segment of our diagram tells us, by means of arcs that were traditionally used to represent ratios *(proportiones)*, that 12 stands to 8, and 9 to 6, in a sesquialtern ratio. On the other hand, 6, 9, and 12 at the upper right display arithmetic proportion: $12 - 9 = 9 - 6$, where the two equal differences *(differentie)* are specified on arcs of three each and the sum of the extremes 6 and 12 is revealed as the double of the mean 9. Next, the main portion of the bottom half of the diagram sets forth the harmonic proportion $12 - 8{:}8 - 6 = 12{:}6$. Here our "picture" mistakenly gives less than required by Boethius' text and at the same time adds something that is not there. We are told that in harmonic proportion the product of the mean and the sum of the extremes is equal to twice (which word is missing) the product of the extremes [that is, $8 \times (6 + 12) = 2(6 \times 12)$]. We are also informed that 8 exceeds 6 by a third part *(tertia)* of 6 just as 12 exceeds 8 by a third part of 12, again a characteristic of harmonic proportions noted by Boethius. The appearance of *quarta,* or fourth part, in the diagram and of 9 as an additional mean is, however, not a representation of something Boethius has just said; apparently it is there simply to draw attention to the corresponding fact that 8 exceeds 6 by a fourth part of 8 just as 12 exceeds 9 by a fourth part of 12. Finally, the very bottom of the diagram relates all that has been pictured above to musical ratios and consonant intervals: two sesquialtern ratios (12:8 and 9:6) are diapente consonant intervals (that is, fifths); two sesquitertian ratios (12:9 and 8:6) are diatessarons (that is, fourths); a double ratio (12:6) is diapason (the octave); and a sesquioctave ratio, here given with Boethius in transliterated Greek as *epogdous* (9:8), is a tone. The arcs intended to represent these ratios and consonant intervals are not properly drawn, however. For other diagrams from this manuscript see Illustrations 100, 104, 214, and 278. And for more on musical ratios, see Illustrations 100–105.

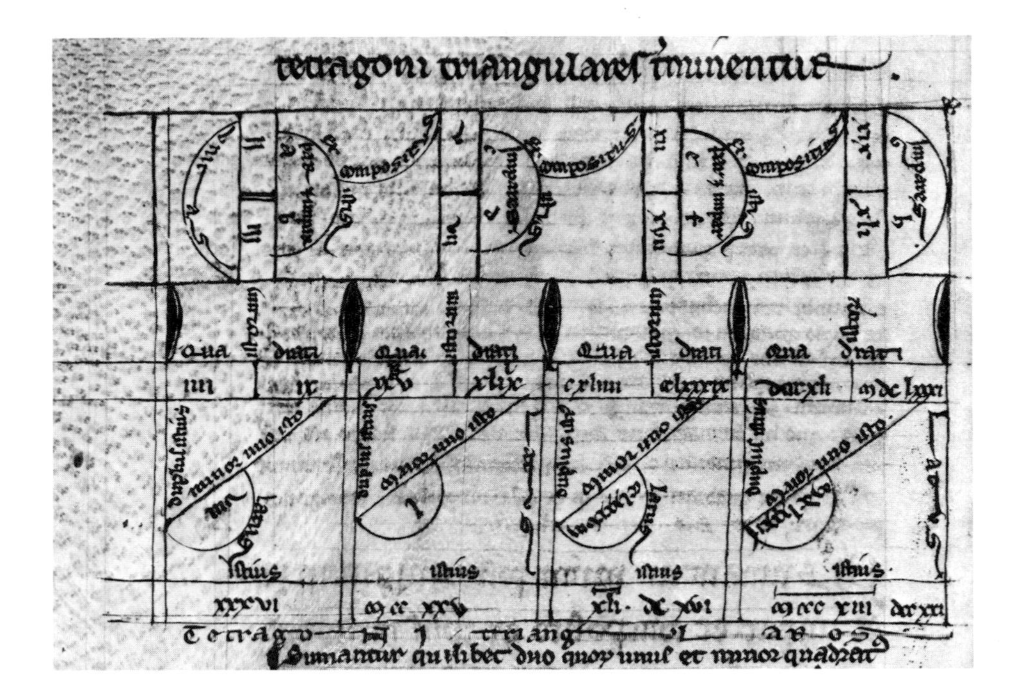

96. Square Triangular Numbers. We have already seen triangular and square numbers as the first two species of figurate numbers represented in Illustration 94. The present diagram, from a thirteenth-century manuscript, depicts the substance of a proposition (VIII, 10) of Jordanus de Nemore's (thirteenth century) *Arithmetica* which tells us that a certain series of numbers will generate pairs of squares whose products will, in turn, generate numbers that are simultaneously triangular and square and which will also reveal the "sides" of the "triangular form" of those generated numbers. The pairs of squares in question are so labeled *(quadrati)* in the center row of the diagram; changing the Roman to Arabic numerals, they are: 4, 9; 25, 49; 144, 289; 841, 1681. Note that we here have pairs k, m such that $k = (m \pm 1)/2$; therefore their products will always be of the form $[n(n + 1)]/2$, which we have seen (Illustration 94) to be that of any triangular number; thus the products of these pairs of squares are the triangular numbers (so-labeled *triangulares*) appearing in the bottom row of our diagram: 36, 1225, 41616, 1413721 (although the last is erroneously written as 1313721). But these numbers are also themselves square numbers (here labeled *tetragoni*), since any number that is the product of two squares is also a square. Inscriptions of 'duplus istius' on the left sides of the four lower squares indicate that the semicircles in these squares contain the doubles of the first square number in each of our foregoing pairs: 8, 50, 288, 1682. The "diagonals" from the semicircles carry labels telling us that these numbers are, alternately, one less or one greater than the second square number in our pairs (thus representing the relation between k and m noted above); 8 and 288 are, however, also the "sides" of the triangular numbers 36 and 41616 (just as 2, 3, and 4 are the sides of the triangular 3, 6, and 10 pictured in Illustration 94), a fact that our diagram represents by a curving *'latus istius'* falling from the semicircles in the first and third of its lower squares. The second and fourth of these squares also contain the words *'latus istius'*, but now connecting the second square number of the pairs above with the triangular numbers below, thus visually presenting one with the fact that 49 and 1681 are the "sides" of the triangular numbers 1225 and 1413721. We have not even mentioned the top half of our diagram, which shows how to derive the pairs of square numbers which have done the yeoman duty we have just mentioned. Beginning at the left with the pair 2 and 3 and moving to the right, the first number of our second pair 5 is composed *(compositus)* of 2 + 3; add to this the first number of our first pair to obtain 7 as the second number of our second pair. Continuing this procedure, we end up with the pairs 2, 3; 5, 7; 12, 17; 29, 41. Squaring these yields our crucial pairs of squares bearing the all-important relation to one another that we have noted above. The manner in which Jordanus generated these numbers guaranteed that their squares would have this desired relation. An interesting historical note can be found in the fact that 2, 3; 5, 7; 12, 17; 29, 41 are some of the "side and diagonal numbers" used in early Greek mathematics to approximate the square root of 2 (something we know through a work of Theon of Smyrna [fl. second century] which was apparently not known by Jordanus). For other diagrams from Jordanus' *Arithmetica,* see Illustrations 97–98.

97. Evenly Even, Evenly Odd, and Oddly Even. A significant segment of ancient Greek, and hence medieval, arithmetic dealt with the theory of odd and even numbers. Part of this theory in turn treated particular kinds of odd and even numbers: (1) even-times even, to use the terminology of Euclid, or simply evenly even *(pariter par)*, to follow Boethius' Latinizing of his source Nicomachus—that is, numbers of the form 2^n: 4, 8, 16, 32, 64, . . .; (2) evenly odd *(pariter impar)*, of the form $2(2m + 1)$: 6, 10, 14, 18, 22, . . .; and (3) oddly even *(impariter par)*, of the form $2^{n+1}(2m + 1)$: 12, 24, 48, 96, 192, ... For Euclid these numbers were slightly different from what they were for Nicomachus, but Boethius, from whose *Arithmetica* in an eleventh-century copy the illustration on the right (above) is drawn, naturally followed the latter scholar. In order to understand this particular diagram, we must first realize the properties had by evenly even and evenly odd numbers. Given any sequence of consecutive *evenly even* numbers a, b, c, d, it is always the case that (1) $ad = bc$, and (2) $ac = b^2$ and $bd = c^2$. Similarly, given any such sequence of *evenly odd* numbers, it is always the case that (1) $a + d = b + c$, and (2) $b = (a + c)/2$ and $c = (b + d)/2$. If we now turn to the present diagram on the right (above), we should begin by noting that its whole purpose is to illustrate the properties of *oddly even* numbers, its central square thus containing just such numbers. Boethius' claim is that (1) any of these oddly even numbers in the *vertical* rows of this central square have the above-mentioned two properties of *evenly odd* numbers even though they are not themselves evenly odd numbers, and that (2) any of these oddly even numbers in the *horizontal* rows have the foregoing two properties of *evenly even* numbers though they are not themselves evenly even. Our diagram illustrates this double claim by the numbers appearing in the arcs surrounding the central square of oddly even numbers. Thus, the three numbers at the extreme left (48, 40, 56) illustrate part (1) of the claim for the four numbers in the extreme left vertical column of the central square (12, 20, 28, 36). That is, $48 = 12 + 36 = 20 + 28$, while $20 = 40/2 = (12 + 28)/2$, and $28 = 56/2 = (20 + 36)/2$. Similarly, the three topmost numbers of the diagram (1152, 576, 2304) depict part (2) of the above claim for the four numbers in the top horizontal row of the central square (12, 24, 48, 96). That is, $1152 = 12 \times 96 = 24 \times 48$, while $576 = 12 \times 48 = 24^2$ and $2304 = 24 \times 96 = 48^2$. The remaining numbers surrounding the central square (among which there are a few errors) illustrate the same claim for the other vertical and horizontal rows. The diagram as a whole falls between two chapters (Book I, 11–12) of Boethius' text and functions as a visual explanation—Boethius calls it a *descriptio*—of the theory set forth in the first of these chapters, while the second chapter serves to explain how the diagram works. Clearly, Boethius felt that this *descriptio* was most important to a proper understanding of his words. Our second diagram on the right (below) is taken from a fourteenth-century manuscript of the *Arithmetica* (Book VII, Prop. 40) of Jordanus of Nemore. The subject is the same as that we have just sampled from Boethius: oddly even numbers. Indeed, the sixteen numbers of this sort in the central square of Boethius' diagram here appear at the lower right in Jordanus' own central square of thirty-six oddly even numbers. Jordanus' claims about these numbers are simpler than those of Boethius, however. To begin with, at the left of his central square, Jordanus has an added vertical row of evenly odd numbers (so labeled on the arc bridging that row): 6, 10, 14, 18, 22, 26, and an added horizontal row of appropriately labeled evenly even numbers at the bottom of the square: 2, 4, 8, 16, 32, 64. The purpose of these additions is merely the depiction of his assertion that the products of each of these evenly odd numbers times each of these evenly even numbers result in the oddly even numbers of his central square. Recalling the general forms mentioned above of these kinds of numbers shows that generally this is true, as Jordanus claims. The remaining two sides of the diagram visually encapsulate the rest of what is asserted by Jordanus' proposition. First, on the top arc we are told that the oddly even numbers in the horizontal rows are continuous proportionals (which is equivalent to one-half of Boethius' claim for his horizontal rows, since $a{:}b = b{:}c$ is equivalent to $ac = b^2$) and, second, the left arc informs us that the numbers in the vertical rows have equal differences, which is again equivalent to half of Boethius' claim, since $a - b = b - c$ is the same as $b = (a + c)/2$. For other diagrams from this manuscript of Jordanus, see Illustrations 98 and 124.

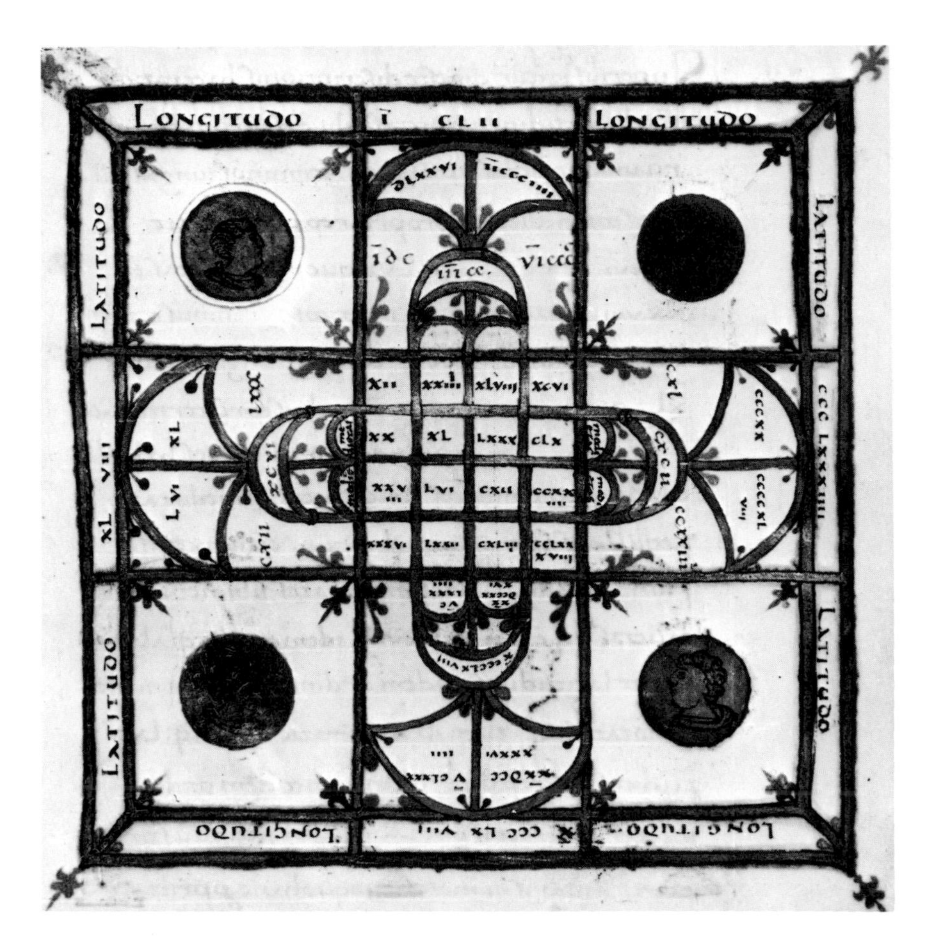
LONGITUDO
CLII
LONGITUDO
LATITUDO
LATITUDO
LATITUDO
LATITUDO
LONGITUDO
LONGITUDO

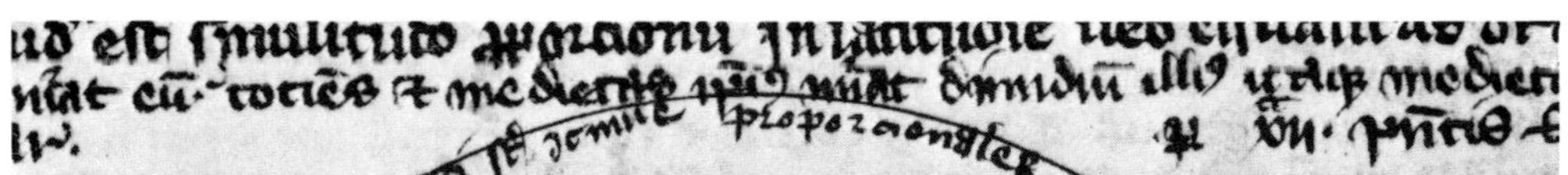

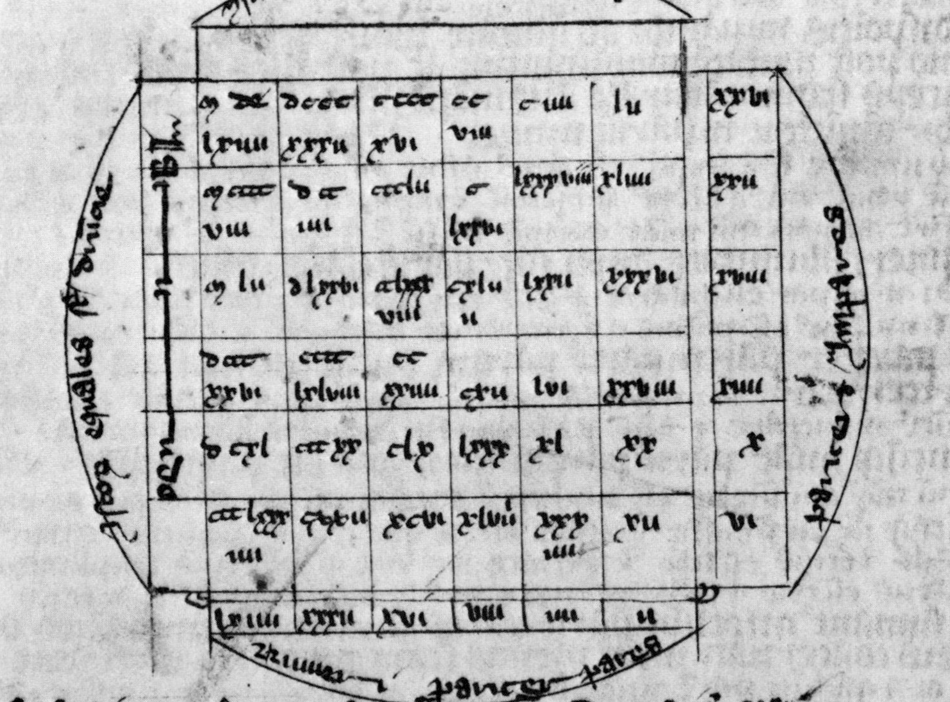

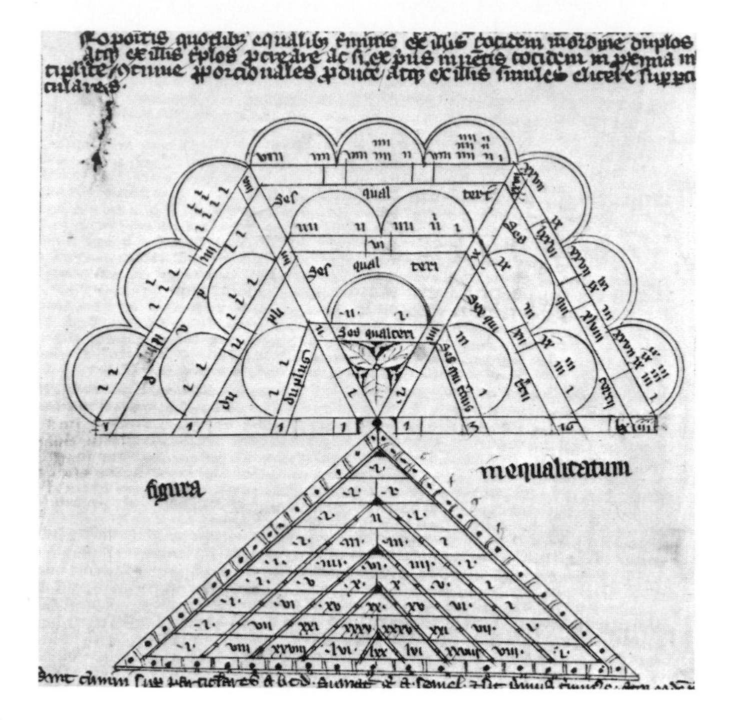

98. Generating Ratios from Equals. This rather curious figure accompanies Proposition 70 of Book IX in a fourteenth-century manuscript copy of Jordanus of Nemore's (thirteenth century) *Arithmetica.* The proposition in question establishes how, beginning with any number of equal numbers, one can on the one hand generate multiple ratios (that is, of the form $nm:m$) from these equal numbers, which multiple ratios will stand in continued proportion to one another (that is, $a:b = b:c = c:d$, etc.) and, on the other hand, also generate from this same base different kinds of superparticular ratios (that is, of the form $n + 1:n$), which again stand in continued proportion to one another. The "secret" of how to accomplish all this is to stipulate a sequence of multipliers that can be applied successively to the equal numbers with which one begins. These multipliers can be set forth, Jordanus claims, in a "triangular form of numbers," and that is what constitutes the lower half of the present figure (although Jordanus tells us how to construct only the top four rows of the triangle). Each horizontal row of numbers in the triangle gives a different multiplier to be applied to different sequences of our given set of equal numbers. Thus, if our set of equal numbers is 1_1, 1_2, 1_3, 1_4 (where the subscripts merely indicate the first, second, third, and fourth equal numbers), then the top row (equals the apex) of the triangle tells us to multiply 1_1 by 1, yielding 1; the second row tells us to multiply 1_1 by 1 and 1_2 by 1, yielding 2; the third row stipulates $(1 \times 1_1) + (2 \times 1_2) + (1 \times 1_3) = 4$; finally, the fourth row gives $(1 \times 1_1) + (3 \times 1_2) + (3 \times 1_3) + (1 \times 1_4) = 8$. The resulting series 1, 2, 4, 8 stand in continued proportion to one another according to a double ratio; Q.E.D., part one. Indeed, just this procedure applied to the same example of 1_1, 1_2, 1_3, 1_4 is illustrated in the left-hand triangle of the trapezoid in the top half of our illustration. Although not illustrated in the figure, in the proof of the present proposition it is made clear that if we apply the *same* procedure of multiplying to our resulting doubles 1_1, 2_2, 4_3, 8_4, we will obtain the continued proportion triples 1, 3, 9, 27. Continuing in this fashion, and always beginning our stipulated multiplication with the smallest term in our series, we can generate all continued proportional n-tuples beginning from unity. On the other hand, if we use the same procedure of multiplication given in the rows of our triangle, but begin with the largest term of the series 8_1, 4_2, 2_3, 1_4 we will generate sesquialtern ratios (that is, of the form $3n:2n$) standing in continued proportion. Thus: $(1 \times 8_1) = 8$; $(1 \times 8_1) + (1 \times 4_2) = 12$; $(1 \times 8_1) + (2 \times 4_2) + (1 \times 2_3) = 18$; $(1 \times 8_1) + (3 \times 4_2) + (3 \times 2_3) + (1 \times 1_4) = 27$; and $27:18 = 18:12 = 12:8$. This procedure is pictured in the center triangle of our trapezoid. The final, right-hand triangle of this part of the figure illustrates a similar multiplication (not mentioned in the body of the proof) applied to the series 27_1, 9_2, 3_3, 1_4, generating the sesquitertian ratios standing in continued proportion $64:48 = 48:36 = 36:27$. Although the body of the proof of the present proposition contains at least one example not given in our figure and cites previous propositions to establish that the results obtained are in fact double, triple, or sesquialtern ratios, etc., in continued proportion, in contrast to the proof, the figure allowed the medieval scholar to comprehend more easily and more adequately just how to generate the ratios in question. Since the crux of Jordanus' proposition was to account for the generation from equals of all multiple and superparticular ratios in continued proportion, the function of the figure was as requisite for the proposition as was its proof, if not more. For other figures from the present manuscript, see Illustrations 97 and 124.

99. A Medieval Number Game. The eleventh century was witness to the creation of a number game of considerable difficulty. It was called by the Greek-derived name *rithmomachia,* literally a number battle, and was played on a chess board (here pictured from a thirteenth-century manuscript) of eight by sixteen squares. The pieces for the game had values corresponding, in modern terms, to: (1) n and n^2, (2) $n^2 + n$ and $(n + 1)^2$, (3) $(n^2 + n) + (n + 1)^2$ and $(2n + 1)^2$. One player's pieces consisted of these values as n ranged over the first four even numbers (here at the bottom of the board), while his opponent's pieces had values of n ranging over the first four odd numbers beginning with 3 (here at the top). In earlier versions of the game, the classes of pieces (1)–(3) were of different colors, but later texts differentiated these classes by different forms (circles, triangles, and squares). Of course, the medieval scholar did not describe the kinds of pieces in the above modern terms, but instead appealed to the kinds of ratios that obtained between these pieces. Thus, to take the four numbers in the lower right corner (45, 81 and 15, 25), the labels running up and down the sides of the squares containing these numbers tell us that the ratio of 81 to 45 is a *superquadripartiens quintas* ratio [that is, of the form $(5 + 4)/5$], and that 25 stands to 15 in a *superbipartiens tertias* ratio [that is, of the form $(3 + 2)/3$]. In the other direction, the "side labels" in these same squares reveal that 45 is *subsuperquadripartiens* to 81 and 15 is *subsuperbipartiens* to 25. What is more, separate syllables of different labels spread out over various squares indicate the genus of the ratios held by the numbers occupying those squares. Hence, taking the four squares at the lower left together with the four at the lower right, we see the hyphenated label *"De ge-ne-re super-par-tien-tium."* This is just to say that the numbers within these eight squares stand to one another in superpartient ratios [that is, of the form $(n + m)/n$, where $n > m$]. Finally, one should note that two of the pieces were most difficult to capture in play: 91 for the even-number player and 190 for the odd-number player. These are both pyramidal numbers since they are both sums of square numbers: $91 = 1 + 4 + 9 + 16 + 25 + 36$; $190 = 16 + 25 + 36 + 49 + 64$. There was, then, a good smattering of medieval number theory in merely understanding the numerical properties of the pieces of the game. The rules for play are too complicated to give anything more than a sample or two here. Capture by sally *(eruptio)* occurs when two appropriately valued pieces are separated by a required number of squares; thus, 5 may capture 45 if they are separated by nine squares (since $5 \times 9 = 45$). Alternatively, capture by ambush (*insidiae,* another military term) obtains when two pieces are abreast a third equal to their sum. Capture of an opponent's pieces was not, however, the ultimate purpose of the game. That was, instead, to end up with pieces whose numerical properties constituted one or another kind of "victory" *(victoria)*. The absolute ultimate *(praestantissima* or *excellentissima)* among victories was to finish in the enemy's camp with an alignment of four pieces (which could be one's own or captured) containing all three kinds of proportion: arithmetic, geometric, and harmonic. Thus, 2, 3, 4, 6 or 4, 6, 8, 12 and so on would constitute such a victory. Yet 4, 6, 8, 12 was the best of all, since all four of the terms stand in geometric proportion and since these numbers contained the basic musical ratios. For other figures from the present manuscript, see Illustrations 120, 124, 133, 139, 141, 149, and 238.

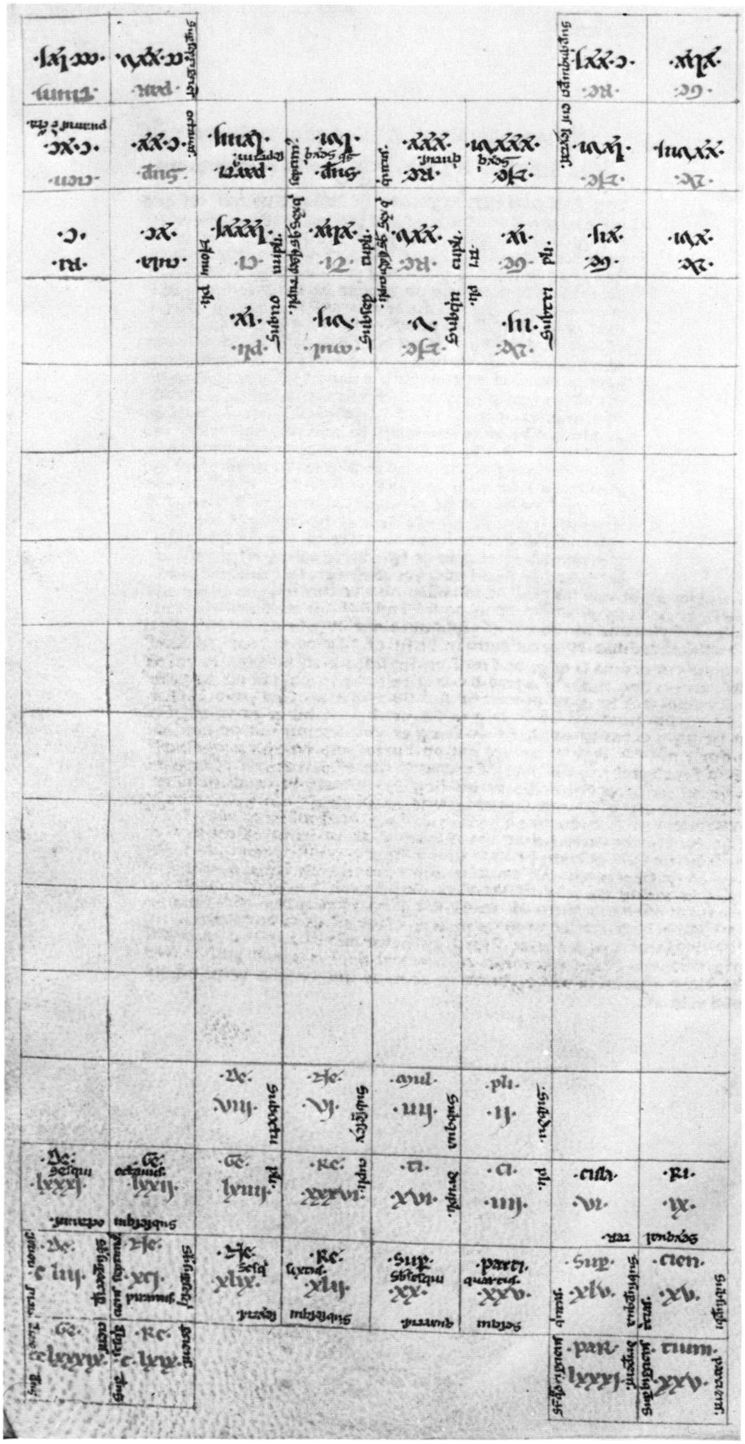

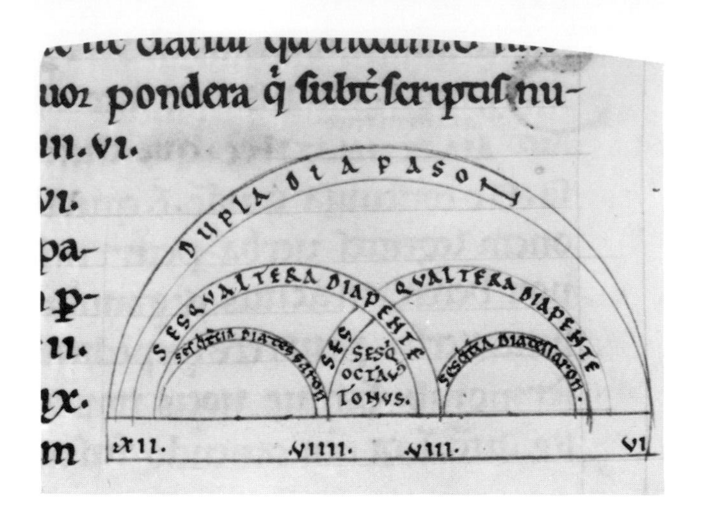

100. The Basic Musical Ratios. There is a strong tradition that ascribes the discovery of the mathematical ratios governing the octave, the fourth, and the fifth to Pythagoreans of the fifth century B.C., although inferences as to just how these ratios were discovered are speculative at best. Yet whatever may have been the details of their early discovery and elaboration, the mathematics of the ratios governing all musical intervals soon became a standard part of Greek musical theory and hence of Boethius' transmission of that theory in his *De institutione musica.* The present "arc diagrams," taken from a twelfth-century manuscript of Boethius' work, revealed the "basics" of this mathematics of ratios. Using the familiar series of numbers 6, 8, 9, 12 (see Illustration 95), the lone diagram above sets forth the four fundamental ratios upon which all else is constructed: the double *(dupla)* ratio 12:6 yielding the octave *(diapason);* two sesquialtern ratios 12:8, 9:6 giving the fifth *(diapente);* two sesquitertians 12:9, 8:6 producing the fourth *(diatessaron);* and a single sesquioctave ratio 9:8 yielding the interval of a tone *(tonus)*. All of this is, of course, relative to the so-called Pythagorean scale that was standard in the Middle Ages and not to the later developments of just intonation and equal temperament. The marginal diagrams on the right begin above with three separate arcs giving the octave, fifth, and fourth in their "lowest" ratios (2:1, 3:2, 4:3) and finish at the bottom with a single arc representing the tone as 9:8 and a more complex figure repeating the octave, fifth, and fourth in terms of 4, 3, and 2. Yet newer material is depicted in the figures between these diagrams. The first gives a visual representation of a twelfth *(diapason ac diapente)*, showing how the triple ratio (6:2) governing this interval is composed of those governing the octave and the fifth. Below this, arcs of two double ratios reveal the quadruple ratio related to the double octave *(bis diapason)*. For other diagrams from this manuscript, see Illustrations 95, 104, 214, and 278.

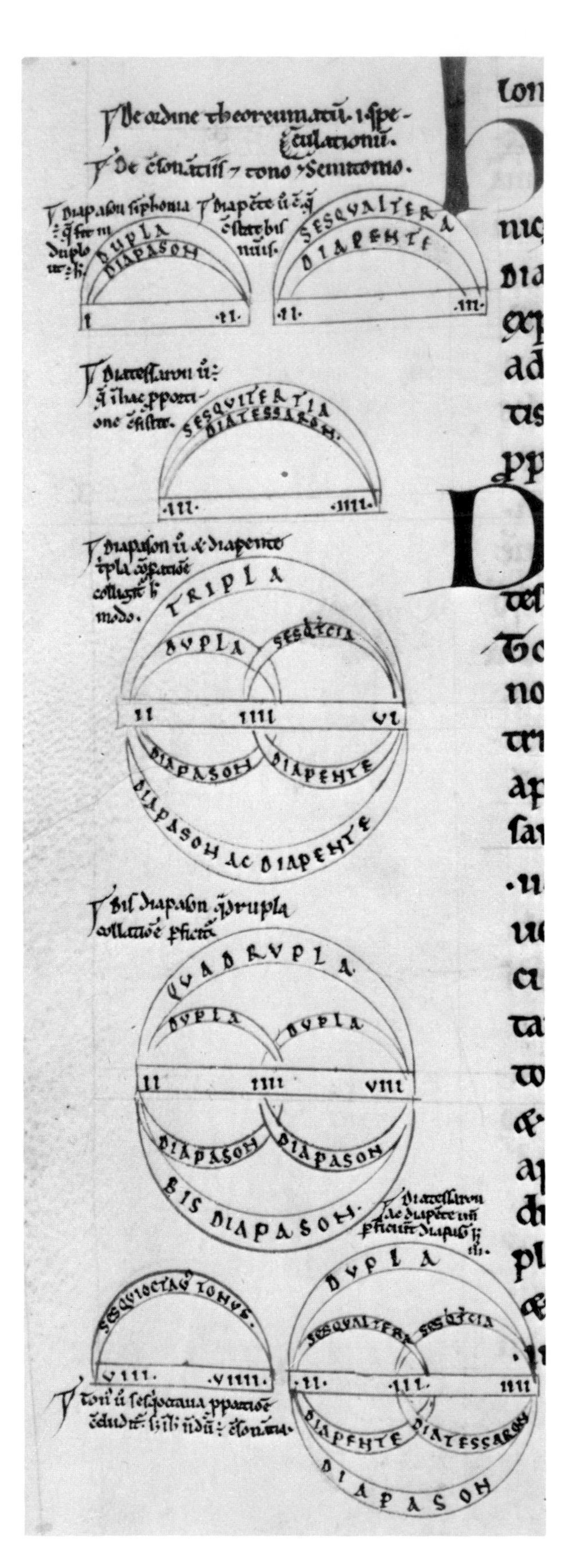

101. Transmigrated Boethian Figures. The utilization of arc diagrams to represent musical intervals and their correspondent ratios became so traditional in the Middle Ages that we find them populating works whose major concern was other than medieval music theory, but which at the same time found occasion to mention at least fragments of that theory. One example of this phenomenon appears here on the right, taken from a fourteenth-century manuscript of an *Arithmetica* which seems to be an alternate version of that usually ascribed to the fourteenth-century Oxonian scholar Thomas Bradwardine. Much the same information is pictured, albeit in a much cruder form, as in the marginal diagrams of the preceding Illustration. Yet in addition to the specification of the pertinent intervals and ratios, an occasional indication is given of the arithmetic difference between the various terms determining these ratios, something that is not especially relevant to the matter at hand. Somewhat more disconcerting is the erroneous information represented in some of the diagrams: For example, the double octave *(duplex diapason)* is held to be composed of 6:4 and 4:2 (where 8:4 and 4:2 is required), the sesquialtern ratio of the fifth is depicted as 4:3 (which is really the sesquitertian of the fourth), and so on. Indeed, even where the information given is correct, the medieval scholar would, it seems, often have had to puzzle his way through the diagram to come up with its "message." An even more uninformative application of the Boethian arc figure can be seen in the diagram below, taken from a fourteenth-century codex of Michael Scot's astrological *Liber introductorius.* Equal arcs are used to represent the intervals of *diatessaron, diapente, diapason,* and *semidiptonus* (the minor third composed of a tone and a semitone). What is more, our diagram implies that the first two and the last two intervals can be had between the same notes (here given in letter notation), although the representation of these same intervals on the staff below the arc diagram corrects this erroneous implication. For other figures from the manuscripts appearing here, see Illustrations 22, 26, 80, 231, 251, and 277.

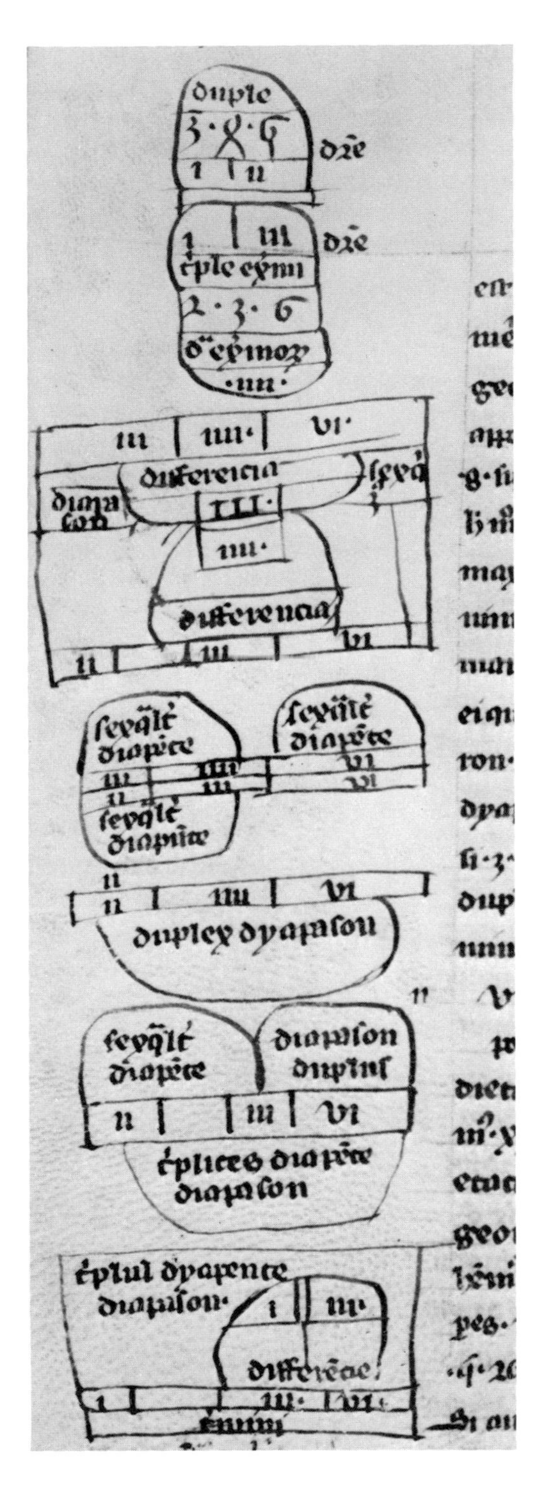

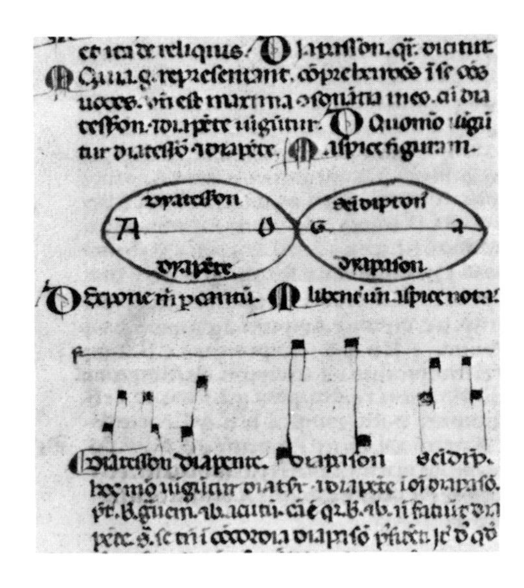

102. Arc Diagrams in Excess. As we have seen, it was a standard medieval practice to represent visually musical intervals and the numerical ratios determining them by appropriately labeled arcs. In fact, the practice was so common in musical treatises that arcs were employed in diagrams where they had nothing to do with the representation of such intervals or ratios; they were, perhaps, just too much a part of drawing music diagrams to be ignored. This twelfth-century manuscript of Boethius contains a typical example of such an automatic appeal to arcs. The purpose of the diagram is the depiction of Boethius' proof (Book III, ch. 13) that the ratio of a minor semitone, namely 256:243, is greater than the ratio of 20:19. The terms of the former ratio appear at the ends of the top horizontal and are labeled *A* and *B*. Their difference *C* = 13 is written in the arc in the central circle. If we divide *C* into *A* and *B* (and round to the nearest whole number) we will obtain, respectively, 20 and 19, duly inscribed on the sides of the triangle under the central circle. But 13 × 20 = 260 and 13 × 19 = 247, the numbers appearing at the ends of the lower horizontal (although transposition of a cipher has erroneously turned CCXLVII = 247 into CCLXVII = 267). But 260 − 256 = 4 = 247 − 243, a common difference noted by the two arcs forming the sides of the diagram. Therefore, 256:243 is greater than 260:247, and since the latter is the same as 20:19, 256:243 is greater than 20:19, a result that is announced by the appearance of "greater" *(maior)* in the top circle of the diagram. A diagram all but identical in form, which is not depicted here, is used on this same page to picture Boethius' proof that 256:243 is less than 19½:18½. For other figures from the present manuscript, see Illustrations 63, 103, and 282.

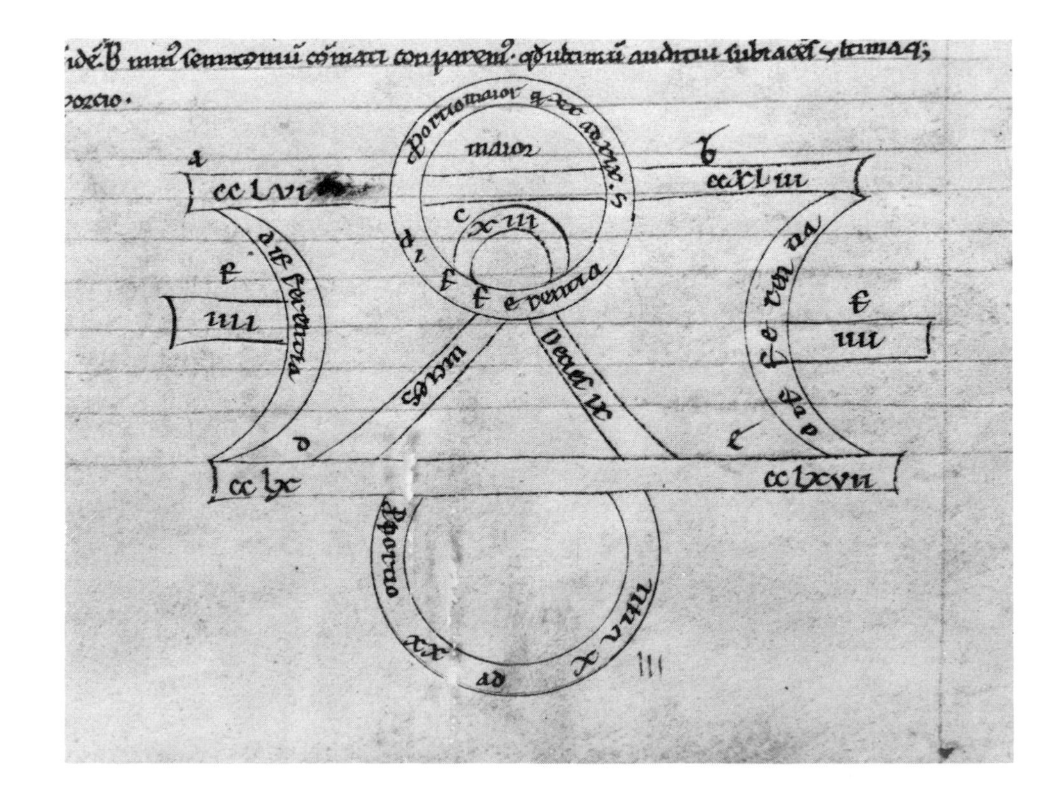

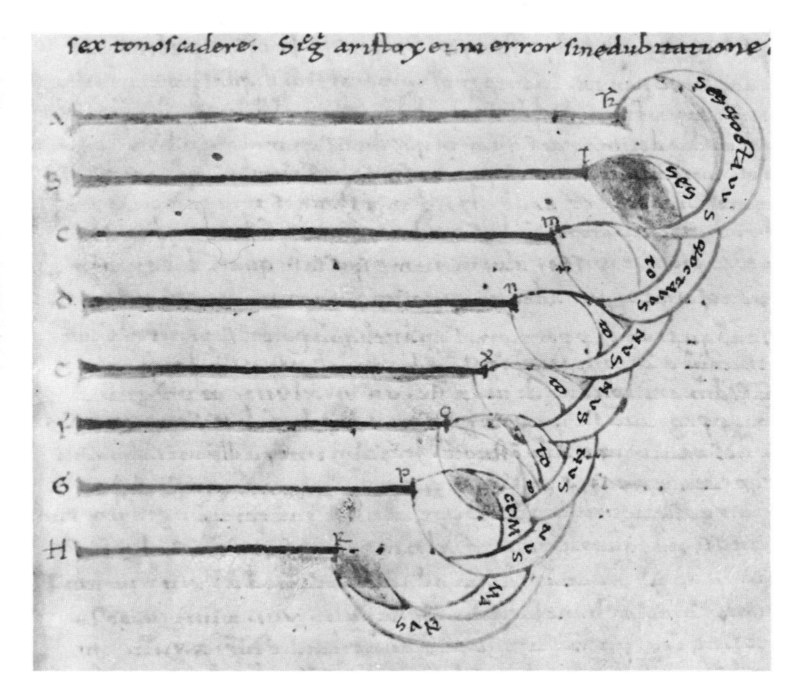

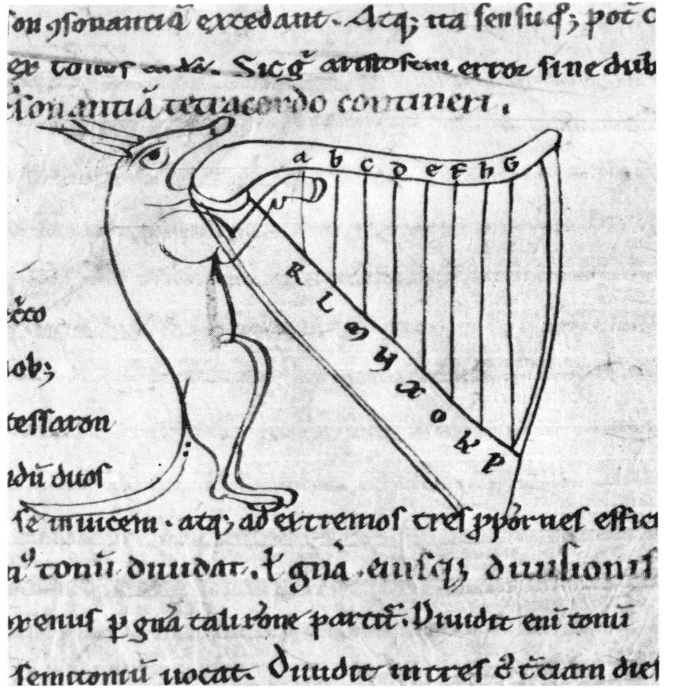

103. More Decorative, Less Sense. The two figures above are intended to illustrate the same chapter (Book V, 14) of Boethius' *De institutione musica;* that on the left occurs in an eleventh-century manuscript, that on the right in a twelfth-century codex. The point at issue is to show that the octave is less than six whole tones and Boethius' proof, taken, he tells us, from Ptolemy, proceeds by the division of an octachord. Allow the first seven strings of the octachord (*A, B, C, D, E, F, G* so labeled at the extreme left in the first diagram) to stand successively to one another in the sesquioctave ratio of a whole tone. Thus, using the letters supplied for both ends of the strings, $AK{:}BL = BL{:}CM = CM{:}DN = \ldots =$ 9:8. But then take a string *HR* such that $AK{:}HR = 2{:}1$, which means that, struck together, they will yield the consonance of an octave. But if one strikes the string *GP,* it will sound slightly sharper *(paulo acutior)* than *HR*. But since the seven strings *AK* through *GP* mark off six whole tones, it is thus made evident to the senses *(sensu)* that the octave is less than six whole tones. The diagram at the left presents Boethius' Ptolemaic argument with but a single error, arcs between the first seven strings stipulating tones or sesquioctave ratios, with a final arc between *GP* and *HR* marked 'comma' (which is the interval between two minor semitones and a tone and hence between six whole tones and an octave, corresponding to the ratio 531441:524288). But the diagram has erred in representing *HR* as shorter than *GP*. How could that be if, when struck, *GP* is sharper than *HR*? The more decorative diagram at the right properly has *HR* shorter than *GP,* but all else is the wrong way round, *AK* being the shortest string and *GP* the longest. Thus, although the first diagram would have been of some assistance in understanding Boethius' argument, the second would have been of no use at all. One is led to speculate that the artist or scribe in the latter instance was much more interested in depicting the shape of the octachord itself and in drawing the little animal plucking it than he was in illustrating the argument at hand. For other figures from this latter manuscript, see Illustrations 63, 101, and 282.

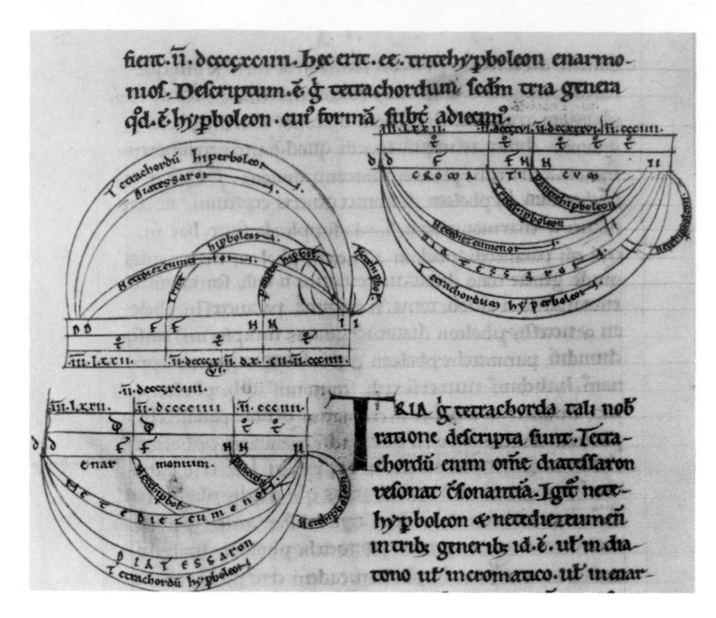

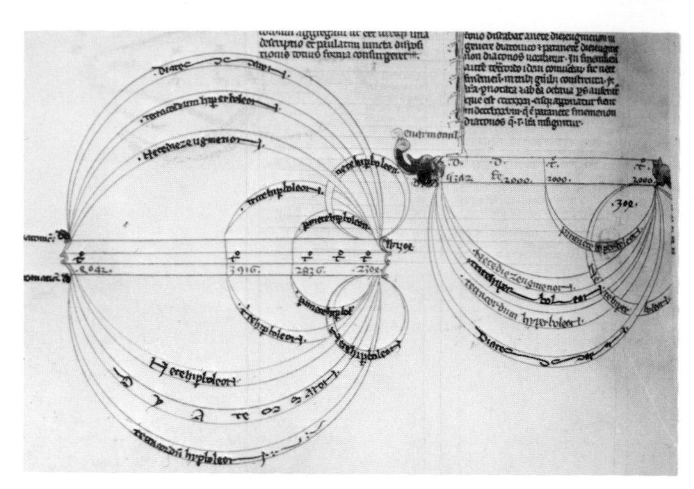

104. The Genera of Tetrachords. One of the fundamental elements in Greek musical theory, and thus also in Boethian medieval musical theory, was the tetrachord, a sequence of four descending tones, four such tetrachords in turn joined in differing ways to form the two-octave Greek scale (cf. Illustration 77). Yet each of these tetrachords could be chromatically modified in a specific way by changing the pitch of its two middle notes, the two outer notes remaining fixed. To delineate these modifications was, in medieval terms, to describe the genera of the tetrachord in question. A description of this sort for the hyperboleon tetrachord (which ranged, in modern notation, from *a′* to *e′*) is that diagrammed here, first from a twelfth-century manuscript of Boethius and then from a fourteenth-century manuscript of the same work. In both figures, two of the arcs bridging the whole of the diagram carry the name of the tetrachord being described *(tetrachordum hyperboleon)* and the interval *(diatessaron,* a fourth) covered by the tetrachord. The remaining arcs indicate the names of the tones involved, from highest to lowest: *nete hyperboleon, paranete hyperboleon, trite hyperboleon,* and *nete diezeugmenon* (the last named represented by the third arc bridging the whole diagram). To understand the remainder of that represented by our diagrams, one must begin with the illustration on the (above) left. Here the central arc figure depicts the diatonic genus (although the appropriate label is missing) of the hyperboleon tetrachord. The middle horizontal row indicates that the interval in question consists (from right to left) of two tones plus a minor semitone. The two outer horizontal rows give the numbers, and the letters standing for these numbers, that determine this tonal arrangement; they are (again from right to left): 2304, 2592, 2916, 3072. Note that Boethius here employs these numbers in place of the usual ratios for a tone and a minor semitone

(Continued)

(namely, 9:8 and 256:243), but, for the diatonic tetrachord, the function is the same since the numbers stand to one another in the required ratios. The only problem with the representation given by this arc figure (and all the others pictured here as well) is that a single symbol [symbol] is used for both the tone and the semitone (whereas the latter should have been indicated by [symbol]). Turning to the upper arc figure in the illustration on the (above) left, we have similar information for the chromatic genus of our tetrachord: Its two middle notes are now modified so that its highest interval is expanded to one of three minor semitones, the remaining intervals being a minor semitone each, and the determining numbers being 2304, 2736, 2916 (here wrongly 2816), 3072. The final arc figure gives the modification entailed by the enharmonic genus of the hyperboleon tetrachord: a highest interval of two tones, plus two intervals of a quarter tone or *diesis* (here symbolized by [symbol]), the relevant numbers being 2304, 2916 (here erroneously 2904), 2994, 3072. The mistakes in these arc figures pale by comparison with those in the diagrams on the (above) right. Here the diatonic and chromatic genera are combined in a single "over and under" arc figure, with no indication at all of the required difference in the middle notes. The same intervals and the same numbers (three of the four wrong at that) are given for both genera. Nor are things any better with the figure at the right representing the enharmonic genus: Although the intervals are correct, all four numbers are wrong, and one of the arcs intended for a middle note bridges the whole. This plethora of errors can most likely be explained by the fact that the copy of Boethius in which these diagrams appear was a presentation manuscript, which placed more value on elegance than on pedagogic value. For other diagrams from both of the present manuscripts, see Illustrations 95, 100, 214, 278; and 44, 59, 60, 62, 178, 241, 252.

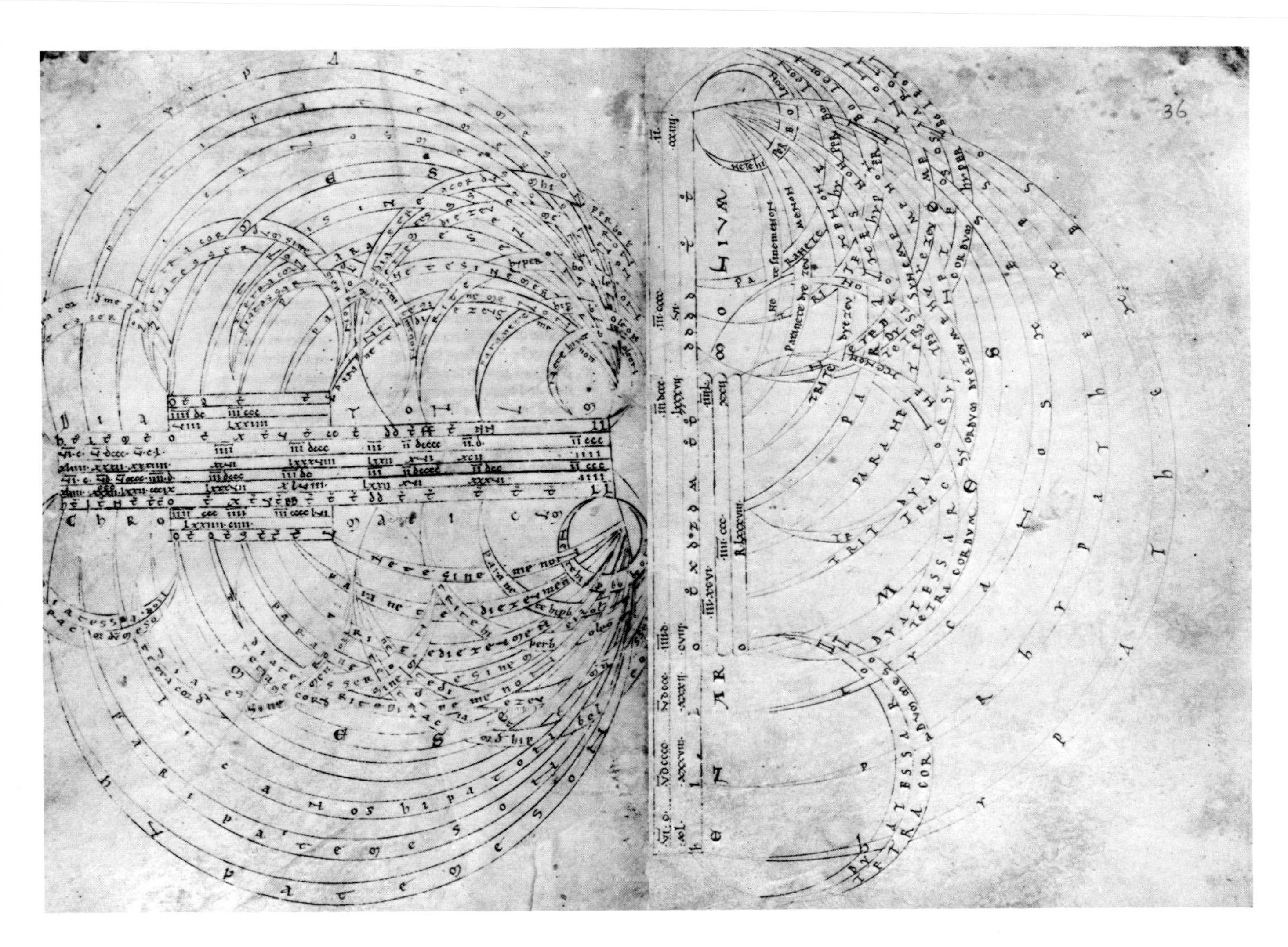

105. Music Theory by Monochord. The teaching device that was unquestionably the most utilized in the Middle Ages for the learning of musical theory was the monochord: a single string stretched over some kind of resonator, the division of the string according to set ratios giving rise to the various musical intervals through its differing vibrations. Relating these divisions to the relevant intervals and notes was one of the central tasks of numerous medieval texts dealing with the *partitio monochordi,* texts which, more often than not, appealed to diagrams to facilitate their comprehension. The double-page diagram above is taken from a thirteenth-century manuscript of Boethius (Book IV, ch. 10) and represents the division of the monochord *per tria genera,* that is, diatonic, chromatic, and enharmonic. The type of information given is parallel to what we have already seen in the preceding Illustration treating of the genera of tetrachords. Indeed, the three monochords which are here divided are composed of three tetrachords (two conjunct with one another, two disjunct), although information is also provided for the added synemmenon tetrachord falling within the range covered by these three. As in the preceding Illustration, the numbers given in the diagram are consistent with the mathematical ratios for a tone and a minor semitone only for the diatonic genus. Since Boethius most likely followed Nicomachus in deriving the chromatic and enharmonic genera from the already divided diatonic genus, the numbers he uses for the former two genera are not consistent with the standard tonal ratios. The top arc figure on the left-hand page gives the required division for the diatonic genus, that on the bottom for the chromatic, while the whole right-hand page devotes itself to the enharmonic division. Taken as a whole, the diagram is so complicated that it could probably be understood only in conjunction with a line-by-line reading of Boethius' text. Even then, tracing one's way through the tangle of arcs must have been a laborious task. If we turn to the diagrams on the next page, taken from a twelfth-century manuscript,

(Continued)

(Continued)

we can witness the monochord put to teaching yet another aspect of music theory. The upper figure, ascribed to the tenth-century Odo of Cluny, gives a pictorial representation of his doctrine of tetrachords, while the lower figure, bearing the name of Guido of Arezzo (whose musical "hand" we have noted before in Illustration 76), depicts the theory of Church modes. Thus, the *Monochordum Othonis* is divided into four tetrachords (the succeeding groups of four notes each receiving the names *graves, finales, superiores,* and *excellentes*) plus two additional notes (labeled *remanentes*). Within each tetrachord, the notes are ordered both by name (*archos, deuteros, tritos, tetrardos*) and, below, by Roman numerals, while the symbols for these notes appear along the upper margin of the diagram. The *Monochordum Guidonis* below this is more complex. The notes are given both in letter notation and solmization names through the center of the diagram, the Roman numerals above and below them representing Guido's technique for delineating modes relative to their "final tones" (thus, modes I and II have *D* as *finalis,* III and IV have *E,* V and VI *F,* and VII and VIII *G*). The arcs above and below these numerals are intended to explain the relation of these *finales* to the scales or *ambitus* covered by their corresponding modes, those above picturing this information for the authentic modes (where the *ambitus* begins with the *finalis* and ends an octave above and where the smaller arcs indicate the *subfinalis* as a frequently admitted added note for the mode in question), those below for the plagal modes (where the *finalis* separates the octave *ambitus* into a lower tetrachord and an upper pentachord). Unfortunately, the upper arcs are carelessly placed, and the interval between the two lowest notes (Γ and A) is improperly pictured, impairing the effectiveness of the diagram as a whole.

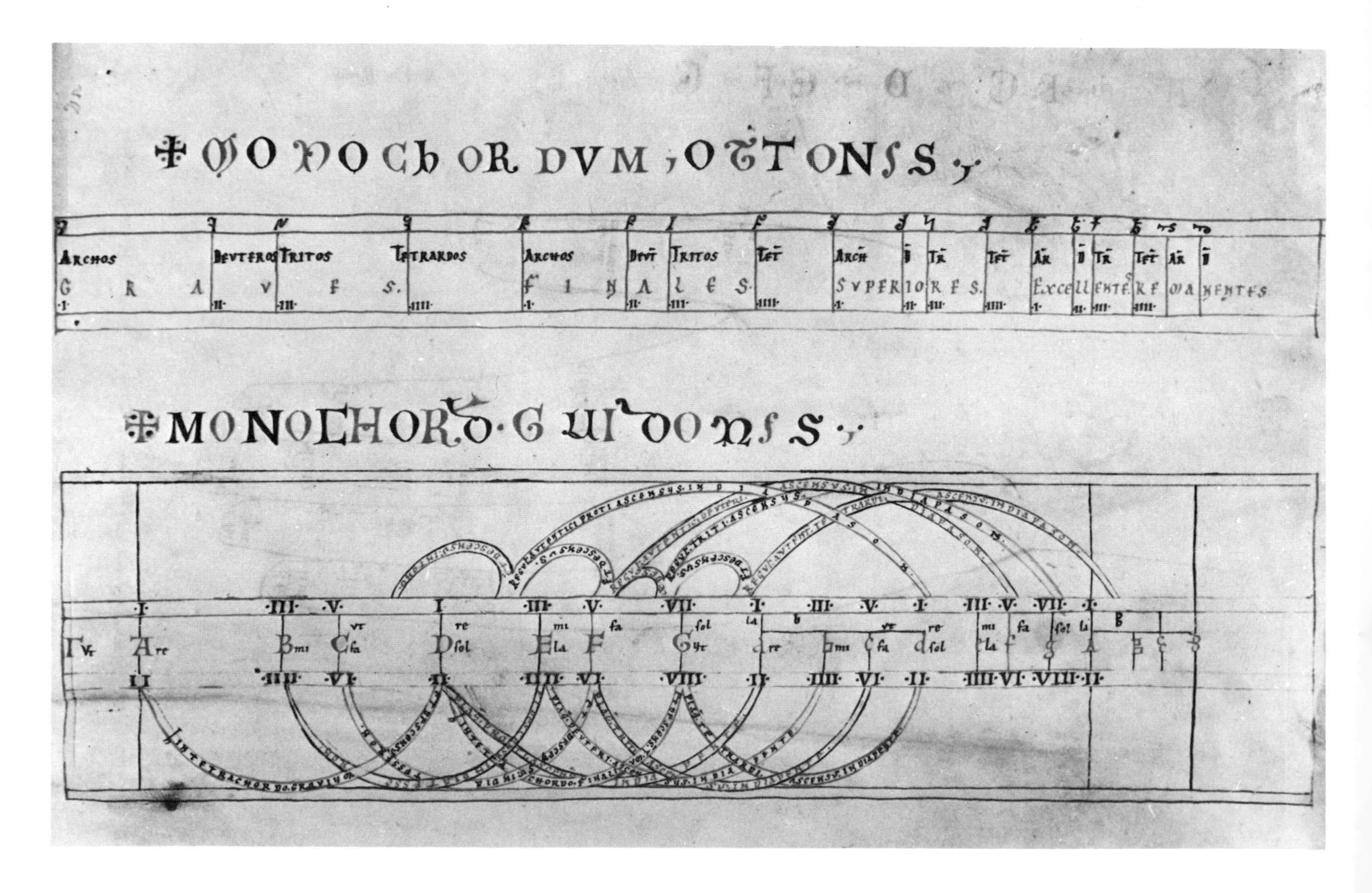

CHAPTER ELEVEN

The Development of the Geometric Figure and Diagram

SOME OF THE EARLIEST evidence we have of Greek activity in geometry reveals the importance of diagrams for that discipline. It may be true that Plato felt that geometers must do more than construct diagrams, must discover what really is, but his very mention of this fact at the same time indicates that, as his student Aristotle was later to maintain, diagrams were significant elements in carrying on the enterprise of geometry in a proper and successful way. There is even later testimony to the fact that the Greek term for diagram had been used interchangeably with that for theorem.

When one follows the tradition of Greek mathematics into the Middle Ages, the awareness of the importance of diagrams is, if anything, even greater. Indeed, it is not surprising that medieval scholars should have been more conscious of their role, since their geometry—which was, by and large, their Euclid—was more didactic, more designed for scholastic use, than the Greek original. The necessity of the compass in drawing diagrams is even mentioned in several proofs of Euclidean propositions, something which would have offended the Greek sense of proper geometrical procedure. Some care is taken to point out that lines are used to represent general magnitudes in Book V of the *Elements* and numbers in Book VII, since this makes things evident not just to the intellect, but to the eyes as well *(non solum intellectum pateat, sed etiam oculis)*. The importance of this visual element in geometry is probably also related to Roger Bacon's claim that through figures in geometry we learn *per experientiam*.

Diagrams are also important for our understanding of the history of mathematics, especially if that history is built upon manuscript copies of the mathematical treatises which are its raw material. Here, variations in diagrams are often every bit as important as variants in the text. They, too, are part of the transmission of a given mathematical work and should be considered fully in the editing of such works, a desideratum that has, until recently, been too often ignored. That the diagrams should not be ignored is also clear from the fact that occasionally a manuscript diagram is better than that produced by modern scholars. Yet beyond the importance of diagrams for the text of a treatise, they often tell us something of how the text was understood, especially when there is evidence that the diagrams were executed at a date later than that of the text they accompany.

The general question of who drew the diagrams and when they were drawn is, unfortunately, extremely difficult to answer. They may have been the product of the scribe who wrote the text, they may have been reproduced by a later scribe, or they may have been introduced by a later reader (although this last alternative is clearly the rarest of the three). Deciding between the first two alternatives is most problematic. There is little handwriting connected with the drawing of diagrams, so a comparison with the hand of the text is often not as informative as it might be expected to be. Furthermore, the differences in the inks employed for figure and text are frequently so minimal that that criterion, too, is often not as useful as one might think, although it is usually more effective than an analysis of hands. As a whole, however, it seems fair to conclude that most diagrams were drawn at the same time as the text was written, or shortly thereafter. And it would also seem that in more instances than not, the same scribe was involved in producing both.

Generalizations concerning the source of the diagrams in geometrical works are easier to make. In a few words, most extant manuscript diagrams were copied from identical diagrams in earlier manuscripts of the same work. This is clearly reflected in the fact that variant versions, or separate redactions, of the same text almost always bear the same variations in diagrams. All of this is, of course, a natural outcome, given what we know of the way manuscripts were produced or, more exactly, reproduced. It is very seldom that a given codex provides us with clues enabling us to hypothesize that the diagrams were constructed through the reading and analysis of the relevant text. At times we see that a reading of the text has caused someone—perhaps even the original scribe—to *correct* an inherited diagram, but even this is not a frequent occurrence.

The choice of examples of diagrams and figures given in this chapter has been governed by the general principle of this whole Album: to concentrate upon different *kinds* of illustrations, whether that difference be, particularly with respect to the mathematical figures at hand, due to form or due to function. As stated previously, all mathematical figures are required for the understanding of the mathematical texts to which they belong. But beyond this, their relation to the text can vary widely. They might, for example, visually reproduce what is stated in the text. This is clearly their predominant function. On the other hand, they might go beyond the text in an attempt to present information that is not there, something that we shall see operative in cases in which the text in question merely provides the enunciation of a geometrical proposition and omits its proof or any further specification. In the medieval period, such cases were not infrequent. Alternatively, a diagram might be employed to cover or summarize a plurality of propositions or, vice versa, a plurality of diagrams might be provided for a single proposition, both when there are multiple cases covered by the proposition at hand and when there is but a single case for which alternate figures are possible. All of these variant relations between diagram and text are exemplified in what follows.

Illustrations are also given of problems and mistakes: erroneous diagrams (when acknowledged as erroneous and when not), cases in which a faulty text has caused a faulty diagram and in which a faulty diagram has (apparently) caused a faulty text, cases of incomplete diagrams, and cases of diagrams without text.

In addition, the later Illustrations in this chapter pay particular attention to areas which, from the point of view of the geometrical subject involved, are inclined to cause greater difficulties in the production of the diagrams required: curves that cannot be constructed by compass alone and the geometry of three dimensions. Medieval scholars frequently explicitly confess their awareness of difficulties with respect to the latter. A body, one translator of Euclid claimed, cannot be depicted on a plane *(in plano depingi non potest)*, and he goes on to remind the reader that the page *(pagina)* is to serve as such and such a given surface. Lines that rise from the two dimensions of the page are, we are told, to be viewed as a stick or bulrush fixed in a lump of wax. Even astronomy is appealed to when certain circles on a sphere are "clarified" by pointing out that they are like equinoctial circles relative to the earth.

Some segments of solid geometry were more problematic than others in the figures they required. The simple prisms of Euclid's Book XI, for example, were much easier to diagram than the cylinders, cones, spheres, and more complex regular polyhedra treated in Books XII and XIII. We shall see not just the variety of techniques used for drawing the figures with respect to some of this later material, but also some of the ways in which the problems involved in providing a diagram for very complex propositions were avoided. In particular, we shall see that a diagram in proper perspective was not always, indeed was quite seldom, the best of all possible worlds.

Note should also be made of the fact that in the case of many of these complex three-dimensional propositions, the diagram in question related first and foremost to the basic construction with which the proof of the proposition begins (for example, the inscription of an icosahedron or a dodecahedron in a sphere, as in Illustrations 121 and 122). Once the construction represented by the diagram had been effected, one had to go on to prove that the faces of the thusly inscribed polyhedron are indeed regular polygons, that each face lies in the same plane, that the vertices of the polyhedron lie on the same sphere, that its edges are of such and such a measure, and so forth. Elements in the diagram were appealed to in proving these things too, but, for reasons of space, they have not received mention in the examples given below.

106. Egyptian Geometry. The Rhind Papyrus, originally uncovered among the ruins of the Ramesseum at Thebes and today bearing the name of the collector who purchased it (see Illustration 83), is the best known and most important document of ancient Egyptian mathematics. Written by a scribe named A^{c}h-mosè in Hieratic ca. 1650 B.C., it contains arithmetical and geometrical materials that are some 200 years older. The two problems pictured here are furnished with geometrical figures which are among the earliest extant in the history of mathematics. That above calculates the area of a triangle, that below of a trapezoid or truncated triangle, of land. The base, or more literally "mouth," of the triangle is given as 4 *khet* (a *khet* usually being 100 cubits), while the number 10 *khet* appears over its upper side. It is not absolutely clear whether this latter value is intended as the side of the triangle or its altitude; in the latter case the calculated result of 20 *setat* or square *khet* for the area of the triangle would be correct and would correspond to the modern formula of $a = \frac{1}{2} bh$, but in the former it would not (unless the triangle were right, which is not apparent either from the figure or the text). The trapezoid below is labeled with a "mouth" of 6, a "cut-off" side of 4, and a height of 20; the calculated result, corresponding to the modern expression $a = [(b + c)/2]\ h$, is 100 *setat*.

107. Figures on Clay. Geometrical figures appear only rarely in cuneiform mathematical tablets. Those pictured here accompany texts written in old Babylonian script and hence date from ca. 1800–1600 B.C. What remains of the tablet shows us that it contained at least fifteen problems (seven more are on the reverse, in addition to the eight pictured here), although in its damaged state it provides very fragmentary evidence for some of these problems. At times we have a text without an accompanying figure, at others a figure without a text, and in both cases often only fragments of either. The problems are given without solutions, as if they may have been intended as some kind of exercise book. In each case, the problem begins with a square whose side is given; other simple geometrical figures—other squares, triangles, trapezia, circles—are then drawn within that initial square, the result being almost more ornamental than geometrical in appearance. Given only this much, the problem then concludes by asking for the surface (= area) of some one or more of the "inscribed" figures.

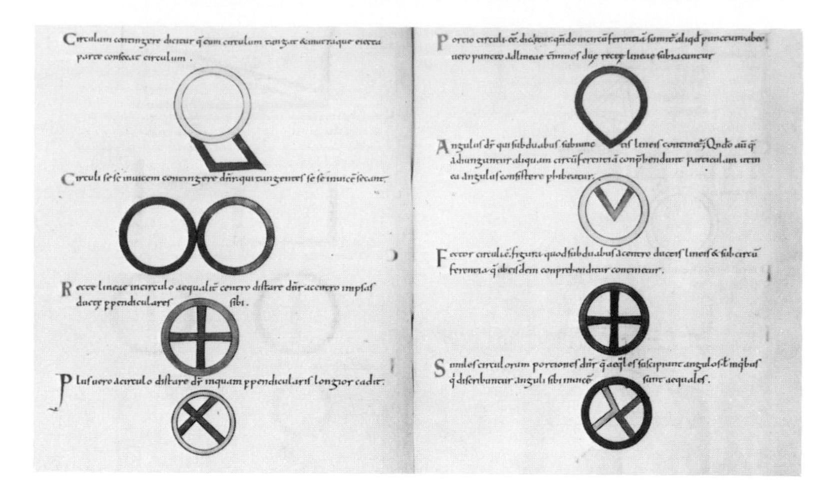

108. Elegant and Useless. The highly decorative eleventh-century English manuscript shown above contains excerpts from an early Latin translation of Euclid's *Elements* ascribed to Boethius. The portion pictured here covers Book III, Definitions 2–11. The figures are presented in thick "lines" of yellow, blue, and red, but the aid they provide in illustrating the definitions to which they belong is wrong more often than it is (within limits) right, although the corrupt nature of the definitions themselves seems frequently to have caused some of the confusion in these figures. Thus, at the top left, the illustration should have been of a straight line touching a circle, but the two words for straight line *(recta linea)* have been detached from the beginning of the matching definition and appended to the last definition on the preceding page (thereby rendering it unintelligible as well) and the stipulation that the straight line not cut *(non secat)* the circle has been mistranscribed as *consecat* ("cut up"), something that, with some imagination, might possibly explain the peculiar figuration under the circle. Similarly, at the top right, the text has conflated the definition of a segment *(portio)* of a circle with that of an angle in a segment, the resulting figure being of no help in understanding either. The two bottom figures on the left are meant to illustrate definitions of straight lines within a circle that are either equally or unequally distant from the center of the circle, the determining factor being the equality or inequality of the perpendiculars from the center to these lines (the first definition again being at fault, since the requirement that the relevant perpendiculars be equal is omitted). The definition of a sector (here erroneously spelled 'fector') on the right-hand page carries a figure which misrepresents the content of the definition, as does the figure at the bottom right for the definition of similar segments of a circle. This leaves only the second definition on each page as being reasonably pictured by their accompanying figures: that on the left-hand page of touching circles; that on the right of an angle standing on *(consistere)* a circumference. The figures in the illustration below are drawn from a thirteenth-century manuscript of a French practical geometry *(Pratike de geometrie)* written in the Picard dialect. Even more decorative in style than those from the eleventh-century Euclid, the diagrams are attached to propositions showing how to calculate the excess *(sorcrois)* of a square over an inscribed circle and of a circle over an inscribed square. This early French geometry was in large part a translation and adaptation of a twelfth-century Latin practical geometry.

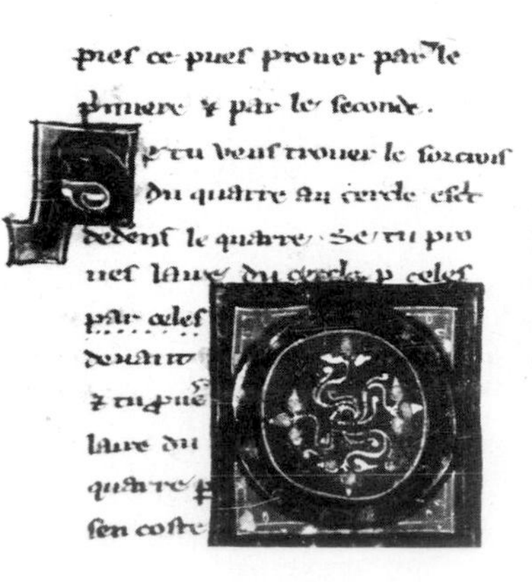

109. Supplementing Enunciations Through Figures. Medieval Latin versions of Euclid were not only extremely numerous, but they often also varied in what they presented of the *Elements.* In particular, some of these versions gave nothing but the enunciations of Euclid's theorems, the result being more like a catalog of geometrical facts than a geometry proper (which would have required the presence of at least some manner of proofs). These "enunciations only" copies of the *Elements* at times resulted from a simple excerpting from Adelard of Bath's Euclid (see the next Illustration), but they were standard procedure in the Boethian versions of the *Elements,* where, at most, only proofs for the first three propositions of Book I were given. Our interest is in the fact that when figures accompanied these mere enunciations, they often provided additional information about the geometrical assertion to which they were attached. Perhaps the figures were in some instances taken from some other version of the *Elements*, but it is at any rate clear that in most cases they presented material that could not have come from the enunciations themselves. They were almost always unlettered, since that information was only to be found in the proofs they lacked. In order properly to appreciate the way in which these figures supplemented what could be learned from the enunciations alone, it is best to concentrate on a single example, Book II, Proposition 10, beginning with a translation of the text of the enunciation in a Boethian version of the *Elements,* while at the same time rendering the import of the example in question more evident by including parenthetic reference to the specifics

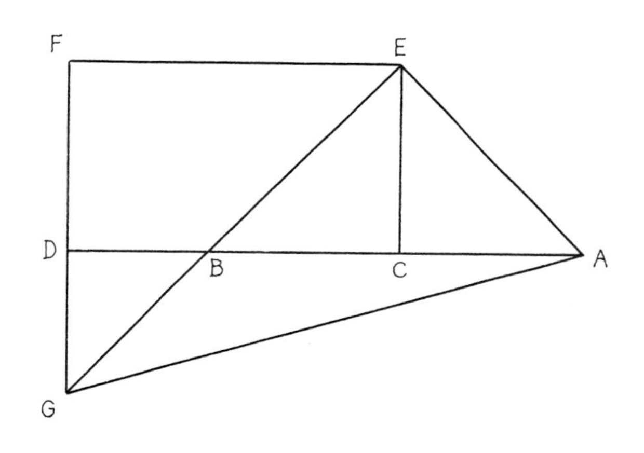

represented in the correct modern diagram given above right: "If a straight line *(AB)* be cut into equal segments *(AC, CB)* and a certain straight line *(BD)* be added to it in a straight line, then the square described on the whole together with the added line *(AD)* plus the square described on the added line *(BD)* is, each square being taken equally, necessarily double the square described on half *(AC)* [of the original line] plus the square described on the line consisting of half [of the original line] together with the added line *(CD)*, each square again being taken equally." (In terms of the modern diagram, the proposition therefore establishes that $AD^2 + BD^2 = 2AC^2 + 2CD^2$.) If we now turn to the three manuscript figures here depicted of II, 10, it is evident that they could not have been derived from a reading and analysis of the enunciation alone. That below, taken from a twelfth-century man-

(Continued)

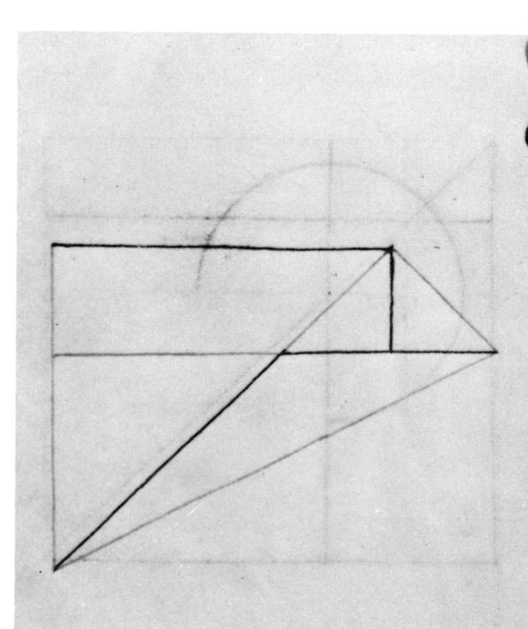

Decimum theorema.

Si recta linea per equalia secetur: eique indirectum quedam linea recta iungatur: quadratum quod describitur a tota cum addita et quadratum quod describitur ab ea que addita est utraque quadrata pariter

(Continued)

uscript of a version of Euclid that mixes the Boethian and the Adelardian (although it is a strictly Boethian enunciation that is given for II, 10), has a marginal figure that is correct (as can be seen by comparing it with the modern diagram); but the triangles it—and the modern diagram, too—contains relate to Euclid's proof, in which the Pythagorean theorem is appealed to. Although it would not be as straightforward as if one had the proof itself at hand, one could reconstruct its essentials given the enunciation plus this figure, unlettered though it might be. The manuscript figure at the top right is taken from a twelfth-century manuscript of the "five-book" Boethian Euclid. It totally misrepresents what is going on and would have been of little or no use in reconstructing anything. The diagrams on the lower right are taken from another twelfth-century manuscript, again of a Boethian–Adelardian mélange, and attempt to supplement the simple enunciations of the text by various sorts of labels and information written within the figures themselves. The diagram for II, 10 appears in the center of the figures enclosed in the "frame" at the right; save for the fact that the angle over the bisected line is acute and not right, it is essentially correct. To begin with, we should note that it labels which parts of the diagram refer to which parts of the matching enunciation. Thus, again using our modern diagram as an aid, it labels what corresponds to *AB* with "the first line cut into halves." Similarly, *BD* carries the word 'additum' and above that "the square on the added line and the half" is specified, while at the upper right there appear the words "the square on the half." More than that, labels are given to parts of the figure that relate to the missing proof. Thus, what corresponds to *EG* is said to be "the base to which the hypotenuse is equivalent" (their squares, of course), while *AG* is inscribed with "this hypotenuse is equivalent to the cathetus—i.e., *EA*—plus the base" (again, their squares). Finally, a series of odd-looking numbers is written in the upper left of the diagram: they refer to which previous propositions, postulates, axioms, and so on are utilized in proving the proposition at hand. Clearly, this figure has been designed to provide a considerable amount of information that is not forthcoming from the enunciation alone. It is naturally not as adequate as a bona fide proof would have been, but it certainly was intended to move in that direction.

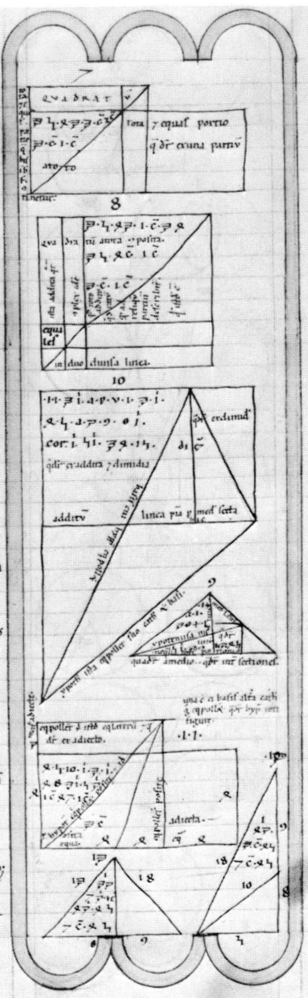

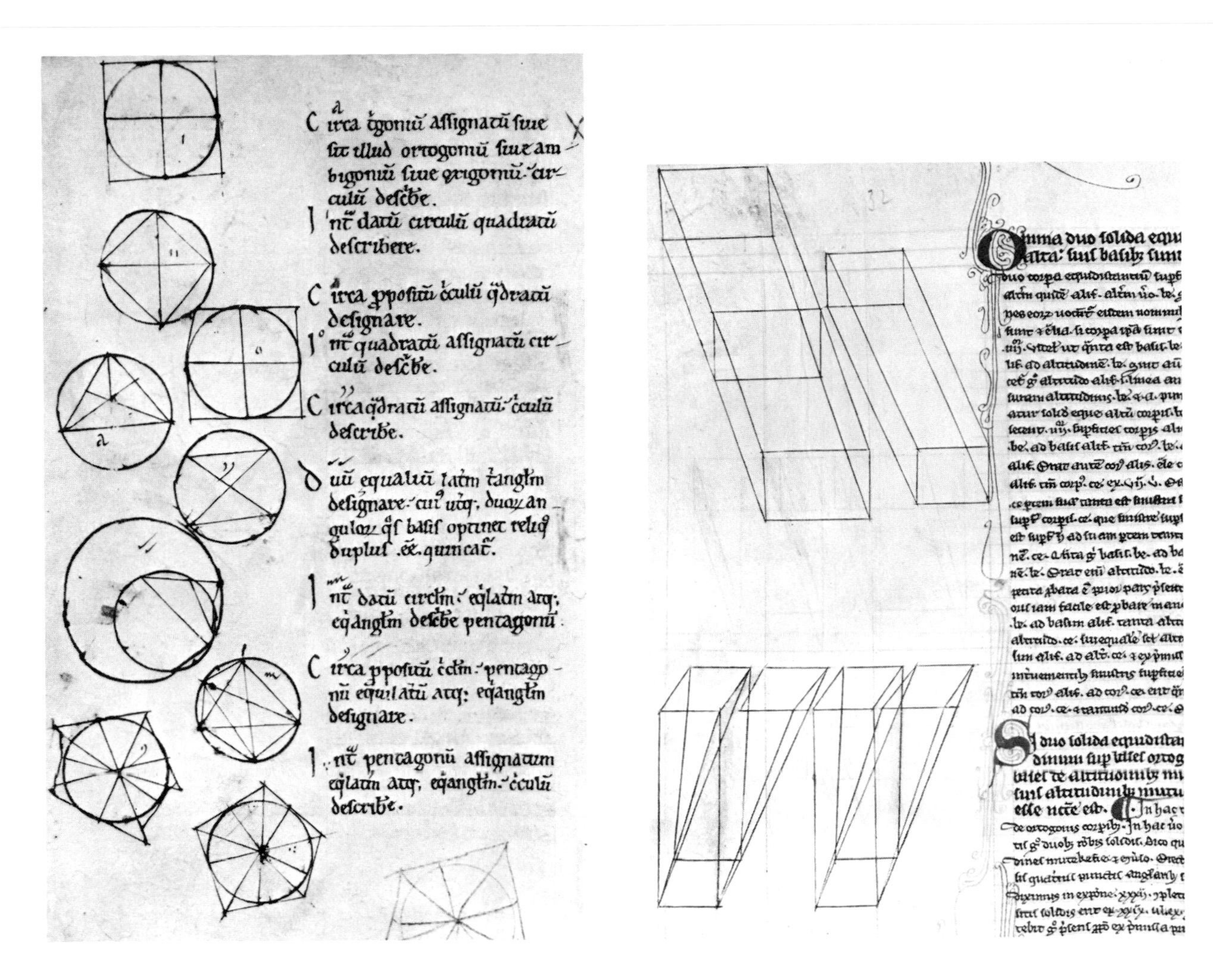

110. Incomplete Figures. The preceding Illustration was taken from manuscripts of Latin translations of Euclid which provided one with nothing more than the enunciations of the theorems of this fundamental geometrical work. The Illustration on the top left is also from a copy of the *Elements* giving enunciations only, but now from a twelfth-century manuscript of the second version of that work made by the twelfth-century English scholar Adelard of Bath. This particular Latin version did give abbreviated proofs, but the present copy has omitted them. As in the preceding Illustration, the added figures fill out the statements made by the proofless enunciations, but now with a new twist. Giving the enunciations and figures for Book IV, Propositions 5–13, this particular example is especially instructive since it has "keyed" the figures to the enunciations to which they belong by employing matching marks of various sorts. They appear first on the figure itself and then over or within the initial words of the relevant enunciation. Thus, the three dots on the pentagon and circle at the bottom of the page relate this figure to the final enunciation dealing with the inscription of a circle in a regular pentagon. Similarly, the 'a' in the circle containing an inscribed triangle "keys" this figure with the first enunciation on the page, the double checks in the "circle within a circle" figure pair it off with the sixth enunciation given (which is concerned with the construction of an isosceles triangle whose base angles are double its vertex angle), and so on. The unlettered figures in the Illustration on the top right are of a different sort. Taken from a thirteenth-century manuscript of the same second translation of Adelard, they belong to Book XI, Proposition 34 (which proves that the bases and heights of equal parallelepipeds are reciprocally proportional). In contrast to the Illustration on the left, the proof is here given, although, in the fashion that was standard for this version of Adelard, preceding the enunciation to which it belongs and not following it. The proof does provide the appropriate letters for both of its figures, but for some reason they have not been filled in. This would seem to be an indication of the fact that this particular medieval copy of Euclid was to have been "processed" in three stages: (1) transcription of the text by a scribe, (2) drawing of the figures, possibly by another person, and (3) lettering of the figures, either by the original scribe or by another. In this instance, stage (3) never came to pass.

111. (Below left) The Value of Figures Alone. This fourteenth-century copy of a unique Latin version of Euclid reveals just what significance figures themselves could on occasion be considered to hold. The two pages preceding that pictured here give the enunciations and proofs for the first six propositions of the first of the *Elements*. Then, most likely tiring of his task after so short a beginning, our scribe—or perhaps the translator himself—apparently decided to finish Book I in as expeditious a fashion as possible: namely, by giving only the figures for Propositions 7 through 47 (a missing proposition in most Latin versions of Euclid accounting for the total of 47 for this Book in place of the 48 one has in the Greek). Just how much these isolated figures could tell one of the substance of Book I of the *Elements* is not clear, but the scholar who prepared this fragmentary version evidently thought they could provide at least some kind of substitute for what the text itself contained. Perhaps he believed that they would serve to recall the content of Book I for one who already had some familiarity with his medieval Euclid.

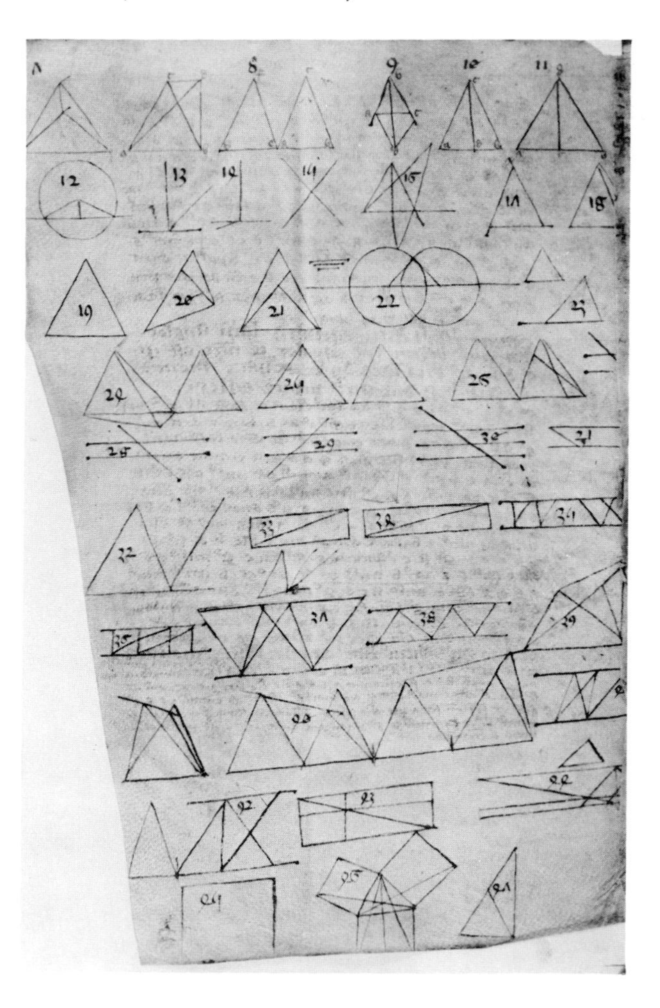

112. (Below right) Figures with Names. A number of Euclidean theorems have traditionally carried proper names and have often been referred to (even in the course of the proof of subsequent propositions) by the "shorthand" provided by these names. Thus, Book I, Proposition 5, came to be known as the "Asses' Bridge" *(pons asinorum)* and was earlier labeled 'Elefuga' ("flight of the miserable"). The Pythagorean theorem (I, 47, or 46 in most medieval versions) was given a variety of names, perhaps the most prevalently used being 'Dulcarnon', apparently of Persian and Arabic derivation, meaning "two-horned." The relevance of this particular name derives from the standard figure for the theorem in question, the squares on the two sides of the triangles sticking up like two horns (see Illustrations 111 and 115). The present Illustration, taken from a thirteenth-century manuscript of Adelard's second translation of the *Elements,* is yet another case of christening theorems with names based on the appearance of the relevant figures. The first, dealing with straight lines falling on the circumference of a circle from a point on a diameter that is not the center of the circle (III, 7), was named "the goose's foot" *(pes anseris)*. The slightest examination of the pertinent figure reveals why. Similarly, the next proposition in Book III, treating of straight lines falling on a circle from a point outside the circle, was called "the peacock's tail" *(cauda pavonis)*, again in obvious reference to the accompanying diagram. In both instances, the scribe of the present manuscript has written the name in question within the figures.

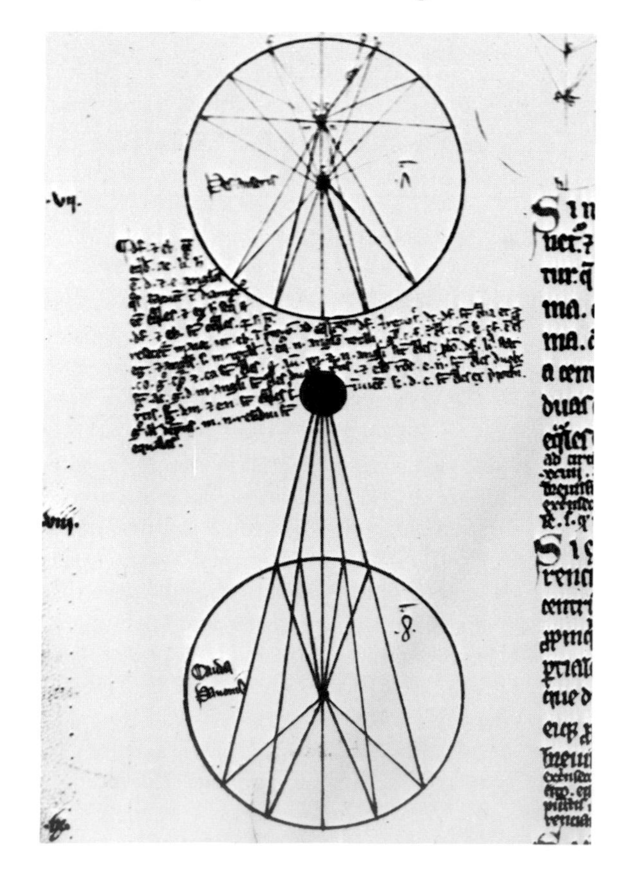

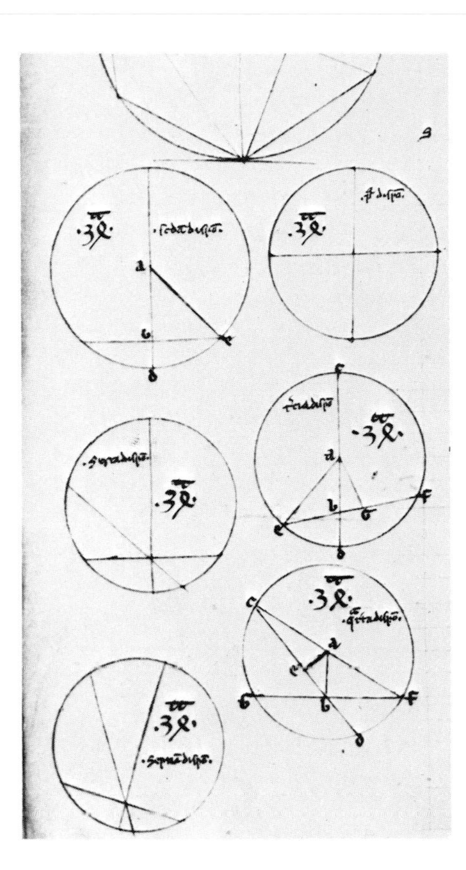

113. (Left) Covering All Possibilities. Whenever there was a multiplicity of cases for a given theorem, Euclid often avoided treating each and every one, an avoidance that was almost never practiced in the medieval Latin translations of the *Elements.* There, completeness was the order of the day. The present instance, taken from a fourteenth-century copy of an anonymous version of Euclid, carries this medieval penchant to an extreme. The theorem in question—Book III, Proposition 34 (III, 35 in the Greek)—establishes that, of any intersecting chords in a circle, the segments of one chord resulting from this intersection determine a rectangle equal in area to that determined by the segments of the other chord. Euclid's Greek treats only two cases: when the chords both pass through the center of the circle and when neither does. The medieval text to which the present figures belong draws attention to all possible cases (labeled *dispositiones*). (1) The two chords may both pass through the center—the *prima dispositio* at the upper right: (2) Only one passes through the center, in which case it may (a) bisect the other—the *secunda dispositio* at the upper left—or (b) divide it into unequal parts—the *tertia dispositio* at center right: (3) Neither chord passes through the center, in which case (a) one may bisect the other—the fourth case at the lower right—or (b) they divide each other into unequal parts—here inappropriately represented by the bottom two figures on the left (which are confused by the inclusion of unnecessary diameters and labeled as the sixth and seventh cases, no fifth case being given). However, as the text itself makes clear, only five *dispositiones* are needed for full coverage. For other figures from the present manuscript, see Illustrations 115, 122.

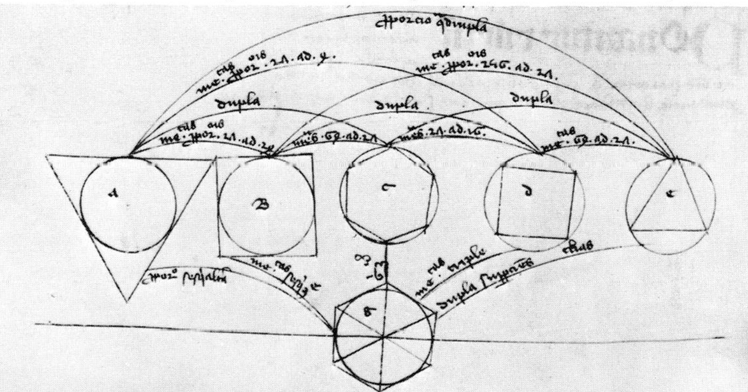

114. (Above) A Summarizing Diagram. Part III of the fourteenth-century scholar Nicole Oresme's *Algorismus proportionum* treated the ratios obtaining between certain regular polygons as inscribed and circumscribed with respect to a constant diameter circle. The present figure, taken from the end of Part III of a fifteenth-century manuscript of the *Algorismus,* effectively summarizes its contents. A variant of the "arc-diagram" we have already seen in Illustrations 95, 100–105, some of the inscribed and circumscribed polygons have areas that are here related according to simple multiple ratios: The label on the arc connecting the circumscribed and inscribed triangles tells us that the former stands to the latter in a quadruple ratio, while the arcs joining the circumscribed triangle with the inscribed hexagon, the circumscribed square with the inscribed square, and the inscribed hexagon with the inscribed triangle all carry the inscription 'dupla', informing one that the first member of these pairs stands in a double ratio to the second member. The ratios of the circumscribed hexagon to the circumscribed triangle, to the inscribed hexagon, and to the inscribed triangle are more complex: The relative arcs tell us that they are, respectively, sesquialtern (3 to 2), sesquitertian (4 to 3), and double superpartient tertian (8 to 3). The remaining relations between the circumscribed and inscribed polygons involve even greater complexity which Oresme expressed in terms of "parts" of ratios. Thus, to cite but two instances, the area of the circumscribed triangle stands to that of the circumscribed square as "half of the ratio of 27 to 16" (*medietas proportionis 27 ad 16*—the single error in our figure here having 24 instead of 16). Since in the medieval mathematics of ratios to take a half part of a ratio is, in our terms, to take its square root, the circumscribed triangle stands to the circumscribed square as the square root of 27 to 16. Similarly, the circumscribed square stands to the inscribed triangle as half of—that is, the square root of—256 to 27. And so on.

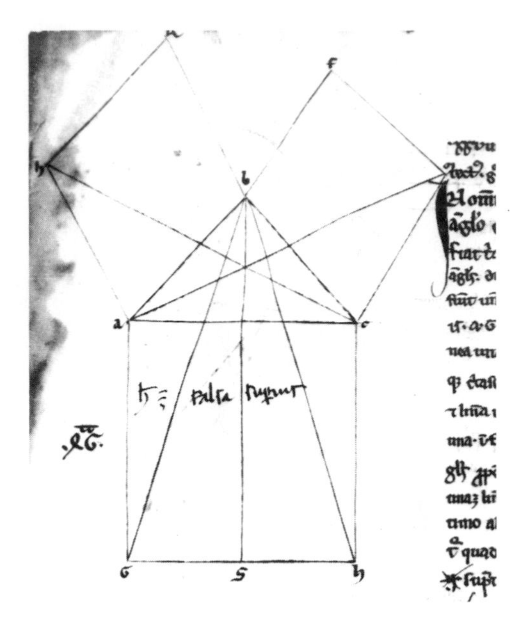

115. Falsigrafia. The Greek, Arabic, and especially medieval Latin manuscripts of Euclid's *Elements* and other geometrical works are for the most part accompanied by quite adequate figures, in some instances drawn with an accuracy and elegance—often with different colors—that rival and even surpass the best efforts of modern printing. In a few cases, however, this graphic excellence is found wanting. Two instances of such *falsigrafia*—a medieval term that covered not only wrongly drawn diagrams but wrongly constructed arguments as well—are given here, both from a fourteenth-century copy of an anonymous version of the *Elements.* That on the top left is the relevant figure for the Pythagorean theorem (Book I, Prop. 46; I, 47 in the Greek). Its problem is that two rhombuses have been drawn above on the legs of the right triangle in place of the requisite squares, a fact that is duly noted by the inscription written across the diagram reading: "this is false above" *(hec est falsa superius)*. The example below (to I, 23) is more involved. Two attempts made by the scribe to provide a proper figure have proved abortive. At the lower right of this diagram we see touching instead of intersecting circles; this is appropriately labeled *falsigrafium.* A second attempt—scratched out and called *falsa* at the top left of the same diagram—erred in drawing the straight lines from the point of intersection of the circles to points *A* and *P,* which are *not,* as they should be, the respective centers of the two circles. They are that in the third, successful, attempt to produce a correct figure, here appearing at the bottom left. For other diagrams from the present manuscript, see Illustrations 113 and 122.

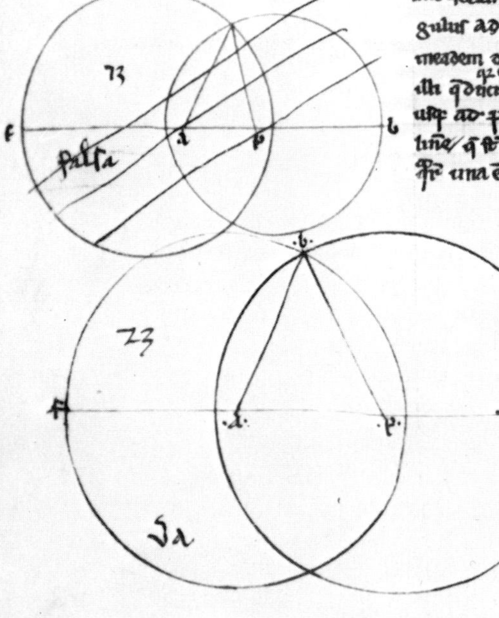

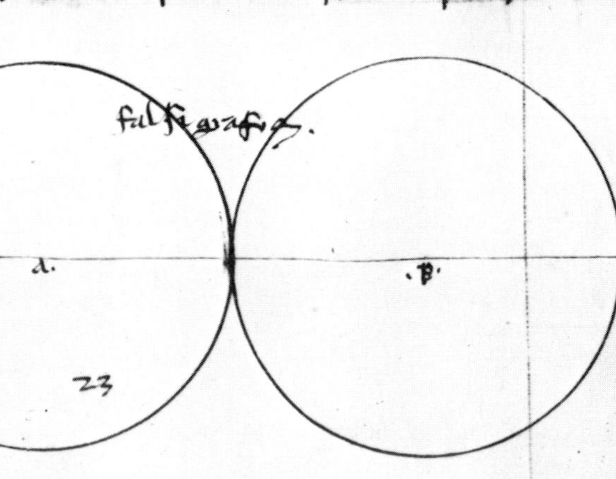

116. Drawing Conic Sections. Medieval scribes always found it problematic to draw curves other than circles or arcs of circles. Manuscript evidence reveals that there were two ways to solve this problem: Either use arcs of circles to approximate the higher curves in question or draw them by hand. The first type of solution—which was that most frequently employed—is here represented by two examples. That directly below taken from a treatise on the measure of paraboloids by the ninth-century Ḥarrānian scholar Thābit ibn Qurra as copied by the mathematician al-Sijzī in 969 A.D., shows three parabolas that have used six circular arcs each in their construction. The drawing of the ellipse in the center diagram has also appealed to circular arcs. The figure is taken from an eleventh-century codex of the Arabic translation made by Thābit ibn Qurra and emended by the Banū Mūsā (9th century) of the *Conics* (Book III, Prop. 38) of Apollonius of Perga (3rd century B.C.). The manuscript itself was written by the eleventh-century Arabic scholar in the exact sciences, Ibn al-Haytham. The ellipses in the bottom diagram have been drawn by hand. They have been taken from seventeenth-century manuscript of the same translation of Apollonius' *Conics* (Book VII, Prop. 8).

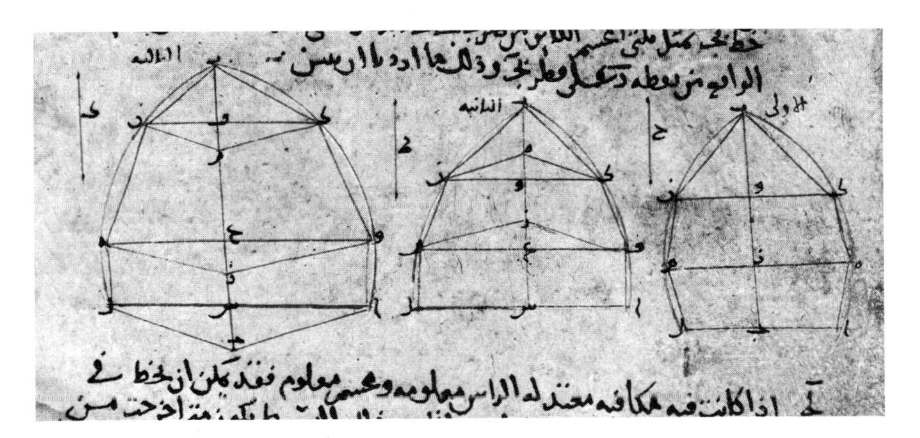

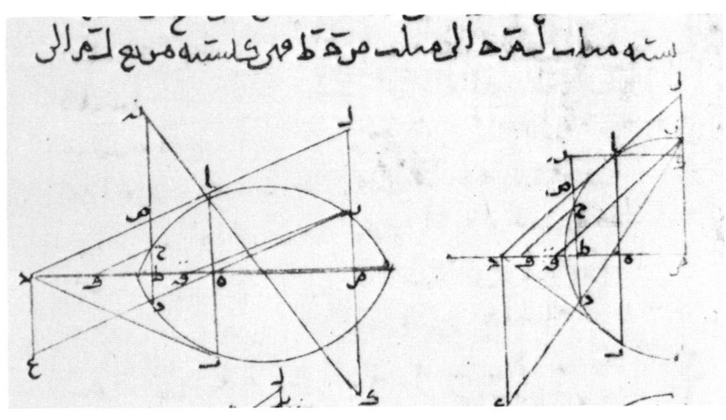

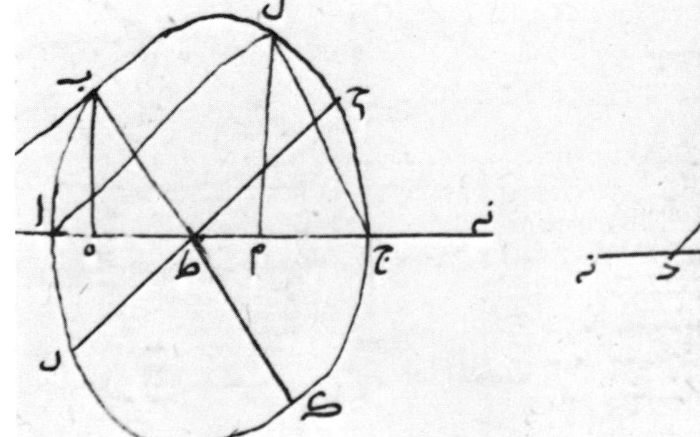

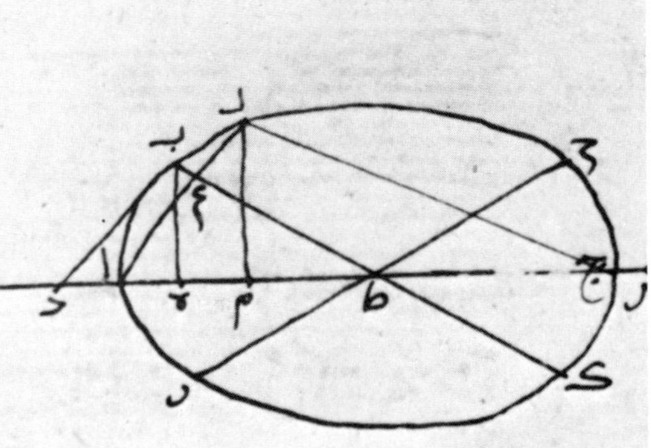

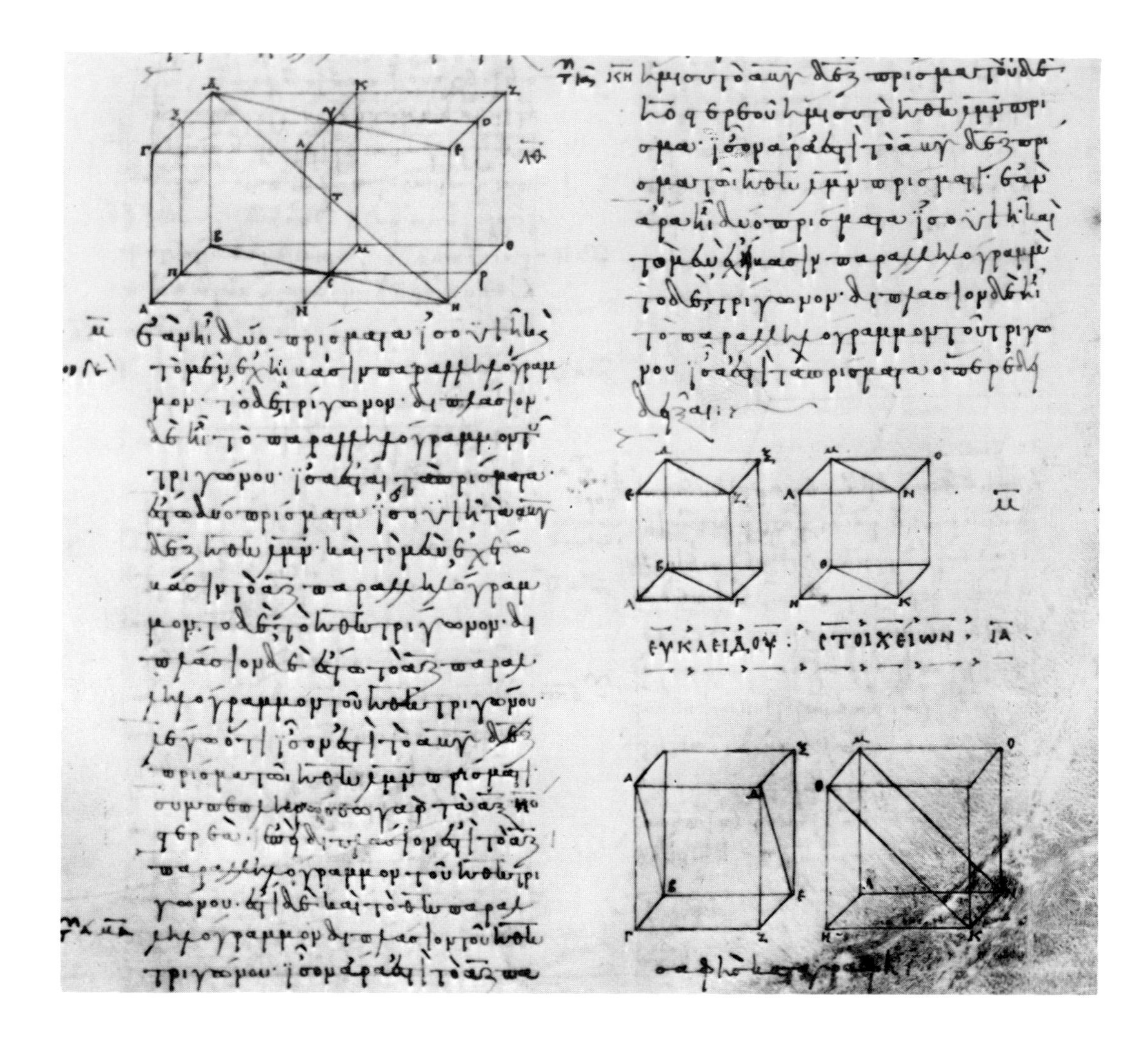

117. Introductory Stereometry. As the basic textbook of ancient and medieval geometry, the *Elements* caused few problems relative to the drawing of figures for theorems that limited themselves to the plane, since neither conic sections nor other higher curves were part of Euclid's charge. It is largely in Books XI through XIII, where the third dimension comes into consideration, that scribes began to reveal their concern with the adequacy of their diagrams and to present one with alternative representations that bespeak fundamentally different approaches to the "picturing" of a given propositon. (Some of this will form the subject of the Illustrations that follow.) The difficulties were fewer in Book XI than in the following two books (although, as the following Illustration testifies, even there the simplest three-dimensional situation could cause concern). On the whole, however, the parallelepiped solids or prisms which formed the substance of Book XI were quite adequately represented and in such a way that their three-dimensionality was intuitively evident. A good case in point is here reproduced from a ninth-century Greek manuscript of the *Elements.* Taken from the very end of Book XI (as the legend **Εὐκλείδου στοιχείων** directly below the first pair of prisms in the right column announces), the figures accompany Propositions 38 and 39 (although they are here numbered 39 and 40—$\overline{\Lambda\Theta}$ and $\overline{\mathrm{M}}$—due to the fact that the manuscript contains an interpolated proposition). Note that not only are the propositions themselves numbered, but so also the figures belonging to them: Compare, for example, the $\overline{\mathrm{M}}$ in

(Continued)

the left margin with the $\overline{\text{M}}$ next to the figures in the right margin. Marginal annotations also reveal which previous propositions are utilized in the course of proof: Thus, Proposition 28 ($\overline{\text{KH}}$) of Book XI is indicated in the space between columns. The figures depicted of the relevant prisms, in addition to being appropriately three-dimensional, are quite adequate for following Euclid's proofs. For the first figure, however, the adequacy is a mathematical one, faithful in representing the geometrically required relations between its constituent elements, but does not also accurately picture the shape of the geometrical solid in question. Indeed, the whole figure in the left hand column is supposed to be a cube. (One is tempted to think that if the figures were added after all of the text was written, not enough space was left to depict this cube with proper shape.) Finally, note should be made of the fact that the right hand column exhibits the not too infrequent practice of providing alternate diagrams for a given theorem: The two lower prisms—appropriately labeled "clear diagram" (σαφὴς, καταγρφή)—belong to XI, 39 as well as the two above. For other figures from the present manuscript, see Illustrations 120 and 122.

118. Specifying the Third Dimension. In the case of many figures drawn of geometrical solids it was intuitively evident (as in the preceding Illustration, for example) that the figure in question was a three-dimensional one. But this was not always so, at least in the eyes of scribes who had drawn the figures. Some additional information was given to make it absolutely clear that the two-dimensional drawing at hand was meant to represent a three-dimensional situation. The present example of such added information is taken from a thirteenth-century manuscript of the second version of Euclid's *Elements* prepared by Adelard of Bath. The subject is Proposition 3 of Book XI, which stipulates that two plane surfaces intersect in a straight line (the curved line in the figure being required by the *per impossibile* proof given of the Proposition). In order to make everything appropriately evident, the scribe has written in the bottom rectangle: "a surface lying in the plane" and on the upper rectangle intersecting that surface: "a surface aloft which cuts that which lies in the plane."

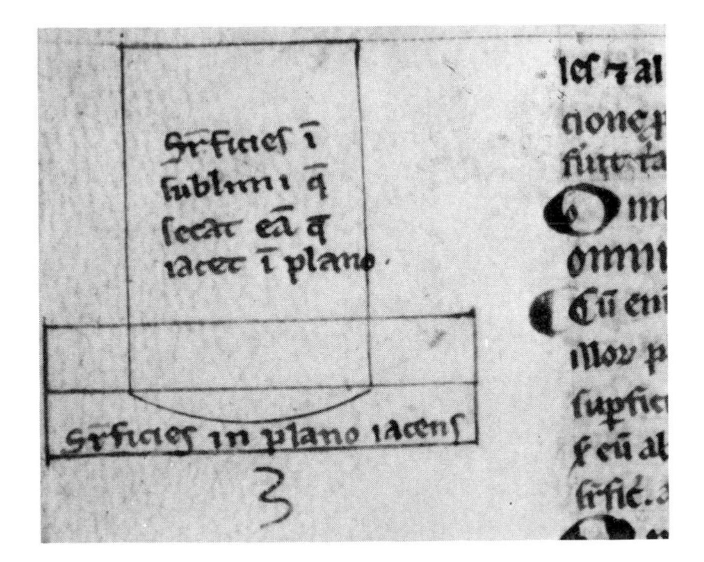

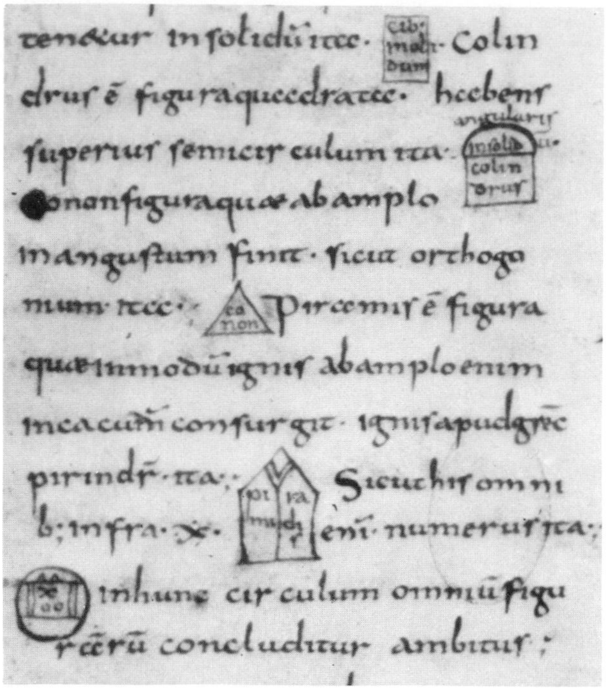

119. A Diagram Creates a Definition. The problems encountered in drawing adequate figures of geometrical solids were evident in the Middle Ages long before much of the content of Greek geometry had been translated and absorbed. An especially interesting instance of one such problem can be seen in the section on geometry in the *Etymologies* of the seventh-century bishop Isidore of Seville. The locus is Isidore's definition (Book III, xii) of a cylinder: "a square figure having a semicircle above" *(cylindrus est figura quadrata, habens superius semicirculum)*. Although Isidore does not say so, it is highly probable that this curious definition was directly derived from some early medieval drawing of a cylinder. An examination of Isidorian manuscripts confirms this hypothesis. The diagrams above occur in a ninth-century manuscript of the *Etymologies,* the rectangle or square plus semicircle depiction of the cylinder directly following the above-quoted definition to which such a depiction likely gave rise. The word 'colindrus' is inscribed in the rectangle, while 'in solidum' appears in the semicircle above, presumably to emphasize to the reader (in a way far more elementary than the added words in the preceding Illustration) that three dimensions are involved. The word 'angularis' written over the semicircle was probably displaced from one of the next two geometrical solids defined, the cone and the pyramid. The last, circular diagram at the lower left is the Isidorian illustration of the fact that, just as every number or digit is contained in ten, so all figures or polygons can be inscribed in a circle. The top illustration on page 127 contains the same definitions and diagrams, but in this instance in a more mature and artistic form as found in a twelfth-century manuscript of the *Etymologies.* The influence of these particular diagrams can be seen in the bottom picture on page 127. They have here migrated from Isidore to the beginning of the chapter on geometry in Cassiodorus' *Institutiones* as in a fourteenth-century manuscript which once belonged to Petrarch. The central feature here is the personification of geometry, the object in her right hand most likely being a depiction of the rod *(radius)* ascribed to her by Martianus Capella (fifth century) which had become a measuring wand *(virga)* in the *Anticlaudianus* of the twelfth-century Alan of Lille. Of greater present interest, however, are the distinctly Isidorian geometrical figures to the left of the picture. Our cylinder appears unchanged at the right in this group of nine diagrams, the "all-inscribable-in" circle in the lower left corner, while lack of space seems to have deformed the sphere in the center and turned the cone at the top right into something like a teardrop. This manuscript was naturally written and illustrated long after more substantial geometrical works such as those of Euclid were translated and well known, but when figures or diagrams were needed, as here, for decorative purposes, older, less "scientific" works like those of Isidore were most often plumbed as sources rather than the mathematical works which were then being studied. For other material from the present manuscript of Cassiodorus, see Illustration 27, 30, 170, and 177, and for more on the personification of the sciences, see Illustrations 172–180.

Chilindrus est figura quadrata
habens supius semicirculum & solidum
Conon est figura que ab amplo
in angustum finit sicut ortogonium;
Pyramis est figura que in modum
ignis ab amplo in acumen consurgit
Ignis enim apud grecos phyr in
appellatur.
Sicut autem infra decem omnis
est numerus ita intra unum
hunc circulum omnium figu
rarum concludit ambitus.

chilin drus.

conon.

py ra mis

De arte geometrica.

Nunc ad geometricam ueniamus que est descriptio contemplatiua
formarum documentum etiam uisibile philosophorum. Quod ut preconijs
efferat Jouem suum in operibus proprijs geometrizare testantur.
Quod nescio utrum laudibus an uituperationibus applicetur. Potest
hec sententia forsitan conuenire ueritati. Geometrizat enim si fas est dicere
sancta trinitas quando creaturis suis quas uno die fecit existentes diuisas
species formulasque concedit. Quando cursus stellarum potentia ue
neranda distribuit & statutis lineis facit currere que mouentur

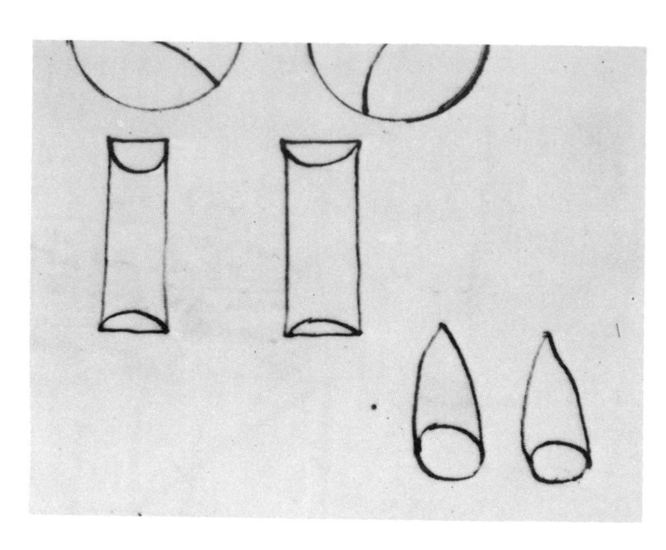

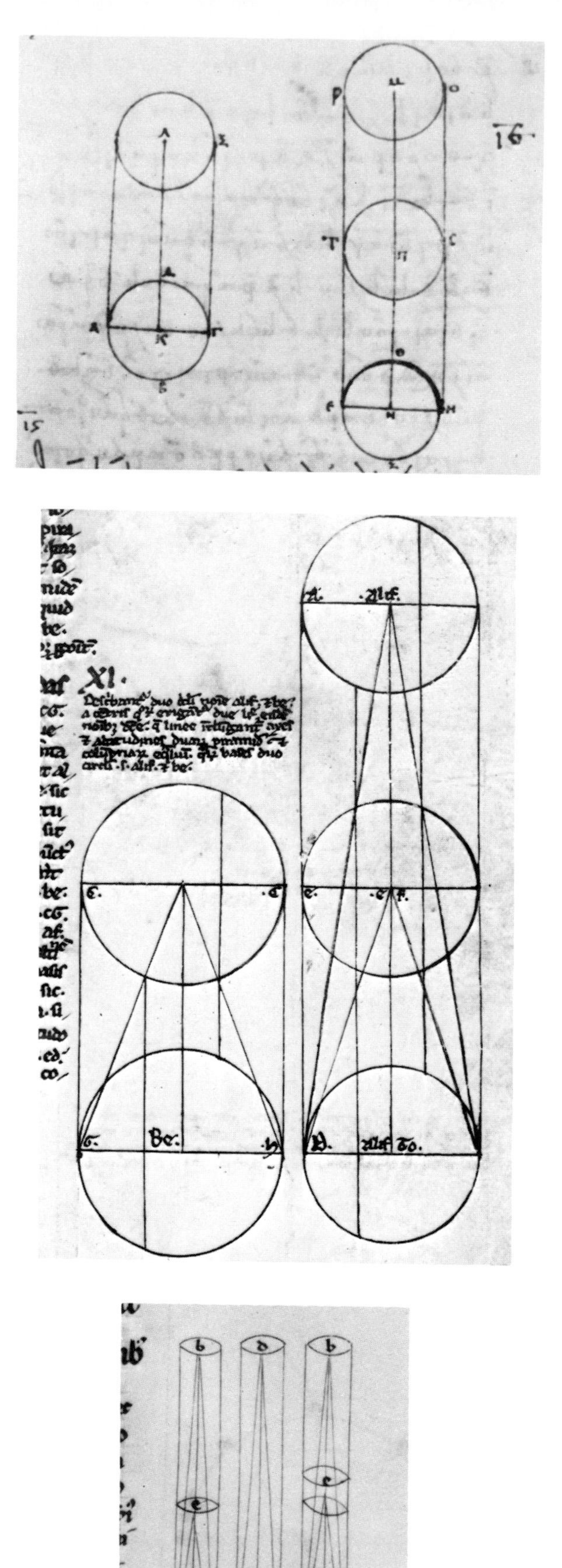

120. The Developing Tradition in the Three-Dimensional. Isidorian cylinders were almost never brought into service as relevant diagrams in proper geometrical texts in the Middle Ages, although something not too far removed from the figures in the *Etymologies* can be seen in the diagrams above left, taken from a twelfth-century manuscript of a Boethian–Adelardian version of Euclid. As in Isidore, the semicircle is called upon to represent the base, but it is now inverted and used for both bases. For the most part, however, semicircles were abandoned in this role. Instead, we most frequently find complete circles employed as the bases or medial sections of cylinders. Two instances of such a practice are depicted here in the top and middle illustrations on the right: one from a ninth-century Greek manuscript of Euclid's *Elements* (Book XII, Prop. 15), the other from a thirteenth-century codex of Adelard's second Latin version of the same. The latter betrays that it is a version prepared from the Arabic since the bases of the two cylinders are referred to as *alif* and *be*. The bottom illustration on the right, from a fourteenth-century manuscript of Campanus of Novara's (thirteenth century) version of Euclid (Book XII, Prop. 14, = XII, 13 of the Greek), employs the arcs of circles to represent the bases of the cylinders, a technique we have already seen (Illustration 116) to have been utilized to depict ellipses (which are, of course, how the bases would appear if pictured in perspective). For other diagrams from the second and third manuscripts pictured here, see Illustrations 117, 122; and 99, 124, 133, 139, 141, 149, and 238.

121. Three Dimensions Better Seen as Two. The primary burden of the final, genuine thirteenth book of Euclid's *Elements* is the inscription of each of the five regular polyhedra in a sphere, the last but one to be treated being the icosahedron (XIII, Prop. 16). Its construction in a sphere is the subject of the present Illustration. To see Euclid's problem and procedure in three dimensions, one can turn to the modern diagrams above. That on the top right simply presents Euclid's goal in perspective; but the diagram at the right center illustrates the substance, if not all of the details, of how he solves the problem at hand: namely, by the construction of pentagons in parallel circles on the sphere. He begins with the lower parallel circle and inscribes a regular pentagon; bisecting the arcs determined by this pentagon in turn yields a decagon and at the same time a second inscribed pentagon. The vertices and sides of this second pentagon give five of the vertices and five of the edges of the icosahedron in question. Next, constructing perpendiculars from the vertices of the *first* pentagon to their points of intersection with the upper parallel circle will determine yet a *third* inscribed pentagon. It yields an additional five vertices and fifteen edges of the icosahedron. Given this, the remaining two vertices, and hence all the remaining edges, can be obtained by finding the poles of the circular segments determined by the two parallel circles. The significant fact to note, however, is that none of the manuscripts (or even printed editions) of Euclid's text utilized three-dimensional figures like those in the modern diagrams above. Instead, they all employ a figure all but identical with that here depicted at the bottom right from a ninth-century Greek codex of the *Elements*. Here three dimensions are "squashed" into two. The circle in this manuscript figure is the lower parallel circle on the sphere, its two inscribed pentagons much more clearly depicted than in the modern, three-dimensional, diagram. The outer, third pentagon is that belonging to the upper parallel circle (which is here not drawn); although equal in size to the first two pentagons, the "squashing" has made it appear larger. What is more, following the two-dimensionality of the manuscript figure, the perpendiculars drawn from one parallel circle to the other are similarly "flattened to the plane." Although the resulting icosahedron does not appear as it does in the modern, three-dimensional diagrams, all of the steps of Euclid's procedure are adequately represented, and in a more easily discernible fashion at that. If we leave the lines drawn within the central pentagon in this manuscript figure out of account (they have to do with the establishment of the poles of the circular segments), we have a figure determined by a kind of projection onto the plane of the lower parallel circle; here intersecting lines all represent real intersections, unlike the modern diagrams where the crossing of lines is often merely due to the three-dimensional representation. Euclid's procedure, if not his finished product, is thus clearer in the ninth-century figure and in all of its similar offspring in other codices and editions. The semicircle to the left of the main manuscript figure is an auxiliary construction allowing one to determine the required radius for the parallel circles from which the main construction begins.

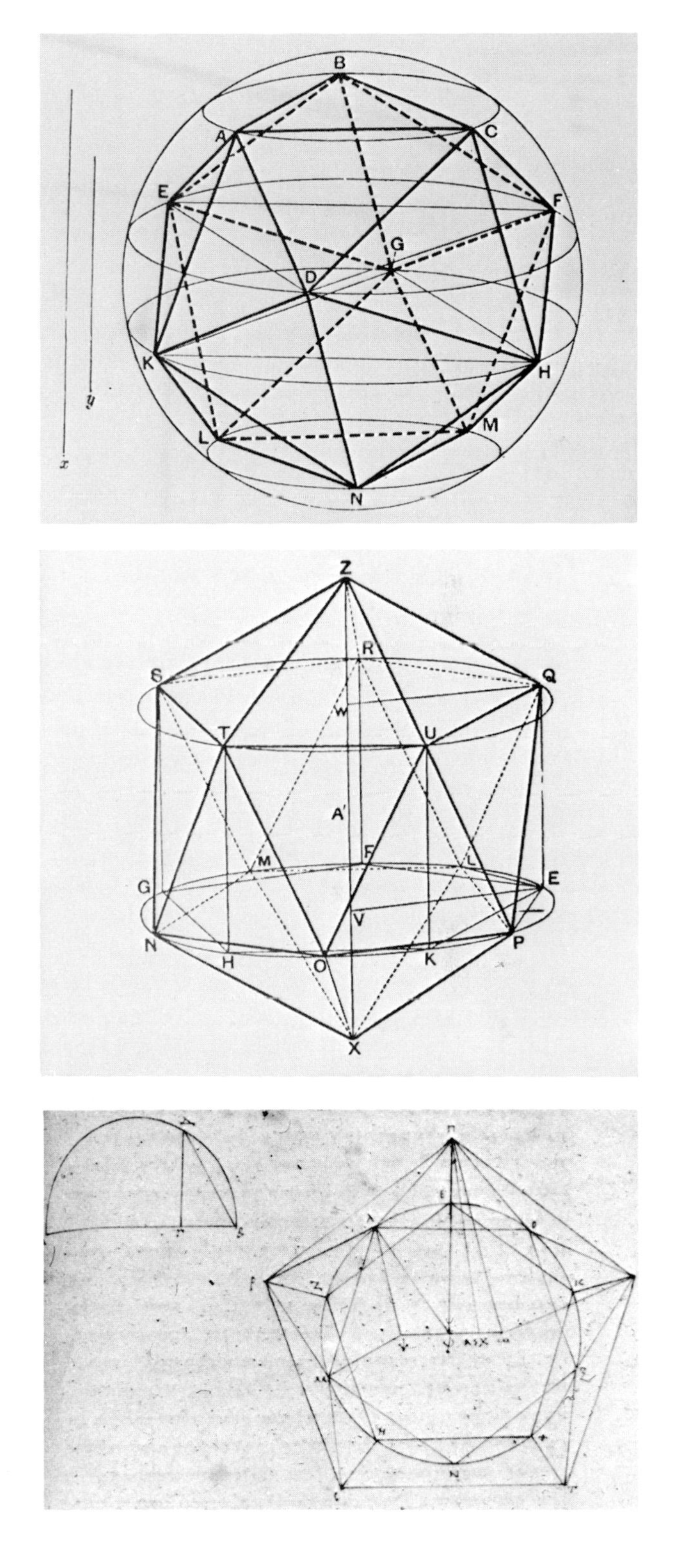

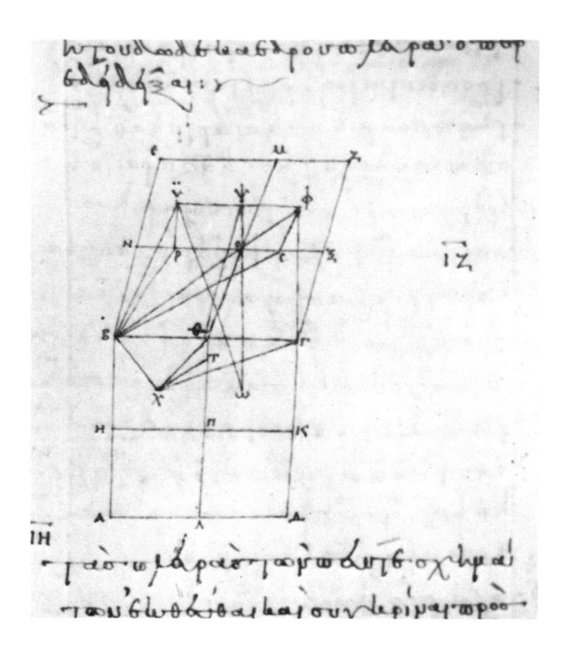

122. Alternate Figures for a Single Construction. The final regular solid that was Euclid's charge in Book XIII was the dodecahedron. In this instance, the standard figure accompanying the text was almost always three-dimensional, but not in the "complete" fashion one might expect. Two modern diagrams may again be useful in revealing just how the Euclidean figures proper for XIII, 17 differ from such an expected one and why. The modern figure on the top left of page 131 shows the desired inscription of the dodecahedron in a sphere. Euclid begins the required construction from a cube that is also inscribed in the sphere (something easily established by an earlier proposition in Book XIII) and the modern diagram on the top right of page 131 shows the relation of the dodecahedron to that cube. But the exact nature of that relation is made much clearer in the traditional Euclidean diagrams for this proposition, which limit one's consideration to only two sides of the cube and a single regular pentagonal face of the dodecahedron. Two such diagrams are given here; the first (top left, page 130) from a ninth-century Greek codex of the *Elements,* the second (top right, page 130) (where the diagram is rotated clockwise 90°) from a fourteenth-century manuscript of Gerard of Cremona's Latin translation from the Arabic. It is far more evident in these "partial" three-dimensional figures than in a more complete one that the inscribed dodecahedron and cube are related insofar as the sides of the cube are diagonals of the pentagonal faces of the dodecahedron. And this naturally determines two vertices of each

(Continued)

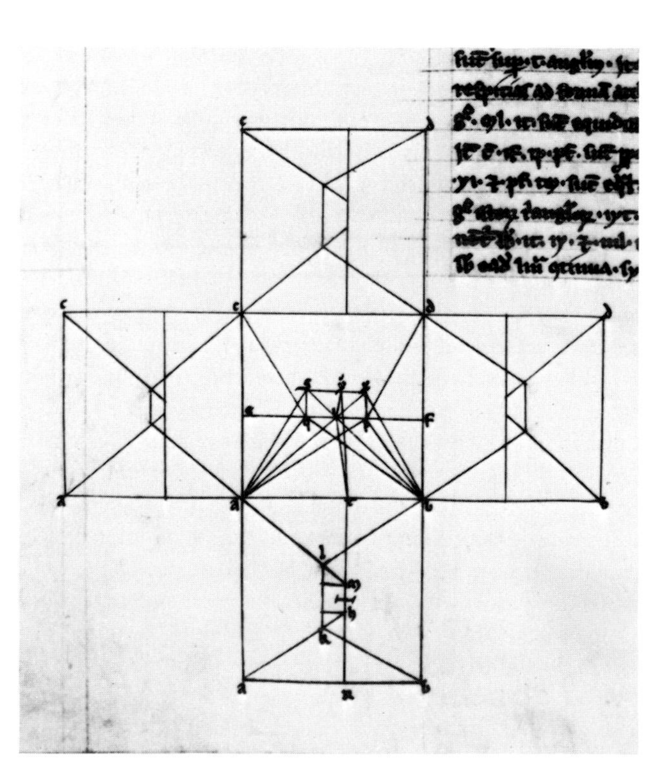

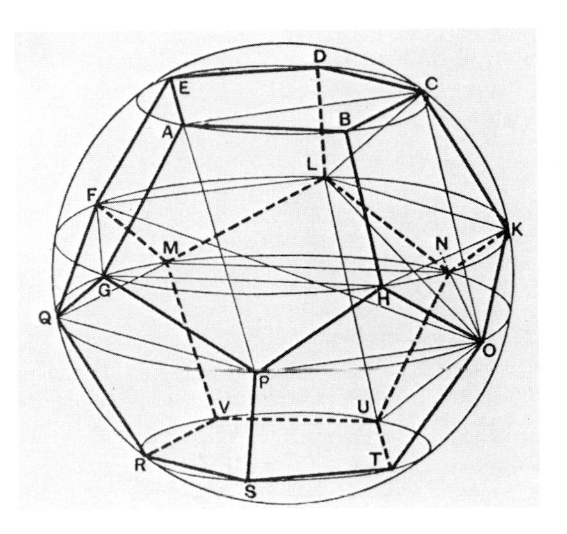

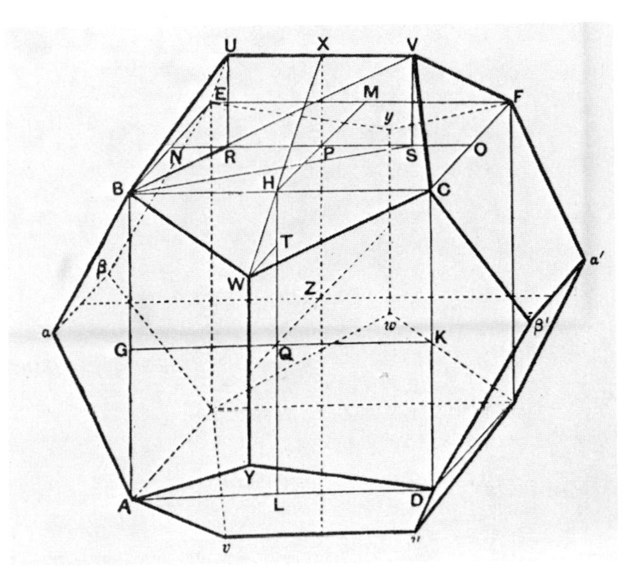

pentagonal face. The remaining three vertices of each face are found by constructing perpendiculars from the faces of the cube, the problem naturally being to establish the precise points on the cube from which the perpendiculars are to be drawn and the requisite lengths of each. Here, again, the procedure is more clearly represented in the "partial" figures in the manuscripts: Halves of the middle parallels in the cube's faces are to be divided in extreme and mean ratio, the resulting points of division providing the points from which the perpendiculars are drawn and the larger segments obtaining from the division being equal to their length. This procedure fully determines the remaining three vertices of each pentagonal face and, considering in a similar fashion the remaining faces of the inscribed cube and the pentagons "attached" to them, all of the vertices of the inscribed dodecahedron. Almost all texts of the elements—original Greek or translations, no matter—appeal to a diagram like the two depicted here to elucidate Euclid's procedure, but one medieval text did not: the third version prepared by Adelard of Bath often referred to as an *editio specialis*. Its quite different solution of diagramming this procedure is here reproduced in the cross-like figure depicted at the lower right on page 130, taken from a thirteenth-century manuscript. Three dimensions are flattened into two. In effect, the "top" is removed from the inscribed cube and its four upright sides are "folded down." The division according to extreme and mean ratio of the middle parallels in all five faces is clearly given and four pentagonal faces are represented (all now in a single plane, but with the relevant relations preserved), even though the positions and the lengths of the perpendiculars mentioned above are given for only one pentagon. Yet whether this two-dimensional technique or that used by the "partial" three-dimensional figure in our other two manuscript illustrations was utilized, the result was a diagram that depicted only that which was essential in Euclid's construction, making that construction more easily comprehensible. For other figures from the first two manuscripts represented here, see Illustrations 117, 120; and 113, 115.

of Euclides Elementes.

In this figure it is more easy to see the former demonstration, if ye erect perpendicularly vnto the ground superficies A G B D, the thre superficieces, GH, K L, and M N, or if ye so erect them that they be equedistant one to the other.

The 16. Theoreme. The 18. Proposition.

If a right line be erected perpẽdicularly to a plaine superficies: all the superficieces extended by that right line, are erected perpendicularly to the selfe same plaine superficies.

Suppose that a right line A B be erected perpendicularly to a ground superficies. The

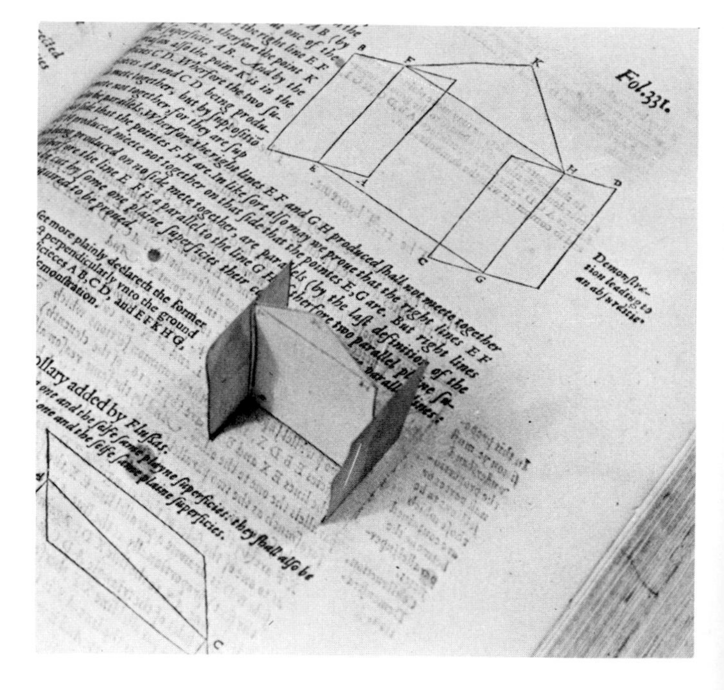

123. "Pop-up" Geometry. Although there are extant medieval manuscripts containing scientific diagrams with movable parts (see Illustration 238), the attempt to make solid geometry truly three-dimensional through the construction of "pop-up" or "stand-up" figures appears to have occurred first in early printed books. The most famous such production is the first proper English translation of Euclid's *Elements*. Made from the Greek and from earlier Latin versions by the Sir Henry Billingsley who was later lord mayor of London (d. 1606), it was published in London in 1570, together with a preface by John Dee, patron and sometime practitioner of the mathematical arts. Our interest in this rather impressive publication lies in its inclusion of pasted flaps of paper that can be folded up to produce three-dimensional models of some of the propositions of Book XI. There are about 34 such creations populating this book, two of which are pictured here. That on the right accompanies XI, 16, which establishes that if two parallel planes are cut by any third plane, their common sections will be parallel. A two-dimensional figure is drawn above the "stand-up" construction corresponding to it. The legend to the right of the upper diagram—'demonstration leading to an absurditie'—refers to the fact that the proof of the proposition proceeds by a *reductio ad impossibile*. The three-dimensional figure at the top belongs to XI, 18, the text of the enunciation of which appears clearly at its left.

CHAPTER TWELVE

The Mathematics of the Heavens

TWO KINDS OF VISUAL artifacts were predominant in the ancient and medieval history of mathematical astronomy: tables and geometric diagrams. The first were related to the presentation of those calculated values—or of the means used to calculate those values—which formed the raw material for that astronomy. Examples of this have been given in the concluding illustrations of Chapter 9. There remain, however, the figures used in conjunction with the geometry of the motions of celestial bodies. These constitute the subject of the present chapter. To be sure, beyond the types of figures sampled in this chapter, there are many other kinds of diagrams (some of them geometrical in one form or another) that were utilized in the early history of astronomy and its ancillary sciences: figures of lunar phases, of eclipses, of the zodiac, and of the "world system" as a whole, just to mention a few. These, however, will be treated later.

The geometrical diagrams of planetary motion which form the nucleus of what follows here almost all derive in one manner or another from the work of Claudius Ptolemy (second century). Some of these diagrams are but repetitions in a simplified form of what one has in Ptolemy himself. Others are developments or emendations of something initially Ptolemaic. And still others are watered down and disembodied parts of what can be found in Ptolemy's *Almagest.* Nevertheless, Ptolemy's position as fountainhead should not obscure the fact that mathematical models intended to explain the motions of celestial bodies are found almost 500 years earlier in Eudoxus of Cnidus' system of homocentric spheres in the fourth century B.C. And between that date and Ptolemy we also have the work of Aristarchus of Samos (third century B.C.) on heliocentric theory, of Apollonius of Perga (ca. 200 B.C.) proving the equivalence of certain eccentric and epicyclic models of celestial motion (see Illustration 125 for the same from the *Almagest*), and, above all, of Hipparchus (second century B.C.), whose work was often used and developed by Ptolemy. Yet almost all the major works of these earlier scholars on the problem of planetary motions either are lost or do not provide a source for the geometrical figures which are our present concern. The history of the diagrams of the mathematics of celestial motions is, therefore, essentially a Ptolemaic one.

On the other hand, if the mathematics of astronomy is taken more broadly to include those elements of spherical geometry that can be applied in astronomy in general, then there is a significant body of diagrammatic material which is not Ptolemaic. The opening illustrations of the present chapter from Autolycus of Pitane (fourth century B.C.) and Theodosius of Bithynia (second century B.C.) are meant as a sample of that material. Other Greek scholars also wrote works devoted to similar aspects of the "mathematics of astronomy," most notably Euclid (fl. ca. 295 B.C.), who wrote a brief treatise entitled *Phaenomena,* and Menelaus of Alexandria (fl. ca. 100 B.C.). In Islam these works, as well as those of Autolycus and Theodosius, were put together with other geometrical and optical writings to form the "intermediate books" *(mutawassiṭāt),* that is, books which could be studied after Euclid's *Elements* but before the *Almagest* of Ptolemy. Indeed, the figures from Autolycus and Theodosius in Illustration 124 are taken from manuscripts containing, inter alia, a similar collection of such "intermediate" works in Latin translation.

The medieval history of Ptolemaic astronomy itself bears a decided imbalance. For the understanding and the development of Ptolemy were far greater in the hands of Arabic scholars than Latin ones. The twelfth century did see Latin translations of the *Almagest*—from both Greek and Arabic—but these translations appear to have been appreciably less utilized in the Christian Middle Ages than the Arabic translations were in Islam. Instead, "handbooks" were written, purportedly providing the essentials of Ptolemaic astronomy.

True, Arabic scholars had their handbooks too—the most important probably being that by al-Farghānī (ninth century)—but they did not play nearly so central a role in the history of astronomy as did those in the Latin West. The latter—of which the two most popular are represented in Illustration 128—became the standard sources from which almost all scholastics learned the rudiments of astronomy, something that is all too evident from the very great number of manuscripts we still have of these handbooks, from the frequency with which plentiful marginalia testify to their extensive use, and from the number of times we find them "assigned" in late medieval university curricula.

At times it seems that even these handbooks were ignored for something still more elementary. Thus, in commenting on the arts of the quadrivium in the thirteenth century, Lambert of Auxerre confesses that, due to its *difficultas* and *prolixitas,* astronomy was not read along with the other arts; instead, one obtained what one needed from a brief treatise called the *Astrologia Marciani,* that is, the chapter on astronomy in the handbook on the seven liberal arts by Martianus Capella (fifth century). Of course Lambert's report must not be taken to represent the mean of medieval Latin comprehension of astronomy. There were scholars who did know and understand Ptolemy, but they were just far fewer than scholars with a similar competence in Islam.

In addition to providing several diagrams taken from medieval Latin astronomical handbooks, the present chapter concludes with a number of illustrations designed to show how two specific elements of Ptolemaic astronomy—the epicycle and the eccentric—were detached from the context of the *Almagest* and made to do duty elsewhere. The history of this detachment begins, at least for the Latin treatises with which we shall be concerned, with the elementary astronomy presented in Roman handbooks. From there, these detached notions or "devices" were reworked and modified, at times with a passing reference to things Arabic, at times not. The result, however, was much the same: an awareness and use of fragments of the Ptolemaic system without any knowledge of the system as a whole. Still, as fragmentary as this utilization of Ptolemaic notions was, it was a fruitful tool of explanation in the elementary works in which it was applied.

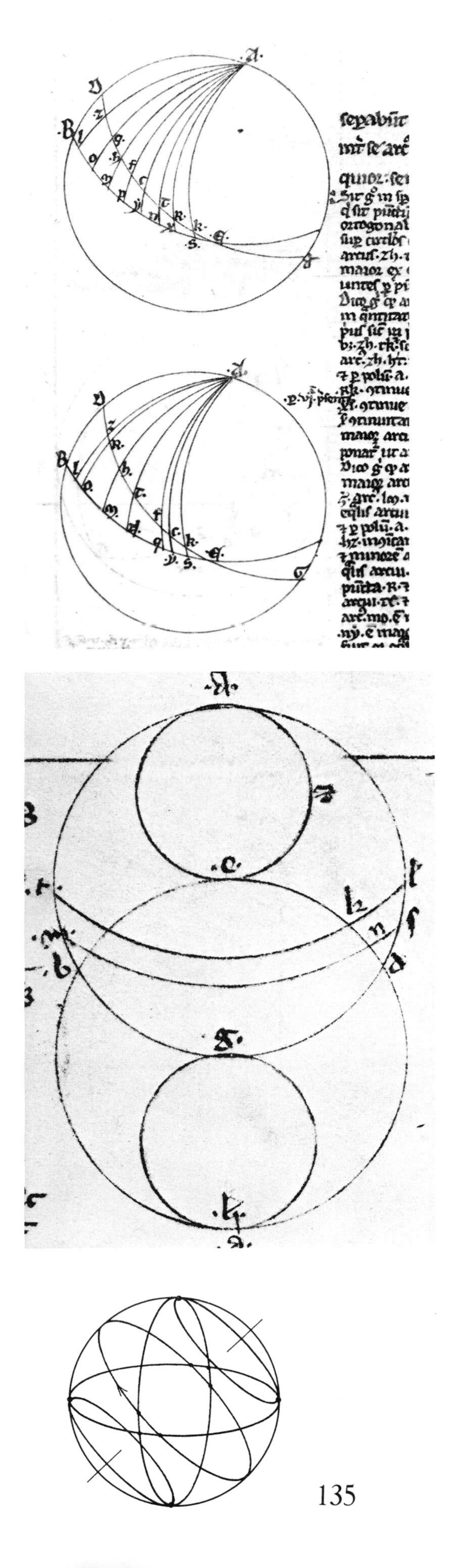

124. The Mathematics of the Celestial Sphere. The figures at the top right are taken from a thirteenth-century manuscript copy of Gerard of Cremona's Latin translation from the Arabic of the *Sphaerics* of Theodosius of Bithynia (second century B.C.). Although this treatise of Theodosius is later than the second of the two works pictured here—*On the Moving Sphere* by the fourth-century B.C. Autolycus of Pitane—the geometry of the sphere it summarizes most likely derives from a time nearly contemporary with Autolycus. The whole of Book I and a good part of Book II of the *Sphaerics* treat aspects of spherical geometry that are not as keyed to the interests of astronomy as the remainder of this three-book work. Yet even in this later portion of Theodosius' work, although many of the elements treated have obvious astronomical relevance, they are not assigned their astronomical names, but simply parade in neutral geometrical garb.

Thus, in the proposition (III, 9) relevant to the figures pictured here, *BE* is an arc belonging to a circle simply described as a great circle on the sphere which is the greatest of the parallel circles relative to the pole *A*, while *DG* belongs to a great circle oblique to these parallel circles. It is evident, however, that *DG* is part of the circle of the ecliptic, while *BE* belongs to the circle of the celestial equator, though the relevance goes unmentioned. Further, the main concern of the present proposition—namely, the relation between the arcs formed on both of these circles by great circles passing through the pole *A*—can in a similar light be viewed as speaking of the relation between ecliptic and equatorial coordinates.

Theodosius' approach to the geometry of the sphere was characterized by John Philoponus (sixth century) as having little to do with "matter" and at the same time contrasted with that of Autolycus as more "physical," more related to "real" motions. Philoponus' estimate of the greater proximity of Autolycus to astronomical concerns is substantially correct and can be illustrated by the eighth proposition of his *On the Moving Sphere,* to which the diagram on the middle right belongs. It is taken from a fourteenth-century manuscript of Gerard of Cremona's Latin translation of Autolycus, and the burden of the proposition (the Arabic-Latin text here differing slightly from the Greek) is to show that for an observer not at the equator, any great circle which, like the horizon *(orizon),* touches the greatest parallel circle of always visible *(manifestum)* stars as well as that of always invisible *(occultum)* stars will, upon rotation of the celestial sphere, come to coincide with the horizon.

Philoponus' view of Autolycus' greater "physicality" is borne out by the fact that the circles in question are referred to by names immediately giving their astronomical relevance. Our interest, however, is in the figure: in place of its using three dimensions like the modern diagram on the lower right, all is flattened into two (as in some of the diagrams in Illustrations 121 and 122). The horizon is the large upper circle in the diagram at the middle right *(ABGD)*, the smaller circle *(AEZ)* within it being the greatest circle of always visible stars, that below and outside it *(GH)* the greatest circle of always invisible stars, while the lower larger circle *(BHD)* is that great circle which, upon rotation of the sphere, will come to coincide with the horizon. This two-dimensional arrangement has the important advantage of always picturing the greatest circles of visibility and invisibility as, respectively, inside and outside the horizon circle to which they are tangent; more than that, they do not intersect the horizon circle—which is observationally true—as they do in the modern three-dimensional figure on the lower right. For other diagrams from the two manuscripts represented here see Illustrations 97, 98; and 99, 120, 133, 139, 141, 149, and 238.

125. Epicycle and Eccentric. Two of the most important mathematical "devices" in the central astronomical treatise of antiquity, Ptolemy's *Almagest,* are the eccentric and the epicyclic models of planetary motion. The diagram pictured here at the top right, taken from a ninth-century Greek manuscript of the *Almagest,* concerns them both. Although these two models of motion were part of Greek astronomical endeavor many years before Ptolemy, they received their most extensive and sophisticated use in the hands of this second-century scholar, and it is from their appearance in his work that they became essential elements in the whole history of mathematical astronomy for nearly 1500 years.

The present figure relates to Ptolemy's proof of the equivalence of the eccentric and epicyclic models for the motion of the sun (*Almagest,* Book II, ch. 3). Understanding it will be facilitated if we again appeal to a modern diagram at the lower right (but this time to one that does not diverge from what we find in the text itself, since it corresponds in essentials to a figure appearing in Ptolemy a few pages before that given here from the ninth-century manuscript). Motions on all circles or orbits are uniform with respect to their centers, the sun at Z moving counterclockwise on the dotted circle with center K, which is distant by an amount e (the eccentricity) from D, the central earth and the center of the second large circle. Thus, the sun moves on a circle eccentric to the earth, being farthest from the earth at H (apogee) and closest at G (perigee). But the same motion of the sun can be accounted for by an epicycle (the small circle) with a radius r equal to the eccentricity e whose center B travels counterclockwise on a circle (called the deferent) with center at D, while the sun at Z moves clockwise about B with the same angular velocity as B does about D. In both the eccentric and epicyclic models, the sun's motion will be uniform about K, but nonuniform about D, a fact that corresponds to the observed variation in the sun's velocity (a variation that is evident from the unequal lengths of the seasons). However, in either model, from D the sun will appear at Z, thus reflecting the equivalence of the two models (although Ptolemy naturally proves this equivalence).

Turning again to the manuscript figure at the top right (where Greek letters correspond wherever possible to the Roman letters in the modern diagram), we have a geometric illustration of the equivalence of the two models when the eccentric circle is either larger or smaller than the deferent circle on which the epicycle travels. Only arcs of the two eccentrics are drawn, HΘ and ΛM. Yet even in these cases the eccentric–epicyclic equivalence is evident, for the sun appearing, respectively, at Θ or M falls on the same straight line ΔMZΘ.

The figure at the upper right relates only to the motion of the sun, but the eccentric and epicyclic models were capable of more general application. For example, in the epicyclic model for the sun, the center of the epicycle and the sun on the epicycle itself rotate in opposite directions but with the same angular velocity. Alternatively, the model for the moon requires that they be of differing velocities, while that for the planets requires that the rotations occur in the same direction. For other figures from the present manuscript, see Illustration 88.

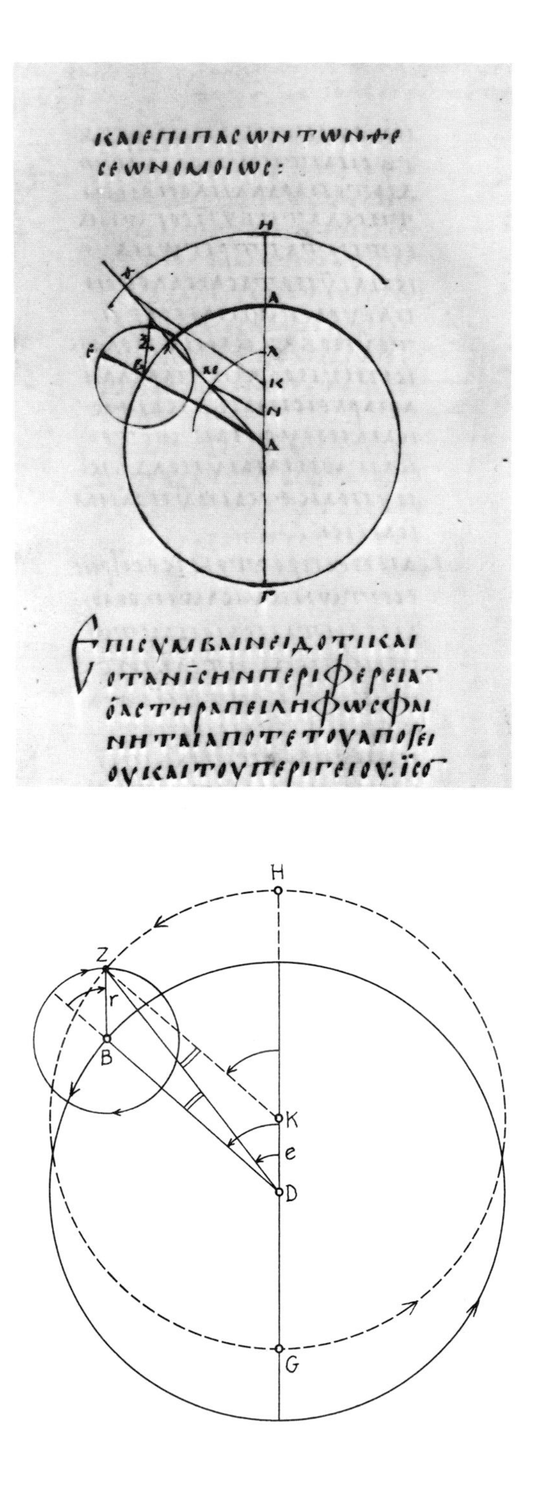

126. Physicalizing the System. One of Ptolemy's later works was his *Planetary Hypotheses,* in the second book (extant only in Arabic) of which he develops physical models for many of the geometrical mechanisms he had set forth in the *Almagest.* A version of one of these physical models is pictured at the right, not from Ptolemy, but from a thirteenth-century manuscript of al-Qazwīnī's *Wonders of Creation.* The three inscriptions in the small circle in the middle of the figure indicate that there are three centers involved in the figure (reading upward from the middle of the small circle): (1) the center of the earth, (2) the center of the turning sphere, (3) the center of the eccentric. Corresponding to these three centers, there are three pairs of concentric circles or spheres: (1) The small circle in the middle with the central earth and the outermost circle, together constituting what is here called the "universal sphere" (that is, the sphere involved in the daily rotation of the heavens). (2) The circle immediately outside, but eccentric to, the small central circle and the second outermost circle, constituting the "turning sphere." This particular sphere is required for the planet Mercury—to which the present figure refers—to account for the retrograde motion of the center of its deferent, an additional mechanism devised in the *Almagest* for this particular planet. (3) The two circles containing the small circle at the top, constituting what is here termed the "eccentric sphere." This final sphere corresponds to the deferent circle on which the center of the epicycle in question moves. Accordingly, the small uppermost circle within the sphere is the sphere of the epicycle *(falak al-tadwīr),* the small circle on its right being appropriately labeled "the body of the planet." The position directly above the sphere of the epicycle is properly indicated as the planet's apogee, while its perigee appears directly below the small central circle. The remaining inscriptions in the figure designate those parts which, respectively, contain or are contained within the various inner and outer spheres.

In Ptolemy's *Planetary Hypotheses,* each of the planetary "shells" here represented by Qazwīnī's spheres consists of some unspecified material, although Ptolemy himself does not appeal to whole spheres, but merely to "slices" of them, resulting in wheel-like bands. Although these nested spheres or wheels are designedly physical, they are not meant to provide an explanation of each individual planet's motion; that is accounted for in terms of a will or vital force ascribed to each planet. For other illustrations from the present manuscript of Qazwīnī, see Illustrations 52, 203, 208, and 253.

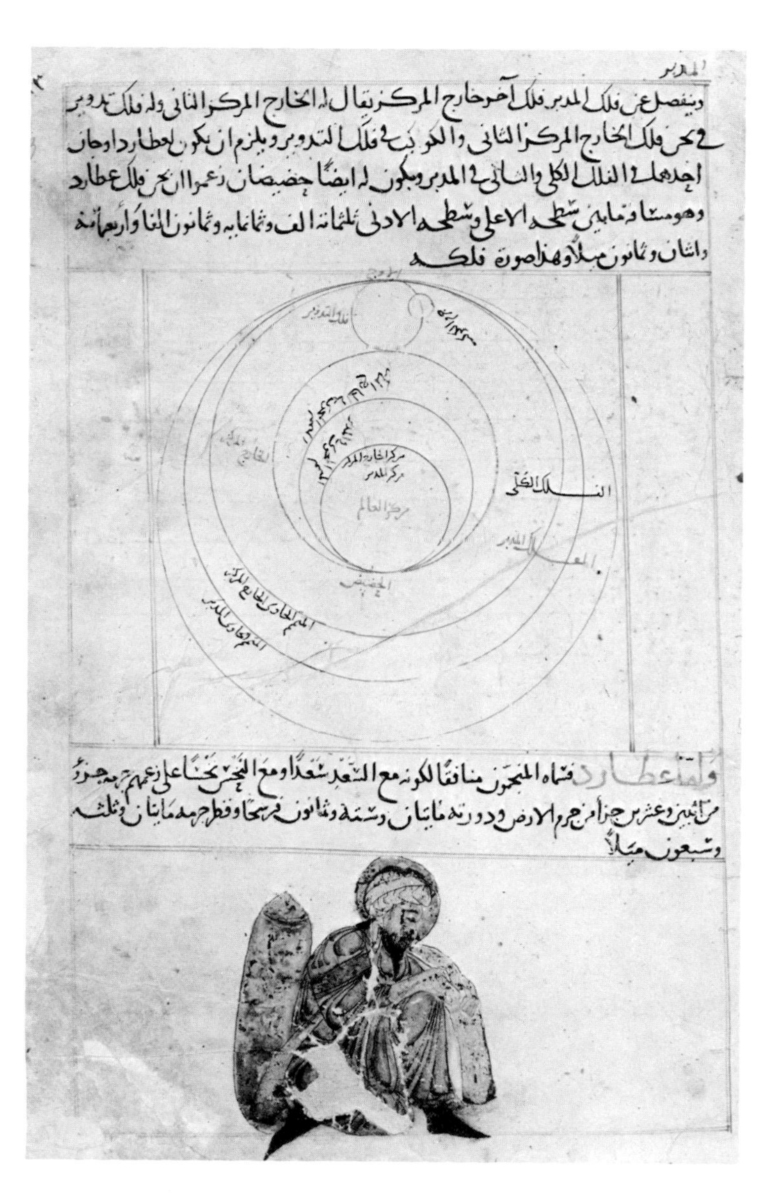

127. The Emendation of the Ptolemaic System. Medieval refinements of Ptolemy's accomplishments in mathematical astronomy belong almost entirely to Arabic, and not Latin, efforts. There were, for example, a number of changes introduced into the Ptolemaic models for the moon and the planets, among the most striking and important being that proposed by the thirteenth-century Persian philosopher and scientist Nāsir al-Dīn al-Ṭūsī. The significance of his innovation is such that it has become known among historians of astronomy as the "Ṭūsī-couple." It is represented here, first separately from a fifteenth-century manuscript of Ṭūsī's *al-Tadhkira fī ᶜilm al-hayᵓa (Memorandum of the Science of Astronomy)*, and then imbedded in a diagram of his lunar model from a fifteenth-century manuscript of a commentary of al-Bīrjandī (sixteenth century) on the *Tadhkira.*

The basic elements of the Ṭūsī-couple are two internally tangent coplanar circles, the diameter of one being half that of the other. This much is quite evident from the top figure on page 139. Both circles rotate uniformly in opposite directions, the angular velocity of the smaller circle being twice that of the larger. Given this, if the point of tangency of the two circles be taken as determining a given point of the smaller circle, then, following the motions of the two circles as stipulated, that given point will move back and forth on the straight line which is the diameter of the larger circle. This fact is incisively brought home by the four circles in the top figure on page 139. The "given point" is so labeled next to or within each of these circles, while the legends above them specify which motions result in the relative positions depicted of the two circles and the given point.

Thus, reading from right to left, the first legend tells us that its "picture" *(ṣūra)* represents the two circles at the beginning and informs us that the small circle will move to the right with respect to the viewer; the larger to the left. The next legend says that it pictures things after a half revolution of the small circle and a quarter revolution of the larger, the following two legends stipulating that the depicted positions are those after a whole revolution of the smaller circle and a half revolution of the larger and, lastly, after one and a half for the small, three-quarters for the large. The larger circle underneath the four above relates to Ṭūsī's proof that the given point in the smaller circle moves up and down the diameter of the larger.

It is significant that Ṭūsī has troubled first to explain and, particularly, to furnish a number of figures for, his device quite apart from its involvement in a model of the motion of some celestial body, a procedure that was not as frequent as one might expect in the early history of mathematical astronomy. However, only a few pages later he effects an application of his device to a lunar model, although the illustration at the bottom of page 139 is taken from a commentary on Ṭūsī's work.

In order to facilitate the understanding of this second illustration, it is helpful to realize that the phenomenon of the moon uncovered by Ptolemy but now known as evection required that the distance of the center of the moon's epicycle from the earth vary. Ptolemy accounted for this variation in distance by an eccentric deferent whose center moved on a "crank mechanism" about the earth which "pulls in" the epicycle (as the moon approaches quadrature).

Alternatively, the oscillating rectilinear motion effected by a Ṭūsī-couple could be used to account for this variation in distance. This is, in effect, what is done in our manuscript figure. Representing the path of the moon's epicycle for half a synodic month, the center of the whole figure is the center of the world. The four small concentric circles represent four different positions of the lunar epicycle (which here takes over the role played by the "given point" in the earlier diagrams), their changing distance from the center of the world being effected by a Ṭūsī-couple "attached" to a radius rotating about that center. Four positions of the Ṭūsī-couple are given, corresponding to the four positions of the epicycle. The path of the epicycle center is the large eccentric curve; clearly, it represents the required varying distance of the lunar epicycle from the earth.

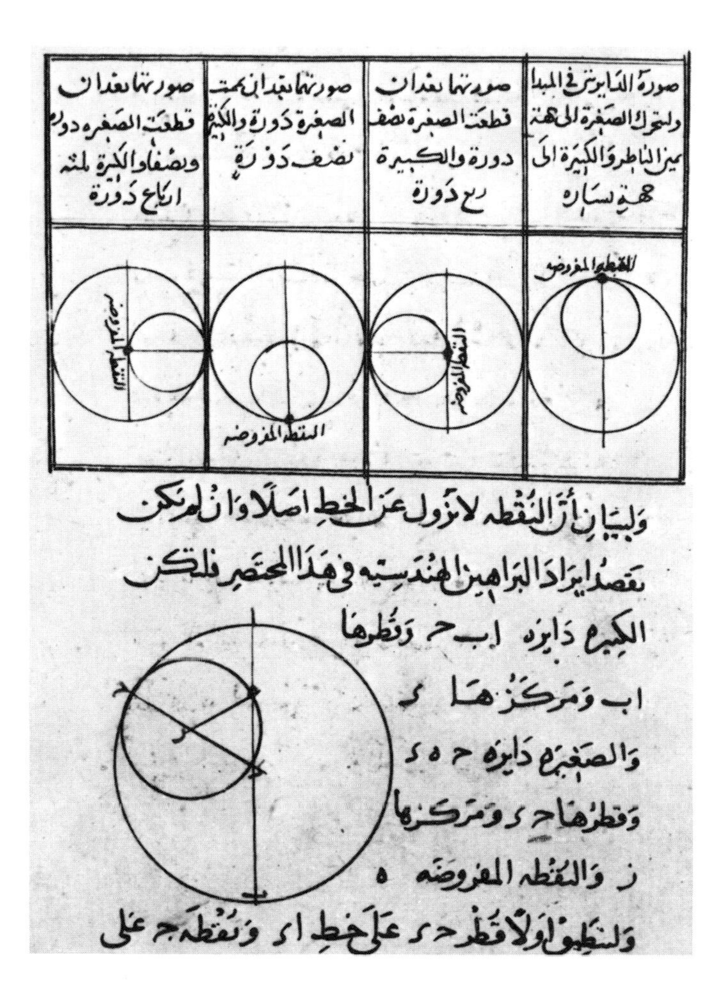

صورة الدايرتين في المبدا والتحرك الصغيرة الى جهة يمين الناظر والكبيرة الى جهة يساره	صورتهما بعد ان قطعت الصغيرة نصف دورة والكبيرة ربع دورة	صورتهما بعد ان تمت الصغيرة دورة والكبيرة نصف دورة	صورتهما بعد ان قطعت الصغيرة دورة ونصفا والكبيرة ثلثة ارباع دورة
النقطة المفروضة	النقطة المفروضة	النقطة المفروضة	النقطة المفروضة

ولبيان ان النقطه لا تزول عن الخط اصلا وان لم نكن
نقصد ايراد البراهين الهندسيه في هذا المختصر فلتكن
الكبيره دايره ا ب ج وقطرها
ا ب ومركزها د
والصغيره دايره ج ه د
وقطرها ج د ومركزها
ز والنقطه المفروضه ه
ولنطبق اولا قطر ج د على خط ا د ونقطة ج على

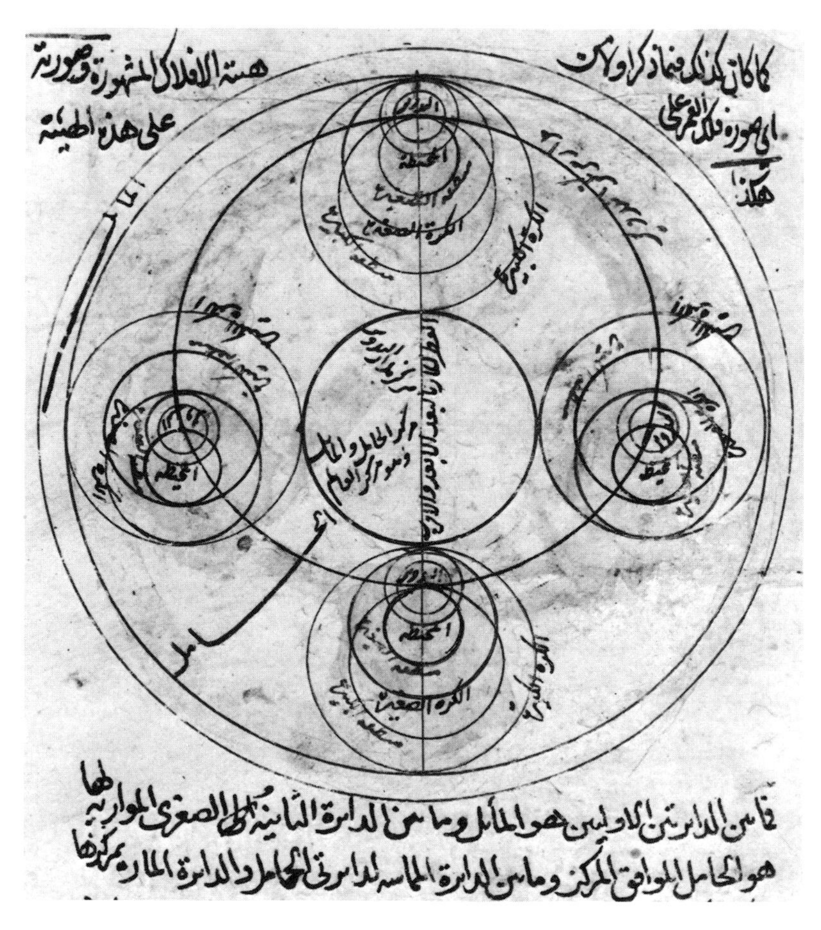

هيئة الافلاك المشهورة وصورتها
على هذه الهيئة
هكذا
فما بين الدائرتين الاوليين هو المائل وما بين الدائرة الثانية والصغرى الموازية لها
هو الحامل الموافق المركز وما بين الدائرة المماسة لدائرتي الحامل والدائرة المارة بمركزها

128. The Handbook Tradition in Astronomy I. The knowledge medieval Latin scholars had of Ptolemaic astronomy was overwhelmingly derived from extremely elementary "digests" of selected parts of that astronomy. The two most popular digests of this sort have furnished the two diagrams reproduced here. That on the top right is taken from a fifteenth-century manuscript of the *De sphaera* of John of Sacrobosco (thirteenth century), as a standard text in many later medieval universities undoubtedly the most commented upon of astronomical works. The first two chapters of this work explain the structure of the universe and the various circles or spherical coordinates needed in astronomy; the third treats such matters as the rising and setting of these signs, the variations in the lengths of days, and the seven climata; while the fourth and final chapter devotes but two pages to the motion of the sun, moon, and planets, and the causes of solar and lunar eclipses.

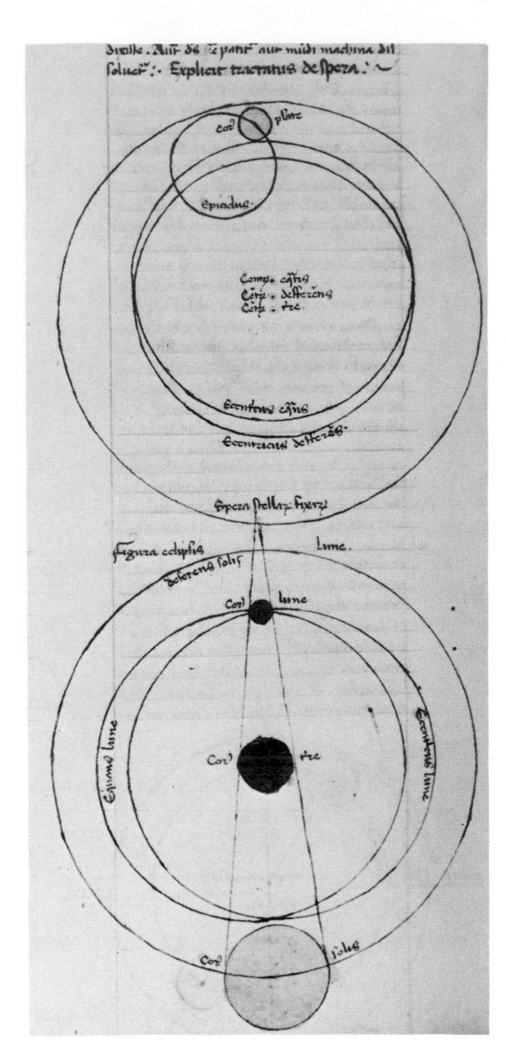

It is at the end of this chapter that the present illustration appears. The lower figure in this diagram is clearly that of a lunar eclipse. That above represents the Ptolemaic model for the motion of a superior planet. The outermost circle is specified as the sphere of the fixed stars, its center *(centrum terre)* being the lowest of the three indicated in the center of the figure. Above the *centrum terre* we find the *centrum deferentis,* namely, the center of the deferent circle on which the epicycle travels. Above that we have the *centrum equans,* that is, the equant point used by Ptolemy in his explanation of planetary motion. There, the center of the epicycle moves on the deferent circle eccentric to the earth, but moves uniformly not about the center of the deferent, but about the equant point, the distance between the equant and the deferent center being the same as that between the latter and the earth. The circles belonging to the first two of these centers are here labeled *eccentricus equans* and *eccentricus deferens.* The epicycle above—which, unlike that for the moon or the sun, rotates in the same direction as its center on the deferent—is also appropriately inscribed, as is the *corpus planete* it carries.

The figure at the bottom right comes from a fourteenth-century manuscript of the *Theorica planetarum,* a thirteenth-century work often erroneously ascribed to the twelfth-century translator Gerard of Cremona, but, if identifiable at all, is possibly the work of Gerard of Sabbioneta (thirteenth century). Whoever the author may have been, it seems clear that this work was written to supplement the extremely inadequate treatment of planetary theory given in the final, fourth chapter of Sacrobosco's *De sphaera.* The fact that so many of its manuscript copies—and there are hundreds of them extant—are joined to the *De sphaera* shows that it functioned most successfully as a supplement.

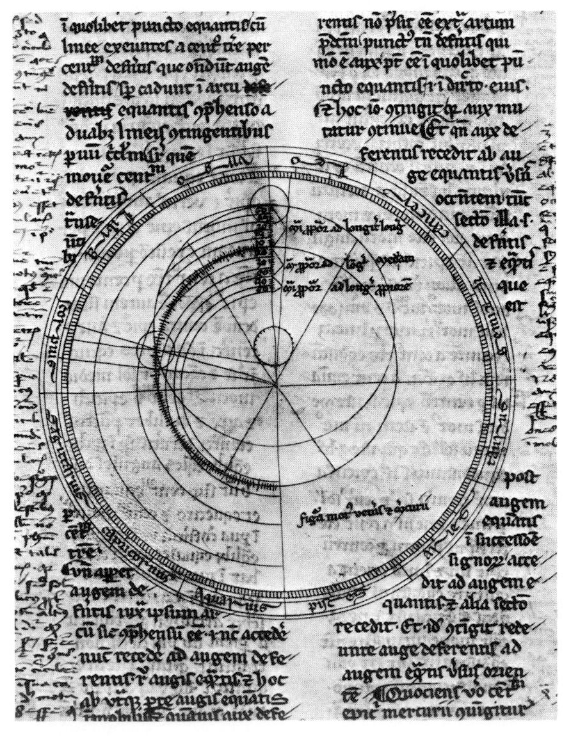

The words in the lower right quadrant of the *Theorica* diagram given here announce that it is a figure of the motion of Venus and Mercury. The epicycle and deferent for each planet are clearly depicted, the small circle in the center representing the added mechanism needed for Mercury (the center of its deferent is not stationary, but moves on this small circle). Three different sets of proportional minutes *(minuta proportionalia)* are indicated and numbered in the top half of the diagram; their function was to enable one to calculate the angle subtended by the planet and the center of the epicycle relative to an observer on the earth (this angle was termed the *equatio argumenti* in medieval astronomy). Finally, note should be made of the inscription of the names of the zodiacal signs on the circumference of the figure.

129. The Handbook Tradition in Astronomy II. Figures for the various models of planetary motion in the Ptolemaic system were not meant to be gathered together in a single all-embracing diagram, yet that is precisely what is done in this thirteenth-century manuscript of a variant anonymous *Theorica planetarum.* Two centers serve the whole diagram: the center of the world *(centrum mundi)* and, directly above it, the center of the eccentrics *(centrum eccentricorum).* Each planet, from the moon through Saturn, has one circle assigned to it relative to each of these centers (those serving as eccentric deferents being labeled on the lower right of the figure). The space between two consecutive concentric circles is that limiting the eccentric deferent of a given planet and is appropriately named "the sphere" of that planet (a hint, it would seem, of a physicalized system such as that in Illustration 126). The apogee and perigee of each planet are indicated by the terms, respectively, 'longior longitudo' and 'proprior longitudo,' while the relevant epicycles and the planets carried by them are also named. All of this leaves one circle unaccounted for: the smallest circle concentric to the *centrum mundi;* it is the "innermost surface of the sphere of the moon" *(intima superficies spere lune).*

Of course, this "total picture" diagram is of no use at all in bringing one into contact with the mathematics of Ptolemaic astronomy. Apart from the problem of drawing an appropriate figure to scale, that depicted here leaves out most of the relevant geometry and misrepresents a good deal of what it tries to cover (the same eccentricity is given for all planets, which is just the opposite of what is in fact the case). Yet in spite of all these difficulties and mistakes, it was deemed worthwhile to have such a diagram. Presumably it was thought to be a way of facilitating the understanding of Ptolemy, at least to the extent required or desired by some medieval scholars.

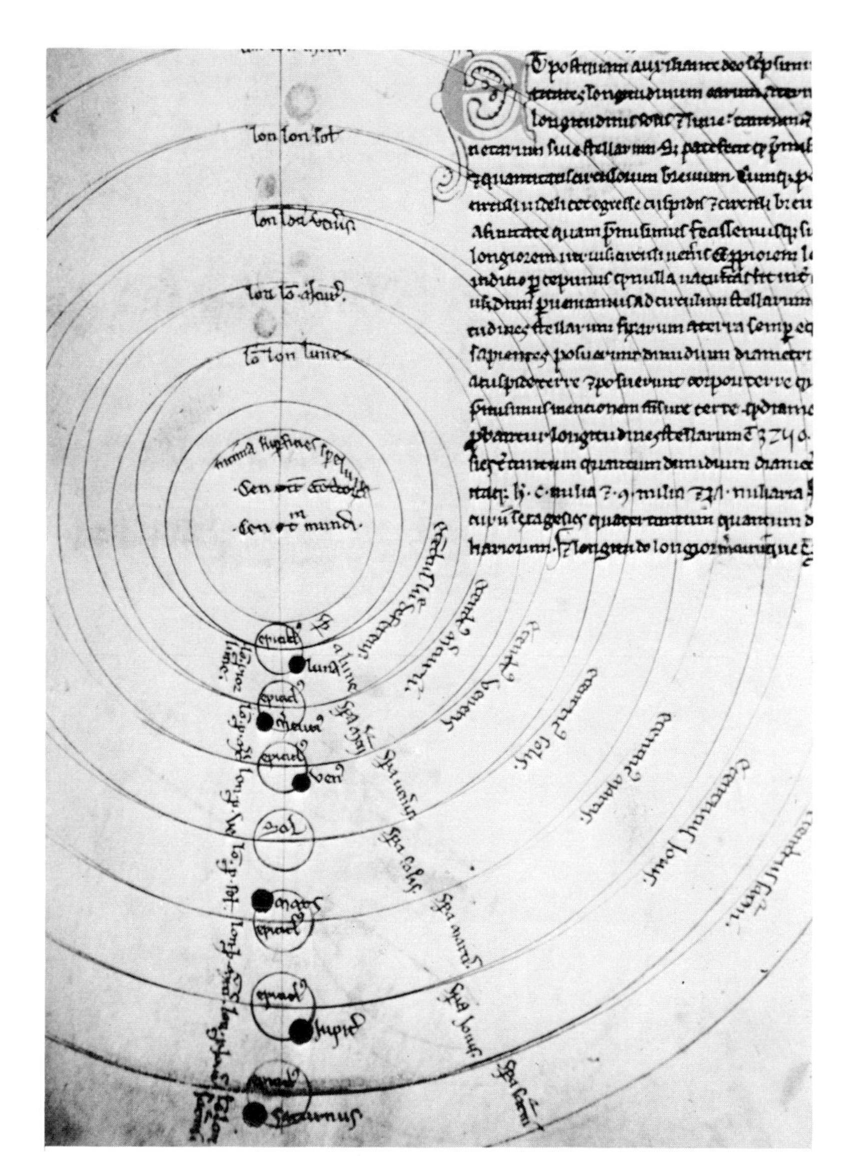

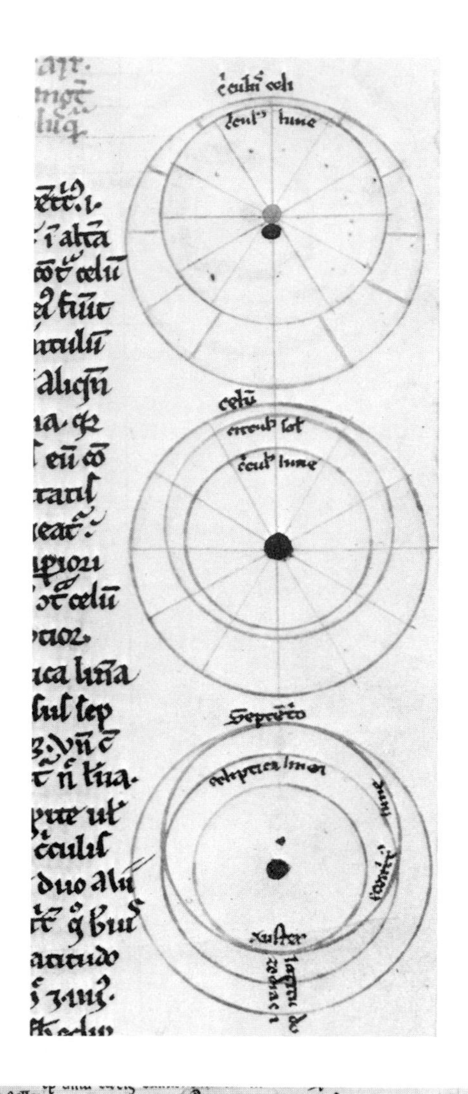

130. Aiding and Adding by Figures. From the eleventh century on, computus treatises devoted to the calculation of the date of Easter and other movable religious feasts began to contain more and more information pertinent to the heavenly motions connected with the variation in these dates. At the same time, they frequently had marginal figures which, if not required for the understanding of the text, certainly provided a visual aid to that end.

The three circular figures at the top left, taken from a late twelfth- or early thirteenth-century manuscript of the *Computus* of the twelfth-century English scholar Roger of Hereford, are an instance of this procedure. The particular text being "aided" is concerned with the inequalities of lunar motion. Thus, the top figure in the diagram reveals the eccentricity of the "circle of the moon" with that of the whole heavens; that directly below it, the eccentricity of the circles of both the moon and the sun to the heavens and to each other. The bottom figure in the same diagram exhibits the motion of the moon in latitude. The innermost and outermost concentric circles represent the limits of the latitude of the zodiac, the central concentric circle being labeled the line of the ecliptic. The inscriptions *septentrio* and *auster* refer to the north and south limits of the zodiacal band. The single eccentric circle in the diagram shows how the moon moves through the latitude of that band. (For other figures representing the motion of the moon and other planets in latitude, see Illustrations 249–250.)

The second set of figures at the bottom left comes from a twelfth- or thirteenth-century manuscript of a treatise entitled *De recta imaginatione spere* usually ascribed to Thābit ibn Qurra (ninth century), although no clear determination has yet been made whether this work is a translation of an Arabic original by that scholar. As a whole, this extremely brief treatise is simply concerned with the description and determination of some of the pertinent coordinates on the celestial sphere. The present figures treat of a matter omitted in the text, but apparently significant enough in the eyes of one scholar or scribe to merit a marginal addition: namely, the distinction between terrestrial and celestial longitudes and latitudes. The figure on the right represents the celestial case, the ecliptic being clearly drawn with ten perpendiculars, each bearing the label 'latitudo.' The text immediately to the left tells us that the longitude of a star is only measured according to the path of the sun; that is, according to the ecliptic or zodiac, beginning from the sign Aries, while stellar latitude is the perpendicular distance from the zodiac. The text at the left in the diagram defines terrestrial *(civitatis)* longitude as the distance from the east or west and terrestrial latitude as that from the equator *(linea equinoctialis)*. The text within the circular figure to which these definitions refer points out that longitudes are there represented by black lines, latitudes by red, while the upper "hatched" *(lineata)* semicircle is the habitable part of the earth; that below, the uninhabitable.

The word 'arim' over the center of the equator refers to a mythical city on the equator, at the midpoint between east and west, which was often regarded as the center of the surface of the earth and, in some works of Arabic astronomy influenced by Hindu sources, at times functioned as the zero meridian. For another figure from the first of the two manuscripts utilized here, see Illustration 252.

131. The Fortuna of Eccentrics. The two fundamental elements, epicycles and eccentrics, in Ptolemy's geometric models for planetary motion often found their way into places in which very little, if any, mathematical astronomy was at stake. The two present figures are examples of such transmigration on behalf of eccentrics. The first, at the upper right, whose circular frame bears the names of the zodiacal signs, is taken from a slim anonymous thirteenth-century handbook of elementary philosophy, cosmography, and astronomy. The function of this figure is the visual explanation of a text (above the figure) in Bede's *De natura rerum* which consists of excerpts from Pliny's *Natural History* (II, 63–64). The topic is the apsides of the planets, here understood as their orbits, and, in particular, the zodiacal signs relative to which the apogees of these eccentric orbits (here called the *absides altissime*) occur.

Unfortunately, the figure is more decorative than accurate. Within a slight margin of error, the apogees are actually drawn at the points on the eccentric circles at which the names of the planets and straight lines from the central earth appear. But these positions do not always match the zodiacal signs stipulated in the text for the *absides altissime.* Thus, the apogee of Mars is to be opposite Leo, not Aquarius; that of Mercury opposite Capricorn, not Aries; that of the moon—missing in this manuscript copy of the text—facing Taurus, not Leo. The text surrounding the figure (ultimately from Pliny, II, 32–44) speaks of the times of the planetary orbits (see Illustrations 48–49) and is not relevant to the present diagram.

The second figure pictured at the lower right was attached to the beginning of a fourteenth-century manuscript containing Greek dramatic and other literary texts. A simple circular illustration of the signs of the zodiac, the interior eccentric circle appears to represent the orbit of the sun, here as the furry-looking figure at the left. In this case, the eccentric is not at all required for the matter at hand; instead, its mere familiarity would seem to have accounted for its presence. For other figures from the first of the manuscripts depicted here, see Illustrations 50, 251, and 280.

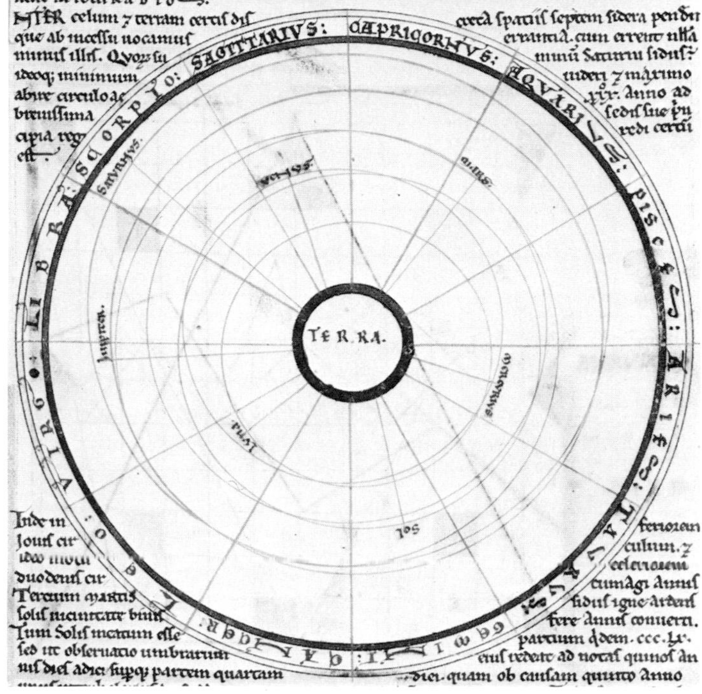

132. The Fortuna of Epicycles. Figures of eccentric orbits were not alone in finding application in all manner of elementary handbooks; in this they were matched, if not exceeded, by the drawing of epicycles. Two twelfth-century Latin works here provide instructive examples of this latter proliferation. That at the bottom left comes from a thirteenth-century manuscript of a treatise dealing with the nature of sublunar and celestial bodies *(Liber de naturis inferiorum et superiorum)* written by the twelfth-century Englishman Daniel of Morley. The subject is the "circles of the planets," which, Daniel claims, is something better learned from Arabic than from Latin sources. His elementary description does not use the terms 'deferent' and 'epicycle,' but employs 'greater circle' *(maior circulus)* in place of the former and 'small' or 'advancing circle' *(circulus brevis vel promovens)* for the latter, although his first choice to designate the epicycle is the Arabic 'elchedwir' (see the *falak al-tadwīr* in Illustration 126) and it is this which is duly inscribed in the present figure.

Using Saturn as an example, Daniel of Morley goes on to explain that the planetary stations occur where the epicycle intersects the greater circle, the course of the planet undergoing *processio* on the epicyclic arc above the greater circle and *retrogradatio* on the arc beneath. Unfortunately, the person who drew our figure did not follow the textual instructions with much accuracy, and has written the words 'statio,' 'progressio' (or 'precessio'), and 'retrogradatio' at places relative to the diagram that are erroneous as often as they are correct.

The figure at the bottom right (page 144) and the one at the top left (page 145) appear on successive pages in a very decorative thirteenth-century manuscript of the *Dragmaticon philosophiae* of the twelfth-century Neoplatonist scholar Guillaume de Conches. The first (page 144, right) provides a model case history of getting things only half right. As in the figure from Daniel of Morley, the subject is the motion of Saturn. With a rather thick zodiac band bearing the names of the signs, there is a central earth with Saturn's eccentric circle drawn—though with precious little eccentricity—about it. The four smaller circles are supposed to represent four positions of Saturn on its epicycle, the epicycle itself unfortunately omitted from the illustration. In a small circle at the left we have *Saturnus in prima statione,* on the right *in secunda statione.* The remaining two small circles tell us that between these stations Saturn is retrograde (here erroneously spelled *rectigradus*). But why should our figure characterize its motion as retrograde both when it is ascending on its epicycle and when it is descending? Although the text in question certainly does not demand such a mistake, it carries the germ for making

(Continued)

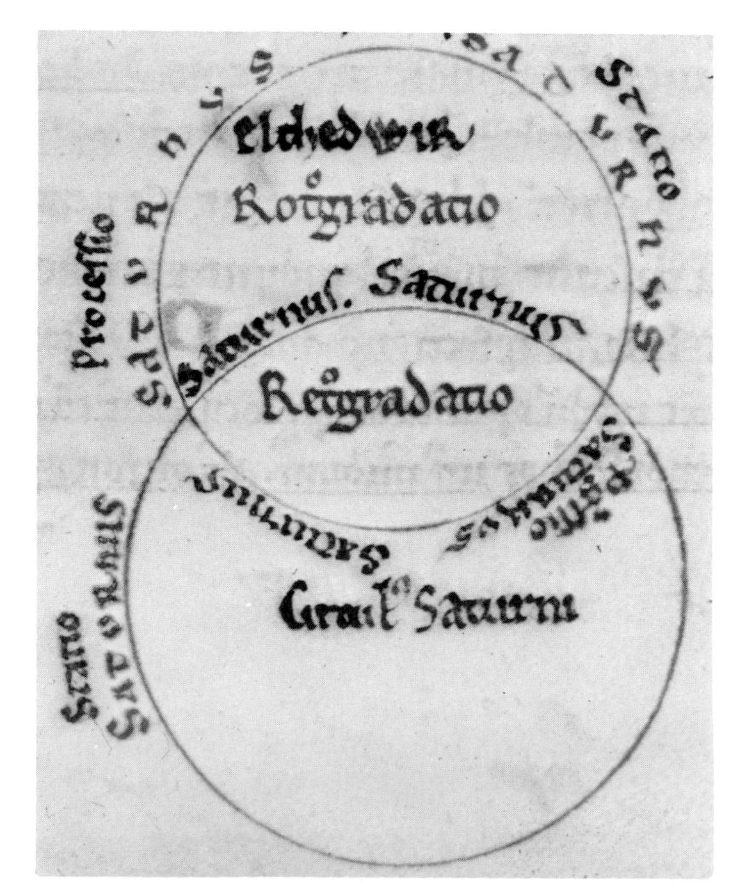

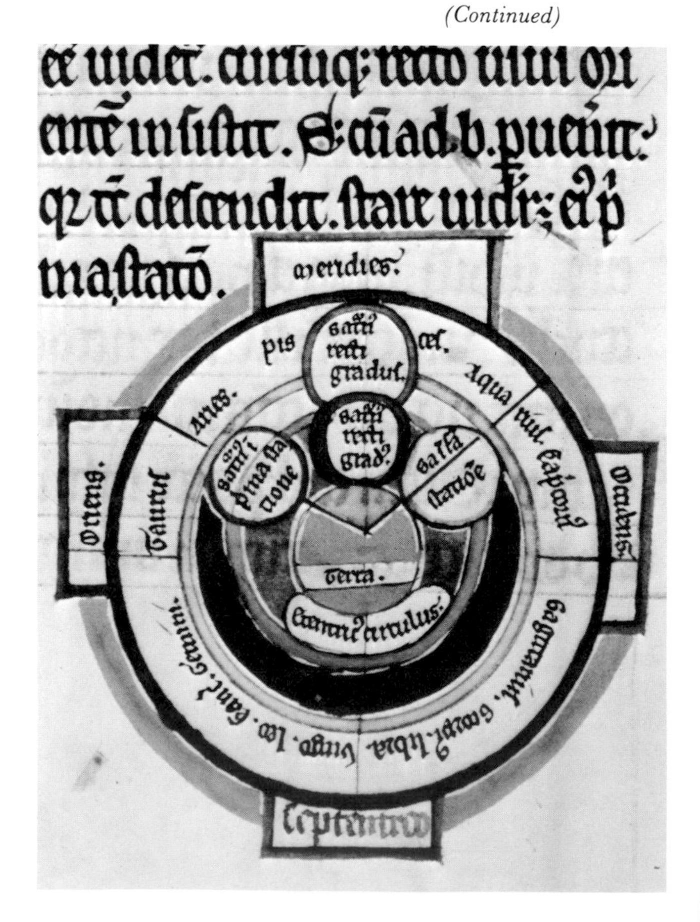

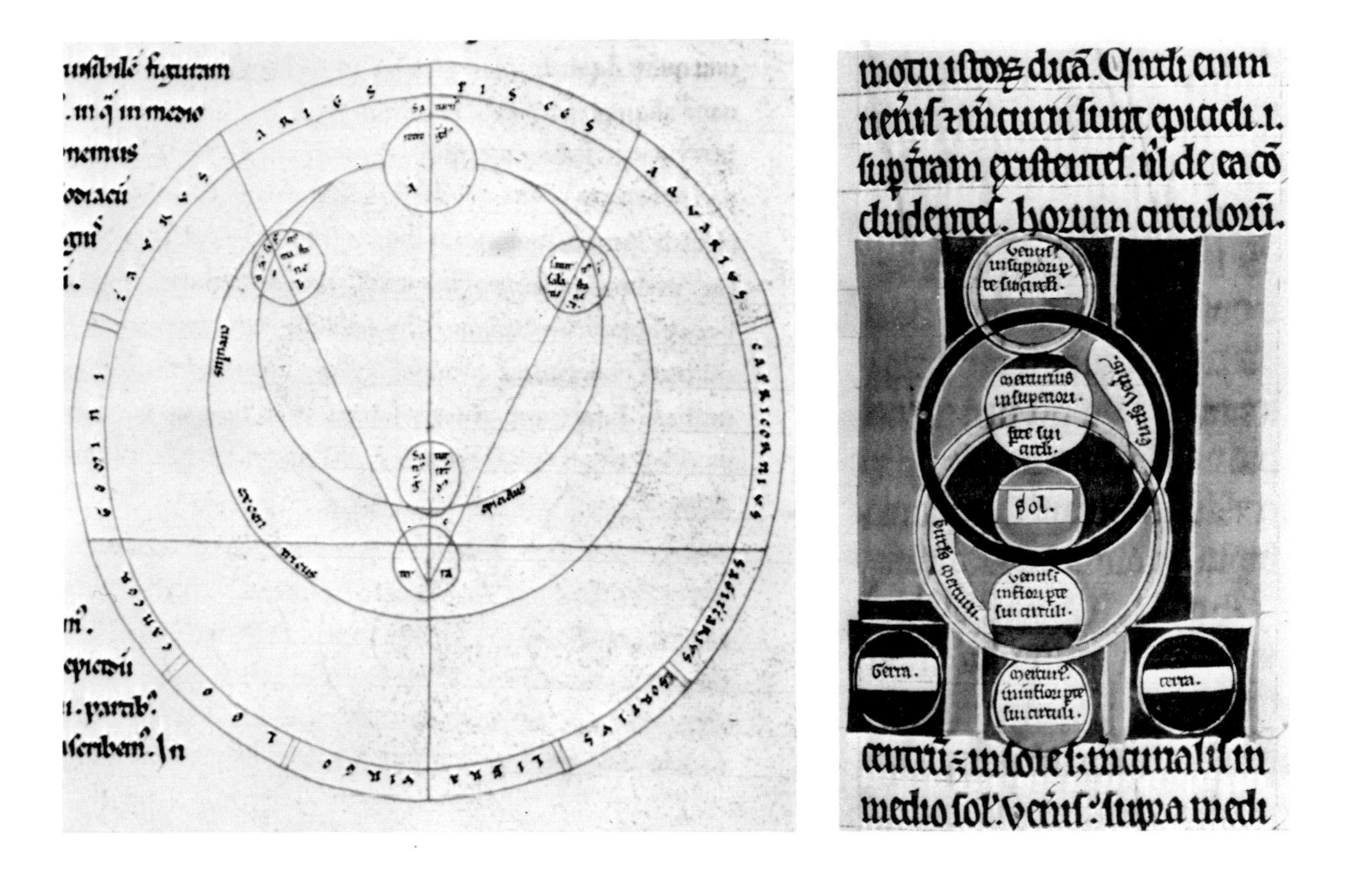

it. For in speaking of the motion of Saturn on the upper part of its epicycle, it says that its path will be directly toward the east *(cursuque recto versus orientem insistit)*, and it is a plausible hypothesis that the word 'recto' in this text led to a misunderstood 'retro' in the figure. Finally, one should note that the stations in the figure are misaligned with the zodiacal signs specified in the text. As a whole, then, the major merit of the figure lies in the striking colors in which it is drawn and not in its instructional value.

At the top left of page 145, the very same figure is taken from another manuscript (fourteenth century) of the *Dragmaticon*. It fares slightly better. Although we again have 'retrogradus'—here properly spelled—for both ascendant and descendant motions, the positions of the stations relative to the zodiacal signs do correspond to those given in the text, the eccentricity of the deferent circle is marked enough to be obvious, and the epicycle itself is drawn (although not with its center on the deferent, but internally tangent to it).

The diagram at the top right on page 145 from the thirteenth-century Guillaume de Conches manuscript employs epicycles for purposes other than Ptolemaic. Evidently elaborating on what he had found in Macrobius (*In somnium Scipionis,* ch. 19), Guillaume has interpreted his statement that the Egyptians held that the sphere of the sun is encircled by the spheres of Mercury and Venus to mean that their spheres are epicycles about the sun which include no part of the earth within their orbit. The center of Mercury's epicycle is that of the sun itself, while that of Venus is above and eccentric to the sun, something that Guillaume explicitly hopes to make clear by a diagram *(quod ut melius intelligatur, figuram depingam)*. Thus, here the upper and darker larger circle is the epicycle of Venus; that below, the epicycle of Mercury; their relation to the sun accurately reflecting the claims of the text. The earth, outside, is represented twice, at the lower left and right. Venus and Mercury themselves are each depicted in two positions on their epicycles: superior and inferior *(in summo et imo)*. Guillaume adds that when both planets are in the former position, Mercury is closer to the sun, the contrary obtaining when both are at their inferior points. This, too, is evident from our figure, if not made explicit by appropriate labels. For other illustrations from this manuscript of Guillaume de Conches, see Illustrations 250 and 285.

CHAPTER THIRTEEN

Geometric Diagrams in Other Scientiae Mediae *and Natural Philosophy*

IN HIS DISCUSSION IN the second chapter of Book II of the *Physics* of how the natural philosopher differs from the mathematician, Aristotle noted that certain of the mathematical sciences were more physical—namely, optics, harmonics, and astronomy. In other places, he filled out the relation of at least two of these sciences to mathematics by claiming that harmonics was subordinate to arithmetic, as optics was to geometry, adding that mechanics was also subaltern to geometry.

In the Middle Ages, both Arabic and Latin, Aristotle's view of this matter appeared not only in discussions about the division or classification of the sciences—al-Fārābī (tenth century), for example, held that optics treated the same things as geometry, but in a different way—but also in the opening paragraphs of treatises devoted to these subaltern sciences as such. One of the central "textbooks" written in the thirteenth century on the science of weights begins by saying that this science is subordinate to both geometry and natural philosophy and that, as a result, the proof of some things will proceed geometrically, that of others philosophically. Alternatively, John Pecham (thirteenth century), author of the most used compendium of optics, tells us that in this science glory *(gloria)* is found physically as well as mathematically, since it is "adorned by flowers" from both these "higher" sciences.

Each of these remarks reflects what was to become almost a commonplace in the latter half of the thirteenth century: that optics, the science of weights, harmonics, and astronomy were all *scientiae mediae* falling between mathematics and natural philosophy. Inasmuch as diagrams relevant to harmonics and astronomy have already been dealt with in earlier chapters, the two "middle sciences" awaiting present attention are those concerned with optics and weights, to which will be added a third "discipline," that known as the "latitude of forms," since in the late Middle Ages it, too, was considered a *scientia media.*

Naturally, it is not possible to survey the content of each of these sciences from antiquity through the Middle Ages, but some indication should be given of the most significant works from which the following diagrams are drawn. In all instances there was a Greek "background" to the science in question. In terms of works that were translated into Arabic and Latin, this is perhaps most striking in the area of optics. Although philosophical and medical writings contained important material on the nature of light and the physiology of vision, Greek optics proper was largely geometrical optics. Undoubtedly there was already activity in this area in the fourth century B.C., but the first important extant work is that of Euclid (fl. ca. 295 B.C.): his *Optics,* in both an original and a later recension, as well as a treatise on mirrors *(Catoptrics)* which came to be attributed to Euclid, although it is clearly not his work. We also possess a *Catoptrics* (extant only in Latin) written by Hero of Alexandria (first century), but it is the *Optics* of Ptolemy (second century) that is clearly the crowning ancient achievement in this field. Ptolemy's work added empirical investigation (especially in the area of binocular vision) to the geometrical procedures that were characteristic of the Euclidean treatises, and it is unfortunate that we possess that work only in a twelfth-century Latin translation made from the Arabic, and even then without its beginning and end.

Euclid and, especially, Ptolemy were fundamental sources for optical writings by Arab scholars, among which one work stands out above all others: the *Kitāb al-manāẓir (Book on Optics)* of the eleventh-century Iraqi, Ibn al-Haytham, known to the Latins as Alhazen. Continuing the combination of the empirical and the mathematical that he found in Ptolemy, Ibn al-Haytham's massive, seven-book work was more comprehensive than any previous treatise in optics, covering a general theory of light and vision, a theory of visual perception, binocular vision, errors of vision, and all manner of problems involved in reflection and refraction.

This work attracted little attention in Islam before the opening years of the fourteenth century, when an elaborate commentary upon it was composed by the Persian Kamāl al-Dīn al-Fārisī (who, like Ibn al-Haytham, also wrote other, smaller works on optical topics). But the major impact of Ibn al-Haytham's work was not among later Arab scholars, but in the Latin West. There, its translation about the year 1200 provided a major source for subsequent speculation, not simply in the Middle Ages, but for later scholars such as Kepler, Snell, Fermat, and others. Most notable in the medieval period of its influence is the ten-book *Perspectiva* of the thirteenth-century Polish scholar Witelo, which to a great extent was a development of much to be found in the seven-book production of his Arabic predecessor.

Yet neither of these voluminous *magna opera* in optics came to be the most utilized of works in that *scientia media*. As to be expected, a more compendious survey served as the standard text, in this case the *Perspectiva communis* of the thirteenth-century Franciscan John Pecham. Later medieval commentaries *de perspectiva* almost invariably relate to this work. Other thirteenth-century authors—such as Robert Grosseteste and Roger Bacon—also paid considerable attention to optical matters, but their influence nowhere matched the extent and consistency of Pecham's.

The Greek tradition in what we now regard as statics, but in medieval terms was called the "science of weights" *(scientia de ponderibus)*, has a history that is quite different from the corresponding tradition in optics. Greek works on statics exhibited two essentially different manners of approaching and treating problems within that discipline. The first was exemplified by treating weights in equilibrium in terms of dynamical considerations; namely, in terms of the velocities weights would have if a balance in equilibrium were set in motion (the velocities interpreted in terms of the arcs swept out in equal times), no strict mathematical proof of the "law of the lever" in question being given. The second way of treating the same problem did give such a mathematical proof, one deriving solely from statical considerations based on the concept of the center of gravity.

The first manner of approach is found in the Pseudo-Aristotelian *Mechanical Problems* (fourth century B.C.); the second received its most sophisticated application in Archimedes' *On the Equilibrium of Planes* (third century B.C.). But neither of these two works appears to have been translated into Arabic, and only that of Archimedes was put into Latin, but then too late (1269) to have been effective in the formation of the medieval Latin tradition in the science of weights. Nevertheless, medieval statics in both its Arabic and Latin dress reveals an integration of the two manners of approach exemplified by Pseudo-Aristotle and Archimedes, a factor made possible by the translation into Arabic or Latin of other Greek treatises (now no longer extant in the original Greek) that embodied these two approaches.

The figures dealing with statics that are reproduced below are all drawn from the thirteenth-century medieval Latin corpus of works on the science of weights associated with the name of Jordanus de Nemore. Unfortunately, the precise identification of Jordanus has hitherto eluded historians and we can with reasonable confidence ascribe only one work to Jordanus—whoever he may have been—among those which traveled under his name. Still, these Jordanian treatises as a whole formed the nucleus for the study and development of *scientia de ponderibus* throughout the Latin Middle Ages.

The Greek background to the final discipline to be considered in this chapter—the doctrine of the latitude of forms—is much less extensive and decisive than in the case of optics or statics. To be sure, Aristotle had noted that qualities—heat or color, for example—could "admit of a more and a less" (that is, could be more or less intense in a given subject). Moreover, some of the Greek commentators on Aristotle (who were not translated for the Latin West to behold) had developed the idea of the range or latitude of such qualitative variations, and similar consideration of such latitudes can be found in Greek and Arabic medical writings.

But the most extensive and significant work in this area was a medieval Latin enterprise of the thirteenth and fourteenth centuries. The first phase of this enterprise concerned itself with just *how* qualities or forms could so vary in intensity over a latitude. Yet whatever answer one may have given to such a question, there remained the problem of how one should best *measure* such a variation, primarily a variation in the intensity of a quality, not over time, but as distributed over a given subject in extension. One answer to this second question was to utilize geometry for the measurement in question, and it is that "answer" whose diagrams make up the concluding Illustrations of this chapter.

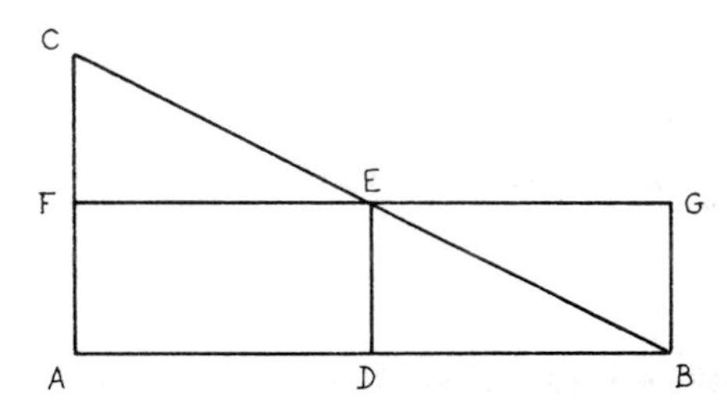

A few preliminary words related to a modern diagram of the latitude of forms will be helpful in understanding these Illustrations. The line *AB* in the modern diagram is to be considered the extension or subject line (the subject, therefore, being considered one-dimensional). All perpendiculars to *AB* are to be considered as variant intensities of a given quality or form (heat, let us say). Thus, the intensity *CA* at point *A* of our subject is twice that of *GB* at point *B*. Correspondingly, the rectangle *ABGF* represents the subject as *uniformly* hot throughout in degree or intensity *GB*. On the other hand, the figure *ABGEC* represents the subject as *difformly* hot.

A more interesting and fruitful case of difformity is given by the triangle *ABC,* which represents the subject as *uniformly difformly* hot, from zero degree at point *B* to degree *CA* at point *A*. But we can measure this same uniform difformity in a more informative way: the subject with varying degrees of heat as represented by the triangle *ABC* is just as hot as—contains the same amount of heat as—if the same subject *AB* were *uniformly* hot in the degree *ED* mean between zero and *CA*. This is established geometrically by the fact that the triangle *ABC* equals the rectangle *ABGF*. Such a procedure is to measure uniformly difform qualities by their "mean degrees."

The same technique was applied when the "quality" in question was motion. Then, however, the subject line represented time, and the triangle *ABC* the case of a body accelerating uniformly (undergoing uniformly difform motion) from rest to degree *CA*. Applying the foregoing equivalence, the geometry involved allows us to state that a body accelerating uniformly from rest to some determinate velocity traverses just as much space as would be traversed if it were to move over the same time with a uniform velocity equal to its mean degree. This is a statement of what is known in the history of medieval science as the mean-speed theorem, something that Galileo was to put to his own use in his investigation of naturally accelerated motion.

The fourteenth-century medieval geometry of the latitude of forms received its most brilliant development at the hands of the Parisian scholar Nicole Oresme, an elementary version of some of his contentions being produced subsequently in Jacobus de Sancto Martino's *De latitudinibus formarum*. It must not be thought, however, that the employment of geometry was the only way qualitative variations were measured. Such measurement was made before Oresme without using geometry by such Oxford scholars as William Heytesbury and Richard Swineshead, the latter of whom is represented in several of the figures given in this chapter because others interpreted parts of his work by geometrical means.

Finally, one should realize that the latitude of forms was brought to bear in measuring things other than qualities and motions. The perfections of radically distinct species (as a whole forming a *latitudo entium*) were so measured, as were such medical notions as the latitude of health. These, too, are represented below.

Mathematical and geometrical figures were also appealed to in areas of natural philosophy other than the latitude of forms, but these will be sampled in Chapter 23. Figures relevant to the latitude of forms can be seen in Illustration 13, in addition to those appearing in this chapter.

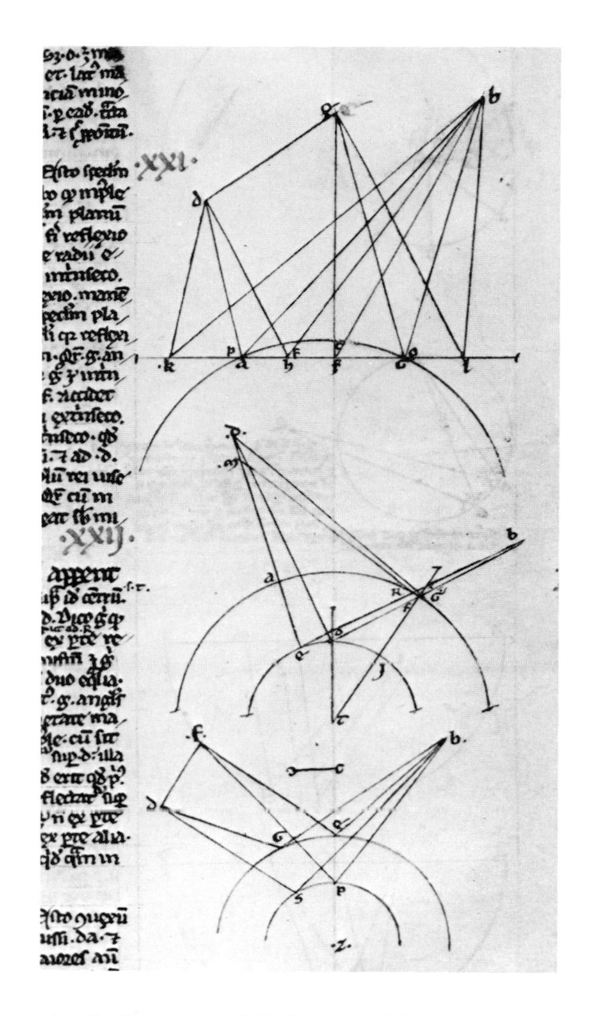

133. Reflection and Refraction. The two Illustrations given here are unremarkable insofar as they are typical representatives of the kind of diagrams found in medieval geometrical optics. The figure on the upper left is taken from a thirteenth-century manuscript of an anonymous translation made from the Greek of the *Catoptrica* (that is, a work on mirrors) ascribed to Euclid in the Middle Ages, although it is in fact a later compilation. The upper diagram in this figure belongs to a proposition establishing that in convex mirrors, the image *(ydolum)* is always smaller than the object seen *(res conspecta)*, while the two lower diagrams in the figure relate to two cases of a theorem claiming that images of the same object will appear smaller in smaller convex mirrors. In all three diagrams the eye is located at point *B*. The symbol >—c above the last of these figures is meant to direct the reader to an alternative figure which is duly found elsewhere on the same page. Note should be made of the fact that in both of these propositions there is no construction of the image in question and that the apparent magnitudes are taken as a simple function of the angle subtended by the visual cone, both factors being characteristic of this pseudo-Euclidean treatise.

(Continued)

The figure on the lower right is drawn from a fifteenth-century manuscript of a chapter in Roger Bacon's (thirteenth-century) *Opus maius* dealing with the formation of images by refraction. In order to cover the full diversity of such image formation, Bacon considers not just plane media, but spherical ones. The diagrams given here relate to the latter. The top diagram in the figure pictures a case in which the eye is located in a spherical medium that is subtler than that of the object seen, the center of medium falling between the eye and the object. The diagram pictures the situation point for point: the eye *(oculus)* occupies the uppermost vertex; the object *(res visa)* is the lowest and longest horizontal line; the center *(centrum)* of the spherical medium, slightly misplaced, is directly above the arc (below which the words *corpus densius* appear); and the image *(ymago)* is given as the shorter horizontal line between this arc and the line representing the object. Since this "image line" is shorter, it properly represents Bacon's conclusion that, in the case at hand, the image is smaller (in absolute terms) than the object. The bottom diagram in the figure presents the case in which the eye is in a denser spherical medium, but now between the center of this medium and the object. As can easily be seen by inspecting the diagram, in this instance the image is larger than the object. Other figures from these two manuscripts can be seen in Illustrations 99, 120, 124, 139, 141, 149, 238; and 142, 143.

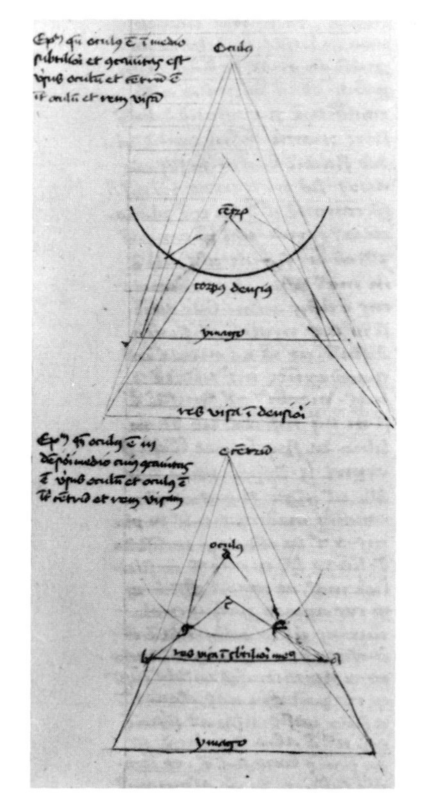

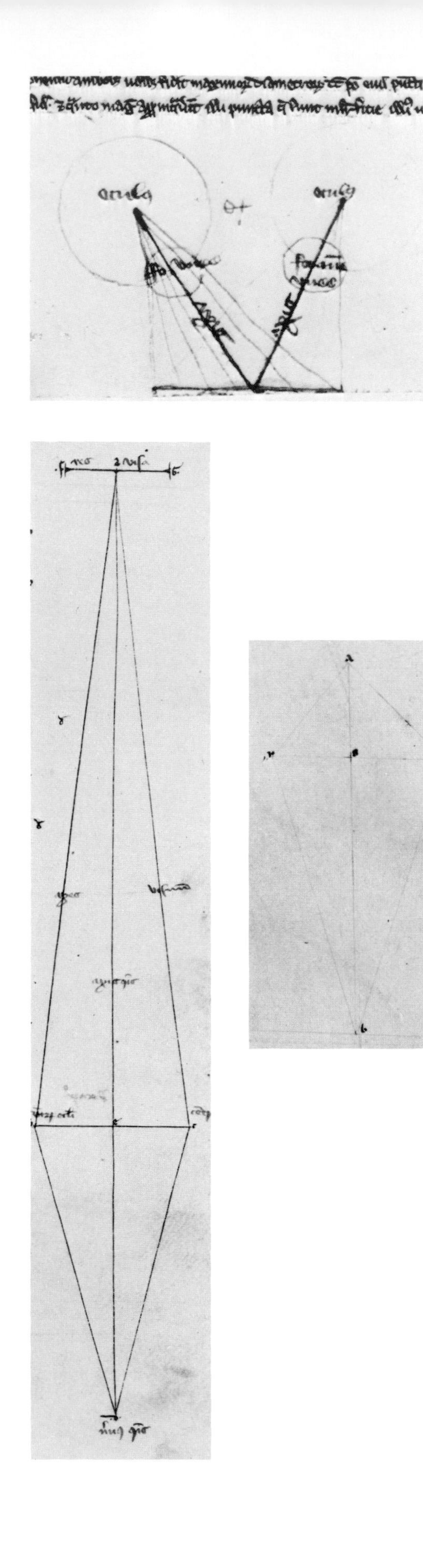

134. The Importation of Figures for Clarification. The *Kitāb al-Manāẓir (Book on Optics)* of the eleventh-century Arabic scholar Ibn al-Haytham was unquestionably the most important medieval work devoted to this particular "subaltern" science. Translated into Latin in the late twelfth or early thirteenth century as the *Perspectiva* or *De aspectibus* of Alhazen (a latinization of al-Haytham), it was of significant influence in the Middle Ages and, especially, in the sixteenth and seventeenth centuries. The feature of this work that is of present interest is not in this influence as such, but rather in the fact that its earlier books frequently do not specify figures or make explicit provisions for them. In particular, Alhazen often did not trouble in these opening books to give the appropriate letters or variables even when some geometrical configuration was clearly at stake. It was natural, then, that in such instances medieval Latin scholars would invent figures or import them from other works in order to facilitate the understanding of Alhazen's contentions.

The figures reproduced here give two examples of that phenomenon. The top figure on the left, taken from a thirteenth-century manuscript of the translation of Alhazen, is intended as a visual clarification of his claim (III, 2) that the two axes of the visual cones proceeding from the eyes will meet at a single point on the surface of the seen object. The little mark between the two eyes in this added figure keys it to that part of the preceding text (in which the same mark appears, here not shown) that is being visually explained. The same person who drew the figure has even troubled in a marginal note to emphasize the importance it, and other figures, have for comprehending the text *(Ecce quomodo visio per duos oculos variatur . . . in figura)*.

The figure itself carries labels for the two axes, has sketched in the relevant visual cones, and has used two circles to represent the eyes, two smaller circles standing for the pupils *(foramen uvee)*. As a whole, the figure could have been drawn by anyone familiar with any number of other medieval Latin writings on optics. This is not true in our second example; there the source of that imported is tolerably clear. The "double triangle" figure shown at the outside left comes from the same thirteenth-century manuscript of Alhazen. Its charge is the clarification of his claim (III, 6–7) that straight lines drawn from the center of the common optical nerve to the ends of the nerves in the eye form the equal sides of an isosceles triangle and that the straight line drawn from the center of the common nerve to the line connecting the nerve centers in the eyes is perpendicular to the latter line. The figure here given in explanation contains more than needed; the lower triangle alone—the *nervus communis* at its inverted vertex *D* and two *centra oculi B* and *C* at the extremities of its base—would have been sufficient. But we also have the upper triangle representing the axes of the two visual cones meeting in a point *A* in the surface *FG* of the *res visa*.

The likely source for this "more than necessary" diagram shown at the inside left is not hard to find. In all probability it is the *Perspectiva* of Witelo, who followed and developed Alhazen on so many points. In the present case, Witelo gives his version (III, 33) of the just-mentioned contentions of Alhazen, but adds consideration of the axes of the visual cones, thus making the second triangle of the diagram relevant. The pertinent Witelonian figure is given here from a fourteenth-century manuscript of the *Perspectiva*. Inverted and differently lettered than the added diagram in the Alhazen codex, it is nevertheless clearly its parent, or at least a close relative. For another figure from this Witelo manuscript, see Illustrations 135 and 216.

135. Degrees of Schematization. The two diagrams given here are essentially illustrations of the same phenomenon: namely, the fact that, in binocular vision, the axes of the two visual cones meet in a single point on the surface of the seen object. The first, on the upper right, taken from a fourteenth-century manuscript of the *Perspectiva* of the thirteenth-century Polish scholar, Witelo, is schematic in the extreme. The two small circles are two eyes with centers *E* and *G*, *A* at the vertex of the triangle being the medial point of the common optical nerve. The rectangle *BCDF* represents the surface of the seen object, *U* and *Z* being two points on that surface at which the axes of the visual cones, passing through the pupil, intersect at different times. It would be difficult, however, to extract this information from the schematized figure without the assistance of the accompanying text.

This is not the case with the center figure taken from a fifteenth-century manuscript of the *al-Baṣāʾir fī ʿilm al-manāẓir (Insights into the Science of Optics)* of Kamāl al-Dīn al-Fārisī, the fourteenth-century Persian scholar better known for his elaborate commentary on Ibn al-Haytham's *Kitāb al-manāẓir.* The greater "autonomy" of this figure for the most part derives from two factors: a certain amount of anatomical detail, and not simply two circles, for the eyes; and, more importantly, the labeling of all the parts constituting the diagram instead of relying, as in the Witelo figure, on mere letters that had to be referred to the text. Thus, within the eye, specific indication is made of the conjunctiva, the uvea and its center, the cornea, the crystalline humor, and the center of vision. The nerves and the axes of the visual cones are labeled, as well as the common axis and the line joining the centers of vision perpendicular to it. Clearly, taken by itself, Kamāl al-Dīn's figure is far more informative than that from Witelo's *Perspectiva.* For other figures from the two manuscripts used here, see Illustrations 134 and 216.

136. (Bottom Right) The Camera Obscura. Another figure taken from the same fifteenth-century manuscript of Kamāl al-Dīn's *Baṣāʾir* here depicts the formation of an image in what has come to be called a *camera obscura* (dark room). The central feature of the figure is the crescent image—perhaps of the sun's crescent—projected on the wall of the *camera* as a screen. It is formed by double cones of light whose vertices all pass through the aperture of the *camera,* the bases of one sheet being the solar crescent, those of the other sheet the inverted image of this crescent. Each point on the sun's crescent also sends its own cone of light through the aperture, the bases of these cones being the circles in the present figure. Those circles surrounding the crescent image all have their centers on the convex and concave arcs of the crescent. It is assumed that the "screen" on which the image is projected is parallel to the solar crescent and that the rays from the latter are perpendicular to the screen. The arcs bounding the circles surrounding the crescent image are labeled as the convex and concave surfaces of the produced light, while the boundaries of the image itself are specified as the convex and concave surfaces of the "middle." Other figures from this manuscript can be seen in the preceding and following Illustrations.

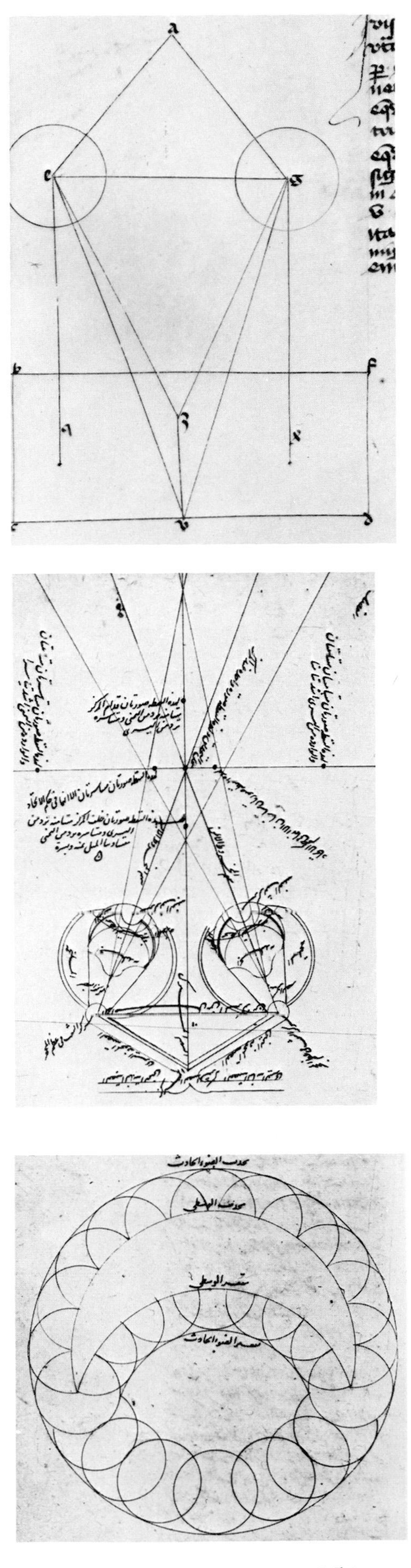

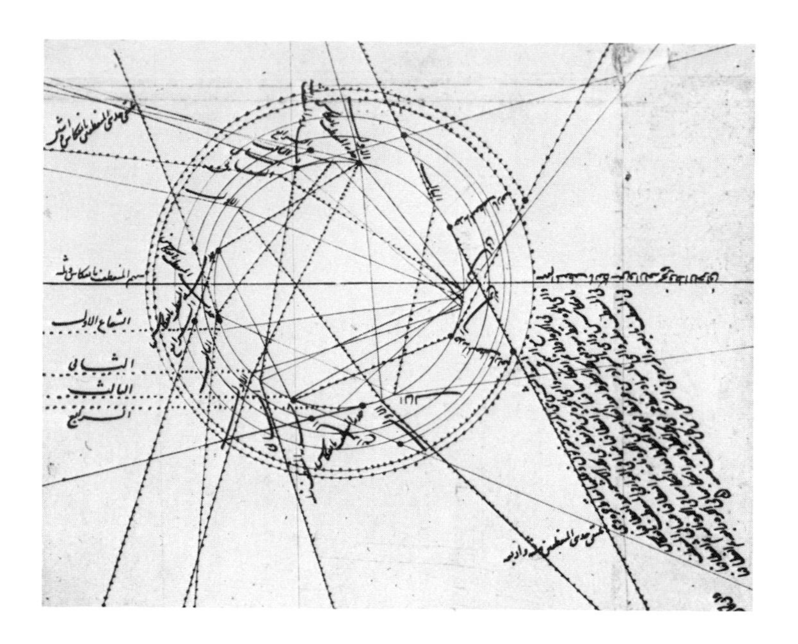

137. (Left) Clarity Through Numbering and Color Coding. This third figure from the same manuscript as the preceding Illustration of Kamāl al-Dīn's *Baṣāʾir* represents the phenomenon of the double refraction and multiple reflection of light incident upon a glass sphere. Kamāl al-Dīn held this sphere to be analogous to the effect of drops of water in producing a rainbow. The complexity of the present Illustration makes it difficult to follow the paths of the rays of light into and out of the sphere. Two techniques are employed to bring this about. First, as the more extensive text in a later hand at the right informs us, the rays coming from the sun are without color, but dotted (four of these rays can be seen at the left, parallel to the horizontal axis of the sphere); the rays are red and without dots after refraction into the sphere, reflection within it, and subsequent refraction into the air (although the figure itself at times does not appear to follow these "instructions" to the letter). Second, the rays are numbered both before they enter the sphere and then upon exit from it. For other figures from the present manuscript, see Illustrations 135–136.

138. (Below) Detail and System. The opening years of the fourteenth century witnessed two essentially successful explanations of the rainbow: that of Kamāl al-Dīn in the East, and that of the German Dominican Theodoric or Dietrich of Freiberg in the Latin West. Both scholars viewed glass spheres as analogous to raindrops and both maintained that the bow involved both refraction and reflection relative to individual drops. The two figures below are taken from a fourteenth-century manuscript of Theodoric's *De iride (On the Rainbow)*. The diagram on the lower left depicts the occurrence of double refraction and single reflection within a sphere as a kind of "model raindrop," a phenomenon which Theodoric held, correctly, to be that involved in the primary bow. The sun, or some other luminous body, is located in a small upper circle at *E*; the eye of the observer *(locus visus)*, below at *F*. The double refraction and single reflection, with the rays crossing after reflection, are clearly depicted within the model drop.

These "internal happenings" are omitted from the larger figure (below right) giving the total "system" of the primary bow. The rays of the sun (which is the small circle *A* at the lower left) are drawn only to one of the drops on the right, but we are to understand their incidence upon the other drops as well. The main purpose of the whole figure is to show that the different colors seen by an observer of the bow (here at *C* on the horizontal base line) come from different drops; the four numbers in the column under *C* match with but one of the four colors coming from each of the four circular drops.

If we take both of Theodoric's figures in sequence, we can say that, just as in Illustration 127, where the Tusi-couple was first explained in detail and then applied to the case of lunar motion, so here the "mechanism" of the rainbow is first given relative to an individual drop and then applied, with the details of this mechanism left out, to the multiplicity of drops "making up" the bow. The different degrees of complexity in the two cases aside, the move from detail to "system," even on the level of figures, is the same.

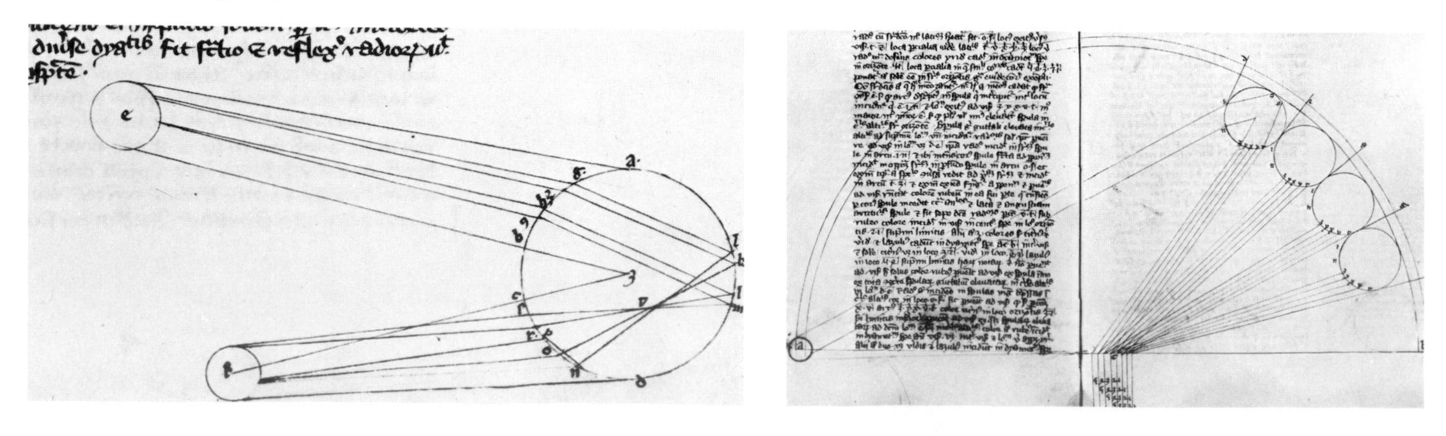

139. Scientia de Ponderibus. The present illustration as well as the two following ones are all drawn from the medieval Latin corpus of writings on the science of weights *(de ponderibus)* associated with the thirteenth-century scholar Jordanus de Nemore. This first group of figures is taken from a thirteenth-century manuscript of the one work in this corpus which can be ascribed with reasonable certainty to Jordanus himself: the *Elementa super demonstrationem ponderum*. The figures themselves, depicting the substance of Propositions 3–6 of the *Elementa,* are accurate and straightforward, their sole concern being the expression of the geometry that is involved in these propositions.

Proceeding from top to bottom, the first diagram relates to the fact that in the case of equal weights on equal arms of a balance, the length of the pendants *(appendicula)* from which the weights are suspended makes no difference. The second, circular figure relates to one of the central operative notions in the Jordanian corpus: that of "positional gravity." The substance of this central notion can be seen in more medieval terms by examining the figure at hand. The horizontal diameter AB of the circle represents a balance in equilibrium with its center at C. If the arm AC of the balance is taken to move upward to position DC or downward to position EC and at the same time we consider one and the same weight to be suspended from A, D, and E, then—our proposition states—the given weight will be positionally heavier *(gravius secundum situm)* at A than at either D or E. This is so because when a weight is in the system of constraint provided by a balance, its positional heaviness is measured in terms of the obliqueness of its descent; the less oblique the descent—here taken in terms of arcs of a circle—the heavier it is positionally and, vice versa, the more oblique the descent, the lighter it is positionally. Further, the obliqueness of "arcal" descent—the arcs in question being of the same circle—is in turn measured by the amount of the vertical intercepted by the arc in question, greater obliqueness corresponding to a smaller intercept of the vertical. This is all neatly "summarized" by the geometry of this second figure. Given equal arcs of descent AZ, DH, and EG, it can easily be seen that the amount of the vertical intercepted by the last two of these arcs—namely, KM and RX—is less than the amount intercepted by the arc AZ—namely, CF. Therefore, a given weight at A is positionally heavier than at D or E. Q.E.D.

The third figure, consisting of two semicircles with different radii, represents a balance with unequal arms, its purpose being related to establishing that if equal weights are suspended from the ends of such a balance, that on the longer arm will tend to move downward. The final diagram deals with equal weights on unequal arms, where the longer arm is bent so that its extremity is the same distance from the vertical as is the extremity of the unmoved shorter arm (in the figure, AC is the shorter arm, CF the longer bent arm, and $AC = FE$). For other diagrams from the present manuscript, see Illustrations 99, 120, 124, 133, 141, 149, and 238.

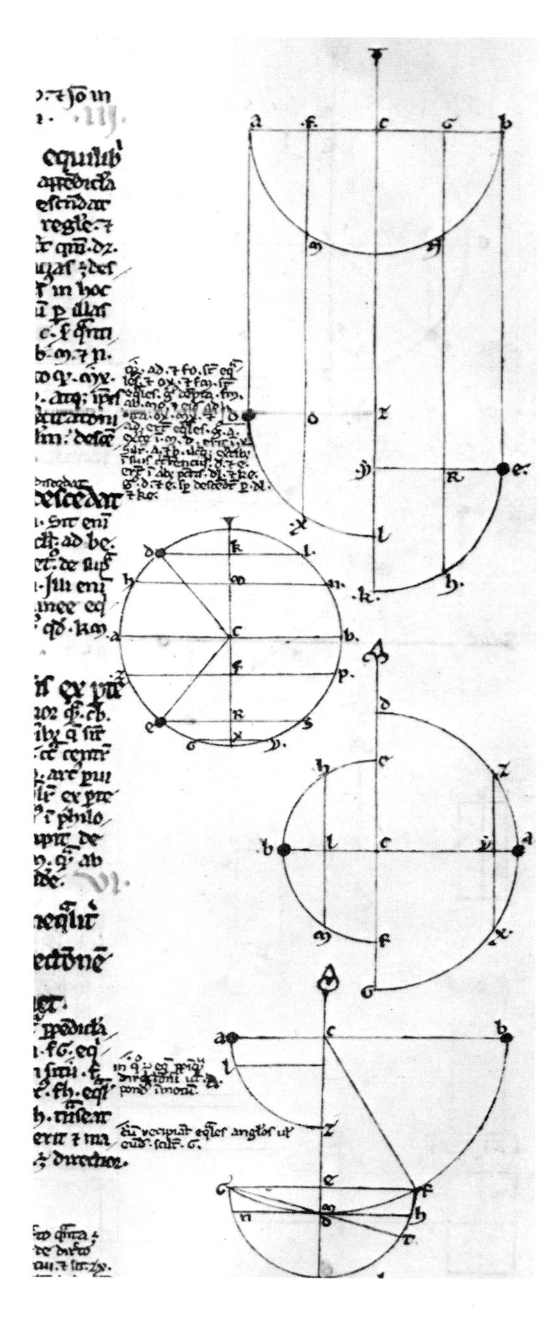

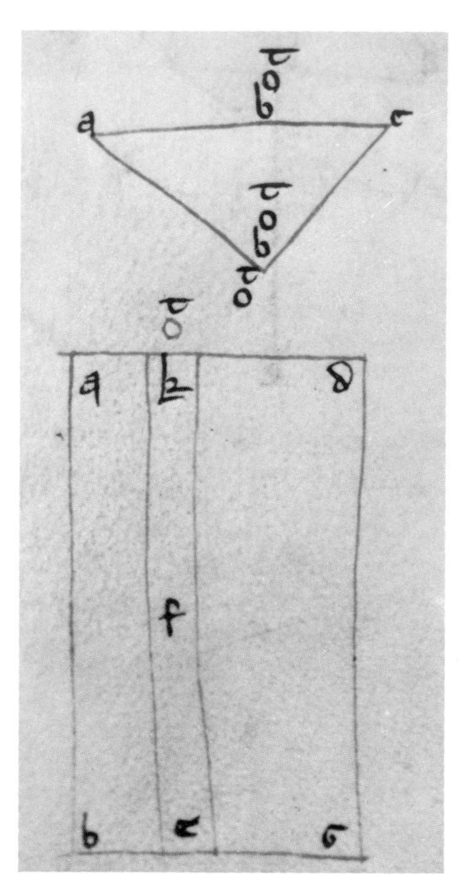

140. Physicalizing the Geometry I. Most of the figures in medieval manuscripts of works on the science of weights are similar to those in the preceding Illustration, if at times drawn with less elegance and accuracy. In a few instances, however, an attempt was made to have figures which also depicted the physical objects involved. Comparative examples of these two ways of drawing figures are here taken from two fourteenth-century manuscripts of the *Liber de ratione ponderis (Book on the Theory of Weight)* belonging to the Jordanian corpus. The center triangular figure at the top right relates to a proposition (IV, 3) claiming that the more cohesive a medium through which a body falls, the more this medium will resist *(quod magis coheret, plus sustinet)*. The surface of the resisting medium is here represented by the line *ABC*, the body falling through it as ᵹ, striking the medium at *B* and moving downward toward the vertex of the triangle, dragging *B* with it. The corresponding figure shown in the second manuscript on the lower right is the semicircular drawing at the top of that diagram. The illustrator has decided to show that the resisting medium is water or some other liquid. Although more "physical" than the matching figure above, this figure's representation of the contents of the proposition in question is less adequate.

Turning to the bottom diagram in the above illustration, the proposition being illustrated maintains that the greater its depth, the denser a medium becomes (due to the fact that it is compressed, not only by its lateral parts, but also by the parts above) and, consequently, that at greater depths the descent of a falling body is proportionally slower. The figure represents this claim by showing that the body (again ᵹ) falling through the medium *ABGD* descends more slowly as it reaches *E* than it does at *F* and more slowly at *F* than it does at *K*. The lower diagram again exhibits the liquidity of the medium, but has misrepresented things slightly by having *K* outside and above the medium and *E* above, and not at the bottom of, the medium as required by the text.

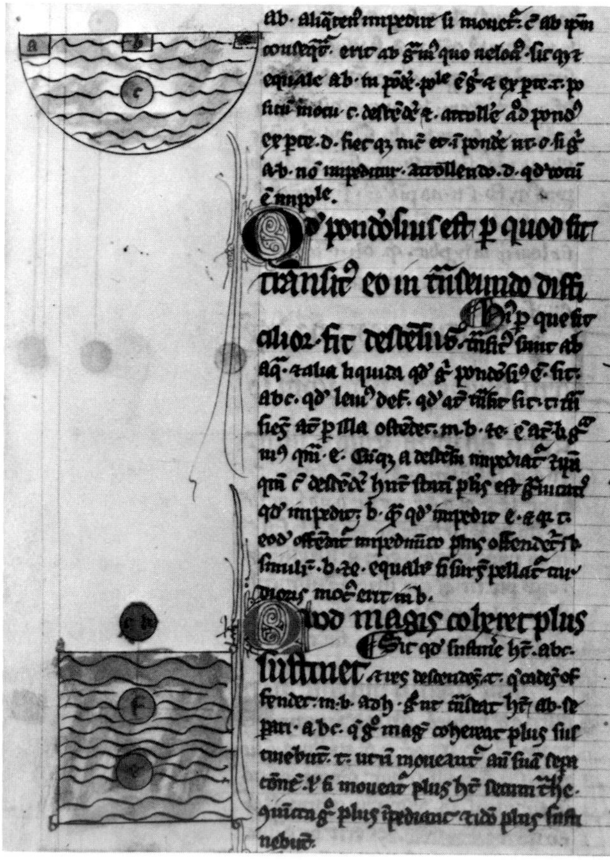

141. Physicalizing the Geometry II. Taken from the same thirteenth-century manuscript as Illustration 139, the figures at the right in one way or another give "physicalized" representations of the three final propositions (IV, 15–17) of the *Liber de ratione ponderis.* The top figure in the upper diagram pictures the case of a body *A* dropping through a medium *B* until it encounters a resisting body *C*, whereupon it will be reflected backward in a straight line *(redit directe).* The text itself says that the medium can be either air or water, here represented by appropriate dotting. It is not clear precisely why the thin black triangular area is employed to represent the fall and rebound of *A*. The simple arrow *BA* below in the same diagram is meant to illustrate the claim that a body with nonuniform weight, if thrown, will always come to a position with the heavier part in front. The figure to the right of the arrow illustrates a proposition having to do with a fact which had already been established in antiquity: namely, that a continuous stream of water will become proportionally narrower the farther it falls. Thus, *AB* is the orifice from which the stream pours and *C* and *E* are two portions of the liquid represented at different stages of their fall, the distance of fall *FZ* being greater than the distance *AF*, and the width of the stream *ZT* being less than its width *FD*.

The single figure below comes from a fourteenth-century manuscript, where it follows the text of the *Liber de ponderibus (Book on Weights),* also a member of the Jordanian corpus. It is not intended to accompany any particular proposition of this work, but apparently simply to illustrate some of the central notions employed in this, and other medieval Latin, treatises on weights. Thus, the circle is said to be that described by weights in equilibrium, the arms of the balance are specified as radii of the circle, the suspended weights are labeled as positionally lighter, etc. However, the most interesting aspect of the figure is its attempt to render all of this more concrete by picturing a hanging balance with "baskets" suspended from its arms in which the relevant weights are to be placed. For other diagrams from the first of the two manuscripts pictured here, see Illustrations 99, 120, 124, 133, 139, 149, and 238.

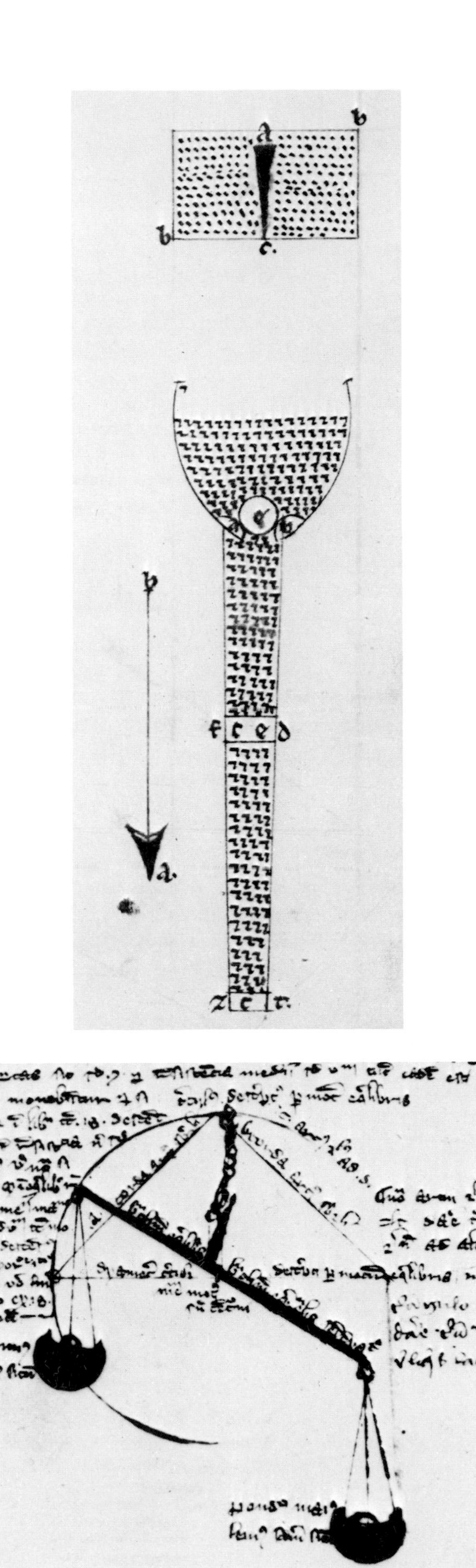

142. Fundamental Elements in the Doctrine of the Latitude of Forms. Geometric diagrams were often employed in proving particular contentions or theorems within treatises dealing with the latitude of forms, but figures were also considered important in introducing the reader to the various geometric entities and configurations that would be utilized in such treatises. In some instances, "introductory figures" of this sort were placed within the body of the text itself. An instance of this procedure is depicted at the top right, taken from a fifteenth-century manuscript of one of the most important, if not the most used, medieval treatises on the latitude of forms: Nicole Oresme's (fourteenth-century) *Tractatus de configurationibus qualitatum et motuum (Treatise on the Configurations of Qualities and Motions)*. Occurring within the text of Part I, ch. 15 of this work, the miniature figures are meant to represent certain configurations of simple, nonuniformly difform qualities. Oresme begins by explaining that qualities of this sort can be represented by various kinds of curves (whose distinction need not concern us here) over a given horizontal "subject line," which curves can be either convex or concave. For this basic distinction Oresme used no more than straightforward textual explanation. But when he went on to explain that these types of configurations can also be differentiated accidentally according to whether the difformity of the quality begins or ends either at some degree or at no degree, he found it more effective to set forth the various cases possible by a few words of text followed by the introduction of a relevant explanatory figure with the words *'sicut hic'* ('as this').

Thus, the first miniature figure in the text of this diagram exemplifies a "convex difformity" that both begins and ends at some degree, while the second gives the case of the same with one extreme at no degree and the other at some degree, and the third an instance of both extremes being terminated at no degree. The remaining two figures in the same text describe the relevant configurations for concave difformities with both extremes at some degree and with one extreme at some degree, the other at no degree. The illustration at the bottom right is a whole page of a more elementary work on the latitude of forms: the *Tractatus de perfectione specierum (Treatise on the Perfection of Species)* by Jacobus de Sancto Martino (fourteenth century). As marginal companions in an early fifteenth-century

(Continued)

manuscript to one of the opening chapters of this work, these figures illustrate some of the most elementary notions to be utilized in the body of the work (rectilinear, curvilinear, acute, right, and obtuse angles, as well as a square and its diagonal and sides).

Appreciably less elementary are the two variant types of uniformly difform latitude, both beginning from *non gradus* but one as an inverted isosceles triangle, the other as a right triangle with its highest degree being labeled *summus* and explicit notation of the equal excess obtaining between the increasing degrees marked off within it. It is worthwhile noting that in other manuscripts of Jacobus' work these very same figures appear within the text in a fashion similar to those depicted from Oresme in the diagram at the top right on page 156. In this instance, however, the scribe apparently believed his figures could be more informative if transferred to the margin, where greater detail and clarity would be possible.

Yet another fashion of presenting introductory figures is seen in the illustration on the right, taken from a late fourteenth-century manuscript of the *De latitudinibus formarum,* most likely also written by Jacobus de Sancto Martino. In the opening pages of this work, the scribe of the present manuscript has provided a good number of marginal figures in illustration of the text. But at the juncture depicted here, he has altered his technique and, beginning with a "heading" reading *incipiunt fygurationes,* has repeated many of the figures he had previously given in the margins and added some new ones as well. Apparently he thought that interrupting the text for a "recapitulation" of figures would be useful to the reader. In most instances he has added explanatory phrases within or next to the figure (for example, the middle triangle at the top bears the legend *ista fygura incipit ab angulo acuto per lineam curvam ascendentem*); in others he has merely referred to an earlier occurrence of the diagram in question by saying (for example) that the "figure is the same as *DE*," where the relevant letters appear inscribed on or about the earlier drawing. There is little doubt that, overall, in the mind of this particular scribe figures were an important part of the treatise he was transcribing. For other diagrams from the first two manuscripts represented here, see Illustrations 133, 143; and 67, 275; for another figure from Jacobus' *De perfectione specierum,* see Illustration 275.

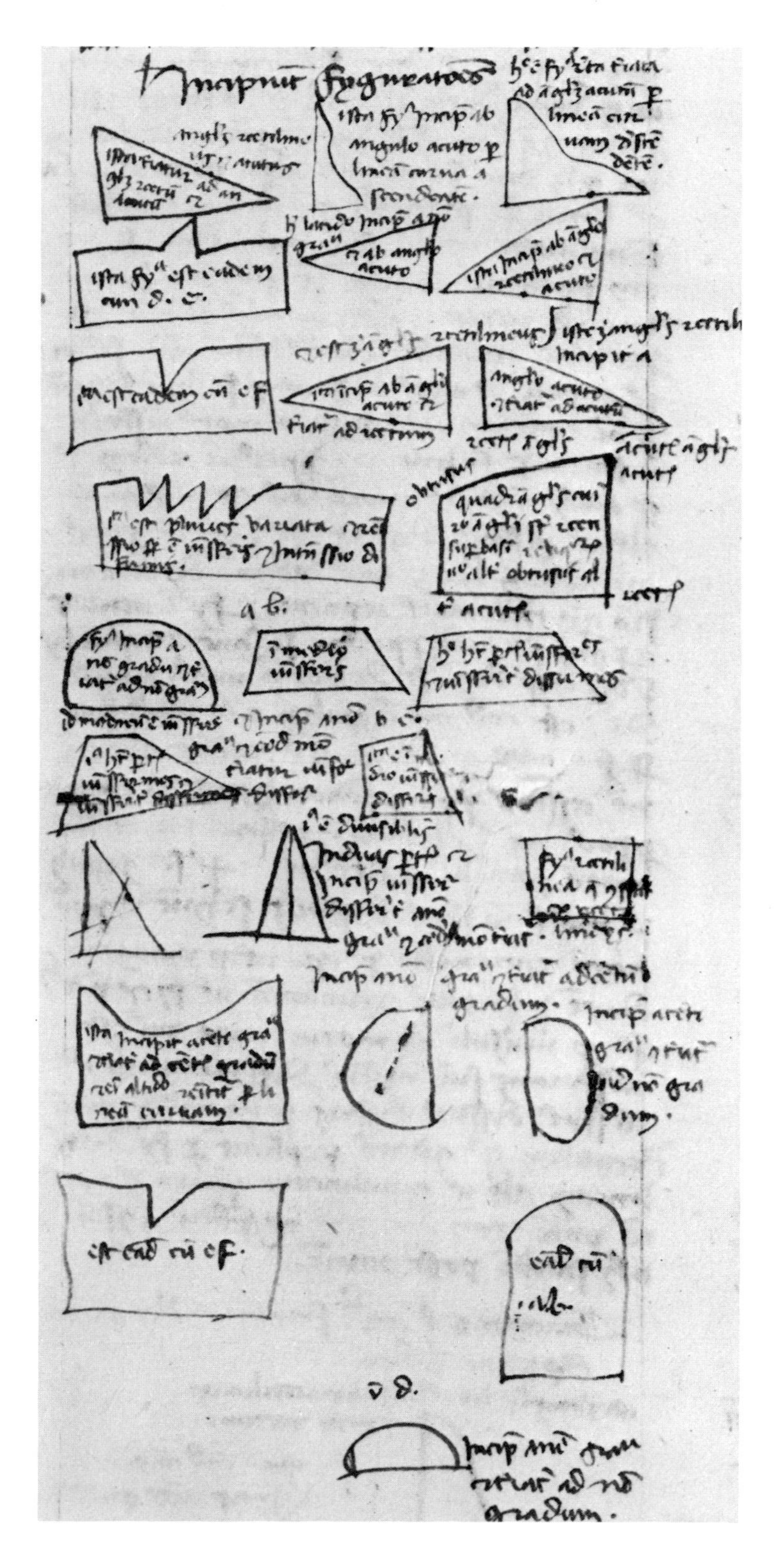

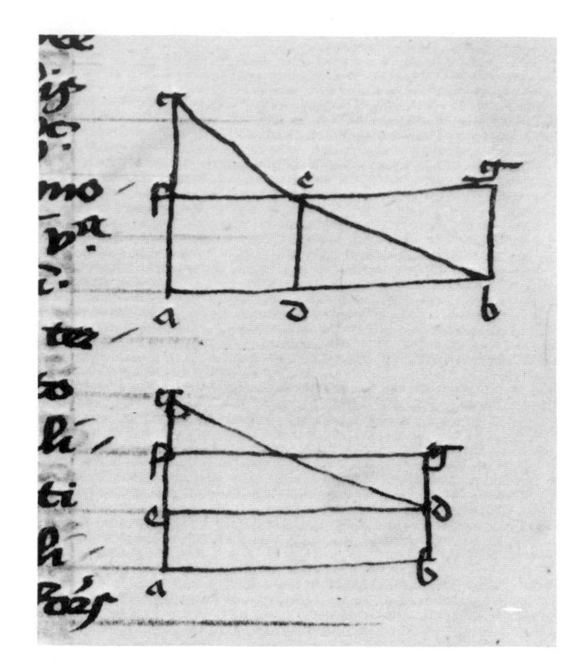

143. The Geometry of the Uniformly Difform. The chapter (III, 7) in Oresme's *Tractatus de configurationibus* to which the diagram above belongs begins by asserting that "every quality, if it is uniformly difform, is of the same quantity as would be the quality of the same or equal subject that is uniform according to the degree of the middle point of the same subject." These words refer, of course, to the "mean degree measure" for uniformly difform qualities that has been mentioned in the introduction to the present chapter. Taken from the same fifteenth-century manuscript of Oresme depicted in the preceding Illustration, the two figures in this diagram relate to the proof of such mean degree measure for the two possible cases of uniform difformity: that beginning from no degree, and that beginning from some determinate degree, both terminating in some determinate degree.

The diagram below, coming from a sixteenth-century manuscript of a commentary deriving from the fifteenth-century Italian scholar Philippus Aiuta on Richard Swineshead's *Liber calculationum,* is of a quite different sort. It does not accompany a proof, but is rather merely illustrative; more than that, it "physicalizes" the geometry involved. The subject at hand is uniformly difform illumination, proceeding from some determinate degree at the source of illumination to zero degree. The luminous source *(luminosum)* is here clearly depicted as a fire, the extent of its illumination decreasing uniformly from degree 8 through a mean degree of 4 to the zero at its termination. It is interesting to note that the text immediately preceding this illustration informs us that the teacher *(preceptor noster,* who is presumably Philippus Aiuta) of the writer of the manuscript did not diagram the topic in question *(non figuraverit)*, but that the reader can now fix his eyes upon such a production *(sed nunc conspice)*. For other figures from the two manuscripts used here, see Illustrations 133, 142; and 13, 144, 255.

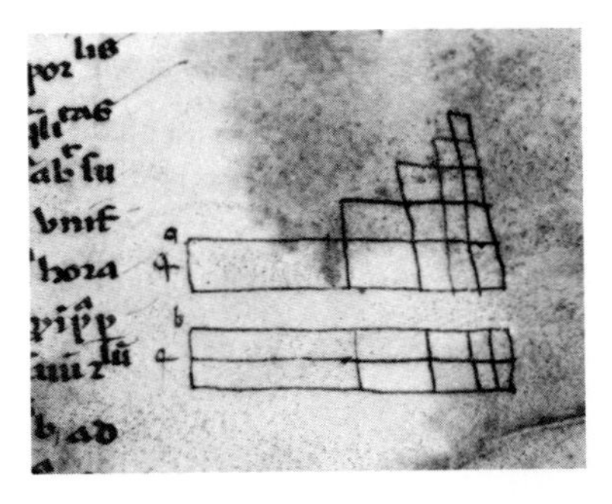

144. Infinite Intensities. It was almost a habit in many medieval treatises dealing with the measurement of qualities to try to establish ways to measure cases in which the quality was infinite, both in intensity and (though less frequently) in extent. The figures above, taken from a fourteenth-century manuscript of Richard Swineshead's *Liber calculationum,* are an example of such measurement of the infinite. Swineshead himself did not use geometry in measuring the distributions of qualities, but manuscript evidence tells us that those who read him, or commented upon him, did.

The first two figures depicted in the above diagram are most likely to be associated with such a reader. Accurately depicting geometrically what Swineshead said, the top figure shows that the topic is a body that is qualified—hot, let us say—in degree 1 over its first half, in degree 2 over its next quarter, in degree 3 over its next eighth, in degree 4 over its next sixteenth, and so on, in infinitum. How hot is this body as a whole? Swineshead asked. The answer was that it is just as hot as "its second proportional part" (that is, the first fourth after the first half), which is to say, as a whole, it is hot in degree 2. (In modern terms, Swineshead has summed a convergent infinite series.)

The lower figure in the same diagram relates to Swineshead's proof of his contention. As a whole, the subject it represents is uniformly hot in degree 2. But it is easy to see that one can rearrange the top "layer" of rectangles in this lower figure and automatically obtain the upper figure representing a quality increasing to infinity over successive proportional parts (merely place the first half of the top layer on the last half of the bottom layer, the first fourth of the top on the last fourth of the bottom, the first eighth on the last eighth, and so on, in infinitum).

The more elaborate diagram below, taken from the same commentary on Swineshead as in the preceding Illustration, consists of a series of "segments," each one of which gives a configuration identical to the upper figure in the above diagram (although now the degrees begin from 4 over the first half). Each of these segments, then, would as a whole be "denominated" by 8. But not only is the infinite increase of the quality over successive proportional parts of each of these segments specified by the words *et sic in infinitum* attached to the uppermost vertical, but the same inscription at the lower right of the whole diagram tells us that we are to understand that an infinite number of such segments are to be given. As a whole, then, the diagram represents an infinite denomination for the quality in question (for us, the infinite series involved would be divergent). It was naturally problematic to diagram anything infinite, but the present two illustrations—especially the last—provide a sample of one frequent medieval technique of accomplishing such a task. For other figures from the second manuscript used here, see Illustrations 13, 143, and 255; for other diagraming of "the infinite," see Illustration 255.

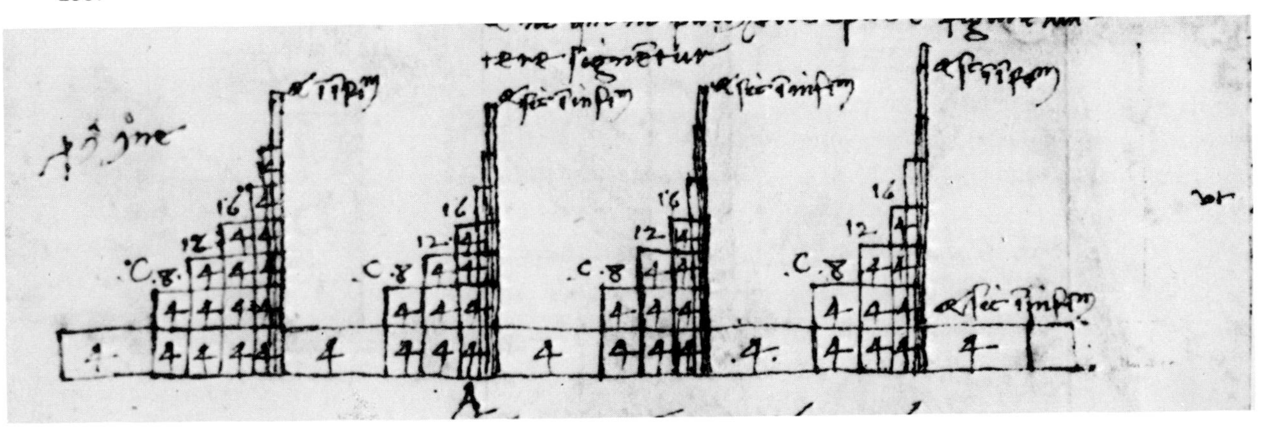

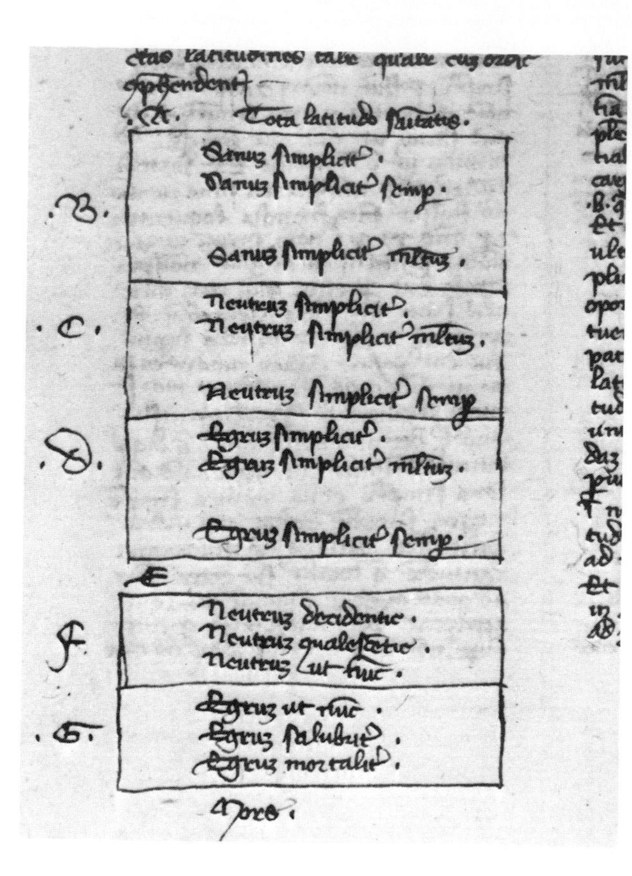

145. Latitudo Sanitatis. The notion that there is a latitude of health *(latitudo sanitatis)* is fundamentally a Galenic notion, but the amount of attention it received increased markedly in the later fourteenth century in the Latin West, reflecting in many respects the interest in natural philosophy in the latitude of forms. One of the results was that diagrams related to this latter doctrine were produced in medical works. Two such diagrams are given here, both deriving from fifteenth-century manuscripts. The first (above, left) comes from a brief work on several aspects of Galen's *Ars parva* by an otherwise unknown Bologninus, and is composed of two rectangles surmounted by the inscription *tota latitudo sanitatis,* which proceeds to death *(mors)* at the bottom. Bologninus tells us that we can take both together as constituting a single latitude which is equivalent to the *latitudo vite,* or we can take them as essentially distinct (a view he believes to be "insinuated" by Galen). On this second interpretation, *A* is a latitude consisting of three subordinate latitudes—and hence rectangles in the diagram—*B*, *C*, and *D*. The bodies falling within these three are, respectively, health *(sanum)*, neutral *(neutrum)*, and ill *(egrum)*, a threefold distinction which itself goes back to Galen. But in all three cases the bodies are healthy, neutral, and ill *simpliciter,* that is, in terms of their simple basic constituents. Thus, a body which is *sanum simpliciter* is one whose basic simple constituents are in a harmonious balance, in *eucrasia,* to use the transliterated Greek term found in medieval medical literature. However, looking at the other members of the rectangle *B*, we observe that a body can be *sanum simpliciter semper* or *multum.* In the former case its *eucrasia* is maximal; in the latter, it falls short of that *eucrasia,* but not appreciably so. And the same distinctions obtained, our diagram tells us, for the *neutra* and *egra* that populate rectangles *C* and *D*. What distinguishes the bottom two rectangles *F* and *G* (which make up the distinct latitude *E*) is that bodies relevant to them are not *neutrum* or *egrum simpliciter.* Instead, they are at best *neutrum* or *egrum ut nunc* (that is, only for the present time). The diagram is brought closer to some of the notions that were operative in the doctrine of the latitude of forms in general, if we note the author's claim that although *A* is a *latitudo sanitatis,* we must not think that it is uniform over all its parts (even though the diagram, he confesses, has been constructed of rectangles). The fact that only *egra* and *neutra* are members of its bottom two elements means that, as a whole, *A* is a difform latitude. And the same is true for latitude *E*. The point is made clearer in the second *latitudo sanitatis* diagram (below right) taken from the comments on Galen's *Ars parva* by the fourteenth-century Italian physician Antonius de Scarparia. Corresponding to the top three rectangles in the first figure, here the top triangle is constituted of the latitudes of *sana, neutra,* and *egra simpliciter.* But this triangle is surmounted by the inscription *summus gradus sanitatis,* the shorter and shorter parallels as one moves toward the vertex denoting lesser and lesser degrees of health. Thus, the diagram itself makes the difformity of the latitude evident, something that was not the case with Bologninus' figure. And the same thing is true with respect to the lower triangle, whose base is labeled as the *summus gradus egritudinis.*

(Continued)

CHAPTER FOURTEEN

Practica Geometriae: *Mathematics Put to Work*

GREEK GEOMETRY AS EXEMPLIFIED in the works of Euclid, Archimedes, and Apollonius belonged to what today one might term "pure" mathematics; it was not an applied geometry in any proper sense at all. Yet one must not let the predominance of such works in the history of Greek geometry as we know it obscure the fact that there was then also a more practical side to geometrical endeavors. The primary extant treatises belonging to this phase of ancient Greek mathematics were the work of Hero of Alexandria (second half of the first century), although there were undoubtedly other writings of a similar sort before those composed by this scholar.

We have no definite record of these "mensurational" writings of Hero of Alexandria having been translated into Arabic, but al-Fārābī (tenth century) speaks of practical geometry in his *Enumeration of the Sciences,* and some historians believe they have detected evidence of Hero's influence in the works of the tenth-century Arabic mathematician Abū᾿l-Wafā᾿ al-Būzjānī, who wrote works on geometric constructions for the "artisan" and on such topics as the determination of the distances and heights of inaccessible objects, a subject of concern, as will be seen in some of the illustrations in this chapter, throughout the whole history of practical geometry.

The Arabic segment of this history is still largely unstudied, and although we know a great deal more about its medieval Latin phase, it too is still in need of considerably more attention. It is clear, however, that the Latin history of practical geometry is relatively innocent of Hero's influence (he was not translated), but finds its point of departure in the writings of the Roman surveyors or *agrimensores* (literally, "field measurers"), sometimes also called *gromatici* (from the *groma,* an instrument used in their trade). The writings of these Roman professionals have come to us in a manuscript of the sixth century—the so-called Codex Arcerianus, a name that derives from its sixteenth-century owner (see Illustration 146). These writings concern not only topics related directly to the division of land (such as methods of division and mensuration, the marking of boundaries, and the making of maps), but also such things as the types and rules of land tenure, land disputes, tables of measures, lists of colonies, and so on.

The influence of this material in the earlier Middle Ages naturally derived in part from later copies of the *agrimensores* treatises themselves, but it also stemmed from the fact that the early Boethian versions of Euclid contained numerous extracts from one or another of these treatises (see Illustration 148). This influence can be seen in the *Geometria* of Gerbert as well as in the *Geometria incerti auctoris* associated with the works of this tenth-century scholar. It is in the twelfth century, however, that the development of Latin practical geometries begins to accelerate. One of the most notable works in that development was the *Practica geometriae* of Hugh of Saint Victor. Hugh divides the whole of geometry into *theorica* and *practica* and tells us that the latter "uses certain instruments and judges by inferring one thing from another on the basis of proportion," a description that is borne out both in his own work and in other practical geometries. As will be clear from a number of the illustrations given in this chapter, the instruments involved were, on the one hand, such things as quadrants, astrolabes, and even mirrors (all of which functioned as sighting devices), and, on the other, gnomons or fixed rods, applied largely in conjunction with the shadows they cast.

Hugh's reference to the importance of proportion in practical geometry is also reflected in what we find in treatises belonging to that genre, principally in the repeated utilization of the properties of similar triangles in determining the measurements in question.

Instruments and considerations of this sort belong primarily to those sections of practical geometries entitled *altimetria,* treating such things as the heights of buildings or the depths of wells. These sections also often contained astronomical material involving the altitude of one or another celestial body. There were, however, two other sections that were standard: *planimetria,* where the task was the determination of the areas of surfaces, and *crassimetria* or *stereometria,* where the volume of solids was at stake. Further, since the mensuration which was the concern of these treatises involved numbers, sections dealing with arithmetical matters were often added.

These, then, were the kinds of issues and procedures that one finds in almost all medieval Latin practical geometries. They appear in treatises devoted to the quadrant (see Illustration 152) and they are very evident in the anonymous twelfth-century Parisian practical geometry (Illustration 149) which served as the Latin original of the first French geometrical tract already sampled in Illustration 108. Similarly, these kinds of considerations formed the basis of the practical geometries composed in the thirteenth century by Leonardo Fibonacci of Pisa, in the fourteenth century by Dominicus Clavasio, and in the fifteenth century by Leonardo of Cremona (Illustration 152). And they are present in any number of vernacular practical geometries as well. Finally, scholars who, like Roger Bacon, did not compose practical geometries themselves, but only reflected on their nature and contents, also realized the importance of these basic concerns.

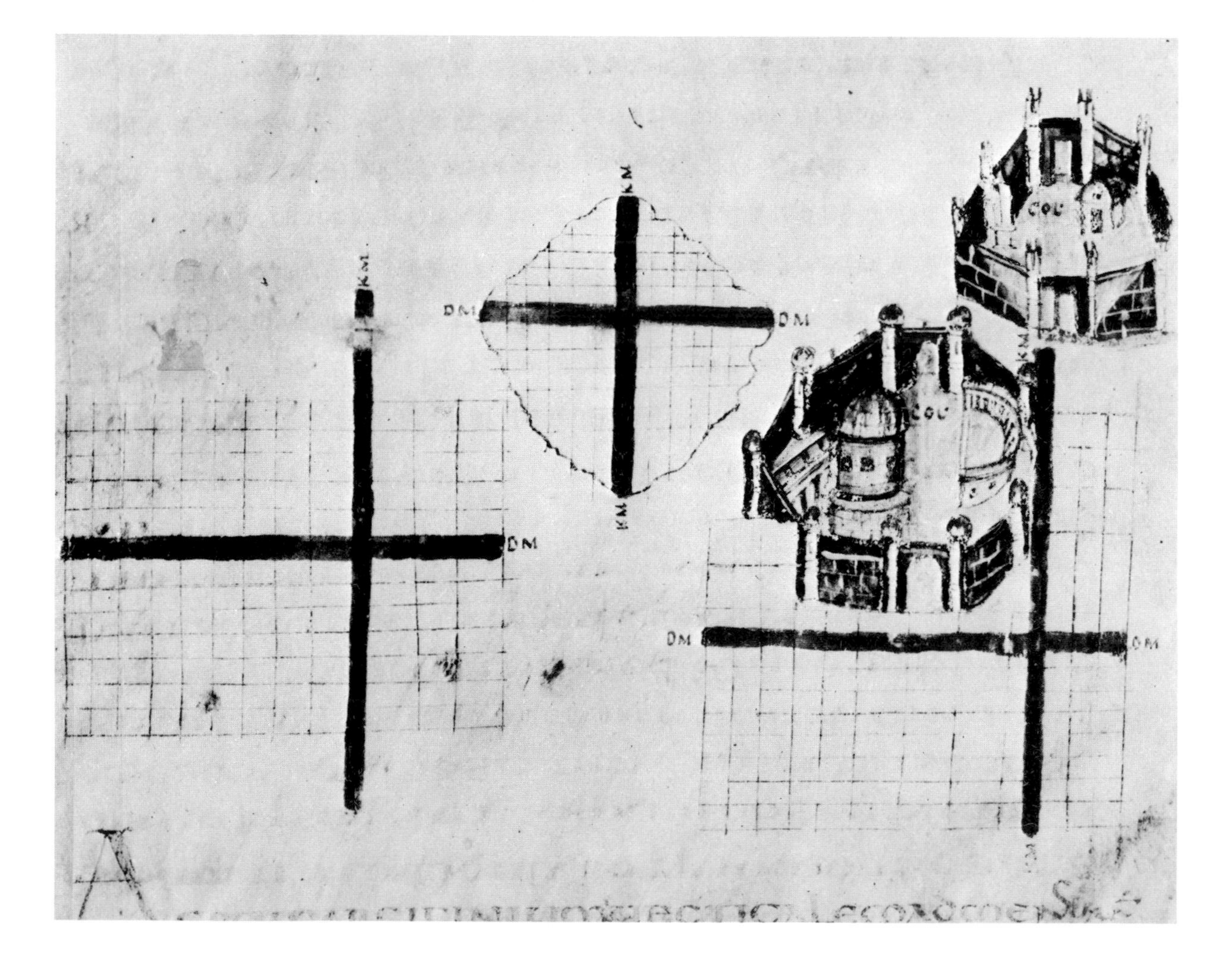

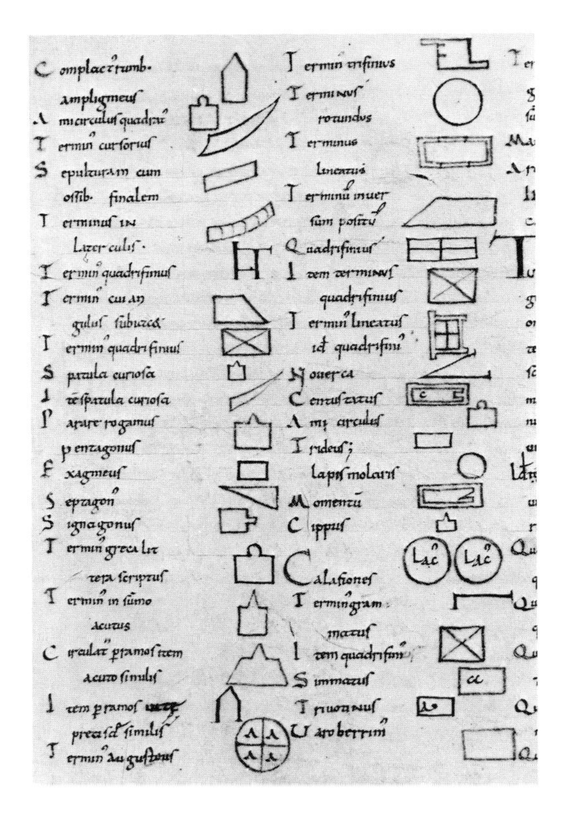

146. The Surveyor's End Product. Of all the kinds of materials populating the corpus of writings of the Roman *agrimensores,* among the most important were those treating of the division of land and the construction of maps on the basis of this division. The illustration (bottom, page 162) of this division and mapping is here taken from the oldest and best manuscript of the *agrimensores,* the early sixth-century Codex Arcerianus. Although the figures do not serve to explain the details of the accompanying text, they nevertheless adequately relate to the introductory words of that text, which announce that "this is how we shall allocate undeveloped land in the provinces." The allocation in question is that called "centuriation," a technique of land partition that was frequently used in Roman colonies. The main unit of land was the *centuria,* a square with sides of 2400 Roman feet (= 710 meters), here clearly represented in the three "checkerboard" patterns. The name 'centuria' possibly derives from the fact that the land unit which was its bearer was at some point of time divided among 100 settlers, each smaller unit being known as a *heredium.* The boundaries or *limites* between *centuriae* were called *decumani* (running east to west) and *kardines* (running north to south), with two main boundaries being designated as the *decumanus maximus* and the *kardo maximus.* These are clearly depicted and appropriately labeled *DM* and *DK* in the three Arcerianus figures. The figures on the page above are from an eleventh-century manuscript of the *agrimensores* corpus and are meant to be representations of the various shapes and markings for the boundary stones (*termini*) which were so much a part of the surveyor's art.

147. (Below) Finding Cardinal Directions. The present figures are taken from an *agrimensores* manuscript written at Luxeuil in 1004 of the treatise *De limitibus constituendis* ascribed to the second-century gromaticus Hyginus (but are probably the work of a later author). Their function is the illustration of the procedure recommended by Hyginus for determining the appropriate bearings for the surveying of land. The upper figure in the diagram relates to his instructions to draw a circle on a flat part of the ground and to erect a sundial gnomon (*scioterum*) at its center; the shadow of the gnomon at times falls within the circle, at others without. (The text does not tell us what the radius of the circle should be, as does a parallel text in the *De architectura* [I, ch. 6, 6–7] of Vitruvius [first century B.C.]: namely, the length of the shadow cast before noon at about the fifth hour.) The lower figure in the diagram gives the specifics for the remainder of Hyginus' procedure. We are advised to mark those points on the circumference of the circle at which the end of the shadow enters and leaves the circle and then to draw a straight line through these two points (although this is not accurately represented in our figure). We are next to bisect this straight line and to draw a perpendicular from the point of bisection; this will give us the required *kardo* or north–south line. Constructing a perpendicular to this line will yield our east–west coordinate parallel to the first line, here labeled *D . . . M* at its ends (that is, *decumanus maximus*). A similar indication for the *kardo* is not given; in its place we are instead informed that this line is "aligned with the meridian" (*meridiano ordinatum*).

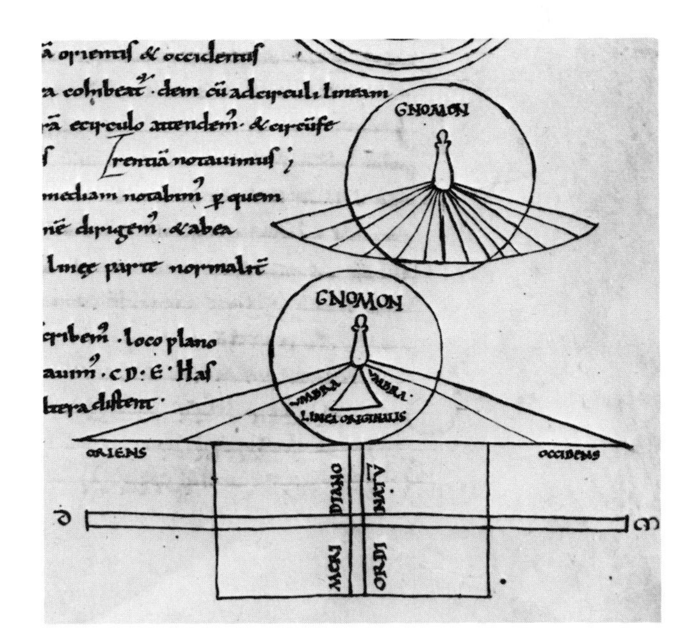

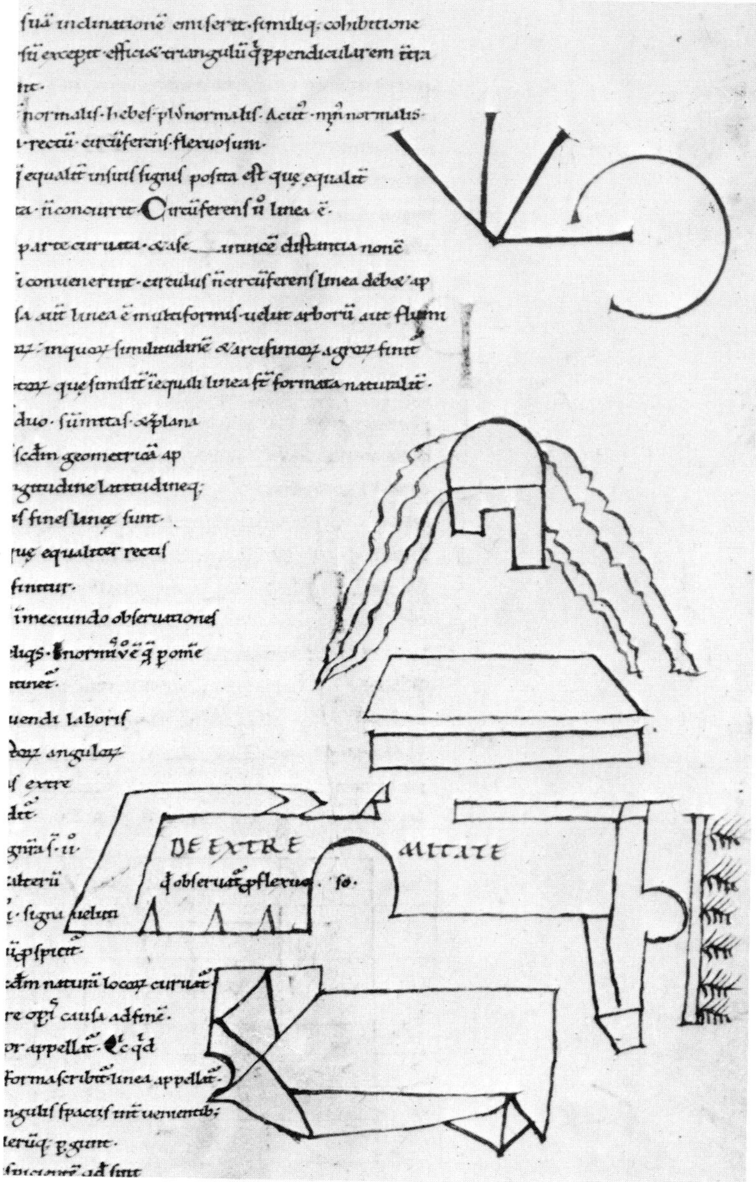

148. Agrimensores Material in Early Medieval Euclid. There were two Latin translations of Euclid's *Elements* ascribed to Boethius (see Illustrations 108–109), both of which included passages from the practical geometry of Roman surveyors' manuals. An example of this incorporation is here taken from a manuscript written ca. 1100 of the so-called two-book Boethian Euclid. After the translation of the definitions, postulates, axioms, and propositions (enunciations only) of most everything in Books I–IV of the *Elements* as well as proofs for the first three propositions of Book I, the introduction of *agrimensores* material begins, deriving largely from the contribution made to this material by the second-century surveyor Balbus. These additions begin with relatively standard geometrical fare: definitions of the different kinds of lines and angles. Indeed, part of this is seen in the topmost figures given in this diagram: right, obtuse, and acute angles and a circular line. But then more "practical matters" are approached. First, the *linea flexuosa,* found, we are told, in such things as trees and rivers. The second figure on this page graphically illustrates this. The next topic is two-dimensional configurations, called *summitates,* the two simplest examples of which are illustrated by the trapezoid and rectangle directly under our figure for the *linea flexuosa.* More complicated *summitates* are pictured by the next two figures, that which is irregular (*enormis*) and, in the lowest figure, that which is oblique (*liquis*). Immediately to the right of the picture of a *summitas enormis,* the small figure with six "trees" sprouting from it again reminds us that this geometry has to do with such practical concerns as land.

149. The Development of Practical Geometry. One of the hallmarks of medieval practical geometries was the utilization of numbers and arithmetical procedures, a factor that was not always reflected in the figures accompanying such works. In the first example on the top right, it clearly is. The ten triangles appear in a twelfth-century manuscript of the *Geometria* of the tenth-century monastic scholar Gerbert d'Aurillac, who became Pope Sylvester II in 999. One of Gerbert's primary preoccupations in this work was calculating the sides and areas of what he called *pythagorici orthogonii*—in our terms, rational right triangles (that is, right triangles whose sides are never incommensurable, but always expressible by integers, fractions, or mixed numbers). Even in the final incomplete chapter of the *Geometria,* the subject is still such triangles, and although the text as we have it specifies no examples, ten of them came to be given in the manuscript tradition by means of figures alone. Unfortunately, although these figures are attractively drawn in the present manuscript, the numerical values given are correct only a dozen times out of thirty-four. A fair number of the errors have arisen due to the greater chance of confusion when transcribing Roman numerals (for example, v has frequently been taken as li). None of the triangles is drawn to scale, either in terms of the ratios between their sides or relative to one another, but such drawing to scale is almost never the case with any manuscript geometrical figures, even when specific numerical values are involved. Six of the ten triangles have "integer sides." (Thus, that at the top left—missing a value for one of its sides—has a hypotenuse of 13, sides of 5 and 12, and an area of 30.) Some of the sides of the four remaining triangles are assigned mixed numbers, the symbols used for the uncial fractions (that is, twelfths) involved being those we have already seen in Illustration 86 as part of the abacist tradition. For example, if we remedy the erroneous and missing numbers for the triangle at the lower left, we have a hypotenuse of $11\frac{4}{12}$, sides of $5\frac{4}{12}$ and 10, and an area of $26\frac{8}{12}$.

The diagram below marks a later stage in the medieval development of *geometriae practicae,* the figures coming as they do from a thirteenth-century manuscript of an anonymous Parisian practical geometry written in 1193. The theorems to which these figures belong announce that their concern is finding the area of polygons (the figures here representing, from top to bottom, the pentagon, hexagon, heptagon, octagon, and nonagon). But the "formulas" given in the text of these theorems belong to arithmetic, not geometry, since they are relevant to calculating polygonal numbers, not the area of polygons. Thus, the equilateral—but not regular—pentagon depicted here belongs to a theorem whose formula for its "area" is, if n is taken as one of its sides, the area is equal to $(3n^2 - n)/2$ [and the text goes on to give an example: if the side (*latus*) is 6, the area (*area*) is 51]. But this is the formula for a pentagonal number, one of which is, of course, 51 (see Illustration 94). Both the text and the figures in this practical geometry therefore misrepresent the actual procedures at hand. One might be misled into thinking that one here has to do with the calculation of geometrical areas, appropriate numerical examples being given for some of the calculations; but in fact the only "geometry" in question in these calculations is that of the "figurate" conception of the numbers being treated. This was to be arithmetical in a manner quite different from that we have observed in Gerbert's practical geometry. For other figures from the two manuscripts depicted here, see Illustrations 109 and 99, 120, 124, 133, 139, 141, and 238.

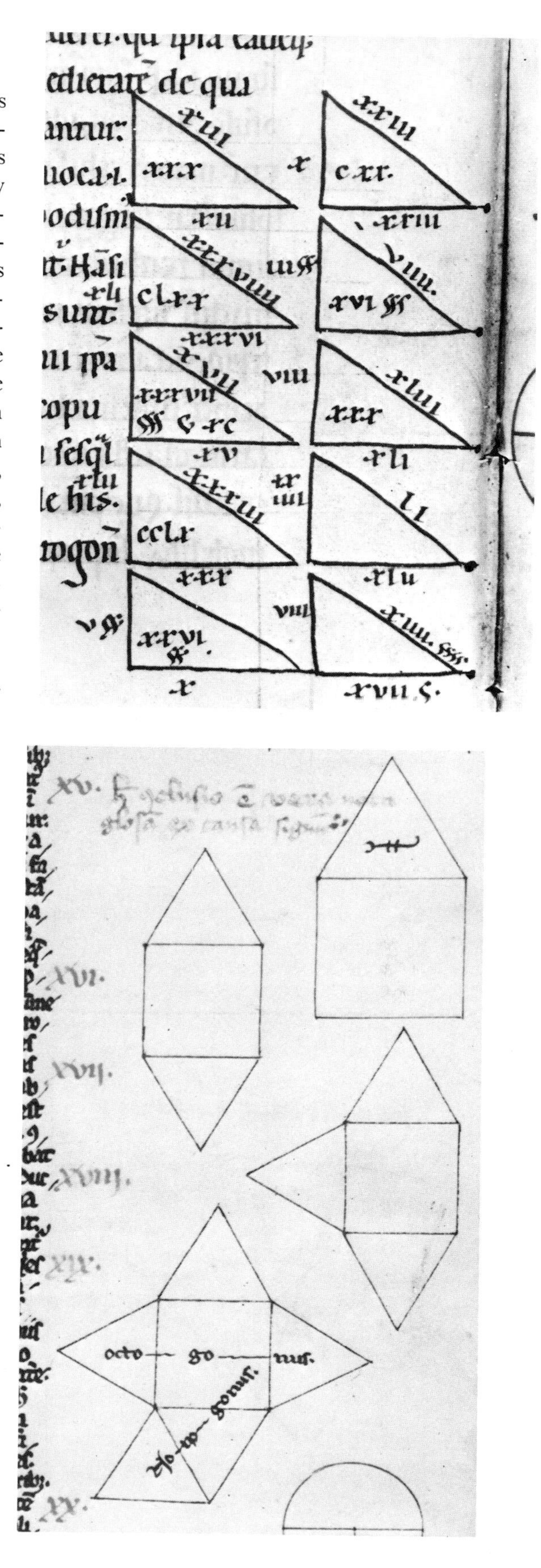

150. Instruments of Measure. A fair number of different artifacts were employed in the measurements often described in practical geometries, but the two most frequently depicted were the compass and the square. Testimony to the great variety of the kinds of works in which the depiction of these two instruments can be found is here given above from an early sixteenth-century French manuscript of a Book of Hours. The compass is being used to measure the orb, or globe, which was a standard medieval symbol of power, whether as an attribute of God the Father, as a sign of the sovereignty of the Son, or, when applied to man, as an indication of imperial authority and dignity. The square and the geometer's wand lie nearby. The square was not only an implement within practical geometry (see, for example, Illustration 153), but also a most important tool for the medieval mason. It is this second role that is behind the depiction of the square at the lower left of the picture on the top right, taken from the tomb of Master Mason Hugues Libergiers of Rheims (d. post 1263). The most interesting thing about this particular representation of the square is the fact that the inner edges of its "arms" are not parallel to their outer edges, although the corner angles formed by these inner and outer edges are both right. This can be seen more easily in the detail depicted on the lower right. The effect of this skewing is that if a "hypotenuse" is drawn from the end corners determined by the outer edges, the two angles thus formed will be 30° and 60°; on the other hand, the angles formed by this same hypotenuse with the inner edges will be 31° 30′ 10″ and 58° 29′ 50″. It has been noted that these angles are not far from the angles (31° 43′ 03″ and 58° 16′ 57″) that are formed by the diagonal and sides of a rectangle whose sides exemplify the golden sections—that is, stand to one another in extreme and mean ratio—although no evidence has been found of how such a "golden section" square (if that is what it is) may have been applied in medieval architecture. Compasses and squares of geometers and masons are also seen in Illustrations 158, 173, 175–6, 178–9, 190.

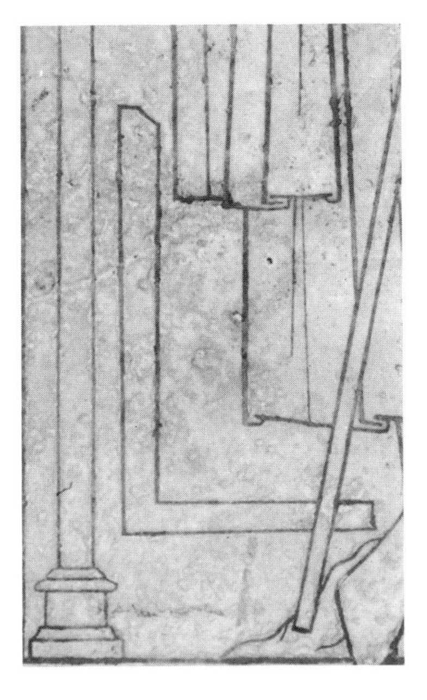

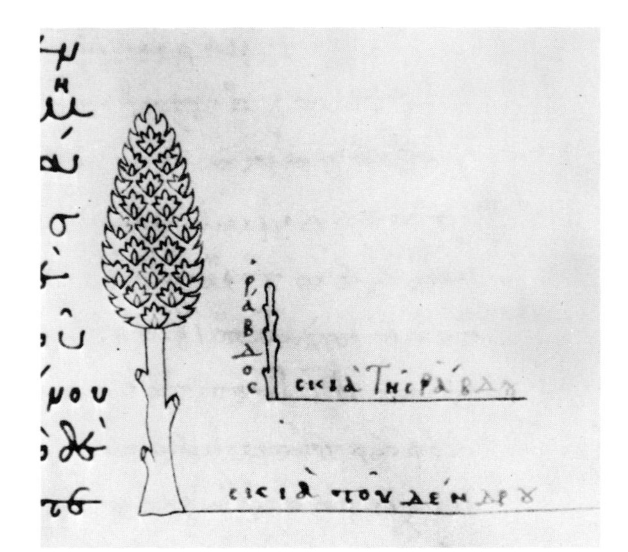

151. Ways of Measuring and Things Measured I. Manuscripts of practical geometries often carried not only figures explaining the techniques of measurement that were applied, but also pictures of the objects measured by these techniques. The first such object depicted above, a tree, comes from an eleventh-century Greek manuscript of Hero of Alexandria's (first-century) *Stereometrica.* If we wish to find the height of a tall tree or column, we can do so, his text informs us, by shadows, preferably between the fifth and seventh hours (that is, of daylight) since the shadows will then be shorter and, presumably, more easily measurable. One can see how this measurement by means of shadows works from the relevant figure alone. A stick or staff (*rhabdos*) of known length is placed in the ground and the shadows simultaneously cast by the stick and the tree whose height is sought are both measured. Given the fact that the ratio of the stick to its shadow is the same as that of the tree to its shadow, the height of the tree is easily calculable. The stick and both shadows are properly labeled in the present illustration, the tree being clear enough from its depiction to obviate any identifying inscription.

The illustration below contains no indication of the technique of mensuration involved, but confines itself to picturing the objects measured: tapestries (*razi*). It is taken from a late fifteenth-century Italian manuscript of Belo Moietta's *Trattato di geometria,* a work written for, and dedicated to, Agostino Barbarigo, doge of Venice from 1486 to 1501. The pictures alone, with the length and width of the tapestries duly inscribed, might lead one to think that determining their area was the primary goal of the theorem being illustrated. The areas are indeed calculated, but only to serve as the basis for discovering how much the larger carpet should cost, given the price of the smaller one (both, we are told, are of equal quality). The relevant calculation is made, incidentally, *per la regula delo 3* (in modern terms, that the product of the mean terms of a proportion equals the product of the extremes). We are also told that the same procedure can be applied to rectangles of wool, silk, and other things.

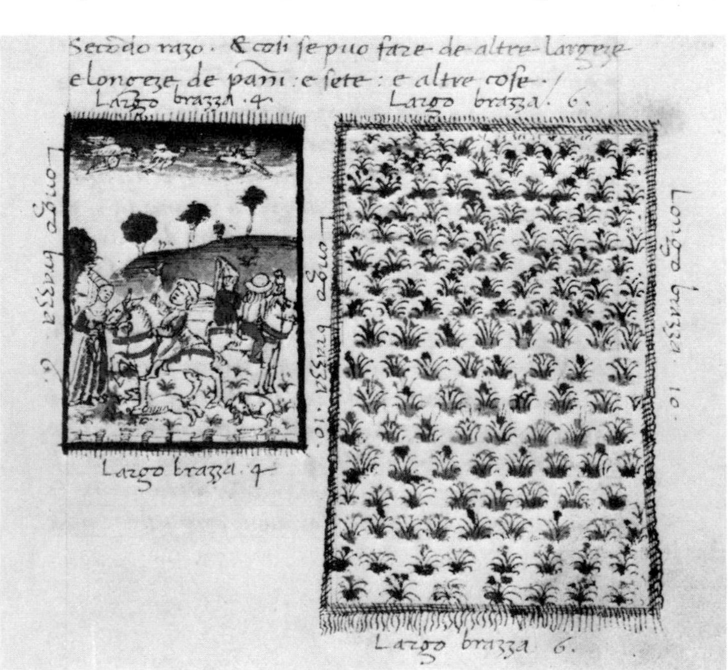

152. Ways of Measuring and Things Measured II. Perhaps the most popular example of measurement in medieval practical geometries was the height of towers. It was standard in any section dealing with altimetry in such works and it is from this kind of locale that these two illustrations are drawn. That on the lower left comes from a fourteenth-century manuscript of a treatise on the "old quadrant" (*quadrans vetus*) ascribed, if not conclusively, to Johannes Anglicus of Montpellier. Yet whoever the author, the treatise was composed in the twelfth or thirteenth century.

The tower's height is here determined by using the quadrant which is the subject of the treatise as the measuring instrument. For present purposes, it is sufficient to note only a few features of this *quadrans vetus* (for others, see Illustration 236). Two pinnule sights are fixed on one of the sides or radii terminating the quarter-circle forming the instrument and a plumb bob hangs from the right angle constituting the center of the circular quadrant. Rotating the instrument while holding the plane of its plate perpendicular to the ground will cause the plumb bob to mark off different angular degrees that are inscribed on the circumference of the quadrant.

The figure at the left on this page depicts the result of the following procedure: While keeping the top of the tower to be measured in the center of the two pinnule sights, one is to approach or back up from the tower until the plumb bob "reads" 45°. At that point, the observer and quadrant are positioned as indicated by the figure, where the observer's eye is at E, his feet at D. Next, mark out a distance (behind the observer) on the ground DC and also measure the distance BD from the foot of the tower to the foot of the observer; $BD + DC$ will give the height of the tower. The basis of the procedure is, of course, that the sighting by means of the quadrant has enabled one to determine an isosceles right triangle, one side of which is "on measurable ground," the other of which is the height of the tower; since the two sides are necessarily equal, the tower is thus measured.

The second figure in the same manuscript represents a case in which the tower's base is not at ground level (*non est in loco plano*), a situation which requires two sightings by the quadrant, one similar to that just described, the other designed to extend a perpendicular to a point on the tower which will determine a height to be added to that obtained by the usual 45° right triangle.

The second illustration on the lower right of page 168 is taken from an early sixteenth-century manuscript of the *Artis metrice practice compilatio* of the fifteenth-century scholar Leonardus de Antoniis de Cremona. The three figures given in this illustration picture different manners of tower measure, that at the top being equivalent to the procedure set forth in the *quadrans vetus* codex, although

(*Continued*)

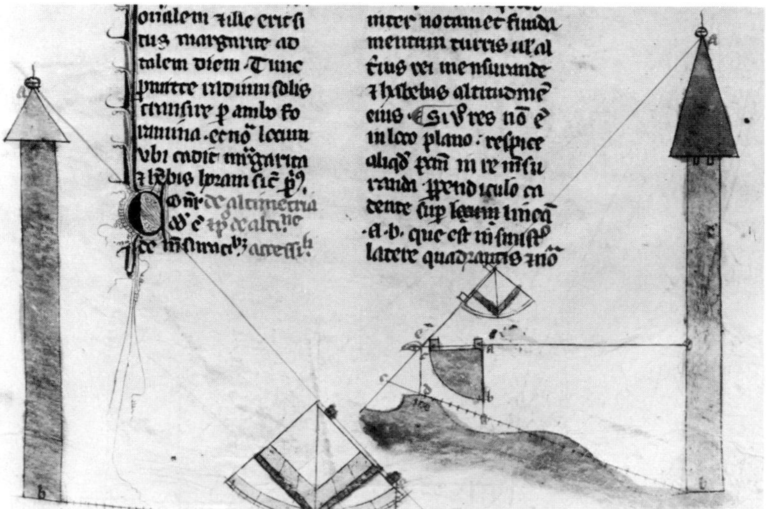

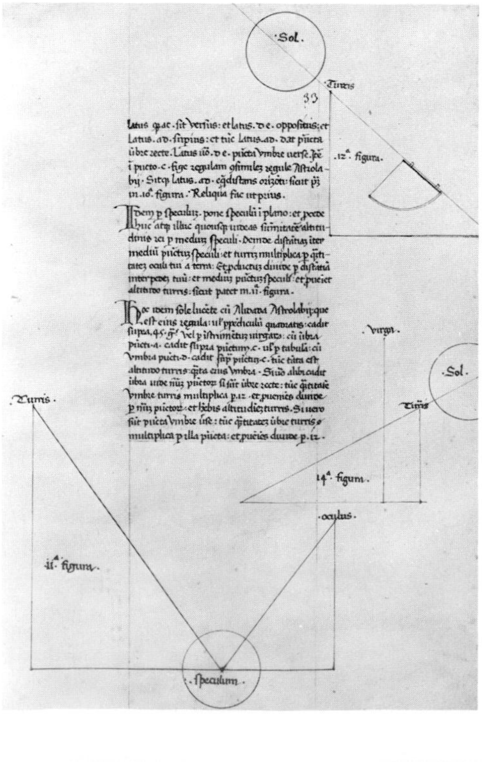

the text to which the present figure belongs also explains variant methods for obtaining the desired result. The middle figure represents a method of measuring a tower by shadows. Given sunshine, the base of the triangle in the figure is a shadow cast by the tower (*turris*); we are then to drive a rod (*virga*) perpendicularly into the ground at such a point that part of it falls within the tower's shadow, part without. If, to simplify matters, we take a as the unknown height of the tower, b as the length of its shadow, c as the part of the rod within this shadow, and d as the distance between the end of the tower's shadow and the foot of the rod (b, c, d all being directly measurable), then, by similar triangles, the height of the tower is obtained by operations equivalent to $a = \frac{(b \times c)}{d}$.

The final diagram in the illustration has to do with a theorem explaining how to measure a tower by utilizing a mirror (*speculum*). The mirror is placed on the ground in such a position that the top of the tower is seen in its center when viewed from an eye (*oculus*) situated at the vertex of the triangle on the right. Given this configuration, since the two triangles in the figure are similar, one can again directly derive the height of the tower by a procedure similar to that indicated in the case of the rod and the shadow. For other pictures of tower measure, see Illustration 82, and for other figures from the first of the manuscripts depicted here, see Illustrations 236 and 238.

153. Geometry in an Architect's Sketchbook. One of the most important works to have come from the pen of a medieval artist is the *Album* or *Sketchbook* of the thirteenth-century architect and scholar Villard de Honnecourt, who was responsible for developing plans for several churches and abbeys. Almost all we know of Villard derives from his *Sketchbook,* written between, roughly, 1225 and 1235, and containing 207 drawings covering an amazing variety of topics. Those pictured here at the right are due to an anonymous contributor to Villard's book and, as the legend in the lower right corner of the illustration announces, are all taken from geometry (*totes ces figures sunt estraites de geometrie*). The type of geometry in question was practical, even though no known specific treatise in that genre can be established to be the source for this page from the *Sketchbook.* A few samples from the text and drawings reproduced here will furnish an adequate idea of the kind of geometrical notions and operations that merited inclusion in this unique compilation. In the upper left corner of the illustration, the legs of a compass span an arc of a semicircle, the accompanying legend telling us that the problem at hand is the measurement of the size of a column which is only partly visible (*la grosse d'one colonbe que on ne voit mie tote*). If we compare this drawing with that immediately to the right in the same illustration, it seems clear that the function of the compass in the first drawing was to determine three points (one by each leg and one in the center of the arc subtended by the legs) through which a circle could be drawn, the circle thus giving the size of the column as a whole. But although the legend to this drawing tells us that it represents the method of finding the center of the circle passing through the three points (*le point en mi on caupe a conpas*), no indication is given of how that point is found.

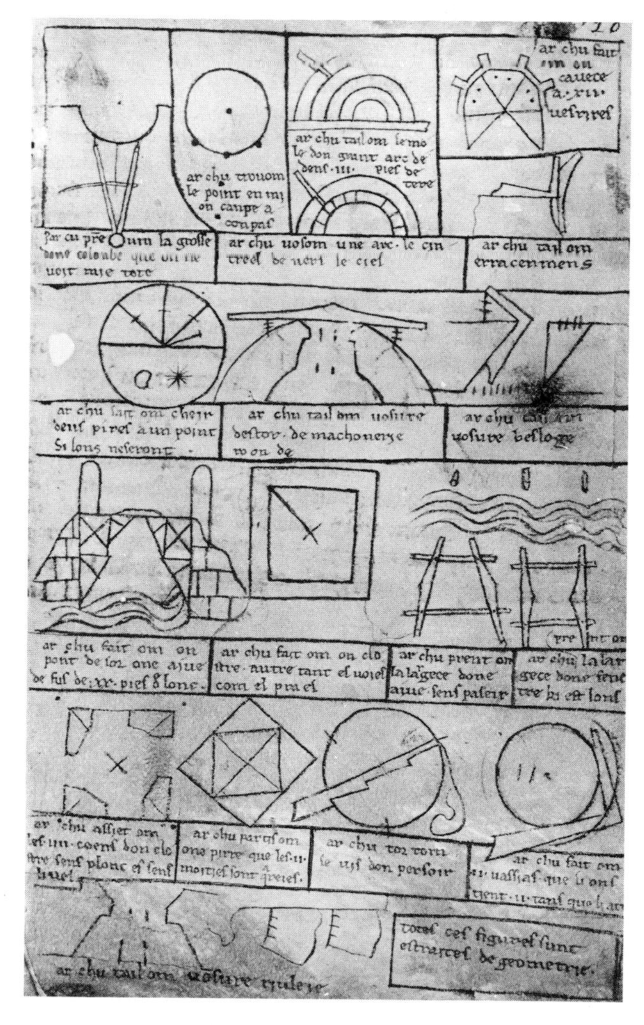

Secondly, turning our attention to the drawing in the lower right corner of the illustration, we immediately recognize the geometer's or mason's square (see Illustration 150) being applied to a circular object. The relevant legend informs us that the problem is the construction of two vessels such that one holds twice as much as the other (*.ii. vassias que li ons tient .ii. tans que li atres*). Just how this is to be accomplished is only too evident from the drawing: Some scribing instrument is placed in the inner corner of the square while it is rotated about the circular object, its two arms always remaining in contact with that object. The resulting second circle—part of which is already traced in the figure—will have an area twice that of the first circle (since its radius will stand to that of the first circle as the diagonal of a square to its side). For another drawing from Villard's *Sketchbook,* see Illustration 203.

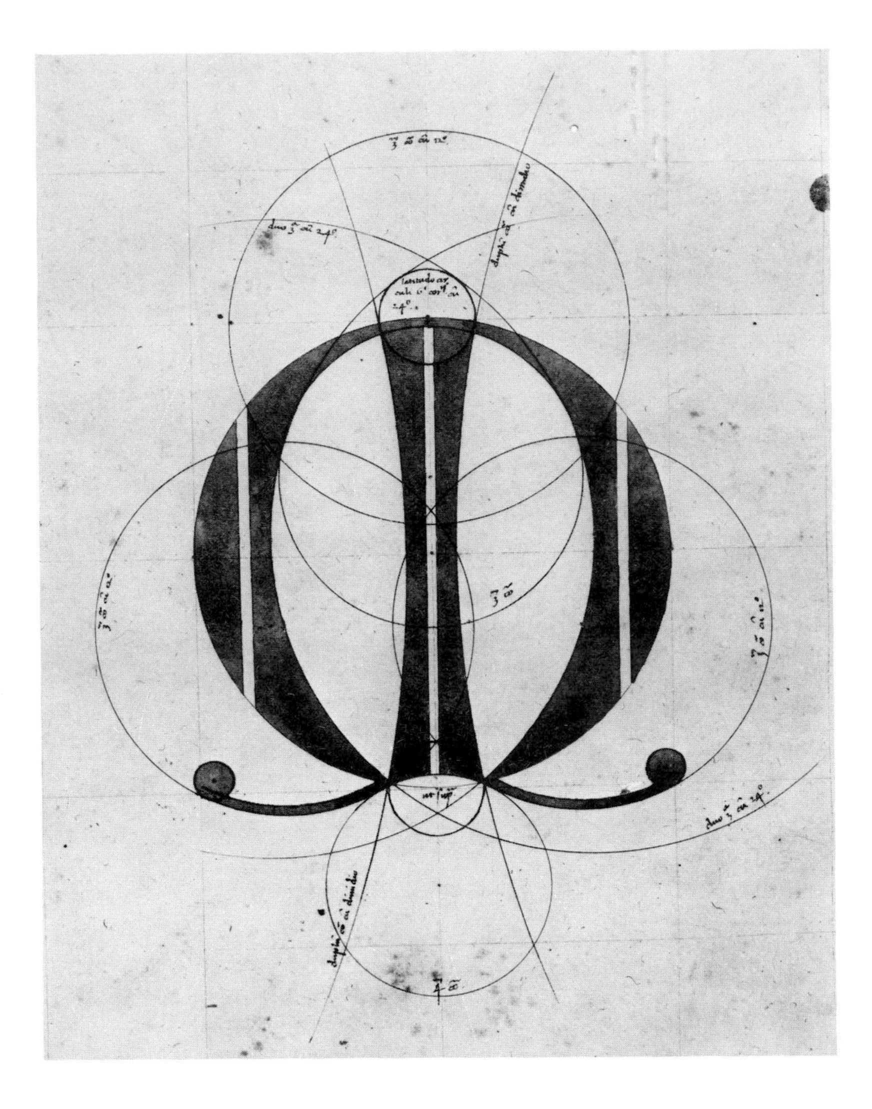

154. Geometry and the Alphabet. In this fifteenth-century manuscript of an anonymous work *de arte scripturali,* geometry has been applied to another domain—that of calligraphy. Although there is no text associated with the letter *M* depicted here (indeed, the whole manuscript consists simply of twenty-odd folios of nothing but figures like the present one), it is clear that save for the three white vertical "stripes" in the *M,* all elements of the letter are carefully constructed by drawing circles with specified centers and radii, the sizes for the latter being appropriately indicated in a very small hand for all circles and circular arcs except the two small circles at the ends of the serifs of the *M.* The sizes of the circles and arcs are all related to a single radius or circle taken as a standard (here termed 'corpus'). Thus, the two arcs forming the sides of the central vertical element of this *M* are two and a half times the standard, the circle bounding this vertical from below is one-fourth the standard, and so on.

Part Four

REPRESENTATION OF THE SCIENTIFIC OBJECT I:

The Scientist and the Sciences

155. Islamic Astronomers. This elaborate miniature is from a sixteenth-century Persian manuscript of a book called *Shahinshāhnāma* (the "Book of the Great Shah"), composed of poems chronicling part of the reign of Sultan Murad III, Ottoman king from 1574 to 1595. Written by ^{c}Alā al-Dīn al-Manṣūr, in addition to recounting expected historical events, the poems celebrate the Istanbul astronomical observatory, its head astronomer Taqī al-Dīn, and even the comet of 1577. The miniature reproduced here depicts the "small observatory," the text at the top giving the initial six lines of four couplets telling us, among other things, that the use of each astronomical instrument involved five wise and learned men: two or three observers, a clerk, and a fifth person for sundry duties. The picture itself is unusually replete with the artifacts of the practicing astronomer. Books are shelved at the upper right (compare the Islamic library in Illustration 185). The table in the center of the room contains all manner of astronomical instruments: hour glasses, what appear to be armillary spheres and plumb bobs of some sort, geometers' squares and triangles, and even a mechanical clock (at the extreme right). Scholars are depicted writing, discussing or sighting with quadrants or other instruments, holding an astrolabe, drawing with a compass, and fixing a tripod. A terrestrial globe is at the bottom center.

THE CONCERN OF THE present and the next parts of this *Album* is the "scientific object." The term is one of convenience, since it is certainly not meant to be limited to covering those physical entities which formed the subjects of investigation for given scientific disciplines. As can be seen from the chapters that follow, the "scientific objects" in question include those devices used in investigating a subject, the places or locales in which the science is learned or carried out, and, as in the present chapter, the scientists themselves.

All these things have been called "objects," not because their depiction in illustrations has in one way or another resulted from their having been observed, but because, unlike concepts or theories, in most instances they were at least capable of being observed. The qualification "in most instances" is needed because when portraying ancient authorities and legends or personifying the "object" at hand, it was not possible for the illustrator to observe his subject. More than that, even when the subject may have been observable—as in the case of a contemporary scientist, the then-extant instruments used in a science, various flora and fauna, and so on—in the majority of cases the relevant observing was not done. Instead, the illustrator constructed his depiction from what was said about the object in a text or, most frequently, from an earlier one found in another manuscript.

In contrast to the celestial phenomena, plants, animals, and instruments that will be the concern of Part Five, the illustrations of scientists and the sciences in the following three chapters were not, for the most part, tied to the substance of a given text. Frequently, as in the case of author portraits, like a frontispiece, they may have been attached to the beginning of a text, but they only occasionally served to explain the contentions of the text. At times such "introductory illustrations" occupied whole pages in a manuscript while at others they were limited to the space of an illuminated initial. Yet whatever their format, they almost always reflected current styles of manuscript illumination which themselves reflected current styles of contemporary painting, a reflection that covered not only the representation of the human figure, but also such elements as background, framing, and coloration. In addition, it was natural that the portrayal of a scientist frequently followed contemporary notions of costume: Arabic garments and headdress, university robes, and clerical garb in general were often considered appropriate for ancient as well as contemporary personages.

Depictions of a scientist or the sciences can be found in sculpture, in wall painting, and in other artistic artifacts, as well as in manuscript illuminations and drawings, but the illustrations given in the following chapters have largely been drawn from the last-named medium, since that is where they are usually more informative, however incompletely, of the actual science involved.

CHAPTER FIFTEEN

The Practitioner of Science

DIAGRAMS WERE ESSENTIAL for mathematical texts, the depiction of plants and animals was standard in materia medica and bestiaries, and other kinds of scientific writings required their own specific type of illustration, but portrayals of the scientist were visual supplements that could be found in manuscripts of works belonging to any scientific discipline. Although most of the extant miniatures and drawings of this sort are medieval, we know that author portraits were part of ancient practice as well. The poet Martial (first century), for example, speaks of a parchment copy of Vergil's works that contained his portrait. Although we no longer possess illustrated codices of such an early date, somewhat later manuscripts allow us to guess that these early portraits may have been medallionlike in appearance with circular borders or distinctly framed with definite rectangular boundaries. An early fifth-century manuscript of Vergil's *Georgics,* for instance, contained such a framed "group portrait"—unfortunately in a considerably damaged state—of the author describing the art of beekeeping to his confreres. Similarly, the sixth-century original—again considerably damaged—of our Illustration 10 is also framed by a broad square border.

Author portraits appear to have had a long history in manuscripts of such writings as the Gospels, but scientific and philosophical manuscripts bore a similar tradition, if not quite as ancient a one, if we are to judge from the claim of at least one scholar who used Greek codices in making translations into Arabic (see Illustration 156). The portrayal of scientists, however, often went beyond what one finds in the usual author portrait in religious or philosophical works. True, scientific texts also contained the simple portrait in various degrees of stylization and idealization. But they also frequently attempted to portray the scientist together with other identifying features. In fact, this seems to have been the predominant practice.

The scientific practitioner might be depicted, for example, together with selected "tools of his trade" (Illustrations 158, 160–163), in which case these tools might be shown in use or simply juxtaposed to the practitioner being represented. Alternatively, the scientist might be pictured "at work" (Illustrations 155, 164–168), sometimes identifiably in the location appropriate to his enterprise (for which see Chapter 17). At times, such portrayals merely revealed the scientist plying his trade, but in others the intent was more didactic (Illustrations 166–168). That is, the picture was meant to show, even if only very sketchily, just how one might perform the particular "scientific task" at hand.

Clearly, in all these cases, that which was depicted was imagined as often as it was pictured from some person or some event seen. In many instances, the subject of the picture could not possibly have been seen. Such is the case, of course, in the mere portrayal of any ancient author in a medieval manuscript. But it also obtained whenever legendary figures or events were involved (Illustrations 157, 163, 169, 170) or where the illustration of satirical and critical judgments of earlier and contemporary figures was the artist's aim (Illustration 171).

In addition to the depiction of the scientific practitioner in the present and the two following chapters, similar representations can be seen in Illustration 11—where the rare occurrence of how an author portrait was executed is given—and in Illustrations 3, 10, 76, 150, 195, 241, 261, and 265.

156. The Medical Professional. Almost any census of the kinds of scholars and scientists portrayed on ancient monuments or artifacts and, especially, in medieval manuscripts would reveal that those most privileged in receiving such a portrayal were connected with the profession of medicine. Three examples of this "majority" are given here. On the top left, a late Hellenistic bas-relief shows a Roman physician reading a papyrus roll, other rolls being visible through the open doors of the cabinet in front of him. Some kind of case containing medical instruments sits on the top of the cabinet. The picture at the middle left is taken from an Arabic manuscript written in Baghdad in the thirteenth century by one al-Ḥasan ibn Aḥmād ibn Muḥammad al-Nashawī. The codex contains an Arabic translation of Books III to V of Dioscorides' (first-century) *Materia medica,* the present illustration depicting the author, book in hand, in proper Arabic costume. We know that the most renowned translator of all from Greek (via Syriac) to Arabic—the ninth-century Nestorian physician, philosopher, and theologian Ḥunayn ibn Isḥāq—worked from Greek codices or rolls that often contained "author portraits," a practice that is here imitated in an Arabic manuscript, even though of a much later date than Ḥunayn. The picture at the bottom left is taken from the initial folios of a fourteenth-century manuscript of the *De herbis et plantis* of an otherwise unknown Manfredus de Monte Imperiali. A fair number of celebrated medical personalities are depicted on these opening pages, Hippocrates (*Ypocras*) and Galen (*Galienus*) being the two "conversing" here. The text between them asserts that "I intend to eat so that I may live, others intend to live so they may eat." As was the case with so many miniatures of Hippocrates, he here holds a book containing the opening lines of his *Aphorisms:* "Life is short, the Art [namely, medicine] long, opportunity elusive, experiment dangerous, judgment difficult" (*vita brevis, ars longa, tempus acutum, experimentum vero fallax, iudicium* [*difficile*]). Facing him, Galen holds a scroll at which he points, presumably thereby emphasizing the Galenic precept for health inscribed upon it: "Whoever wishes to maintain continuous health should take heed of the stomach lest it reject the food necessary to it" (*quicumque vult continuam custodire sanitatem, custodiat stomachum ne, cum sibi necessarium sit, prohibeat cibum*). This particular precept was popular enough, it seems, to have formed the opening lines of a *Practica medicinae* written by one John Bray, presumably royal physician (d. 1381) to Edward III and John of Gaunt, duke of Lancaster. Both Hippocrates and Galen are here represented in typical later medieval fashion, slightly younger and not quite the aging sages one finds in various earlier portrayals.

157. (Above) Patron Saints of Medicine. In many instances, the beginnings of medieval medical writings were decorated with representations of authoritative doctors of antiquity or, in some cases, of the author himself of the work in question. Less frequent was the tradition of opening a work with miniatures of the patron saints of physicians. It is this tradition which is exemplified here in a manuscript dated 1464 of the *Practica* of the early fifteenth-century Italian physician Michele Savonarola, grandfather of the reformer Girolamo Savonarola. Three of the four saints traditionally associated with medicine are present: Cosmas, Luke, and Damian. Luke is depicted with the head of an ox, the standard symbol for this evangelist. Cosmas stands at the left, holding what appears to be either a scroll or a mortar, while Damian is on the right, a urine flask in his right hand. All three are surrounded by "ribbons" inscribed with sayings from Aristotle, Seneca, and Cicero. Almost nothing is known of Cosmas or Damian, apart from their role as medical saints, although the traditional account of this role claimed that they were twins, that they practiced medicine without recompense, and that they eventually suffered martyrdom. The fourth, absent patron saint of medicine was Saint Pantaleon (d. 305).

158. (Below) Contemporaneous Details Reported. Apart from the habits worn by scholars and, perhaps, an occasional artifact or symbol of their profession or station in life, medieval pictures of contemporaries or near contemporaries often did not trouble to represent other specific details connected with the scientist or other individual being depicted. In some cases, however, the trouble was taken, as in the present miniature of the fourteenth-century English astronomer, mathematician, and abbot of Saint Alban's, Richard of Wallingford. Taken from a fifteenth-century manuscript of Thomas Walsingham's *Gesta abbatum monasterii Sancti Albani,* Richard is depicted with one of the symbols of his ecclesiastical office resting on his shoulder, while busying himself with the division of a metal disc. This is a clear reference to his known preoccupation with the construction of astronomical instruments (something the text of the *Gesta* even cites as one of his accomplishments). One is tempted to imagine that the artist intended to represent Richard in the course of designing one of the astronomical instruments with which he is personally connected, the so-called *Albion* (from "all-by-one"), since it involved brass discs in its construction. Further, the blotches on Richard's face are undoubtedly intended to represent the leprosy from which he is reported to have suffered in his later years. These details are such that, were we not informed by the text directly below the picture—*Ricardus abbas de Walingfordia*—of the identity of the person depicted, we would most likely be able to establish that identity on the basis of these details. For a drawing of an astronomical instrument (other than the *Albion*) ascribed to Richard, see Illustration 240.

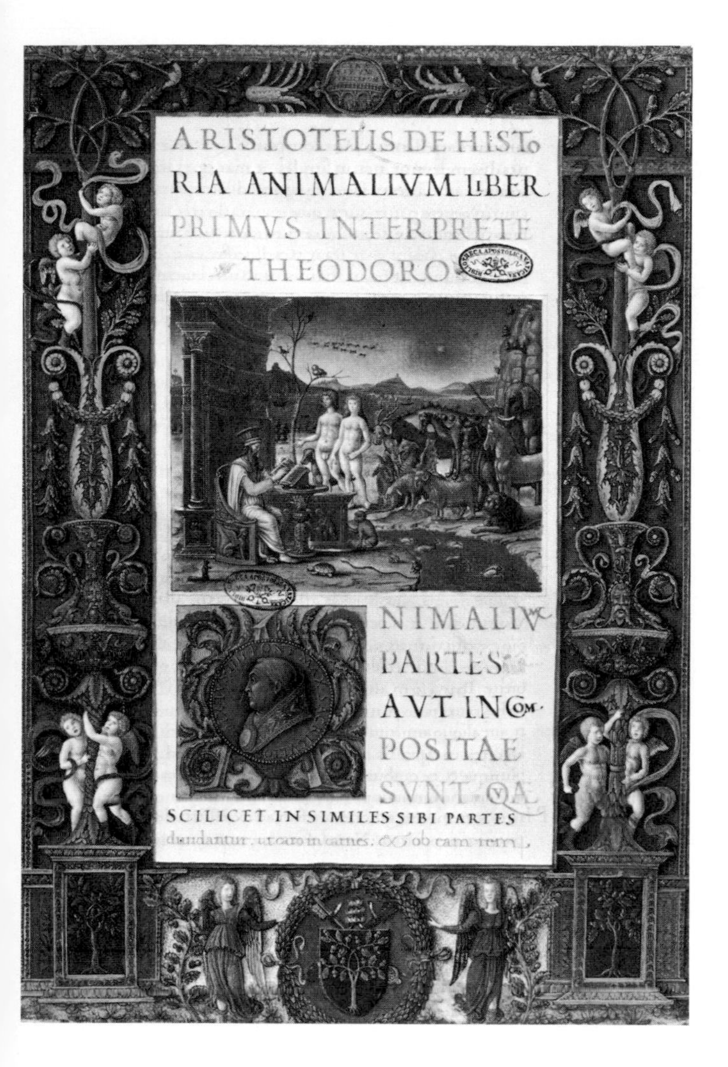

159. (Above) Renaissance Naturalism and Idealization. The present picture is an excellent example of the Renaissance frontispiece to a presentation, or otherwise special, copy of a scientific work. An especially elegant and decorative exemplar of Theodorus Gaza's (fifteenth-century) Latin translation of Aristotle's *Historia animalium,* this codex was produced between the years 1471 and 1484 and was dedicated and presented to Pope Sixtus IV (whose portrait appears in the medallion within the illuminated initial). Surrounded by a marginal frame filled with foliage, angels, and putti, the central miniature shows Aristotle seated at a table writing, while the animals which provide the source material for his work stand "open for inspection" before him. Two naked human beings—presumably Adam and Eve—add a typical Renaissance touch.

160. (Below) An Illuminated Table of Contents. In this fifteenth-century Greek medical miscellany, the scribe or illuminator has seen fit to head the "table of contents" of one of the excerpts making up the manuscript with a pertinent portrait. The excerpt in question (taken from Book VII of the *Epitome medicae* of the seventh-century Alexandrian physician Paul of Aegina) treats pastilles or lozenges (*trochista*) as a part, however minor, of the doctrine of simple and compound medicines. Given such a subject, it was apparently natural for the illuminator to open it with a picture of a pharmacist standing before a table filled with various bottles and phials, while announcing by his gesture "here are my wares." Although without a proper name, he is identified by the inscription above his head as a Jew. For another page from this manuscript, see Illustration 269.

161. Group Portraits. We have already seen Hippocrates and Galen represented together in one of the pictures of Illustration 156. Such "coupling" of important personages was a frequent occurrence in the medieval iconographical tradition, in religious as well as scientific and other contexts. In some instances, the two individuals depicted were master and disciple, in others simply two authorities in the discipline in question. The latter is the case in the first picture given at the top left. Taken from a thirteenth-century manuscript, this "group portrait" serves as a frontispiece for a copy of the *Experimentarius* of Bernardus Silvestris (twelfth century). On the left of the picture, Euclid in medieval garb has been made to serve as a representative of astronomy. He holds what appears to be some kind of armillary sphere in his right hand and is gazing through some type of sighting tube held in his left. On the right of the picture, the figure with an astrolabe hanging from his fingers is identified as Hermannus. This is undoubtedly Hermannus Contractus (that is, Herman the Lame), an eleventh-century German scholar who was influential in the introduction of Arabic astronomical ideas and techniques to the Latin West and who was the author of two frequently consulted treatises on the astrolabe (the *De mensura astrolabii* and the *De utilitatibus astrolabii*). (For other pictorial material of, or related to, the astrolabe, see Illustrations 155, 162, 237–9, 241.) This drawing—like that of Illustration 4—derives from the school of Matthew of Paris, possibly even from his own hand.

The picture at the top right comes from a late fifteenth-century manuscript of the middle-English *Ordinal of Alchemy* of Thomas Norton, the writing of which began in 1477. Here we have not merely two, but four authoritative personalities represented by nothing more than their heads. They are, from left to right, Geberus (who is not the eighth- to ninth-century Jābir ibn Ḥayyān of Arabic alchemy, but the hitherto unidentified author of a number of important Latin treatises in the discipline), Arnaldus (that is, the thirteenth-century Spanish physician and philosopher, Arnald of Villanova), Rasis (the ninth- to tenth-century Arabic scholar Abū Bakr Muḥammad ibn Zakariyyā al-Rāzī), and Hermes (the legendary Hermes Trismegistus, to whom a great amount of literature in the occult sciences was assigned). The works ascribed to all four of these personalities were fundamental within the history of medieval alchemy. Their heads are "sending instructions" inscribed on twisted ribbons to the "students of the art" depicted below. Rasis, for example, "says" that a body will become dry in proportion to the amount it drinks in (*quotiens corpus inbibitur, totiens desiccatur*). The *discipuli* below are shown busying themselves with one or another procedure that was basic to alchemy: grinding with mortar and pestle, tending an alchemical furnace, and stirring a pot. Scholars have hazarded that the second figure from the right is meant to represent Thomas Norton himself. For another miniature from this manuscript of his work, see Illustration 182.

162. Scientists and Instruments. Any number of the pictures of scholars that have appeared on the preceding pages have contained representations of some kind of "tool of the trade" associated with the discipline of the scholar portrayed, a number that becomes even greater if one considers, as one should, books as being such tools. At times, the depiction of these artifacts took on an importance equal to, if not greater than, the portrayal of the scholar himself. Two instances of this are given here. The first at the bottom left is from a fifteenth-century Italian manuscript of the Greek text of Ptolemy's *Geography*. Done by a follower of the Lombard miniaturist Belbello of Pavia, it contains the elaborate framing and architectural detail characteristic of so many Italian illuminated manuscripts of this period. Our interest, however, is in the astronomical instruments accompanying this rather kingly Ptolemy. He holds one astrolabe in his hand while another hangs on the wall in his "study" behind him. In addition to numerous books, his study also contains a quadrant hanging on the window shutter, and what seem to be torquetumlike instruments on the table (the torquetum being an astronomical instrument with equatorial and ecliptic movements).

The much less elaborate miniature depicted at the lower right comes from a fifteenth-century manuscript of Nicole Oresme's fourteenth-century Old French *Traitié de l'espere,* a work which he considered to be propaedeutic to his *Livre du ciel et du monde,* a translation of and commentary on—also in Old French—Aristotle's *De caelo* (for which, see Illustrations 189 and 234—from the present manuscript—and 254, 255, and 278).

163. Science in a Medieval Romance. Medieval vernacular literature contained an astonishing number of versions, in both poetry and prose, of a fantasized biography of Alexander the Great. Originally Greek, the legend came to the West from at least two Latin sources, one of them being the *Historia Alexandri Magni de preliis,* written in the tenth century by the Neapolitan archpriest Leo. Augmented by numerous additions, the *Historia de preliis* served as the base for much of the corresponding vernacular material, including the prose version of the story in Old French, the *Vraye histoire du bon roy Alexandre,* pictured here from a fifteenth-century manuscript copy. After a brief introduction, this *Vraye histoire* begins to follow its Latin progenitor at about the point represented here, the subject being the excellence of the Egyptians among early peoples. They studied things both human and celestial, we are told, and arrived at the certainty of the noble science called astronomy, by means of which they could know things past, present, and future. This, plus the mention of Egyptian knowledge of geometry (*mensura terrae*) present in the Latin but absent from this particular French version of the tale, provided the raw material for the miniature given here. The cleric second from the left gazes heavenward, while the person in front of him points instructively at the armillary sphere in his hands. Next we find two others displaying the instruments of geometry: square, wand, and compass. Between them stands another scholar reading from a book. As fanciful as all of this may be, it is significant to note that the illuminator of this particular copy of the *Vraye histoire* saw fit to supplement the talk about Egyptian competence in geometry and astronomy by including a few of the implements of these sciences which were, by then, well-known symbols of these disciplines. The belief that geometry in particular had its origins in Egypt can be found in numerous ancient Greek sources, going back to Herodotus (fifth century B.C.) at least, but it is unclear in precisely which way this belief became part of the Alexander legend.

164. Two Stages of a Scientist's Work. Written about the year 1200, most likely in Germany, the manuscript depicted here contains a number of standard works on materia medica. Accordingly, it opens with these two full-page miniatures of various phases of activity characteristic of the medieval pharmaceutical profession. That on the left shows one's initial preoccupation in this profession: the gathering of herbs from the field. The roots are being dug out with curious "crutchlike" implements, perhaps somehow related in the illustrator's mind to the instructions in the text that follows not to use "an iron tool" for such a task. This particular stage of work was well described by the original Greek term for such professionals: namely, "rhizotomists" (literally, root cutters). The palm-like tree in the background seems more decorative than instructive and the herbs that have been removed from the earth are not clearly identifiable. In any event, they do not seem to be anything like the bettony described in the text immediately following (the *De herba vettonica* of Pseudo-Musa). The facing page on the right depicts the next step in the herbalist's procedure: the weighing of the herbs collected, here presumably being carried out under the direction of a physician. After being weighed, the herbs would be appropriately processed for use.

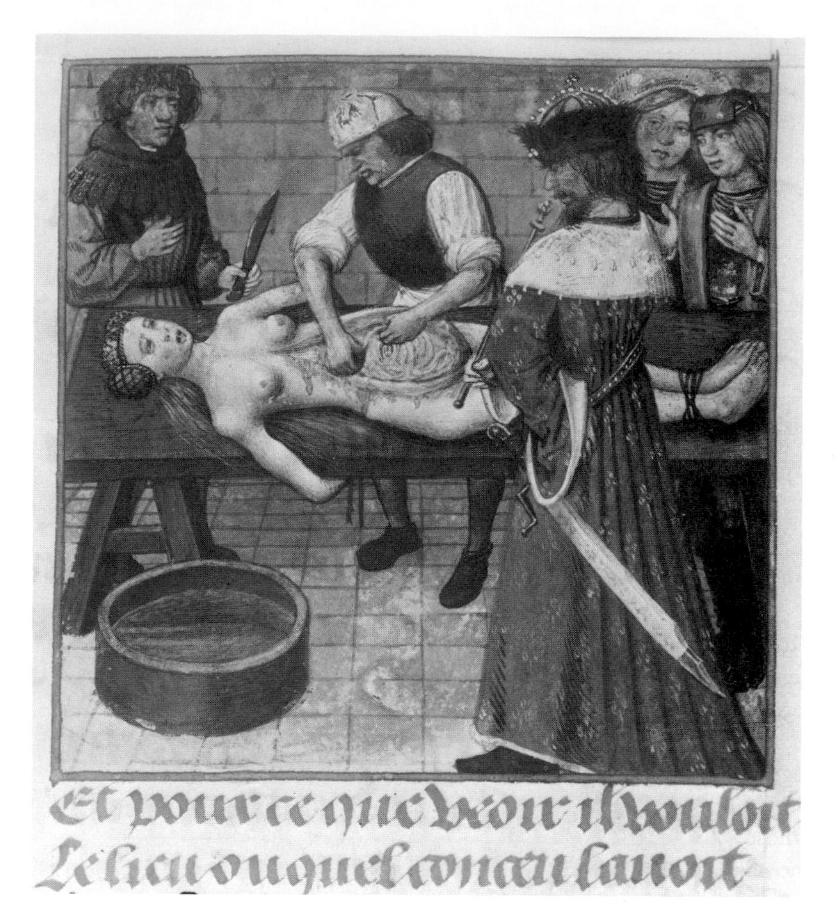

Et pour ce que veoir il vouloit
Le lieu ouquel conceu sauoit

165. Anatomy True and False. The picture at the top left, done by a Flemish miniaturist, is taken from a fifteenth-century manuscript of Gerard of Cremona's Latin translation of the *Canon* of Avicenna (eleventh century), one of the central medical texts throughout the whole Middle Ages (see Illustration 8). The miniature's role is the illustration of a chapter in Avicenna dealing with the nerves (*sermo universalis de nervis proprie*). At first glance, the cadaver lying on the table appears to be a skeleton, but closer inspection reveals that the abdomen still bears an incision through which one can see several intestinal coils. Ermine-robed doctors are clearly in charge of the "neural" anatomy lesson being represented, one of them employing a pointer to indicate the cervical nerves. Although the miniature at the top right might also be thought to be related to some kind of anatomical dissection, nothing could be further from the case. Instead, it illustrates a legend about Nero that was rather popular in the Middle Ages. Here appearing in a fifteenth-century manuscript of the *Roman de la rose* of Jean de Meun (thirteenth century), the topic in question is the horrendous criminal behavior of that Roman emperor. In the course of setting forth the capricious injustice of Fortune, Jean elaborates a passage he had found in Boethius (*Consolation of Philosophy,* II, 6) and recounts not only the familiar burning of Rome, but Nero's dismemberment of his mother so that he could see the place in which he was conceived. It is the depiction of this grisly deed, then, that is behind the "dissection" portrayed and not anything with medical intent, even though the miniature alone may bear some resemblance to some of the pictures that accompanied anatomical or surgical topics.

166. The Preparation of a Medical Draught. This miniature at the top right is taken from a thirteenth-century manuscript of the Arabic Dioscorides illustrated by the so-called Baghdad school. It is meant to picture the preparation of squill wine (Book V, chapter 26), a potion that was held to cure stomach ailments. A funnel, through which the potion is filtered, hangs from a tripod standing between a pomegranate and an orange tree. The physician stands at the right, explaining things to an assistant who seems to find the process of learning his prospective trade slightly perplexing. For more Arabic Dioscorides, see Illustrations 156 and 194.

167. Practitioners Illustrating Surgical Procedures. Occurring at the very end of a twelfth- or thirteenth-century manuscript of texts on materia medica, the illustration at middle right depicts three different surgical operations. The first, at the upper left of the illustration, carries a legend announcing that "hemorrhoids are cut out in this manner" (*emoroida inciditur sic*). The patient stands, bending over, in a gobletlike dish whose function is undoubtedly to catch the flow of blood. In his left hand, the surgeon holds a kind of two-pronged retractor—some variant, one would think, of the Greek *angistron* or hook—in order to extend each pile, which can then be cut by the knife held in his right hand. Although the illustration accompanies no particular text, the procedure it depicts was quite standard and could be found described in Greek, Arabic, and Latin sources. The picture directly below in this same illustration shows the appropriate surgical procedure for the removal of nasal polyps (*fungus de nare sic inciditur*). Again provision is made for the catching of blood, this time by a vessel held by the patient, and again the surgeon wields two instruments: an oddly shaped knife in his right hand for incising the polyp and, in his left, what appears to be some kind of cannula, the exact purpose of which is not clear. The final picture in this group depicts the removal of a cataract (*albule oculorum sic excutiuntur*). A single, needlelike instrument is used by the surgeon in this instance, while he holds the patient himself with his free hand. It is uncertain just what object is being held by the patient, but some historians have conjectured that it is some sort of bandaging material.

The illustration at the bottom right comes from a manuscript of a Turkish medical work: *The Ilkhani Surgery (Jarrāḥīye-i-Ilkhānīye)* of the fifteenth-century physician Sharaf al-Dīn ibn ᶜAlī al-Ḥājj Ilyās, surnamed Sabunju-oghlu (that is, "son of the soap merchant"). Written and presented to the sultan Mahmud II in 1465–1466, this work is a translation and compilation from the more famous *Surgery* of the tenth-century physician Abū ᵓl-Qāsim Khalaf ibn ᶜAbbās al-Zahrāwī (known in Latin as Albucasis, Abulcasis, or Alsaharavius), which in turn depended heavily on the seventh-century Greek work of Paul of Aegina. The particular operation depicted here is that for hydrocele; that is, a watery hernia in the scrotum. Corrective surgery for such a condition was frequently carried out by means of incision, but we also know that cautery could be employed, even though in such a case the cauterizing implement would be sharp enough to cut while cauterizing. To judge from the shape of the instrument being utilized in this illustration, it would seem that it is this cauterization procedure that is represented. For the depiction of instruments from Abulcasis, see Illustration 244.

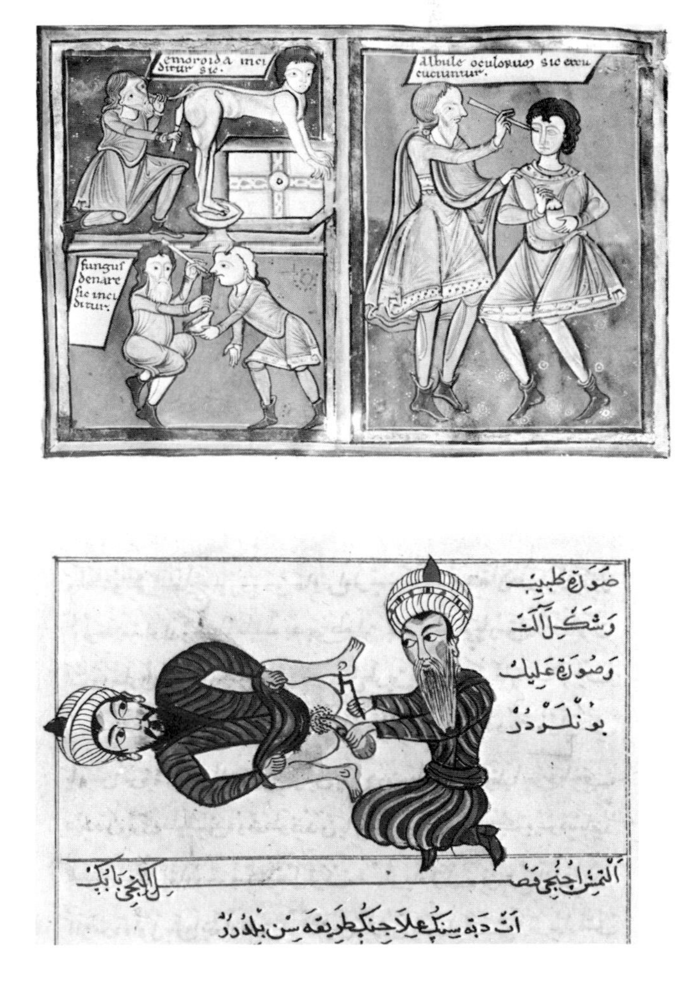

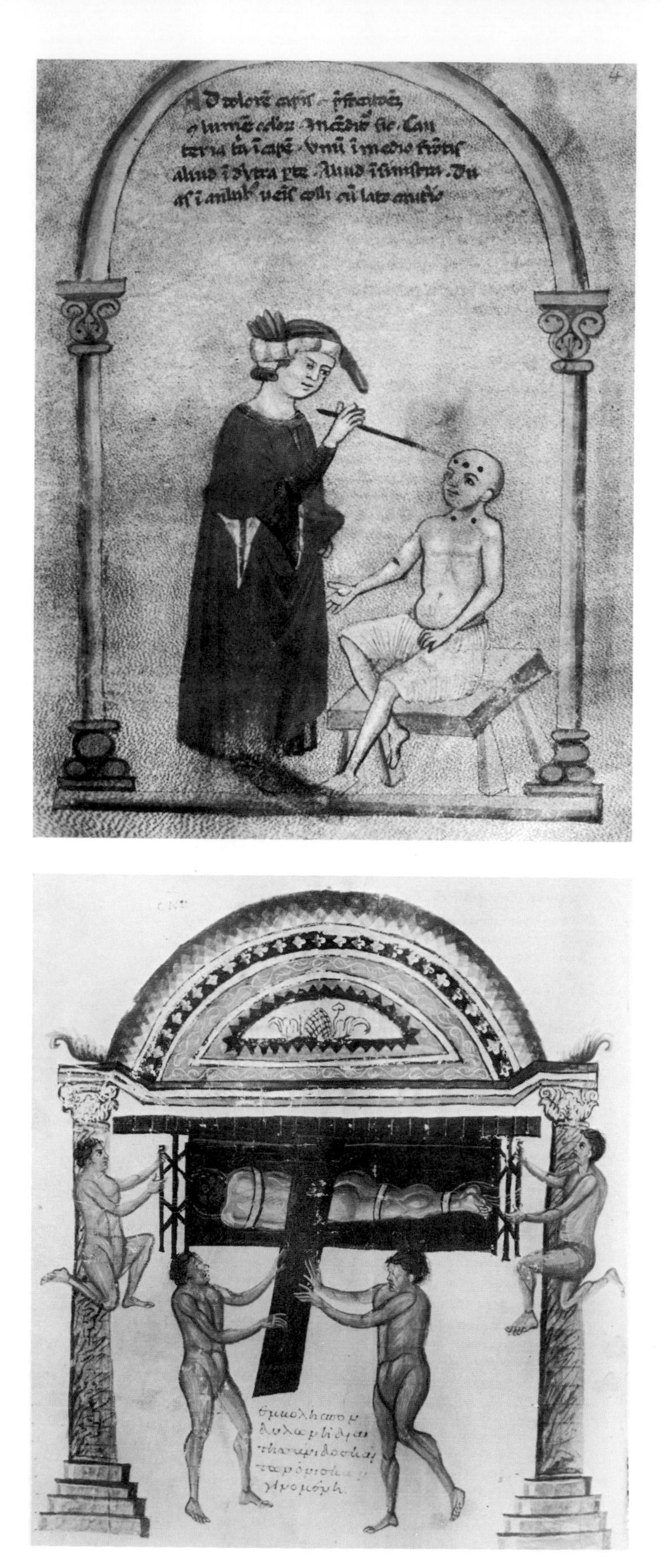

168. Practitioners of Cautery and Orthopedics. The picture at the top left comes from a fourteenth-century Italian medical miscellany and is one of eleven miniatures depicting cauterization, none of which belongs to a specific medical work, although they do contain textual legends of their own. That which heads the present picture informs us that the cautery in question is that appropriate for such things as head illnesses and the impairment of vision. The required technique is clearly represented in the miniature: cauterize three times on the forehead and then twice on the veins of the neck, performing the latter with a somewhat broader cautery iron. The arched frame of the whole picture is rather classical in appearance (compare Illustration 24), but the physician or surgeon is distinctly medieval, his gown clearly indicating that, whatever his status, he has received university training.

The second miniature at the bottom left, in the arcaded presentation so frequent in Greek illustrations, comes from a tenth-century manuscript of Apollonius of Citium's (first-century B.C.) commentary on the *De articulis* (*On Articulations* or *On Joints*) of Hippocrates (fifth century B.C.). It is much less concerned than the cautery illustration with portraying the practitioner; its primary burden is the representation of the orthopedic technique at hand. Here the legend tells us that the procedure depicted is the straightening of vertebrae by means of a plank and a windlass, but the precise text being illustrated makes things more specific. Apollonius is quoting Hippocrates' recommended procedure (*De articulis,* xlvii) for the reduction of humpback. The patient is to lie on a board, adequately bound and attached at head and foot to the windlasslike apparatus, which, through appropriate turning or twisting, will extend the spine. At the same time, the patient is to be fixed in a position such that his spine is slightly above a slot cut in the wall; a plank is then inserted into this slot so that, cushioned by several layers of thick cloth, it rests upon the hump of the spine. With the apparatus so arranged, two assistants apply traction to the spine, while two others pull down on the plank, thereby applying adequate pressure to the hump. That assistants are involved, and not a physician, is indicated by their being depicted as naked. Unlike the preceding cautery illustration, where the legend alone together with the cautery points depicted was sufficient to explain the procedure in question, the present illustration requires reference to an accompanying text in order to comprehend fully what it visually "teaches." For other pictures of cautery and traction machines, see Illustrations 244 and 264.

169. The Archimedes of Legend. Apart from his accomplishments in mathematics and mechanics—which are arguably the most impressive and significant in antiquity—Archimedes (third century B.C.) was also the subject of numerous traditional stories that celebrated his feats as a scientist and engineer. Perhaps the best known of these stories is that told by Vitruvius (first century) of how King Hieron II of Syracuse set Archimedes the task of determining whether a particular crown was of pure gold or had been fraudulently alloyed with silver. Vitruvius goes on to tell us that Archimedes came upon a method of solving the problem with which he had been charged while sitting in his bath, when he noticed that the amount of water which overflowed the tub was equal to the amount by which his body was immersed. So pleased was he with his discovery, we are told, that he leapt from his bath and ran naked through the streets shouting *heurēka, heurēka* ("I have found it, I have found it").

Only slightly less famous is the account of one of Archimedes' many feats in the defense of Syracuse against the Romans: namely, the use of burning mirrors to set fire to the enemy fleet. The earliest record we have of this tale is in Lucian (first century), who omits, however, any reference to burning mirrors and merely says that Archimedes accomplished his incendiary task by artificial means. The first to tell us that burning mirrors were used to focus the sun's rays on the Roman ships is the second-century physician and philosopher Galen. It is this more complete and better-known version of the tale that is depicted at the right as found in a sixteenth-century manuscript of the *De mechanicis quaestionibus* of Matthaeus Christianus. For other pictures from this manuscript, see Illustrations 35, 254, and 281.

170. The Aristotle of Legend. It was not an infrequent practice in the Middle Ages to claim that some ancient scholar had suffered the privilege of having rewarding contacts with Christianity or one of the other "religions of the book." Plato, for example, was held to have visited Egypt and there to have come to know Scripture from the Old Testament. And Ovid was reputed to have converted to Christianity in the thirteenth-century Pseudo-Ovidian poem, *De vetula.* But the most widely disseminated story of such an ilk was to be found in the *De pomo sive De morte Aristotelis* (*The Apple or Aristotle's Death*). Extant in Arabic, Persian, Hebrew, and Latin versions, this curious work is a deathbed dialogue, most likely modeled on Plato's *Phaedo,* which was a similar dialogue relative to Socrates' last day. Although the text of this little work admits that Aristotle did not write it himself, it claims that others did in order to set down "the cause of his joy at his death." As might be expected, a part of this joy is taken to derive from Aristotle's admission of the immortality of the soul and his rejection of the

(*Continued*)

eternity of the world and consequent embracing of its temporal creation. Although these "deathbed beliefs" are quite the opposite of those Aristotle actually held, they were, of course, essential to any believing Muslim, Jew, or Christian. By espousing them, Aristotle could be welcomed as a coreligionist. The miniature given at the bottom left of page 185 is from the beginning of a fourteenth-century manuscript copy of the *De pomo*. Larger than the other interlocutors engaged in his final dialogue, Aristotle lies on his deathbed, clutching the apple of the title in his right hand. Aristotle holds the apple throughout the dialogue, since its fragrance, he confesses, strengthens him and prolongs his life a little. At the very end, Aristotle's "hands began to tremble and the apple which he was holding fell from them."

The second legend about Aristotle, pictured above, is even more remarkable. It is the story of Aristotle and Phyllis. It begins by recounting how Alexander the Great had become overly infatuated with an Indian maiden named Phyllis and had taken her as a wife, whereupon Aristotle cautioned Alexander on the dangers of his overindulgence in this regard. In revenge, Phyllis turned herself toward Aristotle, successfully seducing him into falling in love with her. When Aristotle asked that their love be consummated, she replied

(*Continued*)

that she would do so only if he could prove his love for her by permitting her to ride him like a horse (*sicut equus me portando,* as one version of the legend reads). It is this part of the story that is depicted here in an engraving from the last quarter of the fifteenth century by the so-called Master of the Amsterdam Cabinet. The two individuals leaning on the wall and observing Aristotle with Phyllis astride most likely refers to the end of the whole story. For it concludes by telling us that Alexander in turn asked Aristotle why he was behaving in a manner so contrary to the advice he had so recently given him. Aristotle is said to have replied: "If a woman can make such a fool of a man of my age and wisdom, how much more dangerous must she not be for younger ones? I added an example to my precept; it is your privilege to benefit by both." Although the engraving reproduced here is from the end of the fifteenth century, we know that the legend of Aristotle and Phyllis was already in circulation in the thirteenth, and that a closely related story was included in the *Secreta secretorum,* originally written in Arabic but put into Latin in the twelfth century. And we also know that similar legends existed previously in the East (without, however, Aristotle as a protagonist). For other pictures and diagrams from the manuscript of the *De pomo* used above, see Illustrations 27, 30, 119, and 177.

171. Renaissance and Reformation Criticism of the Middle Ages. Considerable antipathy to scholasticism and medieval learning in general was expressed, at times quite stridently, in writings by Renaissance humanists and others in the fifteenth and sixteenth centuries. Leonardo Bruni (fifteenth century), for example, complained bitterly about English medieval scholars (whom he termed "Barbari Britanni") in logic and natural philosophy; their very names horrified him, and he even called the subject of their preoccupation *suisetica inania,* an obvious reference to the Richard Swineshead we have already met (Illustrations 13 and 144; see also Illustration 256), since his surname frequently appeared as 'Suisset'. By far the greater part of this kind of criticism of medieval science and scholarship is to be found in written sources, but one also occasionally finds it expressed visually. Two examples of the latter are given here, both of them the work of Hans Holbein the Younger (1497–1543). The illustration above is one of eighty marginal drawings made by Holbein (when only seventeen or eighteen) in a copy of a 1515 edition of *The Praise of Folly* (*Moriae encomium* or *Stultitiae laus*) of Erasmus of Rotterdam (ca. 1469–1536). It shows the soul of Scotus (*Scoti anima*) entering the body of Folly. The reference is to the Franciscan theologian and philosopher John Duns Scotus (1265/66–1308), who, along with his followers, was repeatedly the subject of Erasmus' critical barbs, of which the present drawing is meant to illustrate the following: "I think it more suitable while I [scil. Folly] play the divine and tread among thorns to beg that the soul of Scotus, itself more prickly than a porcupine or hedgehog, shall come from his beloved Sorbonne and dwell in my breast for a season; and very soon it shall go back whither it pleases or to the dogs." The added vertical legend at the right of the same drawing tells us to defile logical fools, an instruction rather more graphically depicted in the drawing itself.

The second production of Holbein shown on page 188 is a woodcut representing Martin Luther (1483–1526) as "Hercules Germanicus." Wielding a club and wearing a lion's skin dangling from his neck (possibly in imitation of Hercules slaying the Hydra), he has the pope dangling from his nose, while his left hand is throttling Jacob Hoogstraten (d. 1527), professor of theology at Cologne and Dominican inquisitor. The victims scattered about Luther's feet are more ancient or medieval: Aristotle, undoubtedly as a central authority in medieval philosophy and science; St. Thomas Aquinas (d. 1274), the foremost of all Dominican theologians and philosophers; John Duns Scotus again (at the extreme right with the label 'Schotus'); William of Ockham, the early fourteenth-century Franciscan who was

(*Continued*)

(*Continued*)
considered the chief expositor of the so-called nominalist approach to logical, philosophical, and theological matters; Nicholas of Lyra (d. 1340), a notable Franciscan biblical commentator; Robert Holcot (d. ca. 1349), an English Dominican biblical commentator and theologian with Ockhamist tendencies; and, finally, in the center at the very bottom of the pile, the *Magister Sententiarum* (whose label is curiously given in mirror writing), that is, the twelfth-century Peter Lombard, whose *Book of the Sentences* became the work on which all medieval theologians lectured in obtaining their theological degrees (see Illustration 72). Although it was critical of many of the authors and individuals he opposed, it is likely that Luther himself would not have been overly pleased by Holbein's effort.

CHAPTER SIXTEEN

Personification of the Sciences

THE SEVEN LIBERAL Arts—separable into the trivium of Grammar, Dialectic, and Rhetoric, and the quadrivium of Arithmetic, Geometry, Music, and Astronomy—formed the nucleus of secular learning throughout the Middle Ages, especially in the period before the rise and development of universities, when philosophy as a whole became the focus of the curriculum in the faculties of arts. Most likely related in some fashion to the ancient Greek *enkuklios paideia* (the "circle" of general education needed before higher studies), we know that the seven Liberal Arts were discussed by Varro (second century B.C.) in his lost *Disciplinarum libri IX,* where Medicine and Architecture were added to increase the total to nine. Although other Roman scholars considered other cycles of disciplines, the seven we now regard as the Liberal Arts soon became canonical, to be found above all in *The Marriage of Philology and Mercury* of Martianus Capella (fifth century). Martianus' work came to be not only the most important handbook on the seven Liberal Arts for medieval scholars, but also the first and most significant literary source for the personification of these Arts. His allegorization of them in terms of seven maidens, each bearing attributes intended to characterize the Art she represented, may have suffered changes and undergone development as it was interpreted by its medieval readers (such as Remigius of Auxerre, who wrote a commentary on Martianus' work in the ninth century), but it still remained a basic inspirational element behind almost all later literary and artistic depictions of the Liberal Arts. One has to wait until Alan of Lille's *Anticlaudianus* in the twelfth century to find a document that had similar impact upon this particular tradition of personification. There were, to be sure, other early medieval literary sources that spoke of the Liberal Arts—such as Theodulph of Orléans' (early-ninth-century) poem *De septem liberalibus artibus in quadam pictura depictis*—but none seems to have been as influential as the works of Martianus and Alan.

Theodulph's poem is notable in another respect, since he reveals that his account of the Arts is based upon their depiction—which is indicated by the phrase *in quadam pictura depictis* in the poem's title—on a certain table *(discus)*. Thus, Theodulph provides evidence for a visual presentation of personifications of the Liberal Arts that was contemporary with Charlemagne. And other written sources inform us of yet other early, but no longer extant, representations of the Liberal Arts: paintings that decorated the walls, for example, of the monastery at St. Gall and the palace at Aachen.

The earliest extant representation occurs in a ninth-century manuscript of Boethius' *Arithmetica* (reproduced below as Illustration 172). Since Boethius spoke of the value of the quadrivium in the introduction to that work, it is quite fitting that this ninth-century miniature present personifications of only that "mathematical segment" of the Liberal Arts. Although we do possess other personifications that were executed in the 250 years following this ninth-century miniature, the number of extant representations picks up markedly early in the twelfth century. From that point on, the relevant pictorial sources are quite plentiful. Personifications of the Liberal Arts were part of the sculptural programs of any number of twelfth- and thirteenth-century cathedrals (at Chartres—see Illustration 174—Laon, Sens, Auxerre, Rheims, Clermont-Ferrand, and Freiburg, for example). But mosaics, frescoes—see Illustration 175—sepulchers, pulpits, candelabra, and fountains were also part of this iconographic tradition. It was a tradition that ran healthily throughout the rest of the Middle Ages.

As will be evident below, there were visual representations in which a single Art was personified (in such cases often attached to a "matching" text), in which several were depicted, and in which all seven served as subject. In instances of the last-named sort, one frequently found the Liberal Arts allegorized as "daughters" of Philosophy or Wisdom (see Illustration

173) or placed within the larger context of Christian learning (Illustrations 174–176), a combining of the secular with the religious which mustered considerable support from such authorities as St. Augustine.

The personifications themselves were grounded on Martianus' legacy of the conception of each Art as a maiden, similar to the ancient representation of the Muses. Yet the "persons" of the Liberal Arts were distinguished by the possession of distinct symbols or attributes, again following Martianus' lead. The attributes of the Arts of the quadrivium were in themselves often recognizable enough to enable one to identify the Art in question without much knowledge of the iconographic tradition. Number cords, counting tablets, and finger reckoning were, for example, natural signs for Arithmetic, just as the appearance of a square or a compass was a tolerably sure indication of Geometry. Astronomy could often be identified easily by the presence of an astrolabe, an armillary sphere, or a quadrant, and the representation of any musical instrument all but automatically revealed the fourth member of the quadrivium. To be sure, in some cases the attributes of the quadrivial Arts were not so obvious, but identification was far less straightforward in the case of the trivium. There, one needed more knowledge of the tradition.

This is not a proper place to attempt to list all the attributes used to personify the Liberal Arts. One, Dialectic, may serve by way of example. She is, Martianus tells us, beautifully coiffured and holds a partly concealed coiled serpent in one hand, flowery wax tablets with *formulae* inscribed and a hidden hook in the other. Should anyone be tempted by these *formulae*—perhaps inference patterns of some sort—he could be caught by the hook and drawn toward the venomous serpent. Unlike the attributes of the quadrivium mentioned above, these were in need of interpretation, which was duly furnished by Remigius of Auxerre's commentary on Martianus. Thus, he related the various details of Dialectic's coiffure to fallacious and seductive propositions and to diverse facets of the syllogism. The serpent he interpreted as sophistic cunning (compare the *opponens-respondens* interpretation in Illustration 176), while the *formulae* were held to symbolize the simple proposition, the hook an assumption or a captious conclusion.

It is difficult to say just how much of Remigius' interpretation was accepted and by whom, but we do know that, iconographically, some of Martianus' attributes became traditional, foremost among them the serpent. Often depicted without change, in a good number of instances this serpent was transfigured into the guise of a lizard, a salamander, a dragonlike creature, or the head of a dog. But it was Alan of Lille's transformation of Martianus that was of greatest influence: the *formulae* on wax tablets have given way to flowers (from whence the flowered scepter or branch in many later personifications) and the serpent has been replaced by a scorpion.

This much tells one only of the most frequently depicted attributes for Dialectic. One also finds such things as shears, a cup or goblet, diversely positioned hands (presumably in reference to the relations of propositions within an argument), a tool for boring, an ape, a key, an arrow, the word 'utrum' ("whether"), *ratio* as a bosom-cradled head, writing figures of various sorts, and a balance (which was also a metaphorical representation of logic and logical procedure in written, not artistic, works of antiquity and medieval Islam). Yet no one of these attributes, from Martianus through the end of the Middle Ages, was directly connected with any of the content of Dialectic, as was clearly the case with some of those used to personify the disciplines of the quadrivium. Dialectic's attributes became connected with such content only through interpretation, something that showed itself in visual personifications only when appropriate labels were set into the picture (see Illustration 176). The same was true when a quite different kind of personification of Dialectic invented attributes to cover all the basic parts of the discipline; labels were required then too (Illustration 180).

One other aspect of the medieval tradition of personifying the Liberal Arts bears mention: the custom of furnishing each personified Art with an accompanying representative practitioner. Since the practitioners chosen for this role were, for the most part, scholars who had written works in the discipline in question, there was much more agreement in this matter than in the area of attributes. Here, the majority rule was Donatus or Priscian for Grammar; Aristotle—with a few "votes" for Plato, Parmenides, Zeno, and Zoroaster—for Dialectic; Cicero—and a single voice for Demosthenes—for Rhetoric; Pythagoras or Boethius for Arithmetic; Euclid for Geometry; Tubalcain or Pythagoras for Music; and Ptolemy for Astronomy.

172. A Carolingian Quadrivium. This ninth-century miniature personifying the mathematical sciences is the earliest known depiction of the Liberal Arts. It occurs in a copy of Boethius' *Arithmetica,* composed at Tours (ca. 840) for Emperor Charles the Bald. Music stands at the left, holding some sort of three-stringed instrument. Arithmetic, wearing a crown, stands next to her, a number cord hanging from her right hand while her left reveals an example of finger reckoning (see Illustration 74), although the particular position of the fingers depicted does not seem to correlate with any that were standard in the Middle Ages. Geometry appears next, holding her wand or *radius* (see Illustration 119) over some type of board or tablet on which geometrical figures have been drawn. Supported by a pillar, this tablet may possibly reflect the sand or powder abacus mentioned by Martianus Capella (*abaci sui superfusum pulverum;* VI, 587) as one of Geometry's accoutrements for drawing diagrams. At the extreme right, Astronomy—here called Astrologia—completes the quadrivium. The sun, moon, and stars are barely visible over her head. The significance of the two torches she holds in her left hand has hitherto remained unexplained. For other pictorial material from this manuscript, see Illustration 97.

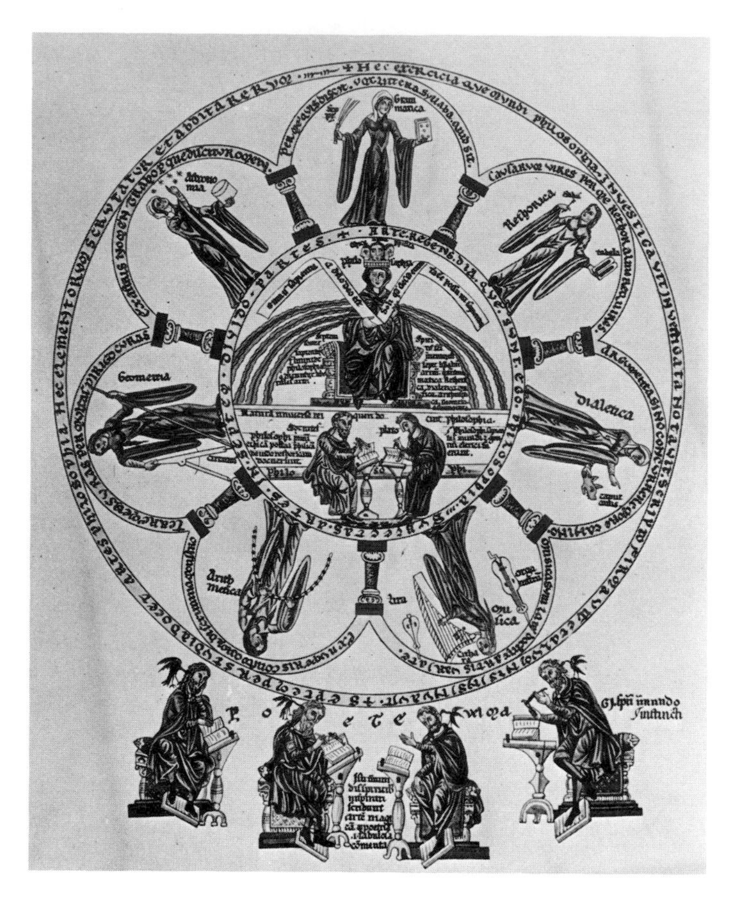

173. The Liberal Arts in a Twelfth-Century Encyclopedic Work. For more than sixteen years in the third quarter of the twelfth century, Herrad of Landsberg (d. 1195), abbess of the convent Hohenberg on Mount St. Odile in Alsace, devoted herself to compiling a massive illuminated manuscript that gathered together, as she put it, "the nectar of the various flowers plucked from Holy Scripture and philosophical works." Tragically, this 324-folio volume containing more than 600 illustrations was destroyed by fire during the bombardment of Strasbourg in 1870, but copies previously made of many of its drawings (and much of the text) provide a source upon which we may still draw. It is such a reproduction that has furnished the illustration of the seven Liberal Arts given here. The *Artes* themselves surround a central circle dominated in its upper half by the figure of *Philosophia*. Wearing a crown with three heads labeled *ethica, logica,* and *physica*—reflecting the so-called Platonic division of philosophy we have already noted in Illustrations 29 and 31—she cradles a scroll in her arms announcing that "all wisdom derives from the Lord God" and that "the wise alone may do what they wish." More informative, however, is the inscription to her right which asserts that "seven founts of wisdom, called the Liberal Arts, flow from Philosophy," a relation of geniture that found frequent expression throughout the Middle Ages and which is here rendered more specific by the legend written on the inner circle informing us that *Philosophia* divides the arts and by them rules all things. Socrates and Plato are seated below *Philosophia* and are identified as those clerics of the Gentiles and sages of the world who first taught ethics, natural philosophy, and rhetoric. Each of the seven Liberal Arts surrounding the central circle holds attributes pertinent to the discipline she represents, some of the attributes being identified by name as well as picture. Dialectic, for example, holds a dog's head *(caput canis)* in place of the more traditional snake. Arithmetic carries the number cord we have already seen in the ninth-century Boethius manuscript in the preceding Illustration, while Geometry now bears a compass as well as her wand. Music has acquired two additional instruments for a total of three attributes: a lyre *(lira)*, a harp *(cithara)*, and a hurdy-gurdy *(organistrum)*. The arches under which each Liberal Art stands all carry verses announcing what that Art can accomplish. Geometry, for example, says that "by many cares I lay out the measures of the earth" *(terre mensuras, per multas dirigo curas)*, and Astronomy reveals her astrological talents: "a sign is learned through the name I draw from the stars" *(ex astris nomen, traho per que* [!] *discitur omen)*, while Dialectic's pronouncement clarifies the dog's head in her left hand: "I permit arguments to join, doglike, in battle" *(argumenta sino, concurrere more canino)*. The outermost circle carries a legend specifying the stages through which *Philosophia* works—investigation, writing, teaching—revealing that she does so through the seven Liberal Arts, exploring the hidden elements of things. The figures at the bottom are poets or magi, characterized as worldly spirits. As a whole, with her elaborate drawing this twelfth-century abbess has personified the Arts and revealed their generation from Philosophy, but at the same time she has put all of this into a framework that establishes the connection between secular learning and the divine.

174. The Integration of Science and Religion. The present personification of Dialectic, dating from the early twelfth century, occurs in the archivolt of the right portal of the west facade of Chartres Cathedral. Surrounding a timpanum and lintel devoted to picturing the incarnation of Christ, the archivolt contains representations of all seven of the Liberal Arts, thereby signifying, as it were, the bringing together of the divine and the secular, the religious and the scientific. Such a rapprochement was especially fitting for the sculptural program of Chartres, since the Cathedral School of Chartres was renowned in the twelfth century for the study of these seven basic disciplines. Thierry of Chartres, one of its most famous scholars and chancellor of the School contemporaneously with the carving of many of these sculptures, composed an immense florilegium, called the *Heptateuchon,* of writings belonging to these Arts. Thierry there declared the seven Liberal Arts to be uniquely indispensable for Philosophy, which, in turn, as the true love of wisdom, was essential for coming to know the truth of all things. More than that, in his commentary on the six days of creation as found in the Book of Genesis, Thierry applied this recommended knowledge of the Liberal Arts, mathematics in particular. One is tempted to believe that Thierry would have interpreted the intellectual gain to be had from these seven disciplines in a manner similar to that expressed in the frequently quoted view of his older brother, Bernard of Chartres, who claimed that twelfth-century scholars were like dwarfs sitting on the shoulders of giants *(nos esse quasi nanos gigantium humeris insidentes)*, being able to see further than their predecessors, not because of their own keener vision or greater stature, but because they were lifted up by the gigantic stature of those who preceded them. Dialectic, as one upon whose shoulders Thierry and others could rest, is here personified by the attributes of a flowered scepter and a dragonlike serpent, the latter looking as if it were a mixture of the traditional snake and the dog's head in the preceding Illustration (even though the manuscript of that Illustration is later than the present archivolt figure). Aristotle, as a representative practitioner of Dialectic, sits beneath her, dipping his quill in an inkhorn and holding an erasing knife in his other hand, while other writing implements are stored on the wall behind him. The two sculptures to the right of Dialectic and Aristotle are the zodiacal signs Gemini and Pisces, appearing here, some historians have claimed, because there was not enough space for them in the archivolt of the left portal of the west facade where the remaining zodiacal signs are depicted.

175. The Program of Dominican Theology and the Liberal Arts. The painting reproduced above is but one in the fresco cycle of the Spanish Chapel of the Cloister of Santa Maria Novella in Florence. Executed in the 1360s under the assistance of Andrea Bonaiuti, these frescoes have been characterized as "a kind of *summa* of ideas and attitudes that were current in the art of the time," a description that is quite fitting to the particular painting given here. Traditionally referred to as *The Triumph of St. Thomas Aquinas,* its title reflects the dominance of the Dominican Order at this Florentine cloister. St. Thomas himself sits enthroned in the center, the three theological virtues of Faith, Charity, and Hope floating overhead, while the four cardinal virtues—Temperance, Prudence, Justice, and Fortitude—flank his throne. The book held by Thomas is inscribed with verses from the *Book of Wisdom* (VII, 7–8) recounting the coming of the spirit of Wisdom and how Wisdom (personified in the gable of Thomas' throne) is to be set before princely things, a sentiment that is expressed by the fresco as a whole. Figures from the Old and New Testaments sit on either side of Thomas, while the heretics Sabellius, Averroes, and Arius appear at his feet. Although, as a commentator on Aristotle (see Illustration 7), Averroes was often used by Thomas himself in his own commentaries, the position assigned him in this fresco most likely refers to the unorthodox views of some of his fourteenth-century followers. Personifications of seven spiritual or theological sciences—with representatives at their feet—are seated in the choir stalls at the bottom left of the fresco, while the Liberal Arts occupy the seven stalls on the right. Enlarged in the detail, the Trivium begins with Grammar at the extreme right, her attributes being three pupils and the door to knowledge. Holding a scroll, Rhetoric is next, followed by Dialectic

(Continued)

bearing a branch and a scorpion as symbols. The Quadrivium follows, headed by Music with a portative organ, then Astronomy holding an armillary sphere, Geometry with a square and a (barely visible) compass, and, lastly, Arithmetic, grasping a reckoning tablet with her left hand while her right hand displays the sign for 400 by finger calculation. Representative practitioners of each science are seated at the foot of their appropriate choir stalls. They are, from right to left, Donatus, Cicero, Aristotle, Tubalcain (hammering on his anvil), Ptolemy, wearing a crown due to the frequent confusion of the astronomer with the kings of the Ptolemaic dynasty, Euclid, and Pythagoras. Finally, note should be taken of the personifications of the seven planets appearing in the gables of the choir stalls of the seven Arts. They are, following the same order from right to left, the moon, Venus, Mercury, Jupiter, Saturn, Mars, and the sun, a correspondence that echoes exactly the matching of the planets and the Liberal Arts given by Dante (*Convivio,* II, 14). It is interesting to note that it is not only this specific point in Dante which renders our fresco a characteristically Florentine production, but also the fact that the iconography of both the Liberal Arts and the planets parallels, with some exceptions, what we find in the reliefs which were produced in the workshop of Andrea Pisano (ca. 1334–1340) for the Campanile of the Florence Duomo. As a whole, this Spanish chapel fresco was one of the most comprehensive representations of the personified Liberal Arts, especially as these disciplines were viewed as related to the broader fabric of Christian learning in the later Middle Ages.

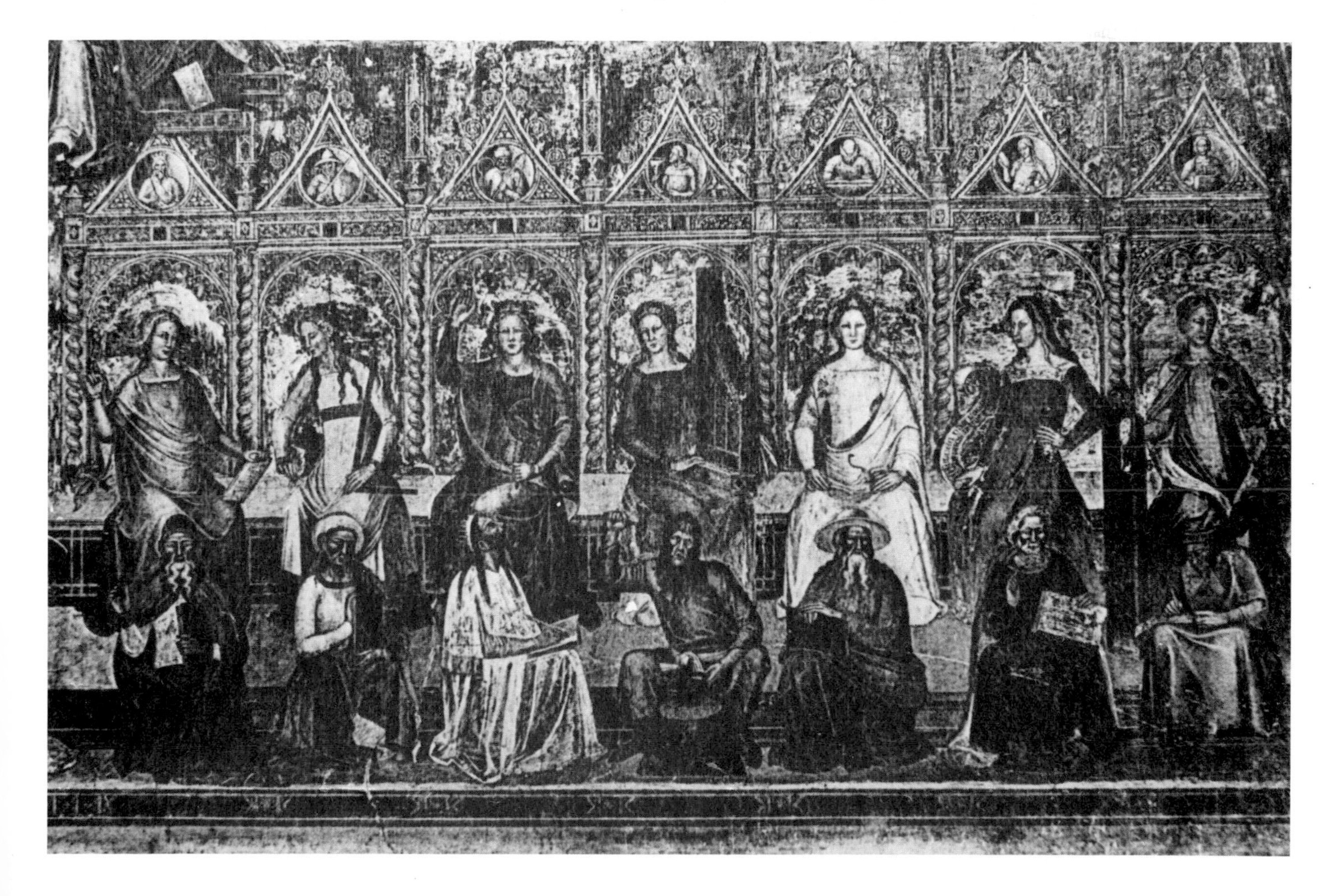

176. Arts, Virtues, and Vices. The symmetry of the number of Liberal Arts, Virtues (Theological and Cardinal), and Vices made their iconographic pairing a natural phenomenon for the medieval illustrator. Such a pairing is reproduced here from a manuscript of Johannes Andrea's *Novella in libros decretalium,* illustrated by Nicolò da Bologna in 1354. Personifications of the seven Virtues occupy the upper row (Cardinal on the left, Theological on the right), the bowed heads of pagan representatives of the corresponding Vices at their feet (for example, "Epicurus voluptuosus" is identified—in a barely visible label—as the prostrate figure beneath *Temperantia,* "Nero iniqus" crouches below *Justitia* at the extreme left, and so on). The seven Liberal Arts populate the lower row, Grammar beginning the Trivium on the left, holding the traditional whip in her right hand while instructing the pupil on her left, who is feeding from her bared breast. Priscian (sixth century) sits below as the representative of the discipline of Grammar. Dialectic—now labeled 'Loicha'—occupies the second position. She here holds two serpents, their relevance specified by the words 'opponens' and 'respondens', while the face of *ratio* looks out from her bosom. Dialectic's representative here is not Aristotle but Zoroaster, as is revealed by his name in the book in which he writes. Rhetoric, with Cicero at her feet, is next, followed by Arithmetic, writing numbers on a tablet, as the first member of the Quadrivium, with Pythagoras at her feet. Geometry is on her left, a square under one arm while she manipulates a large compass with both hands; Euclid is identified by name below her. Music follows with two stringed instruments; she tunes one while the other lies in her lap; Tubalcain sits below wielding his usual hammers. Astronomy completes the cycle at the extreme right. An astrolabe rests at her left while she takes sightings with a quadrant held in her right hand. Ptolemy looks up at her from below, the book in his lap once again indicating the medieval confusion of astronomer and king by the inscription 'rex egiti' following his name. Nicolò da Bologna's illustration of these personified Liberal Arts and Virtues was repeated almost to the letter in iconographic terms in the Italian *La canzone delle virtù e delle scienze,* written and illustrated most likely in 1355 by Bartolomeo di Bartoli, another Bolognese artist who had at times collaborated with Nicolò.

177. Separate Personifications. Drawn from a fourteenth-century manuscript we have already seen on a number of occasions (Illustrations 27, 30, 119, and 170), the three Liberal Arts whose personifications are given here each head the chapter devoted to the Art in question in Book II, *On Secular Letters,* of Cassiodorus' *Institutiones.* All seven of the Arts are so represented in this codex. The personification of Geometry has been given in Illustration 119 and three of the remaining six are depicted here. At the top right, Dialectic stands holding two serpents, two scholars disputing at her left, while another explains some point from a book he is holding. Separately framed, Aristotle is depicted at the left as the practitioner of the personified discipline. The middle illustration shows Arithmetic seated at a desk, giving instruction to four pupils. None of her usual attributes appears to be present, the book before her not being a number tablet (although one can see a Roman XII in its simulated text). The most significant element in this picture is the representation of figurate numbers (see Illustration 94) appearing separately at the left. An attempt has been made to depict triangular, square, pentagonal, and hexagonal numbers (with an added nonrelevant figure thrown in for good artistic measure), although the arrangement of the constituent units is provided only for the square numbers. Astronomy is personified in the bottom illustration. Some type of sphere or disc rests on a table before her, the artist's fondness for decoration rendering any further identification impossible. Similarly, one would expect the object held in Astronomy's left hand to be a quadrant, but again artistic license has altered its shape. The universe is depicted at the right of the illustration, the sphere of each of the seven planets appropriately labeled, while the whole is surrounded by the darker sphere of the fixed stars. The sun and the moon are given special status by being depicted in a fashion that clearly differentiates them from the other planets. At the very center within the sphere of the moon, the earth is represented by a so-called O-T map. Since all seven of the personifications in this manuscript are keyed to the relevant chapters on each *ars liberalis* in Cassiodorus, no attempt has been made to relate any of the disciplines to other parts of Christian learning. More than that, artistic, rather than iconographic, considerations seem to dominate the personifications. Overall, the most interesting feature is the fact that the illustrator has in at least two instances gone outside the traditional iconography for the Liberal Arts to incorporate material drawn from works

expounding or dealing with a given Art itself: the figurate numbers that were so much a part of the Boethian tradition in arithmetic and the diagrams attached to the definitions of geometrical figures by Isidore of Seville (see Illustration 119).

178. Personifications in Illuminated Initials. Geometry is personified at the upper left in the illuminated 'P' which was the first letter of the opening definition (*Punctus est cuius pars non est:* "A point is that of which there is no part") in a Latin version of Euclid's *Elements* made by Adelard of Bath. Occurring in a fourteenth-century manuscript we have had occasion to sample before (Illustrations 44, 59, 60, 62, 104), Geometry here instructs a whole crowd of seemingly eager pupils. She holds a square in her left hand while drawing with a cross-branched compass on a round table before her, the table already exhibiting numerous geometrical figures. The same manuscript also contains the personification of Music at the lower left. Framed within the 'O' of the initial of Boethius' *De institutione musica,* Music strikes her hammers on bells suspended overhead. Other instruments which had traditionally functioned among her attributes have here been transferred to the hands of pupils: one plucks some kind of lute, another strums a zither, and a third bows a rebec. Considering the limited space usually provided in illuminated initials, the artist of the present manuscript has managed quite well in including a sufficient number of personifying attributes.

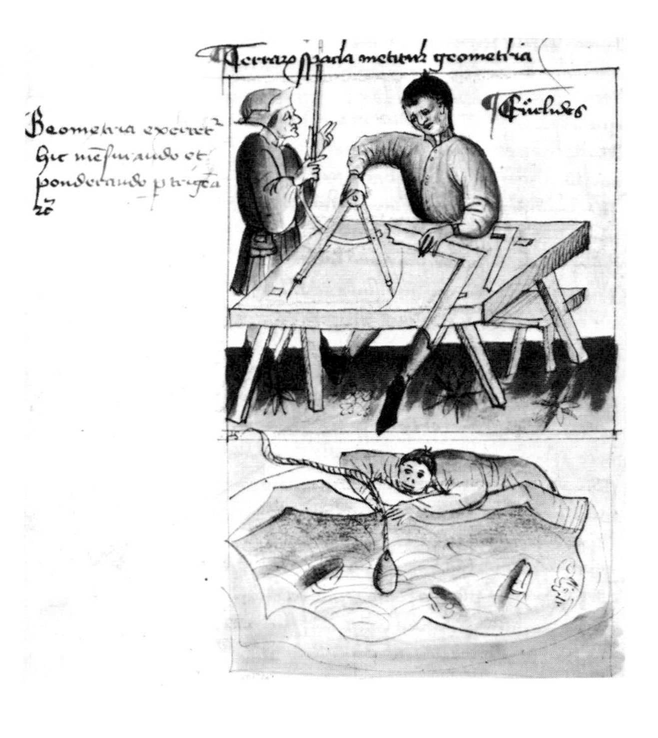

179. Personifications versus Scientists at Work. Although one of the illustrations of Geometry given here is taken from a manuscript and the other from a printed book, their executions were only thirty to fifty years apart. That on the left, above, occurs in an anonymous treatise on the seven Liberal Arts found in a German manuscript of the third quarter of the fifteenth century. Here Geometry no longer receives the usual female personification: the representative geometrical practitioner of tradition, Euclid, has taken her place. The substitution has turned things into a halfway-house between a personification of Geometry and a simple illustration of the geometer at work. The two most frequently occurring attributes in the personification tradition of square and compass are still present and even emphasized. But the presence of working assistants—one alone below taking soundings of some sort—has moved everything in the direction of depicting a practicing scientist. This is confirmed by the marginal annotation at the left, which tells us that "by measuring and weighing by means of triangles and the like, geometry is here being carried out" *(geometria exercetur hic)*.

A similar halfway-house illustration of geometry is at the left, below, taken from the 1517 Basel edition of the *Margarita philosophica* of Gregor Reisch (1470–1525), prior of the Carthusian House at Freiburg and confessor to Emperor Maximilian. In terms of date, we should note that the same illustration already appeared in the Strasbourg edition of 1504 of Reisch's work. The *Margarita* was an elementary encyclopedia, in dialogue form, of the seven Liberal Arts with added sections on natural philosophy, the powers and immortality of the soul, and ethics. The present illustration serves as a frontispiece to Book VI on geometry, both speculative and practical. In contrast to the miniature from the fifteenth-century German manuscript, Reisch has "reversed course" and represented his "Typus geometrie" as a woman. That much alone suggests a return, howsoever slight, to the personification tradition. The implements on the table before Geometry also are consistent with that tradition. But the numerous smaller figures scattered about this woodcut certainly suggest, if not *the* geometer at work, at least geometry being applied, something quite consistent with Reisch's inclusion of practical geometry in this section of his *Margarita*. At the right, two figures are making sightings with an astrolabe and a quadrant (which is wrongly held and whose plumb bob is in a most curious position). The sightings in question could be either of heavenly bodies (since Reisch delineates the importance of geometry for astronomy in his opening paragraphs) or of inaccessible objects that were to be measured (compare Illustrations 82 and 152), since Reisch documents such procedures in his section on practical geometry. At the lower left, a figure seems to be applying some kind of geometrical instrument to the vaulting of a house under construction, for which a stone hangs overhead. Other geometrical implements are scattered on the ground, one figure applying one of them in drawing some type of plan on a board. The word 'jugera' in the inset at the lower right refers to a measure of land that was frequently used in the practice and literature of the Roman surveyors (see Illustrations 146–148). In sum, then, the present woodcut does personify Geometry in the female figure dominating the whole illustration. But its main function appears to have been the revelation of how all the various roles and accomplishments of practical geometry are subsumed under, and directed by, the "speculative" Geometry there personified.

180. Personification with Contemporary Attributes. Save for minor alterations, this woodcut of *Typus logice* from the *Logica memorativa* (1509) of Thomas Murner (see Illustration 72) is a redoing of an illustration apparently first published in the 1503 edition of Gregor Reisch's *Margarita philosophica.* Although Logic is here personified in standard female form, little else is traditional. The familiar attributes of serpents, scorpion, branch, or flowered scepter are all absent. In their stead we have attributes referring to the contents of contemporary logical texts and to the diverse "schools" practicing such logic. To begin with first things first, we should note the horn held to Logic's lips: it bears the twofold legend *sonus* and *vox* ("sound" and "word"), the first two items receiving definition (after, that is, the definition of the term "dialectica" itself) in Peter of Spain's logical textbook, the *Summulae logicales* (see Illustration 70). The two flowers issuing from the mouth of the horn are labeled as the "two premises" required for a syllogism, the latter itself being duly inscribed on the scabbard of the sword hanging at Logic's side. Her left arm is "argumenta," the bow held by her right hand is "questio," and "conclusio" adorns her torso. Her right leg represents the Aristotelian doctrine of categories *(predicamenta),* her left that of the predicables, while the two sabots on her feet symbolize the two basic types of fallacies: *in dictione* and *extra dictionem* (see Illustration 33). All of these (largely Aristotelian) doctrines and conceptions could be found briefly explained in Peter of Spain's standard text. The two dogs in the foreground are Truth and Falsity, the rabbit in front of them "problema." The flora in the lower right corner labeled *parva logicalia* (the "little logicals") refer to that part of Peter's text that dealt with "the properties of terms," a part which carried medieval additions to Aristotle. Two other characteristically medieval creations are the *insolubilia* and *obligationes* represented by the prickly looking bushes just above. Since *insolubilia* were basically variants of the Liar paradox ('ego dico falsum' being the simplest medieval example) and *obligationes* were disputational "games" governed by complicated rules and conditions, the present "thorny" iconography seems appropriate. The "forest" at the upper right represents a different kind of matter. Labeled 'silva opinionum' ("the woods of opinions"), it is intended to symbolize differing views about logic that were then current. Four such different views or "schools" are presented by four different trees: *Albertiste,* for the followers of Albertus Magnus (thirteenth century); *Thomiste,* for adherents to Thomas Aquinas (thirteenth century); *Scotiste,* for followers of John Duns Scotus (d. 1308); and *Occamiste,* for those of the persuasion of William of Ockham (fourteenth century). So numerous and so specific are the "attributes" of Logic scattered throughout this woodcut that their labeling was surely needed to render their identification possible. One is indeed here very far from the more "literary" sort of personification that began with Martianus Capella.

CHAPTER SEVENTEEN

The Locus of Scientific Endeavor: Workshops, Institutions, and Patronage

ANY NUMBER OF the preceding illustrations have shown scientists or scholars in the surroundings appropriate to their professions. But in most instances these surroundings were not central to the illustration. Instead, the portrayal of the scientist, with or without the implements of his trade or discipline, was primary in the illustrator's mind in spite of the depiction of elements beyond the portrait itself (for example, see Illustrations 158–160, 162). Even when a scientist was represented at work, it was often the case that the locale—the physical environs—of that work was unimportant.

The emphasis of the present chapter is, in contrast, on the depiction of that locale or those surroundings. In most instances, at least some of the architectural details of these surroundings are pictured; in others, such detail is minimal or absent, but the assemblage of activities depicted leads us to understand that we are not simply observing a representation of the scientist at work, but rather a scientist at work in the locus proper to his enterprise (the first miniature in Illustration 183 and, in a slightly different sense, the first picture in Illustration 187 are cases in point).

Classified differently, the illustrations in this chapter fall into three groups. The first (Illustrations 181–183) has to do with rather straightforward representations of the locations in which science was done: from the simple study in which scientific works were written, through alchemists' and apothecaries' shops, to clinics and ordinary bedrooms in which the physician worked. The second set of illustrations (184–187) relates not to where science was done, but to where it was learned: that is, academic institutions of any sort.

The final three illustrations of the chapter address a different sort of issue: not the institutions or physical surroundings that provided a place for the study or practice of science to be carried out, but rather the support and rewards that encouraged, in some cases even made possible, that study and practice. In other terms, these illustrations are visual records of the various kinds of patronage that science and scientists received. Often a work was written or some scientific activity was pursued at the request of a patron or patroness, royal or otherwise (so, for example, in the case of Nicole Oresme—Illustration 189—or the master mason in Illustration 190). In other cases, an author dedicated a work to some person of higher social status, not because the work was composed at the behest of that person, but because of earlier friendship or association with that person, an association that may well have carried earlier rewards (the case of Pliny, for example; Illustration 189). Alternatively, it may not have been the dedication of a work itself to some personage, but simply the preparation of a particular copy—usually sumptuous and elegant in form—of some scientific work that was so dedicated (Illustration 188).

181.(Right) A Scholar's Study. As was emphasized in Chapter One of this *Album,* a great deal of ancient and medieval science was carried out in books. It is most appropriate, therefore, that at least one illustration of the "locus of scientific endeavor" be of a scientist or a scholar at his desk. A depiction of Macrobius (fifth century) has here been selected to play that role. It is taken from a 1383 Bolognese manuscript of his *Commentary on Cicero's Dream of Scipio,* a handbook on number lore, astronomical cosmography, and geography that was extremely popular throughout the Middle Ages. Quill and erasing knife in hand, Macrobius is pictured in the course of writing his influential work. Part of his library is stored in a cabinet before him, while other books lie open on a revolving bookrest standing on his desk. The face in the initial below is mostly likely intended to be that of Scipio or Cicero, and not Macrobius, since it stands at the beginning of the Ciceronian *Dream of Scipio* rather than the beginning of Macrobius' commentary. For other pictures from this manuscript, see the Frontispiece and Illustration 15.

182. (Right) The Alchemist's Shop. This miniature is from the same late fifteenth-century manuscript of Thomas Norton's *Ordinal of Alchemy* that we saw in Illustration 161. Here, however, it is not authors who are the illustrator's subject, but the shop of the alchemist, replete with the tools of his trade. The practitioner of the art shown at work at a table is presumably Norton himself. A small vessel sits on the table before him, along with spherical and crescent-shaped objects (possibly, one is tempted to think, as references to the sun and the moon, standard alchemical emblems for gold and silver). Other alchemical vessels and implements can be seen in the open chest beneath the table; but most interesting of all is the balance in a glass case in the table's center, which forms, in a way, the focus of the whole miniature. It is the first known depiction of such an encased balance. Two assistants in the foreground are busy tending furnaces. The one at the left treats a fire that heats a series of stacked alembics, functioning as a kind of fractional distillation tower similar to that later named a "hydra" by authors like Giambattista della Porta (*De destillatione,* 1609). On the right, the other assistant stokes a furnace holding a vessel similar to a "pelican," yet another type of distillation apparatus. The inscriptions on the scrolls or ribbons associated with each individual in the miniature describe various alchemical phenomena. Most notable is that belonging to the assistant on the left: it quotes the better part of the seventh dictum of the Hermetic *Tabula smaragdina (Emerald Green Table),* a central work in medieval alchemy: "Separate earth from fire, the subtle from the thick. . . ." *(Seperabis (!) terram ab igne, subtile aspisso. . . .).* The other two inscriptions may also be quotations (if not nearly so well known) from one or another alchemical tract. That over the head of the other assistant claims that one should observe the colors in a vessel held in the hand *(Mane prope vas, nota colores)*—although it is difficult to determine whether a vessel is really depicted in the assistant's hand—while the instructions over the alchemist in charge simply direct one to compose the philosopher's stone "without contradiction" *(Compone lapidem absque repugnantia).* There is an interesting seventeenth-century reproduction of this particular miniature in a printed version of Thomas' *Ordinal;* it can be seen in Illustration 184 of the second volume of the *Album of Science.*

183. At a Clinic, Bedside, and Apothecary Shop. The miniature at the left, taken from a fourteenth-century Greek manuscript of an antidotary entitled *Dunameron* by the thirteenth-century Byzantine physician Nicolas Myrepsos, presents a veritable bustle of medical activity. The physician in charge at the left is seated on cushions in front of a drawn curtain which was perhaps used for examining patients. A patient stands on crutches immediately in front of him while he scrutinizes a urine flask held in his left hand. Next, in the center of the miniature, stands an assistant holding what appears to be a basket for the transport of urine flasks. A mother sits behind him, cradling a child in her arms, but in pointing to her own head she seems to be indicating that she is the one who is ill. The next figure to the right is another assistant, a book (or perhaps an abacus) under his left arm, a partly opened covered basket containing medicines or flasks of some sort in his right hand. Finally, at the extreme right, a more youthful helper sits, grinding medicines by "double-pestle" action—a rarity in medical illustrations—some of the medicines themselves stored on shelves above him. Note that with the exception of this last "junior helper," all those who are engaged in being "practitioners" at any level are wearing identifying triangular hats. Although this miniature is surrounded by a decorative margin, its "space" is flat, lacking the architectural detail and perspective seen in the miniature on the right, which appears in a fifteenth-century manuscript of a French translation of the *De proprietatibus rerum,* an extremely popular (nineteen-book) encyclopedic work by the thirteenth-century Franciscan, Bartholomaeus Anglicus. Book VII of this voluminous work was devoted to medicine—largely gleaned from writings ascribed to Constantine the African (eleventh century)—and this illustration heads its second chapter on therapy for maladies of the head *(des remedes de la douleur du chief),* although it is intended only as a frontispiece and not as a visual explanation of the text. The main room depicted in the miniature shows a pharmacist weighing medicine that has been prepared for a waiting woman client. Jars most likely storing other medicines are on shelves over a Gothic window, while the table over which the pharmacist works is fitted with numerous shallow drawers containing pills or other medication. A large urn and a sack full of some kind of medicinal herb lie in front of the table. Through the large window or door at the left, a physician can be seen in the house across the road in the course of diagnosing by uroscopy the illness of a bedridden patient.

184. (Above) A Locus of Greek Learning. This mosaic from a Roman villa was discovered in 1897 in a *suburbium* called Civita outside the Pompeii city walls. Most likely dating from as late as the first century A.D., it has usually been taken to represent Plato's Academy, even though this interpretation cannot be definitively established. It rests largely upon the identification of Plato as the figure seated in front of the tree on the basis of some resemblance between this figure and known portraits of the fourth-century B.C. philosopher. In any event, the central figure appears to be using a stick to draw on the ground, possibly a reference to the legendary importance of mathematics in the Academy (an importance that is emphasized in a late story recounting that Plato had an inscription placed over the Academy's doors reading: "Let no one unversed in geometry enter"). Presumed scientific reference can also be seen in the sundial at the top of the central column and in the globe resting in a box at Plato's feet. Plato founded his school in the early fourth century B.C. in a park or gymnasium on the outskirts of Athens called the Academy. Although there undoubtedly were other academic institutions in Athens and elsewhere in Greece before that time, the Academy is the earliest of the more renowned ones, such as Aristotle's Lyceum, Epicurus' Garden, and the Stoa. Plato's role in his Academy was, to quote a later ancient source, "to act as architect and set the problems." Although he was clearly not a mathematician himself, we know that others who were associated with the Academy in the fourth century B.C. were—such as Eudoxos of Cnidus and Theaetetus of Athens. As a center of philosophical activity and learning, the Academy continued to exist, though with some change of locale, until the sixth century.

185. (Below) An Arabic Library. The illumination of medieval Arabic manuscripts was markedly less frequent than that of Latin codices in the Middle Ages, but one Arabic work that often did have the privilege of such artistic treatment was the *Maqāmāt (Assemblies)* of Abū Muḥammad al-Qāsim al-Harīrī of Basra (1054/1055–1122). This work was a collection of fifty picaresque tales, each one a *maqāma* (a séance or assembly) told in terms of the adventures of a main protagonist named Abū Zaid. Although Harīrī himself intended his stories to serve a moral purpose, their fame rests not on content but on the elegance, ingenuity, and other *tours de force* of their rhymed prose, something that fits well with the fact that Harīrī was also the author of several grammatical tracts. The present miniature is one of ninety-nine that adorn a manuscript of the *Maqāmāt* written at Baghdad in 1237, illustrated by one Yahyā ibn Mahmūd al-Wasītī. The first séance occurs in the library at Basra pictured here. Apart from the figures at the bottom involved in recounting this initial tale, the miniature is notable for its depiction of the storage of books stacked flat in numerous cubbyholes. One historian believes that this representation indicates that they are unbound, though they may also be boxed or have wraparound covers serving as bindings.

186. Spiritual and Academic Life at a Medieval College. A number of colleges at Paris in the Middle Ages were "grammar colleges," the students ranging in age from seven or eight to sixteen and the focus of their curriculum being the first member of the Trivium, together with a smattering of some of the other Liberal Arts. One of these institutions was Ave Maria College, founded in 1339 by John of Hubant. The set of drawings shown here is taken from a rarity: a fourteenth-century manuscript cartulary of the statutes of a college with accompanying illustrations. All the drawings bear captions, most of them in Old French. The one on the left in the top row illustrates the required tradition of selecting a student each week, called the hebdomadary *(septimanarius; semeniers)*, to light a lamp every evening before the altar of the Blessed Virgin. The picture directly below shows a lighted candle before the same altar, illustrating a student's obligation to place a candle there during the daily saying of the Matins and Vespers of Our Lady. In the upper right-hand corner we see a student standing in the door of the college while offering a quart of wine to an old woman, the intended reference being to a statute prescribing such a gift *(quarta vini danda pauperibus)* on certain feast days. Directly below we have a picture designed to remind the hebdomadary of another of his duties: cleaning the chapel. According to the legend above, the bottom row depicts the duty of the first student to arise to ring the bell and say the Ave Maria *(Commant li premiers des enfanz qui s'eveille de nuit doit sonner la campane et dire Ave Maria)*. Directly to the right the concern is the pet goldfinch *(chardenreau)* kept—with the founder's permission—in the College chapel: every day the hebdomadary was to feed the finch and on each solemn feast day clean its cage. The drawing occupying the second row from the bottom deals with a more academic matter: library regulations. The particular regulation pictured is that of the weekly inspection *(ostensio)* of the books, done by the master and chaplain and also by the outgoing and incoming hebdomadaries. The most notable feature about all these drawings is simply the fact that it was thought to be good pedagogy to provide students with such visual aids, the better to remind them of the statutes of their college.

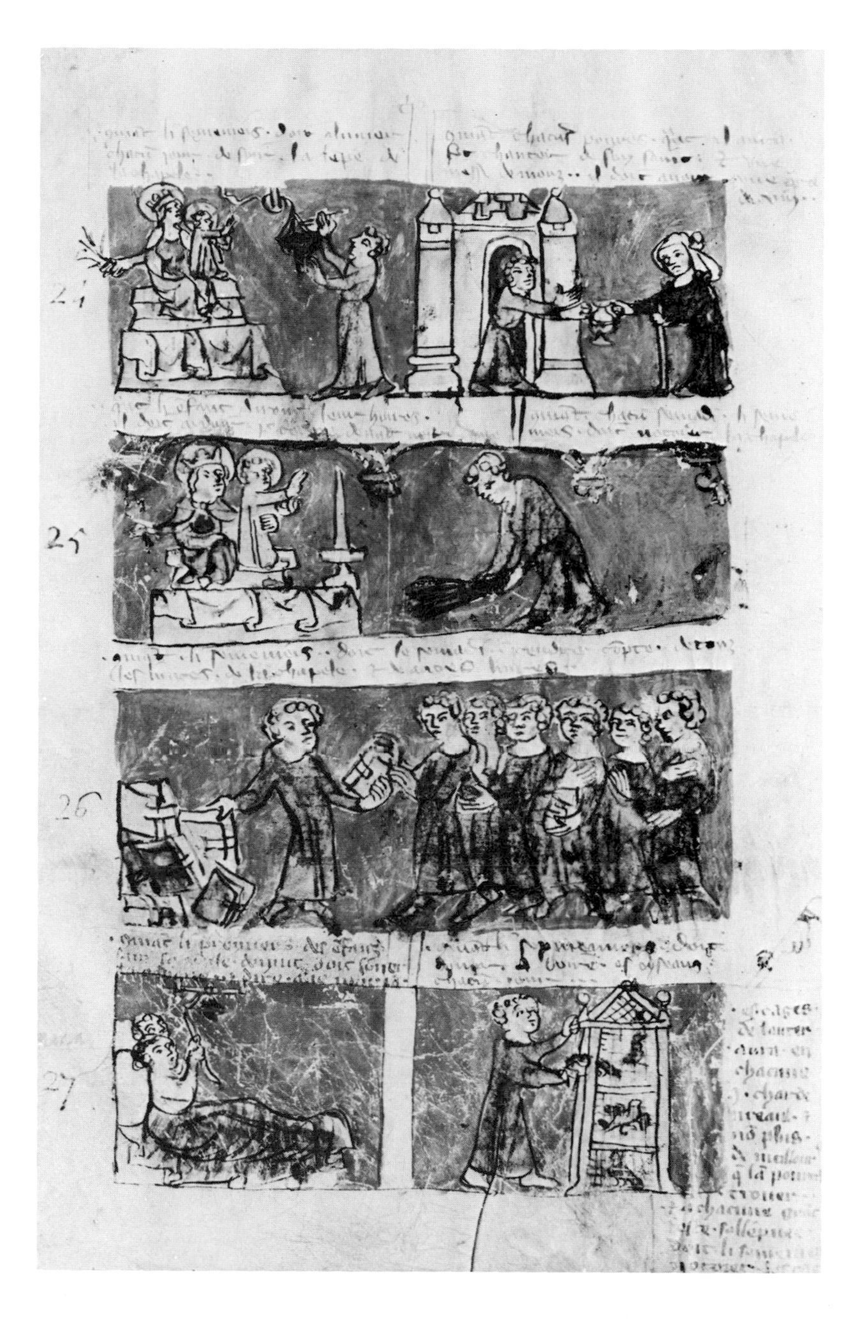

187. Learning: Ordinary and Royal. Both pictures reproduced here pertain to grammar school, not university, education. That above is found in a fourteenth-century manuscript of the *Image du monde,* written in Old French in the thirteenth century by an otherwise unknown Maitre Gossouin or Gauthier of Metz and covering theology, cosmography, geography, natural history, and many other corners of medieval learning. This work was immensely popular, particularly in its three versions in verse form, the first (consisting of 6,594 verses) being completed in 1246. The version depicted here, however, is in prose. The present picture occurs at the beginning of the chapter detailing the seven Liberal Arts (the incipit below reads: *La premiere des .vii. arz si est gramaire. . . .*). The master in this miniature, in clerical dress, is clearly not teaching the subject at hand "without tears." He is glowering at his pupils and pointing determinedly at the open text held by one of them; eventual comprehension of the contents of the texts is further assured by the club in his other hand. Since learning by rote and punishment for failure to do so were common features in this phase of medieval education, the miniature is not far from the mark. The second illustration, below, is taken from a manuscript prepared at the very end of the fifteenth century for Maximilian Sforza, son of Lodovico il Moro, Duke of Milan, and Beatrice d'Este. Executed by Gian Pietro Birago, this miniature shows Maximilian in the course of lessons that are far less arduous than those represented above. He is even guarded against any discomfort from the heat by having a dwarf fan him while he works with his tutor, Gian Antonio Secco, Count of Borella. Other youths are at play, while Maximilian attends to his studies. A Latin grammar based on that of Aelius Donatus (fourth century) is the central text given by the manuscript and is most likely the discipline on which Maximilian works.

188. A Codex with a Mark of Patronage. One of the earliest extant and most famous illustrated scientific manuscripts—indeed one of the earliest extant illustrated manuscripts in any genre—is a Greek copy of the *De materia medica* of Dioscorides made for and presented to Princess Anicia Juliana ca. 512 A.D. In addition to the illustration of some 400 plants on pages facing the text (an example of which is below in Illustration 193), this codex contains a number of prefatory miniatures. A descendant of one of these was given above as Illustration 10. Another is given here from the sixth-century manuscript itself. It documents the fact that this manuscript was created expressly for Anicia Juliana in what is the oldest dedication miniature that has hitherto been uncovered. The princess herself sits enthroned in the center gesturing to a book—probably the dedication copy—held by a putto on her right. Personifications of Magnanimity *(megalopsychia)* and Prudence *(phronesis)* stand at her sides, while another, smaller personification (just below the putto), identified as "gratitude of the arts," appropriately bows with humility before her patroness. Still more material from this manuscript may be seen in Illustration 202.

189. Dedication Miniatures for Ancient and Medieval Works. In contrast to the preceding illustration, the pictures given here are not from the dedicatory codices of the works themselves. Instead, they are depictions of the dedications announced in the text to which they are attached. The first, above, is from a twelfth-century Danish codex of Pliny's *Historia naturalis.* Pliny the Elder (Gaius Plinius Secundus, A.D. 23/24–79) compiled his 37-book *Natural History* from the writings of nearly 500 authors, both Greek and Roman. Treating cosmography, terrestrial phenomena of all sorts, animals, materia medica, geography, and much else, the work was completed in A.D. 77, dedicated to Titus Flavius Vespasianus (emperor, A.D. 79–81), but published only posthumously. It is the dedication to Titus that is pictured here. Although Titus here wears a crown, we know that Pliny's dedication to him occurred two years earlier than his accession, since in the prefatory letter to the *Historia* Titus is referred to as having served as consul six times, a fact that dates the completion and dedication of the work at some point in 77. Pliny had served in Germany with Titus, and one of the reasons behind the dedication may well be their long, close association on this and other occasions. Pliny himself died only several months after Titus succeeded his father Vespasian; observing at too close a range the eruption of Vesuvius that engulfed both Pompeii and Herculaneum,

(Continued)

he was overcome by noxious fumes. The second illustration, bottom right, is of a medieval dedication. It is taken from an early fifteenth-century manuscript that once belonged to Jean, Duke of Berry, of Nicole Oresme's *Le livre du ciel et du monde,* a translation and commentary in Old French on the Latin version of Aristotle's *De caelo.* Oresme had provided similar vernacular translations and commentaries on other Aristotelian works (the *Ethics,* the *Politics,* the *Economics*), all of them at the behest of King Charles V, whom he had already known well when Charles was dauphin. Completed in 1377 as the last of his translations, the *Livre du ciel* opens by confessing that it was composed "at the command of the all-powerful and most excellent Prince Charles, the fifth so named, by God's grace King of France, seeker and lover of all noble wisdom." Testimony of Charles' appreciation of Oresme's translating activity and, in particular, of his *Livre du ciel,* is found at the very conclusion of this work, where Oresme confesses that while he was engaged in its composition, Charles rewarded him with the bishopric of Lisieux. The present miniature is found at the beginning of this fifteenth-century manuscript. At the left, Oresme presents his work to Charles, while three attendants stand behind the king. The right half of the picture provides a representation of the starry heavens surrounded by a crenellated border. The center of this heavenly sphere is occupied by a smaller sphere divided into three parts, the lower half depicting water, the upper left quadrant fire, and the remaining quadrant a wall, trees, and various buildings. The whole presumably refers to at least three of the four elements. For other miniatures from the present manuscript, see Illustrations 162 and 234; for various diagrams from another codex of Oresme's *Livre du ciel,* see Illustrations 254, 255, and 278.

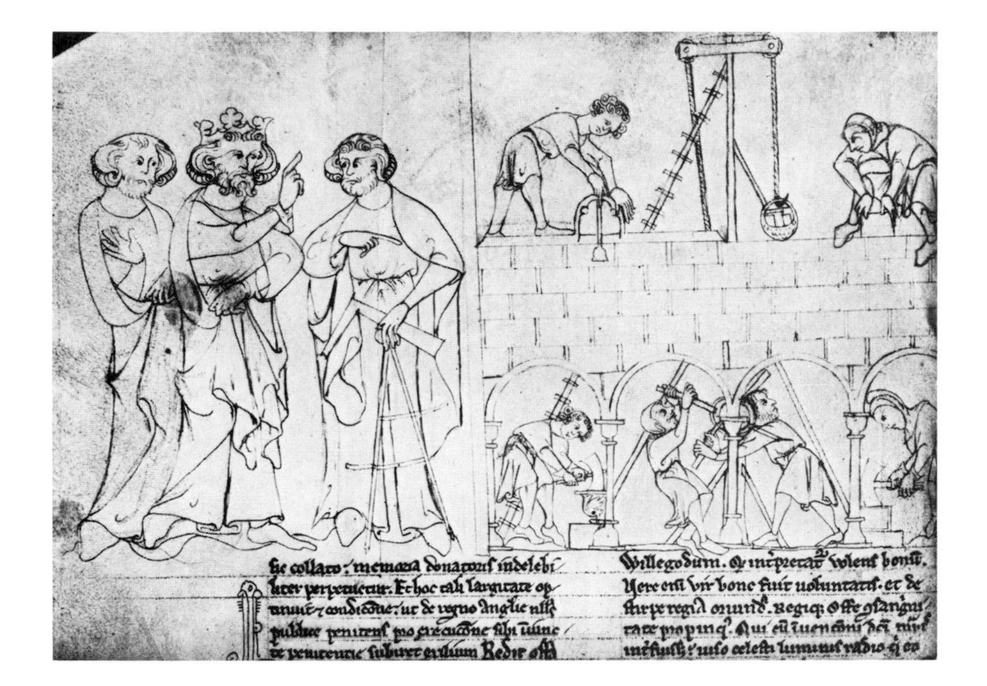

190. Another Kind of Royal Patronage: A King and His Master Mason. The drawing reproduced above is taken from a manuscript that was written ca. 1250, although the drawing itself is a later addition to the codex. The manuscript contains a number of works of the thirteenth-century English chronicler, hagiologist, historian, and artist, Matthew of Paris, the present picture occurring in a copy of his *Vitae Offarum* (that is, his account of the eighth-century Anglo-Saxon kingdoms of Offa of Angel and Offa of Mercia). Since the drawing is of a later date, it cannot be one due to Matthew or his school (for that, see Illustrations 4 and 161). The central figure in the left panel of the drawing is clearly a king. He is shown giving instructions to a master mason standing at his left, appropriately identified by the usual emblematic square and compass. Work on a cathedral or some other church architecture—which is presumably the subject of the conversation between the king and his master mason—proceeds apace in the right panel. One attempt to identify the figures in this drawing has opted for King Henry III (1216–1272), the fourth of the Plantagenet line. This would fit thirteenth-century history rather well, since Matthew of Paris had met Henry on a number of occasions and we know that Henry was instrumental in supporting the construction of church architecture in thirteenth-century England. In view of these facts we could even propose that the master mason depicted might be Master Henry de Reyns, who worked extensively under King Henry's direction on Westminster Abbey from 1245 to 1253, or perhaps Master John of Gloucester or Master Robert of Beverley, both of whom did similar work under such royal patronage during the succeeding years of Henry's reign. Nevertheless, although this supposition fits nicely with political and architectural history, attention paid to the content of the text to which the present drawing is attached and, above all, to a bit of the history of the illustration of Matthew of Paris' works, clearly shows that the identification of King Henry III and any of his master masons cannot be correct. The text at hand is, as was mentioned above, Matthew of Paris' *Vitae Offarum*. Now, one part of this history of the Offas recounts the (legendary) promise of Offa of Angel to found a monastery and the subsequent fulfillment of that promise by Offa of Mercia through his founding of what was Matthew of Paris' own monastery at St. Albans. Thus, the suggestion is that the king in our drawing is Offa of Mercia and that, consequently, the whole subject of the illustration is the construction of St. Albans under his aegis. This suggestion is nicely confirmed if one notes, first of all, that substantially the same tale of Offa of Mercia's founding of this monastery is told in Matthew's *Vie de Seint Auban;* second, that a thirteenth-century illustrated manuscript of that work contains two separate pictures representing the tale; and third, that the drawing reproduced here was clearly based on these two earlier illustrations. Thus, Offa of Mercia must be the king in our drawing, a fact that is rendered absolutely certain by the label 'Offa Rex' in the thirteenth-century original from which it was derived.

Part Five

REPRESENTATION OF THE SCIENTIFIC OBJECT II:
Nature and Artifact

191. Flora, Fauna, and the Heavens. Taken from a twelfth-century manuscript of the *Herbarius* of Pseudo-Apuleius, the picture at the top left shows the plant basilisca (*polygonum bistorta;* adderwort or snakeweed). The plant is reported to grow where the basilisk snake—reputed to be the king of serpents—resides; three such reptiles appropriately surround the plant's roots. The roots themselves, we are told, sometimes assume the form of a bear's paw, a feature again depicted by the manuscript miniature. Among the presumed virtues of this herb is the fact that one who possesses it will be protected from all manner of evil serpents. The bottom left illustration is from a thirteenth-century English bestiary. The subject: the unicorn. The story: that this most famous of mythical beasts cannot be approached by hunters without the aid of a virgin. By baring her breast, she renders the unicorn tame, which then puts its head on her lap and falls asleep, whereupon the hunters can attack it. The allegory: Christ, as a spiritual unicorn, became incarnate in the womb of the Blessed Virgin, and hence was taken by the Jews and crucified. The illustration at the bottom right is taken from a tenth-century codex executed at Fleury of a Latin translation of the *Phaenomena* by the third-century poet, Aratos of Soli. This scientific poem was immensely popular among Roman scholars—Cicero is credited with one translation of it—and throughout the Middle Ages, especially for its account and, through its illustrations, depiction of the constellations. That shown here is Aquila. Framed in the classical style that was standard for manuscripts of the *Phaenomena* at this time, it does not represent the constituent stars of this constellation at all. For other pictures from the first two manuscripts used here, see Illustrations 167, 201, 202, and 264.

IT HAS ALREADY BEEN emphasized that the locution "scientific object" in the titles of the immediately previous and the present parts of this *Album* is meant to refer only to those objects of scientific concern that permitted of observation, not necessarily objects that were observed. Even then, this "capability of observation" needed qualification because in some cases it would not have been possible to observe the particular object being depicted. All of these factors are relevant to the following three chapters as well as to the previous three, though there is an important difference in the manner in which the last qualification applies.

Thus, the pictured "objects" in Part IV that could not have been observed were nevertheless almost always depicted in terms of entities that could have, and often were, observed. An ancient or legendary scientific authority, for example, could be represented in terms of perfectly familiar contemporary scholars or authorities. And the same applies to those "persons" involved in personifications. In the present part, however, the "never seen" scientific object is often quite different, especially in the case of monstrous beasts and monstrous humans. Here the antiquity and legendary character of that being depicted involved the mythical, and thus a different kind of imagination on behalf of the depicting artist—even though his imagining would be "pieced together" from objects that could be, or were, seen. Of course, the picturing of scientific objects of any sort, whether they were observable or not, was almost always guided or affected by the text or previous picture at hand.

The concluding parts of this *Album* are separated from the present and the preceding ones by the distinction of object versus theory. This distinction was not, to be sure, one that was necessarily in the mind of the ancient or medieval illustrator of scientific texts. More than that, it is far from being a rigorous one. Naturally, in many instances the purpose of the representation of objects is the illustration of some point of scientific theory. But the "overlap" in the other direction is of greater relevance to the immediately following chapters. For here, in such instances as illustrations of anatomy and of scientific instruments, the depiction of a given object entails the simultaneous depiction of some aspect of theory. By no means, however, is the overlap of object and theory limited to illustrations in these areas of scientific endeavor. The drawing of a plant in an ancient or medieval herbal, for example, in some cases might well be nothing more than a purely descriptive depiction of the appearance of the plant in question, whether that appearance is based on actually seeing the plant or not. Nevertheless, in many instances some theory was added to this "mere appearance," especially when some effort was made to depict the plant's medicinal use (see Illustrations 195 and 196). A similar phenomenon occurred even more frequently in the depiction of animals. There, some belief about the animal, the real as well as the mythical, entered into the illustration, a belief that went beyond mere description and partook of theory.

The division of the pictorial material into three chapters for this final section on the scientific object is naturally one of convenience. Still, if one leaves out of consideration those related pictures taken from non-scientific sources, each chapter can be seen as a genus with respect to actual species of scientific works, especially relative to such works as were written in the Middle Ages.

CHAPTER EIGHTEEN

The Population of the Sublunar World

THE ILLUSTRATIONS INCLUDED IN this chapter can roughly be divided into those representing flora and those representing fauna, with a special section of the latter devoted to humans. Unlike the many scientific illustrations, the purpose of which was simply to facilitate or improve the reader's understanding of a scientific text, the kind of picture that is of concern here was, more often than not, essential to the comprehension of the science at hand.

The function of the depiction of plants in ancient and medieval herbals, for example, was to enable one to identify the plants in question. The earliest herbals—such as that of Diocles of Carystus (fourth century B.C.) or the final book of the *Historia plantarum* of Theophrastus (ca. 370–288/285 B.C.)—as far as we know were not accompanied by illustrations. That apparently first occurred about a century later. We have, for instance, the testimony of Tertullian (ca. A.D. 160–240) that the second-century B.C. poet, Nicander of Colophon, who composed a work on herbal and animal remedies (see Illustration 196), painted as well as wrote *(scribit et pingit)*, but this testimony most likely refers to copies seen by Tertullian and not to Nicander's own enterprise. Yet we do have good evidence of the existence of an illustrated herbal contemporary with Nicander: it was the work of Crateuas, physician to Mithridates VI, Eupator Dionysus (120–63 B.C.), who had a keen interest in herbalism himself. Not only does Pliny (A.D. 23/24–79) tell us that Crateuas painted the likenesses of plants (*pinxere namque effigies herbarum: Nat. hist.*, XXV, 4), but the early sixth-century Anicia Juliana codex of Dioscorides (see Illustrations 188 and 193) informs us that some of the material was derived from Crateuas. Thus, since the work of Dioscorides in the second century was the major source, in both doctrine and picture, for medical botany throughout the Middle Ages, Crateuas may be regarded as the fountainhead of this illustration tradition.

Treating more than 800 plants and other medicinal substances, Dioscorides' five-book Περὶ ὕλης ἰατρικῆς *(De materia medica)* was not only translated into both Arabic and Latin, many copies of which bore illustrations, but also served as a most important source for other writings on medicinal simples.

Herbals other than those of Dioscorides also played a role in the tradition of plant iconography. Undoubtedly the most important of these alternative works was the *Herbarius* of Pseudo-Apuleius, most likely composed in Latin in the fourth century and not translated from an earlier Greek source (see Illustrations 193–195). Existing in more extant manuscripts than any other medieval herbal, this work covers each medicinal plant in a separate brief chapter (of which there are 130), its various names and uses cataloged in as precise a manner as possible.

The importance of the pictorial tradition for herbals is especially evident in the *Herbarius* of Pseudo-Apuleius. Except for instances in which we find interpolations from the Latin translation of Dioscorides, the text of Pseudo-Apuleius does not include a description of the plant depicted. Such depictions were, however, necessary if one was to be able to identify the plant in question in addition to knowing its names and uses. It appears, therefore, that this requirement was satisfied by an iconographic tradition that was quite separate from the tradition of the text.

Herbals may well have served as the most important

part of the ancient and medieval history of the illustration of plants, but such illustrations naturally also occurred in other kinds of written works and in media other than the book. A few of these depictions are also included in this chapter (see Illustrations 198 and 199).

If the herbal provided the focal point for floral illustration in the Middle Ages, the bestiary did the same for fauna. The origin of the bestiary was a Greek, but Christian, treatise called the *Physiologus* (see Illustration 200). The place and date of its composition are uncertain, but Christian authors of the fourth, and possibly also the third, centuries were aware of its existence. Equally uncertain is the date of the first Latin translation, though some evidence suggests a date as early as the fourth or fifth century. The oldest known copies of a Latin version are of the eighth century, but it was undoubtedly translated considerably before that time.

In any event, the history of the Latin *Physiologus* or bestiary is an appreciably complex one. There were, as to be expected, bestiaries in almost every major vernacular tongue. Additionally, bestiary material found its way into other medieval works, both Latin and vernacular, a phenomenon that was most natural in light of the fit of the bestiary with Christian concerns. It was a fit that derived basically from the essential formula of the bestiary: to describe and characterize an animal or some other object or event in nature and, on the basis of that description and characterization, to elucidate a moral lesson allegorically. Such a procedure was most congenial to the early Christian penchant for viewing nature symbolically. Indeed when, at the end of the twelfth century, the English scholar Alexander Neckham confesses in his *De naturis rerum* that he "is not going to enter into abstruse inquiries concerning the natures of things, but to collect together a quantity of known facts, and treat them morally," he is clearly in harmony with the bestiary tradition. Further, if this moral treatment anywhere within that tradition could relate what was said to specific items of Christian doctrine or could cite fitting passages from Scripture, so much the better.

Yet all manner of works other than the bestiary can be found in which the drawing of an animal is related to some aspect of the text or is present for a strictly decorative purpose. This will be only too evident in many of the following illustrations. However, one particular nonbestiary work should be singled out for special mention: the treatise on falconry, the *De arte venandi cum avibus,* by the thirteenth-century emperor Frederick II, revised in part by his son Manfred. Although its later books relate to the details of the art of falconry, the initial part of this remarkable work constitutes a veritable treatise on general ornithology. And the illustrations found in its extant manuscript copies—both the original Latin and in translation—are equally remarkable (see Illustration 202).

Another segment of fauna whose depiction is frequently found outside the bestiary tradition is that of the monstrous (see Illustrations 206–208). The bestiaries did treat their share of monstrous beasts, but the monstrous humans who seemed to harbor an even greater fascination for the medieval mind found most of their artistic representation outside of the bestiary work. They occurred in accounts of travels supposedly made to fabulous lands, or simply in geographical reports of remote or unvisited places in general. Since the description of the monstrous often occurred side by side with the recounting of more "normal" natural history and geography, pictures of this genre have also been included in this chapter.

The final pictures in this chapter are medical. In accordance with the express intent of concentrating on the scientific object, they are drawn from anatomy. Even here, however, the illustrations are much more "theory laden" than the others making up the chapter. Although the anatomical details pictured were most likely seen at one time or another, there seems little doubt that elements of anatomical and physiological doctrine guided at least part of what was finally depicted. Indeed, there is so much theory involved in the illustration of one part of human anatomy—the brain—that its consideration has been postponed until Chapter Twenty-three, in which the theoretical side of medicine will be the central concern.

Note should be taken of the fact that representations of plants, animals, and medicine have also appeared above: in Illustrations 10, 16, 18, 19, 20, 164–168, and 183.

192. (Right) A Papyrus Herbal. The earliest extant fragment of an illustrated herbal comes from a papyrus roll (ca. second century of the Christian era) that was discovered during the excavation of Tebtunis in Egypt. The date of the papyrus illustration given here is somewhat later (ca. A.D. 400, found in Antinoë), but the illustrations (there are only two) are slightly less fragmented. It is noteworthy that we have here part of a leaf from a papyrus codex, not a fragment of a papyrus roll. The initial word in the text immediately below the illustration has been deciphered as ΣΥΜΦ[ΥΤΟΝ], that is, *symphytum* or comfrey. Although the plant depicted does not seem to match the descriptions given of symphytum (or of plants for which it is a synonym) in Dioscorides, there is some resemblance between the present papyrus illustration and the illustration of an herb called *simfitum album* in several Pseudo-Apuleian manuscripts. Precisely what the connection was, between this papyrus and these manuscripts, is impossible to say. Nevertheless, it is interesting to note that a connection can be made, if one takes into consideration both textual and illustrative factors.

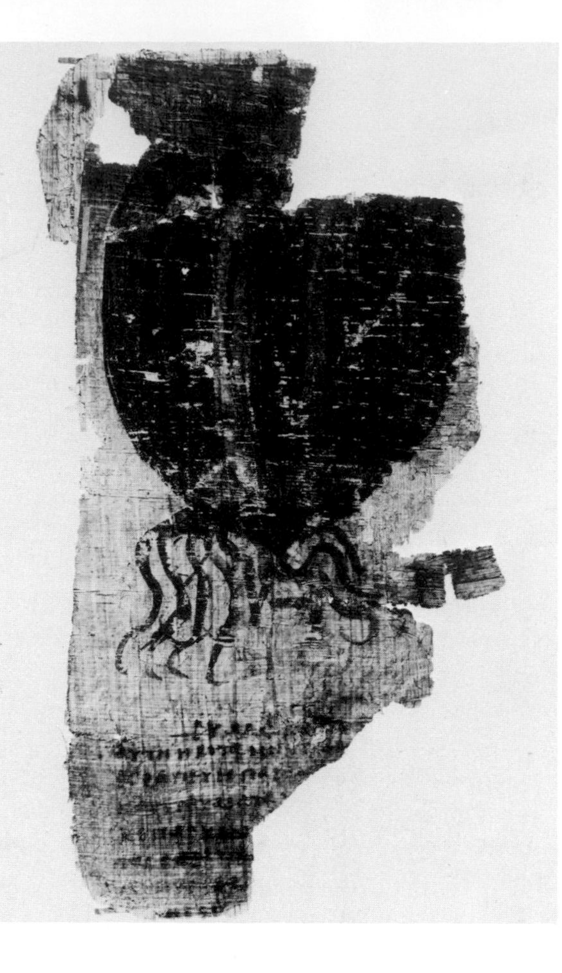

193. One Herb, Six Centuries. The single herb shown here in three different representations is dracontea (*δρακόντιον μικρόν* in Dioscorides' Greek, the modern *arum maculatum* or cuckoopint). With a single exception, all of these representations are concerned merely with the description of the plant itself, not with its medicinal properties. In terms of the text that accompanies the two pictures in the left-hand column on the facing page (both Dioscorides: one Greek, one Latin), the salient elements of this description are few and straightforward. Dracontea's leaves are ivylike (*κισσοειδῆ*; *cisso similia*) with white spots (*σπίλους λευκούς*; not in the Latin). Secondly, there is an upright stem (*καυλός*; *virga*) two cubits high, serpentlike in form (*ὀφιοειδῆ*; *simile serpenti*) with purple spots; the seeds on the top of the stem are like clusters of grapes (*βοτρυοειδῆς*; *votruo simile*). Lastly, the root is round and bulbous (*στρογγύλη*, *βολβοειδῆς*; *vulbi habens rotunditatem*). Just this much of Dioscorides' text in one way or another formed the basis for all three of the present depictions of dracontea. The first is, appropriately, from the earliest extant codex of Dioscorides: the Anicia Juliana of about A.D. 512 (see Illustration 188). Its representation of the herb's features described in the text is reasonably accurate, even though the clusters of seeds and specified colored spots are absent. The second illustration of dracontea is from a ninth- or tenth-century manuscript written in Beneventan script of an anonymous Latin translation made of Dioscorides, most likely in the sixth century. As can be seen from the few Latin words from this translation given above, it contains a fair number of transliterated Greek words; the present codex is also the only illustrated one of this Latin translation. Although much cruder, this illustration is surely related to the one in the Anicia Juliana manuscript. However, it adds the spots that were missing in the earlier manuscript, and the stem, taken together with the bulbous root, now looks slightly snakelike. Most significant, however, are the man and the serpent to the right of the plant. This appears to be the first extant occurrence of the depiction of a human "patient" in an herbal. Its function is to illustrate a particular therapeutic use of dracontea that is mentioned later in Dioscorides' text: namely, that if one rubs oneself with the root of dracontea one will become immune to snakebite *(si ex ea radice quisquis se unexerit, tutus ab incursu serpentum servatur)*. The remaining dracontea illustration, at the

(Continued)

far right, is not from a Dioscoridean text but from a seventh-century copy of the *Herbarius* of Pseudo-Apuleius. Here, the leaves have assumed a quite different configuration, though the grape-cluster seeds that were mentioned in Dioscorides' text seem to be present at the top of the stem. The intriguing fact is, however, that the only description of the plant given in the text of Pseudo-Apuleius says that its root is like a dragon's head. One is urged to conclude, therefore, that in the case of this illustration the depicted form and features of the plant as a whole derived from an illustration tradition that was independent of the text. (The information—much of it referring to Dioscorides—that appears immediately to the right of the plant in the present picture is not part of the text, but is a later sixteenth-century addition, as are the two German words [*Schlangen kraut;* snake plant] below this addition.) It is notable that, with a partial exception made for the first illustration from the Anicia Juliana codex, these representations are quite two-dimensional. One is tempted to think that, just as in certain instances of the diagraming of geometrical solids (Illustrations 121, 122, 124), such a "flattening" minimized overlapping and preserved actual size so that all features depicted would be made as evident as possible.

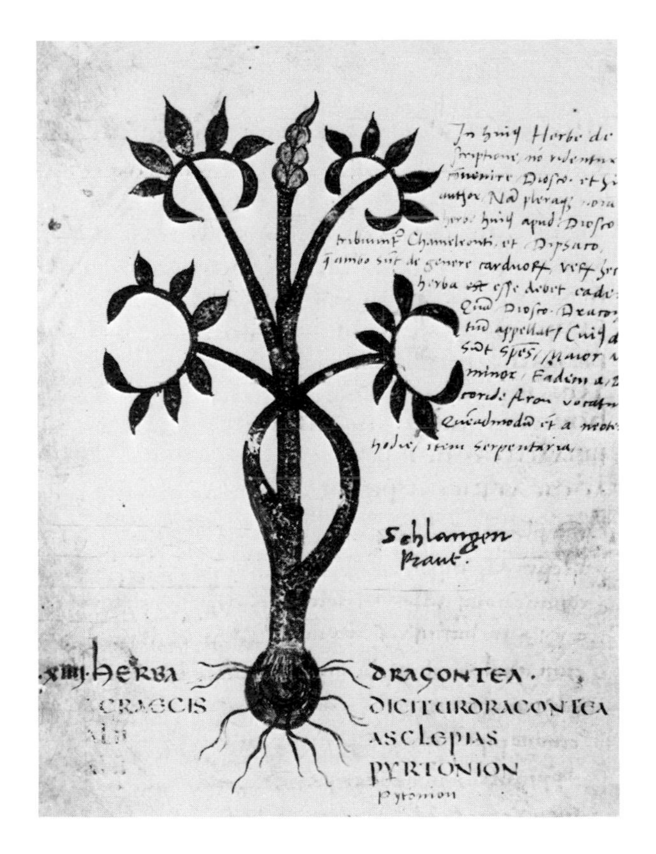

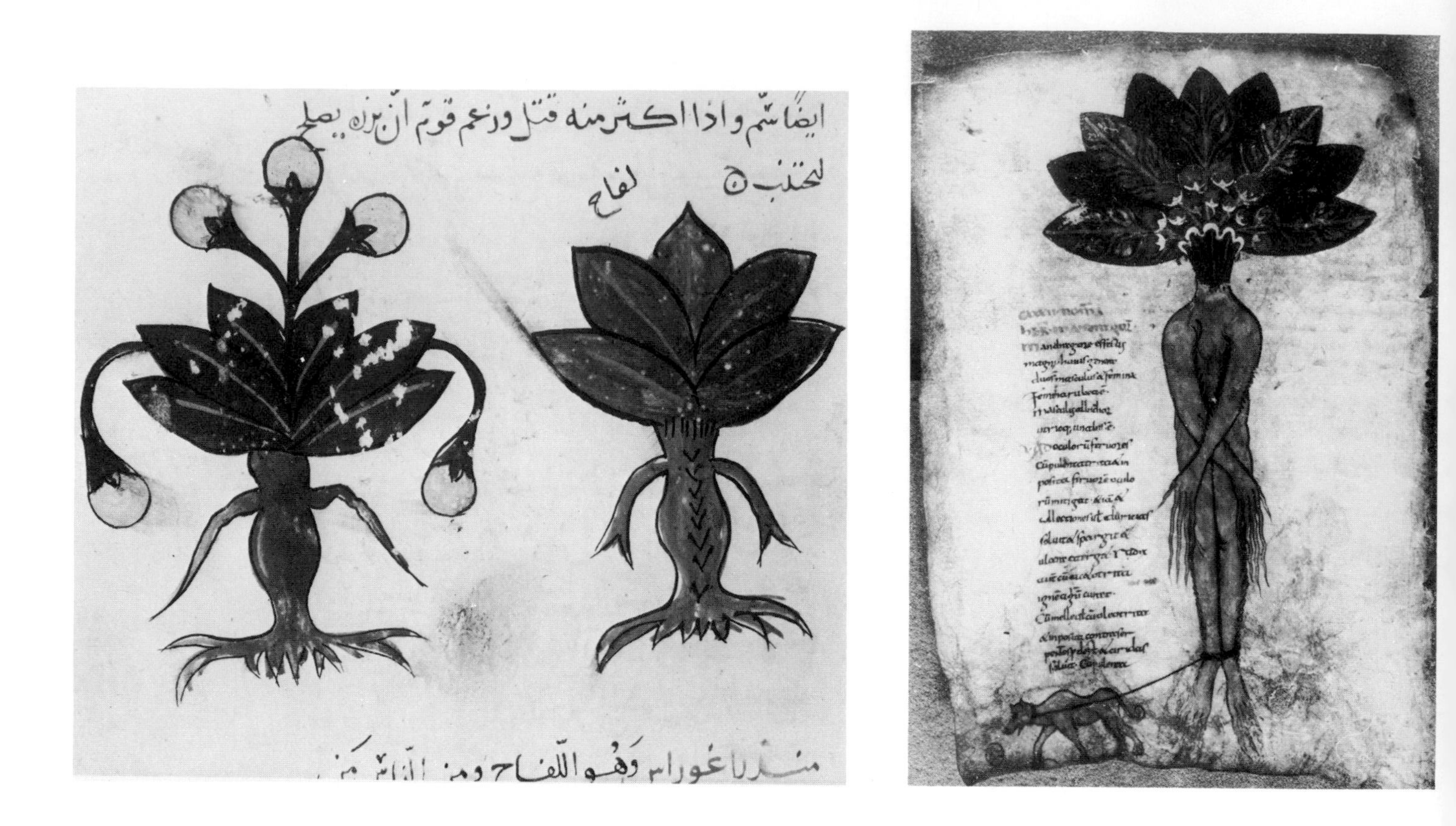

194. An Anthropomorphic Plant. No herb has more legendary and fantastic elements in its history than the mandrake *(mandragora officinarum)*. The most notable of these elements was the belief in its human form, a belief most likely deriving from the two leglike branches of its tuberous roots. Dioscorides distinguished two types of this plant, one of them female, black with leaves smaller and more narrow than those of lettuce and having a heavy, foul odor. The male, on the other hand, he claimed, has larger and broader white leaves, is sweet smelling, and has berries or "apples" twice the size of those of the female variety. Both, he said, lack a stalk. Dioscorides went on to tabulate the medicinal uses of the mandrake, but we should realize that such uses were part of the medical tradition much earlier, since some of them (for example, for wounds, gout, erysipelas, sleeplessness) are cataloged by Theophrastus (ca. 370–288/285 B.C.) in his *Historia plantarum* (IX, 9, 1) and are mentioned earlier still in several writings of the Hippocratic corpus. Even Aristotle saw fit to mention its soporific effects. Of the four pictures of mandrake given here (none of them including any of its medicinal uses in its representation), only one is Dioscoridean: that at the top left of page 218, from an Arabic translation of the *De materia medica* as found in the 1239 manuscript that has already provided the portrait of Dioscorides in Illustration 156. Two names for the herb appear below their depiction, one in transliterated Greek: *mandhraghūrās,* the other the more usual Arabic term: *luffāḥ* (which also appears above the plant on the right). The fact that the top berry or apple of the left-hand plant has been painted over part of the last letter of a line of text is direct evidence that the illustrations in this manuscript were fitted into blank spaces in the text that were provided for them. The next two illustrations are taken from copies of the Pseudo-Apuleius *Herbarius,* the first (top right, page 218) from a ninth- or tenth-century manuscript that contains portions of text from the Pseudo-Dioscorides *De herbis femininis* as well as from Pseudo-Apuleius. The anthropomorphic form is even clearer here than in the two Arabic representations. More worthy of note, however, is the dog leashed to the plant's roots, a feature that was present in any number of medieval representations of mandrake. As is made clear in the text of Pseudo-Apuleius, this refers to the legendary caution connected with the digging of mandrake. One is to draw a circle on the ground about the plant with an iron implement; excavate the root, not with the iron tool but with something ivory; then tie the root with a cord the other end of which is fastened

(Continued)

about the neck of a hungry dog. Then, food is to be placed—here as the ball at the lower left of the illustration—before the dog so that his straining for it will uproot the mandrake. These rather fantastic instructions were not a medieval creation, since we have records of related notions in antiquity. Pliny (A.D. 23/24–79), for example, tells us (*Nat. hist.* XXV, 148) that diggers of mandrake should avoid facing the wind and trace three circles around the plant with a sword. Aelian (ca. A.D. 170–235) reports in his *De natura animalium* (XIV, 27) that peony is to be uprooted in similar fashion by a dog leashed to its roots. And Theophrastus refers to the drawing of circles with a sword as well as a number of other strange directions for the digging of other herbs, all of which he considers fictitious and farfetched (*Hist. plant.* IX, 8, 6–8). The next Pseudo-Apuleian illustration (bottom left, page 219) is from a late-twelfth-century manuscript. Highly stylized like the previous pictures, this artistic representation is undoubtedly of the female mandrake. The final illustration of a mandrake, at the bottom right, is taken from an herbal compiled in a 1419 manuscript by Benedetto Rinio, physician at the University of Padua. Culling material from almost sixty earlier writers on a variety of aspects of medicine and natural philosophy, Benedetto had as his avowed purpose not to record the usual medicinal uses of herbs, but to set forth new applications and to correct errors that had previously caused the misidentification of plants. The manuscript was illustrated by one "Andreas Amadio, venetus pictor." The depiction of mandrake here is much more realistic, though the positioning of the split root still carries anthropomorphic suggestions. However, we should not take this greater realism to imply that the illustrations in this herbal were made from plants in nature, since there is conclusive evidence that many of these illustrations were copied from an earlier codex. (One of the pictures from this earlier codex is given below in Illustration 197.) The legendary human form of the mandrake as well as other fanciful notions associated with it were part of tradition and popular culture well beyond the Middle Ages. The English *Herball* (1551–1568) of William Turner, for example, speaks of "the rootes which are conterfited and made like litle puppettes and mammettes, which come to be sold in England in boxes, with heir, and such forme as man hath are nothyng elles but folishe feined trifles and not naturall. . . . I have in my tyme at diverse tymes taken up the rootes of Mandrag out of the grounde but I never saw any such thyng upon or in them."

195. Herb Plus Therapy. The upper and middle pictures at the right are taken from manuscripts of the Pseudo-Apuleius *Herbarius,* and both attempt to depict the use of the medicinal plant in question in addition to, or instead of, simply depicting the plant itself. That at the top, from a thirteenth-century manuscript, is of Artemisia leptofillos (*Artemisia arborescens* or *abrotanum;* southernwood). Since, as is the case with almost all plants in Pseudo-Apuleius, the text lacks a description of the form of the plant itself, the depiction of artemisia at the top of the page derives either from other texts or, more likely, from some illustration tradition. Instead, the portion of the text given in this illustration is concerned with where the plant grows (hills and ditches), with the fact that its crushed buds yield an odor of oil of marjoram *(samsucum),* and most significantly with two of its therapeutic uses: for stomach illness and for nervous tremors. The first of these uses is depicted here: if a poultice *(malagma)* is made from crushed artemisia and oil of almonds and applied to the patient's ailing stomach, a cure will be effected in five days. (The text given concludes with the additional information that if artemisia is hung over the entrance of a house, no one will inflict injury on its occupants.) The second illustration, middle right, comes from a twelfth-century Pseudo-Apuleius manuscript. The medicinal plant in question is plantago (plantain), but in this case there is no representation at all of the plant itself. Instead, we have an illustration of two of the roughly two dozen uses of this herb listed by Pseudo-Apuleius: for the bite of a snake or of a scorpion. For another picture from the last manuscript, see Illustration 203.

196. (Bottom Right) Therapy Revealed by a Poetic Tale. This illustration is taken from a tenth- or eleventh-century manuscript of the *Theriaca* of Nicander of Colophon (most likely second century). The *Theriaca* is a Greek didactic poem in hexameters that provides an account of snakes, spiders, scorpions, and other poisonous creatures together with the remedies for the maladies they may inflict. The protagonist in the present illustration is Alcibius (whose name appears at the upper left). The medicinal herb that will come to his aid is echium (probably viper's bugloss), described by Nicander as thick with prickly leaves, slender rooted, and topped by flowers like violets. The story recounts how Alcibius was bitten in the groin by a male viper while asleep. Awakened brutally by the bite, he arose and pulled the echium root from the ground, breaking it into small pieces and sucking them while applying its rind to his wound. The plant, the viper, and Alcibius tearing the herb from the ground are all duly pictured here.

197. (Right) Herbal Realism. We have already noted a quite realistic portrayal in the last representation of mandrake given in Illustration 194. Indeed, in the nineteenth century the pictures in the manuscript from which that mandrake was drawn were felt to be so accurate and realistic by John Ruskin that he declared them to exhibit a "Venetian respect for law." But we now know that many of these Venetian pictures were not "real-life" drawings but were copied from an earlier herbal. The text of this herbal is a yet uninvestigated Italian translation (possibly from a Latin translation) of a treatise originally written in Arabic by Serapion the Younger (Ibn Sarābī, most likely twelfth century). The manuscript of it used here was written and illustrated about 1400 in Padua for Francesco Carrara the Younger, the last lord of Padua. The illustration of melons at the top right is taken from this codex. Not only are all parts of the plant depicted with greater accuracy and realism than in other medieval herbals, but the "flattening" for didactic purposes has disappeared and the overlapping of elements within the picture is no longer avoided. The "realism" of the illustration at the lower right, taken from a fifteenth-century manuscript, is of a quite different sort. Flattening is at an absolute maximum, but realism is preserved because the pictures of plants in this manuscript were created by coating real leaves and stems with something like tempera paint and then "printing" their images by pressing them on the page. Unfortunately, the person who created this unique fifteenth-century product did not have as much botanical knowledge as "artistic" ingenuity, since many of the plants thus represented are misidentified. Those appearing here (duly labeled with Italianate names) are: ameos *(ammi maius)*, corrigeola marina (*convolvulus soldanella;* bindweed), basilicus (*origanum marjoram;* sweet marjoram), and liquiricia (*glycyrrhiza glabra;* Spanish licorice).

198. (Left) From Pattern to Nature: The World of Plants in Medieval Cathedrals. Floral decoration was frequently used on Gothic capitals in medieval cathedrals, so much so that historians of art have devoted numerous articles, even whole books, to the problem of the evolution of such decoration. Thus, it has been maintained that the flora on cathedral capitals move from stylization and extreme simplification with little detail in the twelfth century, through "naturalistic exactitude" in the thirteenth, to the extreme complication of the fifteenth-century flamboyant style in which details are multiplied and emphasized while naturalism lessens. Examples of the first two stages in this evolution are given here. That above left is of the capital of a pillar, constructed about 1190, in the north aisle of Notre Dame in Paris. Simple and without much detail, the leaves have an abstract, unreal look, despite the presence of seedpods protruding through several of them (a feature that has occasioned the remark that they give an impression of a plant like Aaron's rod). The remaining leaves, however, strike one as without even a suggested mate in nature. The second example, below left, is of a capital from the West Gallery in the Naumburg Cathedral in Germany. This is now the second, thirteenth-century, stage of floral decoration that bespeaks naturalistic exactitude. No longer is there merely an impression of some plant, with some similarity of the leaves on the capital to those of a known species. Now one can identify the leaves as those of a specific plant: namely, *artemisia vulgaris* (mugwort). In medieval sculpture such real-life "portraits" clearly antedated their occurrence in manuscript herbals, even though the consequent identification of the sculptured plants may not be as definite and exact as some historians would like to believe.

199. From Nature to Pattern: An Utrecht Psalter Tree. In the preceding illustration the simpler, more stylized and patterned representation preceded the more naturalistic in time. Here the reverse obtains. Both of the present illustrations are taken from copies of the Utrecht Psalter and both depict the first Psalm. The man seated writing in the edifice at the top of each picture is the godly man whose blessedness is declared in the opening line of the Psalm. Of greater present interest, the tree below him is a reference to the opening words of the third verse: "And he shall be like a tree planted by the rivers of water, that bringeth forth his fruit in his season. . . ." The illustration at the top left, however, precedes that at the bottom right by more than 300 years. It comes from a manuscript written at Hautvillers about 820. To be sure, the
(Continued)

tree in this illustration is quite distant from the naturalism of the Naumburg capital in the preceding illustration; but it is much less stylized and patterned than the tree in the second picture given here. This comes from the Eadwine Psalter, written at Canterbury about 1150. Not only has the tree become a stylized, symmetrical ornament, but even the land has turned into a mass of decorative scallops. What there was of nature in the ninth-century psalter has been sacrificed for artistic elegance of a quite formal sort.

200. Two Early Bestiaries. Extant manuscripts of the Greek *Physiologus* are later than the earliest existing copies of the Latin bestiary versions based on it. The same situation exists for the earliest illustrated manuscripts of these Greek and Latin works. Thus, the picture at the right comes from a ninth-century Latin codex, that at the left from a Greek manuscript written about 1100, both being the earliest extant copies of their genre. The Greek codex was in the possession of the Evangelical School in Smyrna until 1922, when it was lost during the burning of the library by the Turks. The present illustration, taken from plates made from the manuscript in 1899, accompanied the *Physiologus* entry on the fox (ἀλώπηξ). A crafty and deceitful creature, when in want of food, we are told, it lies still on its back upon the ground while staring upward and holding its breath. Thinking the fox dead, birds sit upon him in order to peck at his flesh, whereupon the fox jumps up, seizes the birds, and devours them. Later medieval Latin bestiaries (which add to the tale by saying that the fox rolls in red earth to give the appearance of being covered with blood while feigning death) frequently carry but a single picture in representing this entry: simply the prostrate fox with birds perched upon him. The present Greek illustration is notable in adding two further "picture strips": one in the middle showing three birds in flight and another below completing the tale by depicting the fox successfully capturing his prey. The accompanying "explanation" (ἑρμηνεία) likens the fox to the devil in deceitfulness, warning that whoever wishes to partake of his flesh (taken to mean fornication, avarice, lustful pleasure, and murder) will die. The second picture, on the right, is from the earliest extant illustrated Latin bestiary, written by a certain Egbert (Haecpertus) slightly after the middle of the ninth century. Its subject is the panther, many-colored (that is, spotted) like Joseph's coat (*Genesis,* XXXVII, 3), and said to be the friend of all animals save the dragon. When it has eaten its full, we are told, it sleeps for three days and upon awakening emits a sweet-smelling roar. Other animals who hear the roar draw near so that they might be filled by the sweet aroma. In the present miniature, the panther is a light yellow-green with red spots; he has presumably just emerged from the cave behind him. A deer and what seem to be a boar and a fox are the animals the panther has attracted, while a serpent (most likely as a depiction of the dragon which is the panther's only enemy) sits coiled on the rocks above, spewing venom toward his foe. The accompanying allegory opens by referring to Christ who, on the third day, rose from the dead, whence all from near and far were replenished by the sweetness of faith. (The attractive odor emitted by the panther is mentioned as early as Aristotle, who recounts that it serves him to lure his prey.)

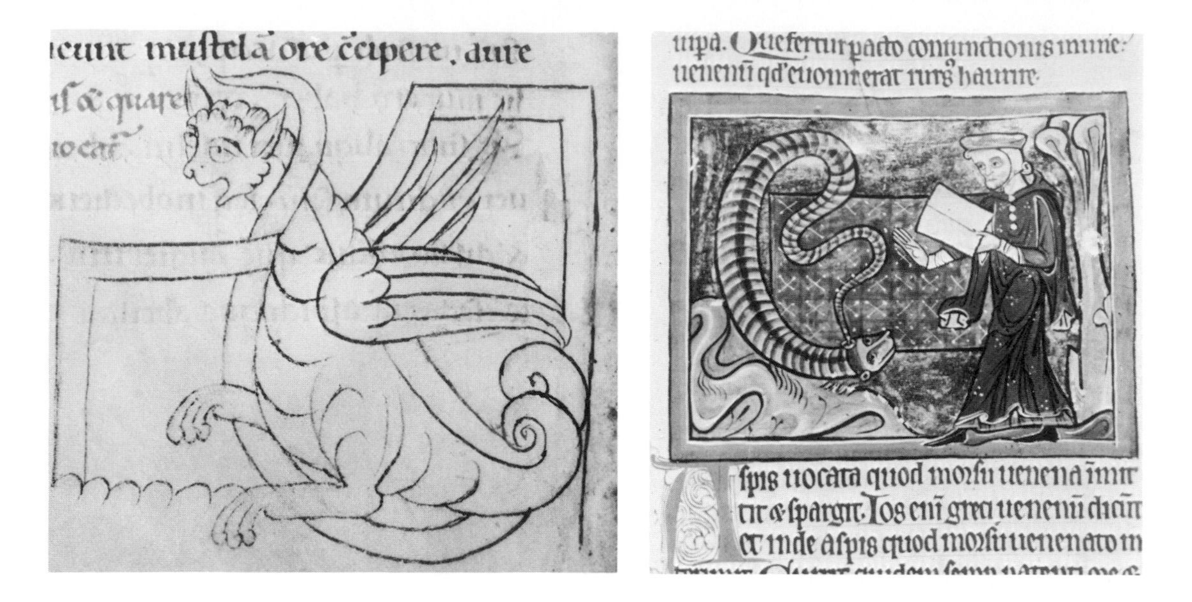

201. Variants in a Bestiary Illustration: The Asp. The central feature of the account of the asp in medieval Latin bestiaries is that when faced with the enchanter *(incantator)* who by song tries to charm it from its lair, it frustrates the attempt by pressing one ear to the ground while plugging the other with its tail. Deriving from a twelfth-century English manuscript, the first picture of the asp at the top left omits all details of this particular story, limiting itself to a simple depiction of the asp itself (in the winged-dragonlike, and not very snakelike, appearance frequent in illustrated bestiaries). The second picture, top right, from a thirteenth-century English codex, presents the asp without wings in a more serpentlike form and properly represents the central details of the bestiary account: the enchanter is there, presumably singing from the book held in his hands, while the asp has one tail-stopped ear with the other hugging the ground. The third depiction of the asp, below, is taken from another thirteenth-century English bestiary codex. Wings are once again present and the asp is appropriately represented with closed ears. But the enchanting has been replaced by a quite different scene which is not part of the bestiary text. Protected by a shield, a man wields a stick to prod the tail closing the one ear. This new element is found in a number of other illustrated bestiaries, the stick being replaced at times by a type of hook, the intention in all instances presumably being to dislodge the tail stopping the asp's ear. One source of the whole account may well have been biblical, for the fifty-eighth Psalm (verses 4–5) speaks of those who are "like the deaf adder (*sicut aspidis surdae* in the Latin Vulgate) that stoppeth her ear, which will not harken to the voice of charmers." The allegory connected with this bestiary account of the asp does not, however, relate to this psalm, but instead to the wealthy, who press one ear to earthly desires and whose other ear is closed to the voice of the Lord. Only men and asps, some bestiary versions declare, refuse to listen. For other pictures from the last two manuscripts used here, see Illustrations 191, 202, 206, and 273.

202. Birds Inside and Outside Bestiaries. Throughout the Middle Ages the depiction of birds was almost always closer to their natural form than the depiction of other animals, no matter what kind of written work these pictures accompanied. Comparison of the three examples given here and on the next page with this chapter's illustrations of other animals should provide evidence for this fact. The first miniature, at the top left, of three eagles is taken from a thirteenth-century English bestiary that we have had occasion to note before (Illustrations 191, 201). The text immediately below opens with the claim (deriving from Isidore of Seville) that the eagle's name *(aquila)* comes from the sharpness of its eyes *(ab acumine oculorum)*. In fact, the greater part of the bestiary account of the eagle relates to its eyesight. Thus, we are told that its vision is so keen that it can see fish swimming in the sea far below. Once a fish is sighted, the eagle dives into the sea, capturing his prey and carrying it off on the wing, a part of the account appropriately represented in this miniature by the fish-bearing eagle on the left. Yet such keenness of sight is not a luxury retained without event by eagles. As they grow old, their wings grow heavy and their eyes become darkened with a mist. To remedy this condition, the eagle goes in search of a fountain. When he finds it, he flies high above it into the rays of the sun, which singe his wings and consume the mist from his eyes (here all duly represented by the eagle and sun at the top of the picture). After this happens, he descends to the fountain and, as the lowest eagle in our picture shows, submerges himself in its waters three times. Immediately he is completely renewed; his wings regain their strength and his eyesight is revived with ever greater splendor. The allegory attached to this bestiary account claims that those whose eyes of their hearts are clouded with mist should seek out the spiritual fountain of God (a reference to baptism) so that they may turn their heart's eyes to the Lord and, like an eagle, renew their youth. The second illustration, at the top right, is taken from a copy (ca. 1300) of a French translation made for Jean de Dampierre and his daughter Isabel of the treatise on falconry of Emperor Frederick II, the *De arte venandi cum avibus*. In keeping with the carefully descriptive and naturalistic character of this treatise, the birds here are pictured with reasonable accuracy. The subject being illustrated is that of birds' crests; the red comb of a cock with its wattles at its throat "like a beard" and the owls with tufts of feathers to both right and left are easily identifiable at the upper left and the lower right. Less obvious, but equally precise, is the depiction of the red excrescence *(une choze)* between the nose and the forehead of the two birds directly below the cock on the left. The text identifies these two as an aquatic bird *(oisel aquatique)* and the male of certain red-beaked swans *(cignes)* who carry this protuberance like a nut *(noisette)*. The final picture, reproduced on page 226, comes from a paraphrase of the earliest extant illustrated treatise on birds: the *Ornithiaca* or *Ixeutica* of an otherwise unknown Dionysius (first century B.C.). Probably composed at some point

(Continued)

(Continued)
in the fifth century, this paraphrase appears as a kind of appendix in the sixth-century Anicia Juliana manuscript of Dioscorides (see Illustrations 188 and 193). Perhaps the most notable feature of the illustration reproduced here is that, unlike what is to be found on the earlier pages of this codex of *Ornithiaca* paraphrase, the birds are not depicted separately and dispersed throughout the text in their proper places, but are "pulled from the text" and collected together on a single, full page, resulting in twenty-four framed "avian portraits." Only the first of six rows on this page is reproduced here, the birds represented identifiable as (from left to right): an ostrich *(struthio camelus)*, a great bustard *(otis tarda)*, a spotted crake *(porzana porzana)*, and a common partridge *(perdix perdix)*.

203. (Facing Page) The Many Medieval Faces of the King of Beasts. Illustrations of any number of animals can be found in medieval manuscripts of works other than bestiaries, but no animal was depicted more frequently or in more different kinds of works than the lion. Some notion of this variety of works, and of how the illustration of this one animal followed the changing styles of manuscript painting and drawing, can be seen from the five examples—from the seventh through the thirteenth centuries—given here, none drawn from a bestiary proper. The first, at the top left, is from the Book of Durrow from the Columban monastery of that name, the earliest (ca. 680) of the Hiberno-Saxon Gospel books. Without mane, this doglike lion appears not as the symbol of St. Mark but, according to the pre-Jerome ordering, as the symbol of St. John (usually represented by an eagle). The second, middle left, is from a twelfth-century English manuscript (seen also in Illustration 195) of the *Liber medicinae ex animalibus* of Sextus Placitus (fourth century?), a work similar in structure and purpose to the *Herbarius* of Pseudo-Apuleius, but treating of medical preparations of an animal origin. Thus, the fat of the male lion—here with a distinctly unlionesque face—can be useful for curing neck, foot, and ear trouble as well as nervous disorders and snakebite, while a segment of the womb of a lioness (here on the right) is claimed to provide an effective contraceptive. Third (top right), from the sketch-book (ca. 1225–1235) of Villard de Honnecourt (see Illustration 153), is one of the most striking and famous medieval drawings of a lion. The legend at the top right tells us that "Here is a lion seen from the front." It is clear that a compass has been used to draw its rather human-looking round face, something that is rather intriguing in view of the final words of the legend, which instruct one to "Note well that it was drawn from life" *(sacies bien qu'il fu contrefais al vif)*! A porcupine *(porc espi)* appears at the lower left. Next (bottom left), from the thirteenth-century manuscript of al-Qazwīnī's *Wonders of Creation* we have sampled before and will meet again (Illustrations 52, 126, 208, and 253): this depiction of a lion is more accurate than any we have met above. The text tells us that all run from the lion's roar except the donkey. The last depiction, bottom right, comes from a manuscript most likely written shortly before 1437 of the *Book of the Fixed Stars* of the tenth-century Arabic scholar ᶜAbd al-Raḥmān al-Ṣūfī (for more of which see Illustrations 227–28). Less realistic than the picture from Qazwīnī, the lion here naturally represents the constellation Leo.

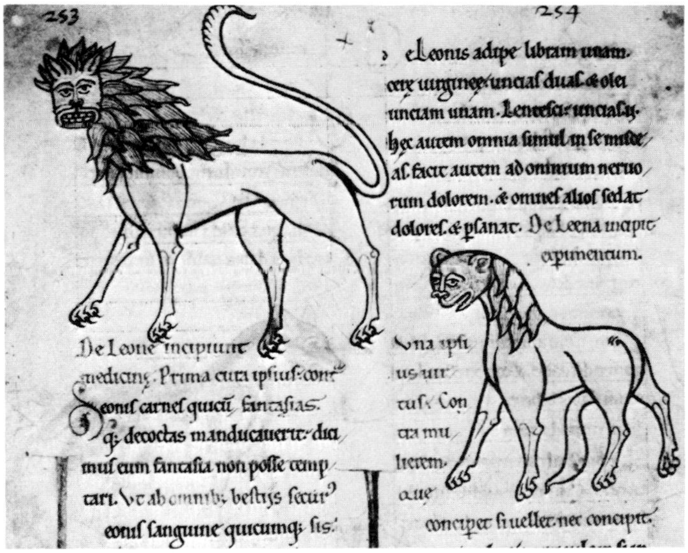

LEO

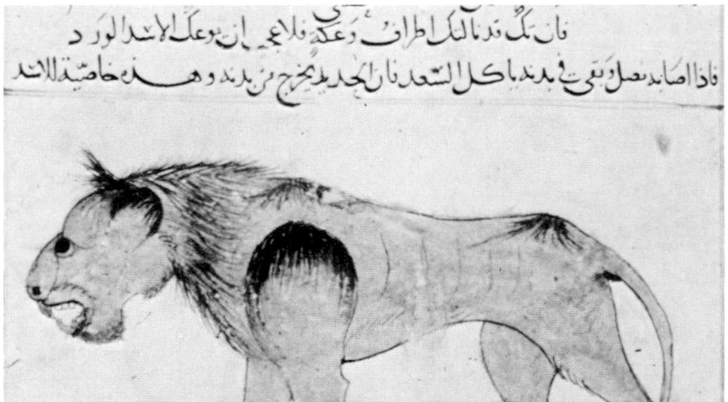

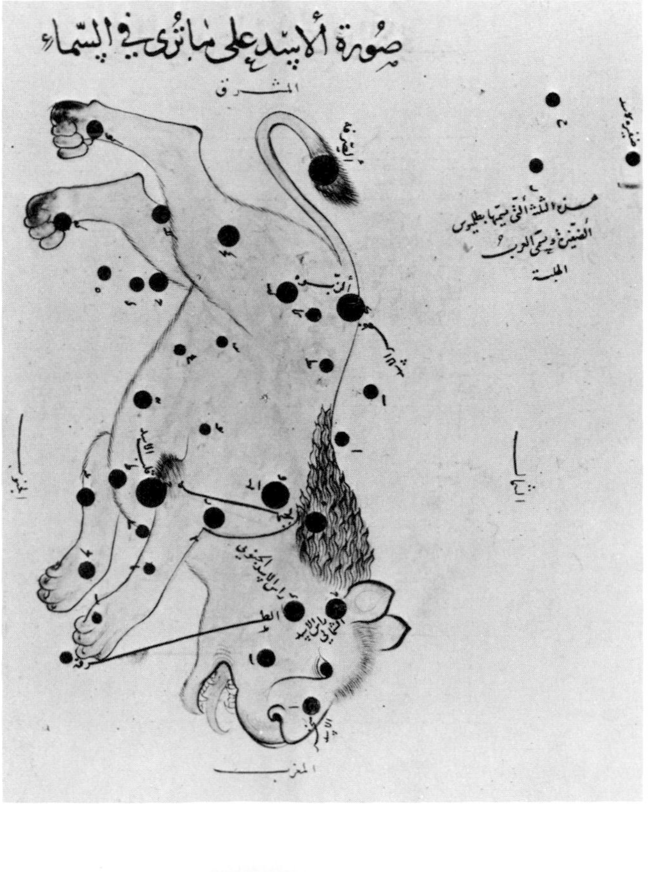

Veez ci i lion si com on le voit par devant & saches bien qu'il fu contrefais al vif.

Veez ci i porc espi; c'est une bestelete qu'lance se soie quant ele s'corece.

253

De Leone incipiunt medicine. Prima cura ipsius: contra fantasias. Leonis carnes quicumque decoctas manducaverit: dicimus eum fantasia non posse temptari. Ut ab omnibus bestiis securus ... Leonis sanguine quicumque sis ...

254

... e Leonis adipe libram unam. cere virginee uncias duas. & olei unciam unam. Leneresci uncias ... hec autem omnia simul in se misceas. facit autem ad omnium nervorum dolorem. & omnes alios sedat dolores. & psanat. De Leena incipit experimentum.

Una ipsius virtus: Contra mulierem. que concipet si vellet. nec concipit.

صورة الاسد على ما ترى في السماء

المشرق

المغرب

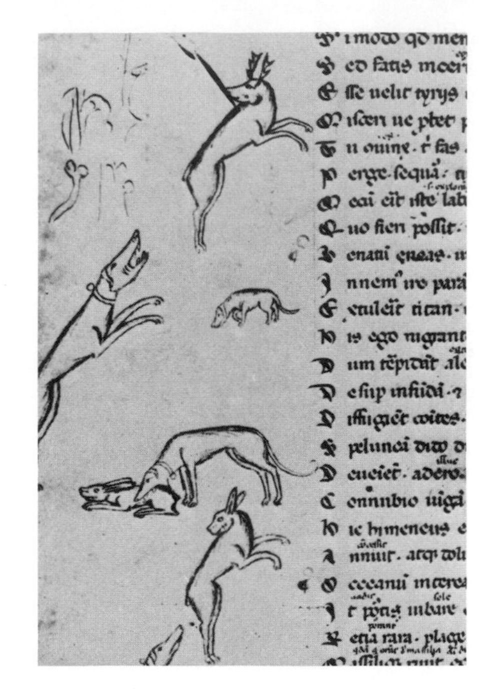

204. (Right) Only Marginal Difference Between Animals. These marginal animals are found in a fourteenth-century Italian manuscript of Vergil's *Aeneid,* their ostensible purpose being to illustrate the account of the hunt of Dido and Aeneas (IV, 117). All of the animals are drawn in standard, unpretentious profile poses. Those standing on their hind legs are quite similar in overall form. But the two standing animals nearest the text are not merely similar, but almost identical. All one needed to do to transform the deer at the top into the hare below was to replace the antlers with ears. One is tempted to think that the artist employed instructions from a model book (see Illustrations 16 and 18) that explained how to draw any number of animals with minimal change.

205. (Below) Belles-Lettres Zoology. The most monumental of the writings of the ninth-century Mu^{c}tazilite theologian and polymath, ^{c}Amr ibn Bahr al-Jāḥiẓ (that is, "the goggle-eyed") of Baṣra, is his *Kitāb al-ḥayawān (Book of Animals)*. Using much from Aristotle, Jāḥiẓ's zoological compilation is more a work of literature *(adab)* than of science: his accounts of the various animals he treated proceed as much through anecdote and digression as through the reporting of tradition. Toward the beginning of this seven-volume work, Jāḥiẓ reveals a conviction about the value of books that provides an excellent description of the whole work on which he was about to embark: "A book is a receptacle filled with knowledge, a container crammed with good sense, a vessel full of jesting and earnestness. . . . It will amuse you with anecdotes, inform you on all manner of astonishing marvels, entertain you with jokes, or move you with homilies, just as you please." The two illustrations given here occur in a fourteenth-century Mamluk manuscript of the *Kitāb al-ḥayawān.* The first shows a white cock copulating with a black hen, the whole scene framed by stylized plants. "Moreover," we are told, "the cock mates with the hen without recognizing her, and this in spite of his lust for hens and his craving for coition." The picture of the lizard and cat occurs on the same page. Here we are treated to two Arabic proverbs, one of which may have suggested the depiction of the cat and the lizard together: "More regardless than a lizard" (since this animal eats its young), and "More pious than a cat and more reckless than a lizard." Whether the expression of the cat betrays piety is open to speculation.

206. Bestiary Marvels and Monsters and Their Progeny. Although the bestiary animals represented in the foregoing illustrations were always reported to have some characteristic or to indulge in some behavior that was much more fantasy than fact, they were at least almost always identifiable as actual animals. But the medieval bestiary also included mythical aspects of nature within its bailiwick. The first example given here, at the top left, comes from a thirteenth-century English bestiary with illustrations by the school of Matthew of Paris (for another picture from this manuscript, see Illustration 4). The subject is firestones *(lapides igniferi)*. They are to be found, male and female, in the mountains in the East. When some distance from one another, they are inert, but if brought closer together, flames immediately burst forth, setting the whole mountain ablaze. The moral is that men should battle against lust and keep themselves remote from women lest the resulting flames consume the good they have derived from Christ. Adam, Samson, and Joseph are cited as "object lessons." The second picture, top right, from another thirteenth-century English bestiary (also seen in Illustrations 201 and 273), is of the manticore, held to be born in India. Bearing the body of a lion the color of blood, with gleaming eyes, three rows of teeth, and a tail with the stinger of a scorpion, it lusts for human flesh. Here the human face is not the artist's anthropomorphizing of the animal, but a characteristic expressly required by the text *(facie hominis)*. Most of these features can be found in Pliny (*Nat. hist.*, VIII, 30) and, more importantly, in his source, Aristotle (*Hist. anim.*, II, ch. 1), who in turn cites Ktesias of Knidos, a fifth-century B.C. scholar who wrote on India, as his source. Aristotle's word for this mythical eastern beast is *martichora,* a Greek form of an Old Persian word for the tiger, which literally means "man-slayer." Bestiary influence of the manticore can be seen at the bottom of the page, in Conrad Gesner's depiction of a *monstrum satyricum.* It appeared in the *Historiae animalium* of this sixteenth-century Swiss naturalist, and the claim is made that such an animal was captured in 1530 in a forest in Saxony.

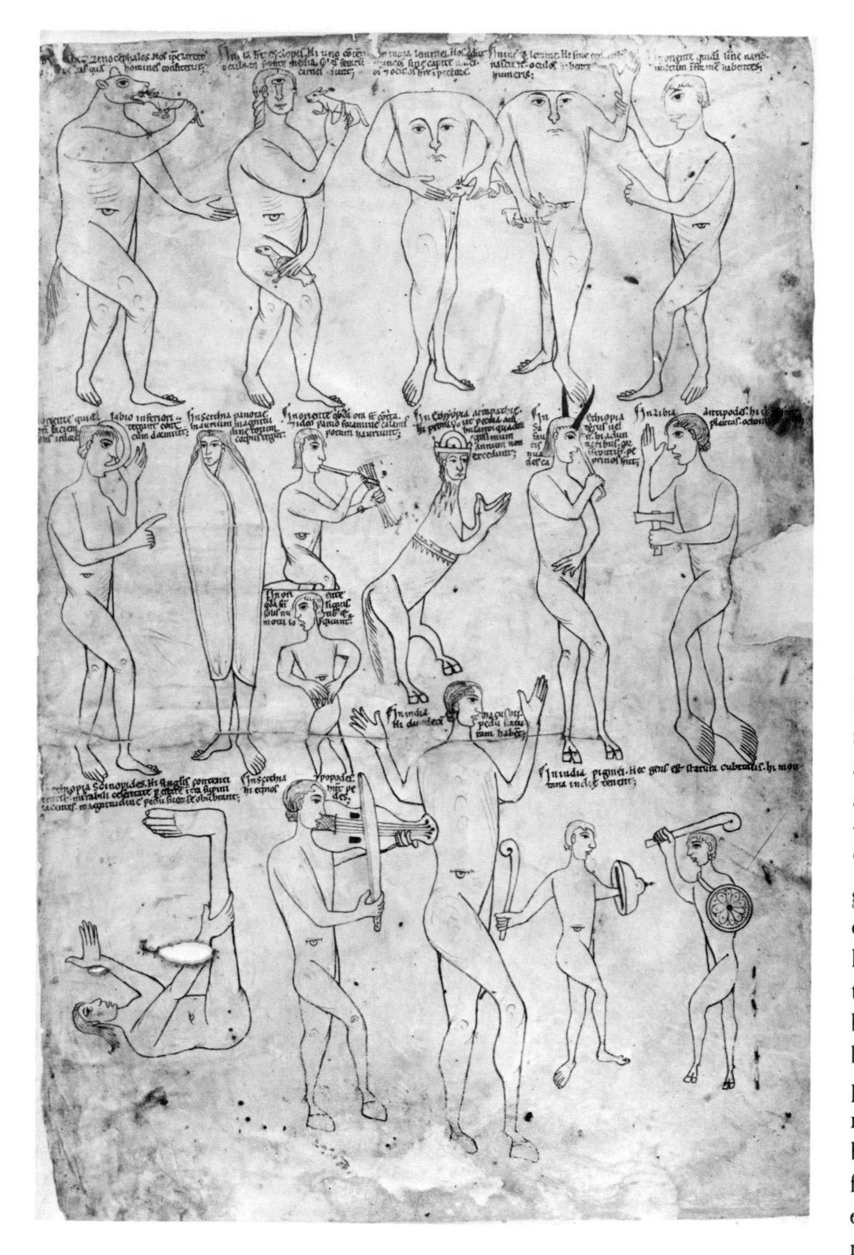

207. A Medieval Catalog of Monstrous Races. Pliny's *Natural History* was a phenomenal collection of information, much of it concrete and factual, but also quite generously populated with the curious and fantastic. In this latter category, his influence as a source—often through various intermediaries—was especially marked among later medieval scholars. The Plinian fabulous races provide a particularly good example of this phenomenon. The single page reproduced here from a twelfth-century German manuscript furnishes a kind of chart of these various monstrous humans, their depictions serving as models for others who might wish to represent them in other works or places. Beginning with the top row we see, from left to right, one of the Dog-heads, or Cynocephali (correcting the spelling of the present manuscript); next, a Cyclops or Round-eye, part of ancient culture as far back as Homer; then two Blemmyae, headless humans with faces on their chests or shoulders (the latter variety sometimes distinguished by being called Epiphagi); followed by a noseless man. The second row begins with a member of Amyctyrae, a creature whose lower lip is so large that it can shade itself from the rays of the sun when it sleeps; next, a specimen of the Panotae ("all ears") since their ears are so large that they can cover their whole body; then a representative of the "straw drinkers," since they are without a proper mouth; and, directly below, the "speechless ones," who communicate only through gesture. The next three races represented in this middle row are the Artibatirae, who walk on all fours; the horned humans with hooked noses and goats' feet; and the Antipodes or "opposite footed." The bottom row begins with a Sciopod ("shadow foot") who, though but one-legged, is extremely swift, and its single foot provides an umbrella guarding it from the sun when it reclines; next, a Hippopod ("horse footed"); followed by a twelve-foot giant who also has horses' hooves for feet; and finally two pygmies, only a cubit tall. A legend accompanying each picture also reveals (again deriving the information ultimately from Pliny) just where such monstrous races can be found: in Africa or the East; Scythia, Ethiopia, Libya, and India being the specific locales mentioned. This pictorial "catalog" may not cover all of the most significant Plinian races—cannibals (Anthropophagi), the Astomi ("mouthless") or Apple smellers, and the Amazons ("without breasts") are omitted, but most of those which were favored for depiction elsewhere (on Gothic capitals as well as in books and other kinds of painting) are present here.

208. Monstrous Races at Home. Following a tradition that was already established in antiquity, most medieval discussions of monstrous humans did not occur together with the treatment of monstrous beasts in bestiaries or other zoological treatises, but rather in expositions of geography and in books of travels. An account of a particular distant locale all but automatically led to the consideration of its more fabulous inhabitants. Two examples of this are given here. That above comes from a fifteenth-century manuscript of a *Merveilles du monde* compiled by Harent of Antioch. The country that this illustration depicts is one that was a traditional catch-all of the unusual and fantastic: Ethiopia. It is, the text tells us, a place filled with venomous beasts such as dragons, unicorns, and basilisks (the king of serpents, able to kill either with its odor or with a single glance). Appropriately, a number of such beasts are represented in the present picture. More numerous, however, are the monstrous humans. Many of them, both male and female, are hairy creatures. Some are identifiably Plinian, such as the two headless Blemmyae just to the left of center and the Panota with the immense ears at the upper right. Others, however, are either not Plinian or are elaborations or transformations of races found in this Roman author (such as the two lizard eaters at the lower left, who may perhaps derive from Pliny's fish eaters or Ichthiophagi). The second illustration, on the right, comes from the same thirteenth-century codex of Qazwīnī's *Wonders of Creation* cited elsewhere (Illustrations 52, 126, 203, 253). Citing both Ibn al-Faqīh (a Persian geographer, fl. ca. 903) and the renowned philosopher, physician, and alchemist Muḥammad ibn Zakarīyā al-Rāzī (d. 923/924), Qazwīnī is here in the course of telling us of the inhabitants of the island of al-Rāminī in the China Sea. There are found barefoot, naked men and women with hairy bodies who live in trees. Their speech is incomprehensible since they communicate with whistling sounds. They are only four "hands" high, so one may assume that they are some sort of pygmy. Although they are not represented in these pictures, Qazwīnī also informs us that on this remote island one can find snakes 200 cubits long; they eat each other and swallow elephants.

209. Medieval Anatomy Men. The most striking medieval anatomical illustrations are to be found in a series of five pictures—hence christened the *Fünfbilderserie* by the German historian of medicine, Karl Sudhoff—appearing in a number of medieval manuscripts. At times the number of pictures in the series was more or less than five, but the standard and most interesting group consisted of a bone man, a muscle man, a nerve man, a vein man, and an artery man, all of them depicted in a more or less squatting position. The first three of these five have already been seen (in two copies each) in Illustration 20, the first of the manuscripts there utilized being the earliest codex containing this *Fünfbilderserie.* One of the remaining two of this standard series is given here from a late thirteenth-century English manuscript, the "squat" much less pronounced than in most other copies of the series. The picture at the top right is of a vein man. In addition to the venous system, the alimentary canal is also represented, the esophagus (crossed by three cervical vertebrae) leading to a strikingly circular stomach with an equally circular intestinal coil directly below. The small circle through which various veins pass in the center of the stomach is most likely intended to represent the diaphragm. A five-lobed liver appears at the left, a gall bladder in its middle, while the spleen, shaped like the sole of a shoe, is on the right. The major veins are traced throughout the body, although the labeling present in other examples of the *Fünfbilderserie* is absent here. The next picture, bottom right, comes from a late-fourteenth-century manuscript of an illustrated Persian anatomical work, the *Tashrīḥ bil-taṣwīr,* written by Manṣūr ibn Muḥammad ibn Aḥmad . . . Ilyās (fl. ca. 1396–1423). One of six such illustrations in this manuscript, its interest is not simply in the evidence it provides for the existence of such a series of anatomical pictures outside the Latin West, but also in its addition of a depiction of pregnancy to the group. Here we have an artery woman, the digestive tract again clearly represented, the liver hugging the stomach on the left, the spleen to the right (but now rather heart-shaped), with two kidneys visible just below the stomach. The rather well-developed fetus is connected directly to the heart by an artery. It should be noted that squatting anatomical figures of the sort given here also exist in Far Eastern sources. They can be found in Siamese and Tibetan medicine, where in some instances they were used in teaching as late as the nineteenth century. For other pictures from the manuscript of the first figure reproduced here, see Illustrations 211, 212, and 271.

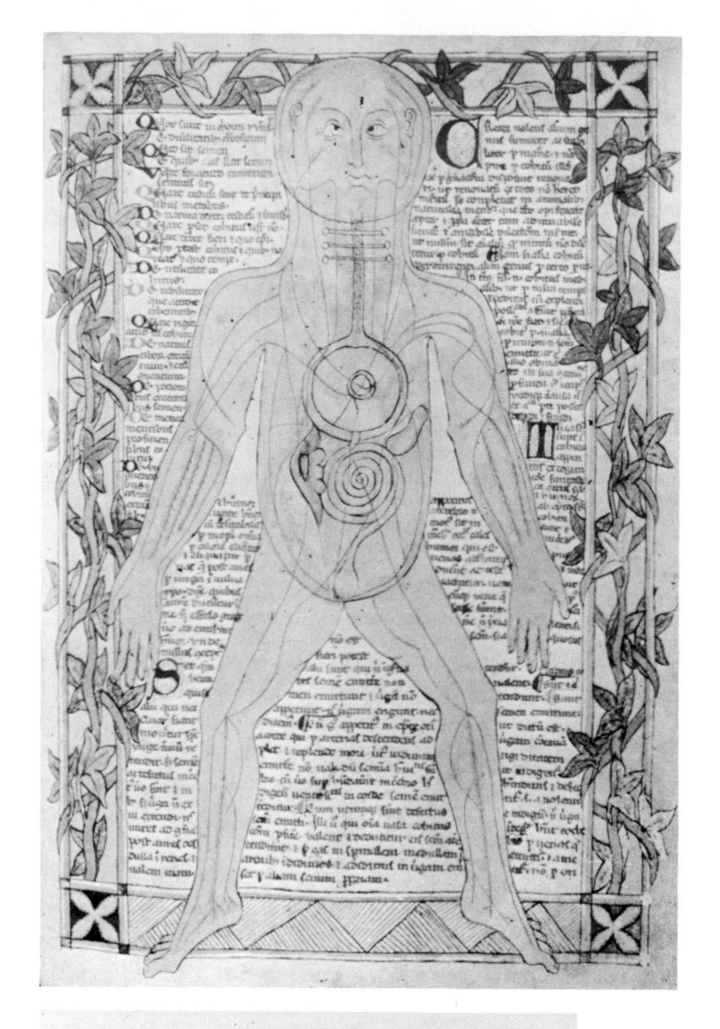

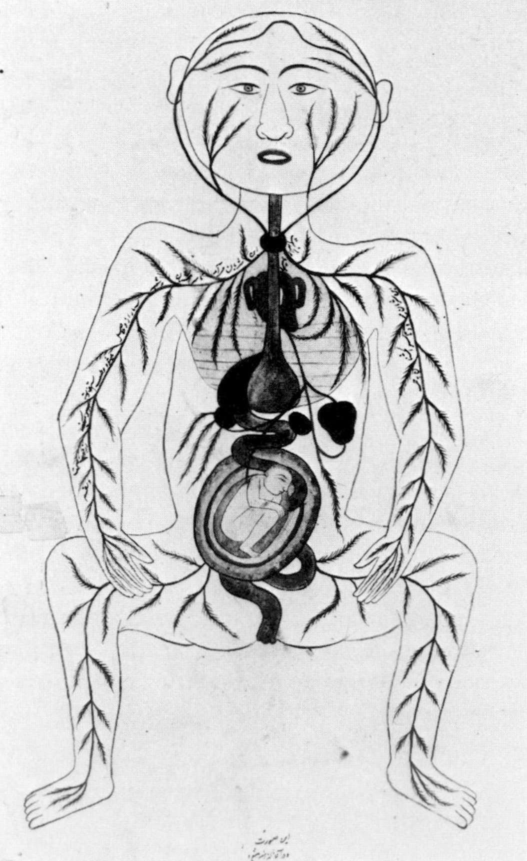

210. Other Anatomical Figures. Although the series of anatomy pictures that has been the subject of the preceding illustration is perhaps the best known of medieval efforts in this regard, many other types of such pictures can be found in manuscripts of the Middle Ages. Just two examples of these many alternative types are given here. That at the top right comes from a rather unusual early-fifteenth-century manuscript: twelve skins of vellum sewn together to form a scroll over seventeen feet long. It contains the text of the *De arte phisicali et de cirugia* of the fourteenth-century English surgeon, John Arderne, liberally decorated with marginal and intercolumnar drawings relating to an amazing number of aspects of later medieval medicine. The drawing reproduced here depicts the contents of the skull and face and the thoracic and abdominal viscera by an artistic device unusual, if not unique, in medieval illustrations: splitting the body along a ventral medial line and folding it open for view (a similar dorsal splitting occurs in another drawing in the manuscript). The added touch of the figure here holding his own body open reminds one of a similar artifice used in later printed anatomical works (see, for example, Illustration 282 of the second volume of the *Album of Science*). Some of the organs revealed are more easily recognizable than others, but careful inspection will uncover (from top to bottom) cranial nerves, the larynx, trachea, vertebrae of the upper spine, lobes of the lungs, the heart, a three-lobed liver with the stomach and spleen in sequence, the intestines, two kidneys (well down in the pelvis) with bladder between, and the penis. The objects between the legs and to the left and right of the figure are surgical instruments, while the text flanking the whole sets forth the proper remedies for various ailments. The second illustration at the bottom right, in and beside an illuminated initial, is quite different in style and intent. It does not even come from a medical work, but rather from an elegant fourteenth-century manuscript of the voluminous *De animalibus* of the thirteenth-century Dominican philosopher and theologian Albertus Magnus. We are here provided with the depiction of an anatomy lesson. The physician stands on the right instructing two students, one behind him, the other at the left, presumably clarifying some issue for himself by referring to the "supplementary" anatomical figure in the margin. Despite the small size of the two "anatomy men," one can still clearly observe and identify a number of organs. The figure inside the initial reveals veins, esophagus, heart, stomach, a four-lobed liver, two kidneys, and a bladder, while the dorsal view of the marginal man again shows veins and the heart, plus the spinal column. For other pictures from the second manuscript used here, see Illustration 262; for more of John Arderne, see Illustration 269.

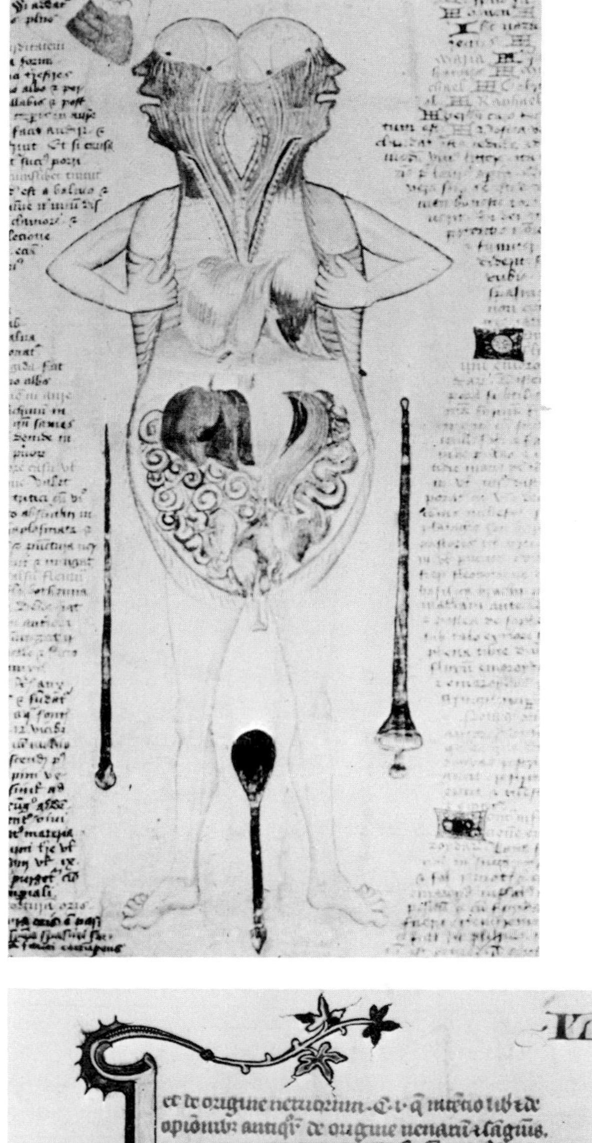

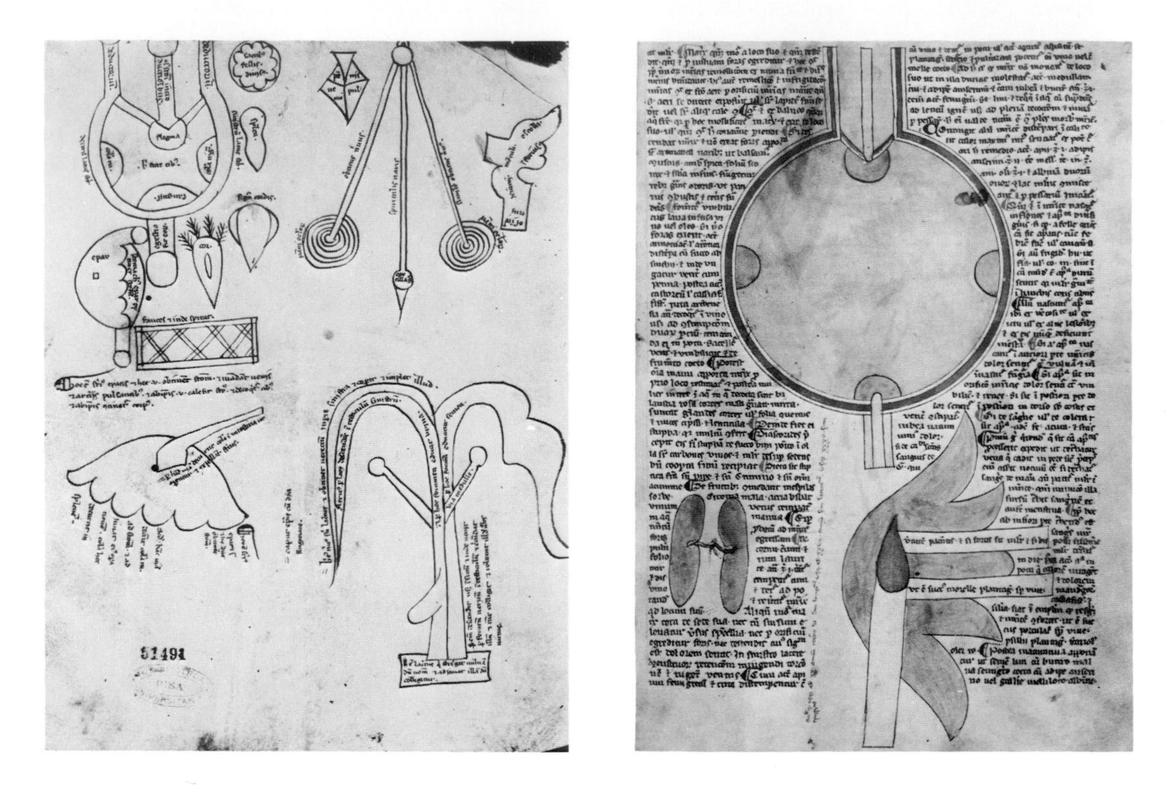

211. Anatomy Organ by Organ I. By far the greatest share of medieval anatomical illustrations displayed their subject within depictions of the whole human body. On occasion, however, we do meet with pictures or diagrams of separate organs unrelated to any illustration of a more complete torso. Two instances of this rarer phenomenon are given here. The first, top left, is from a thirteenth-century codex. A circular stomach occupies the upper left on this manuscript page, four semicircles in its interior stipulating the four humors: phlegm *(flegma)* at the top, blood *(sanguinis)* at the bottom, with red bile *(colera rubea)* at what is labeled as the stomach's right side, and black bile *(colera nigra)* at the left side. The entry of the esophagus into the stomach is appropriately labeled above *(os stomachi introductionibus)*, as is the pyloric exit below *(egestio fit eorum)*, while food in the course of digestion is indicated in the center *(hic stat cibus)*. The two *deductoria* on the left and right above are probably intended to represent some kind of support for the stomach. The circle with internal scallops directly to the right of the circular stomach is the gall bladder *(saculus fellis divisus)*, while the spleen is depicted directly below it, followed by two drawings of the heart (to the right below the stomach). To their left on the other side of the intestinal end of the stomach is a diagram of the liver *(epar)*; bearing five lobes, it appears to be represented as lying over another figure of the stomach. The rectangle with crosshatching immediately to the right has been held to represent the trachea (largely on the basis of its inscription, *fauces et inde spirat)*. To describe the remaining items on this page in less detail, at the lower left there is another liver diagram, this time with six lobes, while at the top center, the small diamond-shaped figure schematically represents the lungs and their lobes. The large tripodlike illustration to the right of this covers the eyes and nose. The partially visible diagram at the far right is identifiable only through the inscriptions within it: esophagus *(os colli)*, jejunum *(ieiunii)*, and intestinal orifice *(os intestini)* suggest that the digestive tract is in question. The final figure at the lower right is presumably of the uterus. Aristotle himself had evidently diagramed the uterus in his lost work *Dissections* (ἀνατομαί), and it is represented here in the two-horned form he claimed for it (*Hist. anim.*, I, ch. 17; III, ch. 1). The second manuscript page, top right, also of the thirteenth century, is a copy or "cousin" of several of the figures just examined. The same circular stomach appears at top center, but without the appropriate labels identifying its relevant parts. Similarly, the five-lobed liver represented in the second manuscript at the lower right is a descendant of the liver diagrams in the first illustration (the two inscriptions on this diagram belong to the surrounding text and not to the diagram itself). Finally, at the lower left is a depiction of organs absent from the first manuscript: two kidneys. For other figures from this second manuscript, see Illustrations 209 and 271.

212. Anatomy Organ by Organ II. Although similar in content, the figures taken from the two manuscript pages reproduced in the previous illustration are quite different in style: those from the first manuscript are freer and more informal; those from the second are more rigid and stylized. Other examples of the latter kind of figure are given here, the first (bottom left) from the same codex that provided the second set of diagrams in the preceding illustration. That depicted is the reproductive system of the human female. The vertical column extending upward from the bottom center is the vagina, the various constricting muscles being properly labeled at its lower end. Its upper end bears a muscular "cover" *(cooperimentum)*—apparently a reference to the cervix—while the uterus above holds an oval-enclosed fetus. The inscription surrounding the oval explains that it is here that the semen collects and that the infant is nourished and grows. Two arcs, descending on the right and the left, end in two circular ovaries (called testicles) from whence the female semen flows. That on the right contains an additional feature in the second circle above it: here is the locus of the blood *(statio sanguinis)*, which during menstruation flows down a tube *(via sanguinis menstrui)* to the ovary below. The two inverted teardrop figures on the lower half of the page are labeled as muscles, although without any indication of function. Perhaps they have been mistakenly taken from another diagram, as most likely is the case with the two vertical columns at the lower left and right, both identified as the "neck of the womb" *(collum matricis)*. It is interesting to note that the illustrator of this manuscript page has adopted the labor-saving device of fully labeling items almost always only on the right of this symmetrical diagram; the matching elements on the left simply bear the inscription "same here" *(similiter hic)*. The second diagram presented here (bottom right) is taken from an early printed edition of the voluminous commentary on Avicenna's *Liber canonis* (see Illustration 8) composed by the fifteenth-century physician Jacques Despars (Jacobus de Partibus). The subject is the exposition of the bones of the upper jaw (*Canon,* Lib. I, fen 1, doctr. 5, summa 1, cap. 4). Here, Avicenna informs us, the arrangement of the relevant sutures creates two triangular bones under which are two other bones. It is this "geometry" of skeletal anatomy that is depicted here by Jacques Despars. The upper two triangles within the larger equilateral triangle are Avicenna's two triangular bones, while the bones below them are glossed here as being quadrangular. Almost all of the sutures mentioned by Avicenna are duly labeled, as are the canine and incisor teeth (below the base of the larger triangle), functioning as locators for the sutures. The trapezoid above the main triangle is a diagrammatic gloss of Avicenna's additional remark that the bones in question form right angles with the sutures longitudinally dividing the whole palate, acute angles with the roots of the canine teeth, and obtuse *(expansus)* angles with the nostrils.

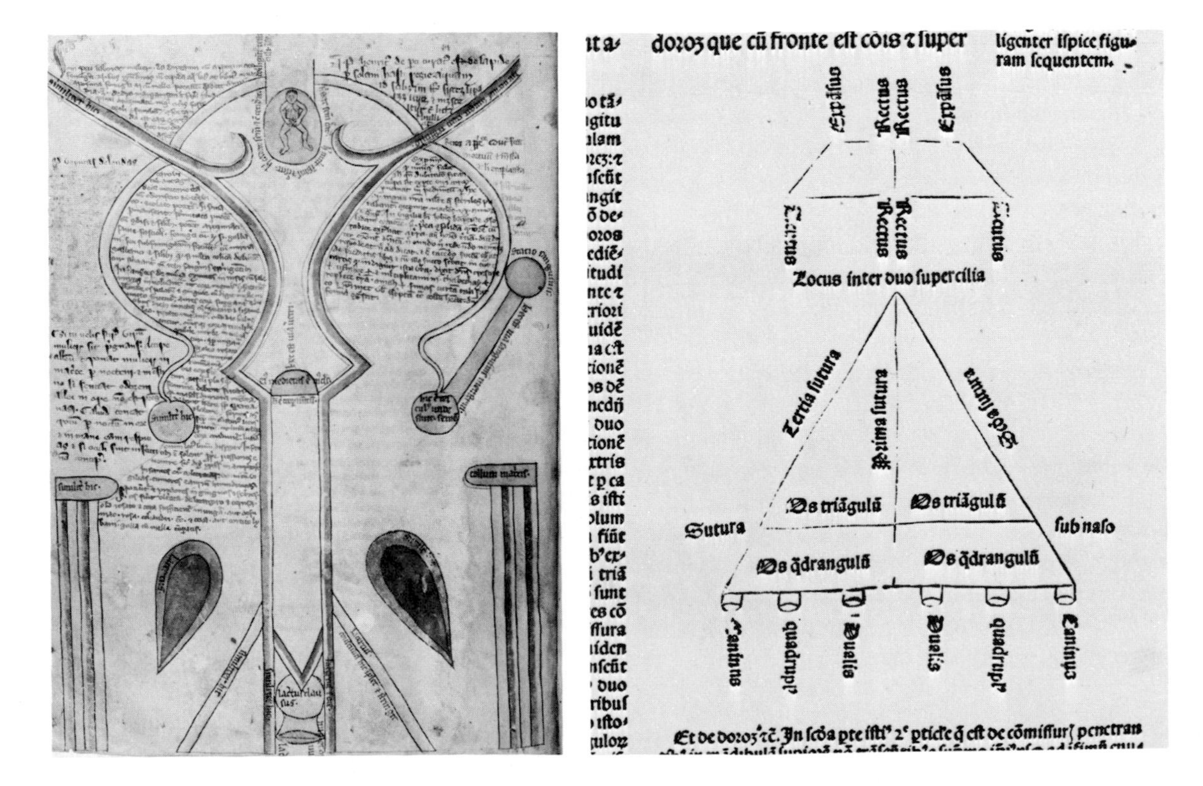

213. The Eye: Greek into Arabic. The present two figures are taken from the *Ten Treatises on the Eye (Kitāb al-ᶜashr maqālāt fi'l-ᶜain)* ascribed to the ninth-century translator, physician, and theologian Ḥunayn ibn Isḥāq. Although this work is the earliest extant "primer" on ophthalmology, its Greek ancestry is clearly revealed in its subtitle: *On the Structure of the Eye, Its Diseases and Their Treatment, According to the Conception of Hippocrates and Galen.* The manuscript of this work used here was composed in Syria in the twelfth century, copied from an older codex written in 1003. Ḥunayn refers to diagrams in his text, so it is certain that figures such as these were part of the original and not the addition of some later scribe or reader. The figure at the upper right displays the structure of the eye itself. The outer lozenge represents the two lids of the eye, the small circle at the center the lens, described in the text as an icelike humor (from the Greek *κρυσταλλοειδές*, says Ḥunayn), not of perfectly spherical form (as seems to be the case in the present diagram), but round and flattened. The notion of the lens at the eye's center is Greek, here faithfully reproduced as part of Arabic doctrine. The pupil is depicted by the small circle inside the crescent (which is the uvea or iris) toward the bottom of the diagram. The front of the eye is at the bottom, while the hollow or optic nerve (here not labeled) exits at the top. The circle between the pupil and the lens contains the visual spirit. The line functioning as a diameter of the larger central circle while curving in front of the lens is the arachnoid membrane; it divides the cavity of the eye into an anterior part containing the aqueous humor and a posterior part containing the vitreous humor. There are three coats or membranes surrounding the eye: the outermost conjunctiva (with two curving oculomotor nerves attached at the left and right), the sclera together with the cornea, and, as the innermost membrane, the retina. The second figure, at the lower right, occurs a few pages later in the same manuscript of Ḥunayn's work. Its purpose is to facilitate understanding of the eye's muscles, of which there are nine, according to Ḥunayn. The three pictured (with labels in the form of vertical lozenges) at the top of the eye surround the optic and oculomotor nerves and serve to hold the eye in place. Another exterior muscle at the top of the eye (now with a horizontal lozenge label) moves the eye upward, while that at the bottom moves it downward. The two exterior muscles at the left and right move the eye toward the temple or the nose; two interior, oblique muscles—here depicted as label-bearing lozenges at the top and bottom within the eye—account for its rotation. Ḥunayn also explains that there are three muscles for the motion of the upper lid (the lower lid lacking muscles, since it is motionless); one of these three muscles moves the lid up, the other two move it down. This is presumably diagramed by the semicircular figure at the upper left. Besides the fact that the "conception" of the doctrines represented by these figures is Greek, it is notable that Ḥunayn's text acknowledges this debt through his repeated citation of the original Greek terms representing the central elements in these doctrines. It is also notable that although the features of the eye's structure that appear in both Ḥunayn's text and the accompanying diagram receive a fuller explanation in the former, the figures occasionally do go beyond the text by their inclusion of features that go unmentioned in Ḥunayn's discourse. For another picture from this manuscript, see Illustration 65.

214. The Eye: Arabic into Latin. It has already been noted in the caption to Illustration 134 that the most important Arabic contribution to medieval Latin optics derived from the Latin translation of Ibn al-Haytham's *Kitāb al-Manāẓir,* so it is not surprising to learn that the views of this eleventh-century scholar concerning the anatomy of the eye were also influential in the Latin West. The question of the influence of the figures representing this anatomy is, however, more complicated and problematic; for the figures of the structure of the eye found in the Arabic manuscripts of the *Kitāb al-Manāẓir,* as well as those that occur in codices of the commentary of Kamāl al-Dīn al-Fārisī (see Illustration 135) on that work, are different from those that appear in manuscripts of the Latin translation of Ibn al-Haytham's magnum opus. And it is these latter figures that were influential in the construction of figures in the optical writings of medieval Latins. Their traces are clearly present, for example, in the figures of the eye one finds in manuscripts of the *Perspectiva communis* of John Pecham, the *Perspectiva* of Witelo (see Illustration 216), or the *Opus maius* of Roger Bacon. In fact, a fourteenth-century manuscript of the latter work is the source for the present figure. Although not identical to corresponding figures to be found in manuscripts of the Latin translation of Ibn al-Haytham, it is clearly a descendant of one or more of the latter. The small circle at the top is the pupil *(foramen uvee),* the uvea itself being the innermost of the three tunics or membranes labeled at the bottom of the figure, where it is followed by the cornea and the *consolidativa* or sclera. The intersection of the two central circles determines the *anterior glacialis* or crystalline lens, separated here by a straight line from the *posterior glacialis* or vitreous humor from which the optic nerve descends, cutting through the bottom of the diagram. Bacon was acutely interested in the centers of the circles or spheres determining all of these elements, and the inscriptions on the right under the *anterior glacialis* specify just where these centers are: the highest, and consequently the closest to the pupil, is the center of the vitreous humor, followed, as one moves deeper into the eye, by that of the uvea, of the cornea, of the *anterior glacialis,* of the whole eye together (since their circles are concentric), and finally of the *consolidativa.* The most curious feature of the whole diagram is the straight line tangent to the pupil. The inscription above and below it tells us that this line is the side of a square that is inscribable within the sphere of the uvea "according to the view of some" (whose identities go unmentioned by Bacon). This seems to be an extremely oblique way of saying, in terms of the geometry of the figure, that the center of the uvea would be the same as the center of the vitreous humor.

(Continued)

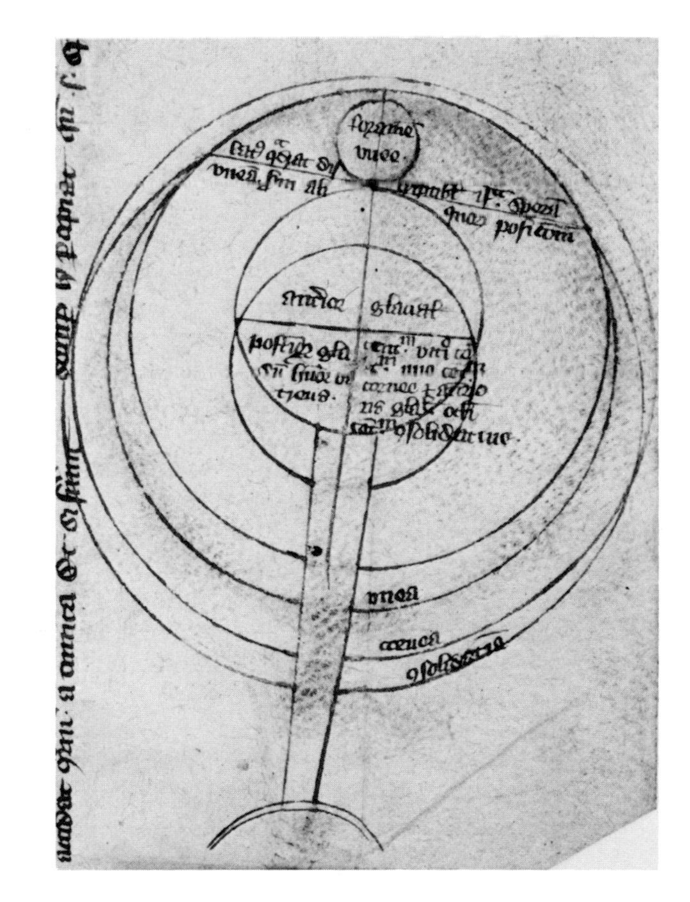

215. (Right) The Eye: The Figure as a List of Parts. The curious figure reproduced here is taken from a fourteenth-century manuscript of a *Liber de occulis* called *Salaracer* by an otherwise unknown Macharias (though this treatise does have some as yet undetermined connection with a twelfth-century *Tractatus de passionibus oculorum, qui vocatur Sisilacera* by a magister Zacharias). Labeled a "figure of an eye in a human head," the central eye itself is clearly decorative, as are the other facial parts depicted. What is important is simply some kind of visual indication of the ten components of the eye: three humors and seven tunics or membranes. All ten of these elements are "listed" on concentric semicircular bands in the center of the head (except that the crystalline humor or lens has the innermost full circular band all to itself). These semicircles bear little or no relation to the actual structure of the eye (even with allowance made for the errors that usually occurred in medieval eye diagrams); their function, rather, is simply to act as "slots" on which to enter the names of the eye's components (the order of names being roughly similar to the actual order of components). Thus, beginning at the forehead, the semicircular bands carry labels for the conjunctiva, cornea, uvea, *albugineus* (that is, aqueous) humor, *tunica aranea* (that is, the interior capsule of the lens), and crystalline humor or lens itself at the center. Working upward from the "ear position," we have the sclera, the *tunica secundina* (that is, the choroid membrane), the retina, and the vitreous humor. All of these humors and tunics are listed again in a classificatory arrangement to the right of the central figure, but now with an indication of what diseases are suffered by which tunics. Finally, note should be made of the fact that, in addition to affording a pictorial vehicle for a catalog of the elements of the structure of the eye, the back of the head provides a place for displaying similar matter concerning the skull. The outermost semicircular band here carries the label "craneum," followed by "dura mater" and "pia mater." The crescent remaining between these bands and the rear semicircles of the eye bears the inscription "cerebrum."

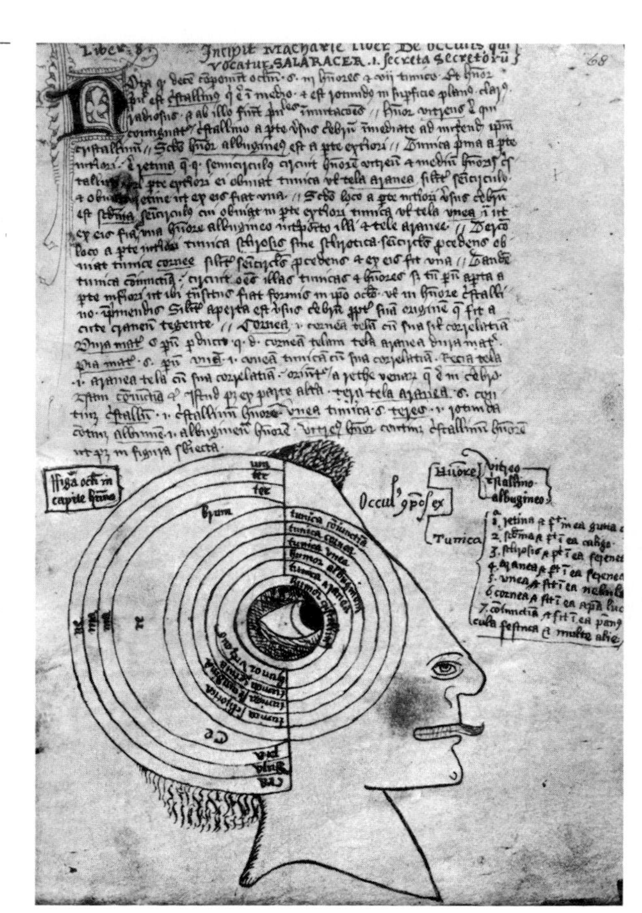

216. The Eye: Constant Text, Changed Figure. The subject of the figures on the facing page is the fortuna of a single eye diagram: that belonging to the *Perspectiva* of the thirteenth-century Polish scholar Witelo (see Illustrations 134 and 135). The diagram itself, again influenced by that which accompanied the Latin translation of Ibn al-Haytham, is quite standard in form and content. Thus, looking at either of the first two examples of the diagram (left and right, above) and reading from top to bottom, one has the usual small circle for the pupil cutting through the cornea and uvea. Below this, the upper crescent of the central circle shows the aqueous humor *(humor albugineus)*, the anterior capsule of the lens *(tela aranea)*, the lens itself, another *tela aranea* (which is possibly the ciliary zonule), the vitreous humor, and, probably, the posterior capsule of the lens (again *tela aranea*). Continuing downward, we meet the *concavitas nervi optici* or the retina, with a cone determined below by the central optical axis and four interior and exterior tunics of the optical nerve. The "box" at the base is labeled *punctus medius nervi communis,* which is to say the center of the optic chiasma. The triangle veering off to the left of this "box" is most likely meant to refer to the optical nerve of the other eye. To consider each figure reproduced here separately, then, one should begin with that at the upper left, taken from a fourteenth-century manuscript of Witelo's *Perspectiva.* It is all but identical with the second figure at the upper right, taken from the first printed edition (Nuremberg, 1535) of Witelo's work. A later printed edition of this work was produced in Basel in 1572 under the editorship of Friedrich Risner (who included an edition of the Latin of Ibn al-Haytham's work in the same volume). The text of this edition is the same as that first printed in 1535, but the relevant figure, here lower left,

(Continued)

is quite different in form even though the difference in content, though present, is not very great. Just why and how this variation in the figure came about is immediately evident on comparison with the figure at the bottom right. It has been taken from the 1543 Basel edition of the *De humani corporis fabrica* of Andreas Vesalius. The inescapable conclusion is that it was Vesalius' eye diagram that dictated the change in the Witelo figure, without there being any change in the text. And precisely the same thing occurred when Risner prepared a corresponding figure for the Latin Ibn al-Haytham of this 1572 edition.

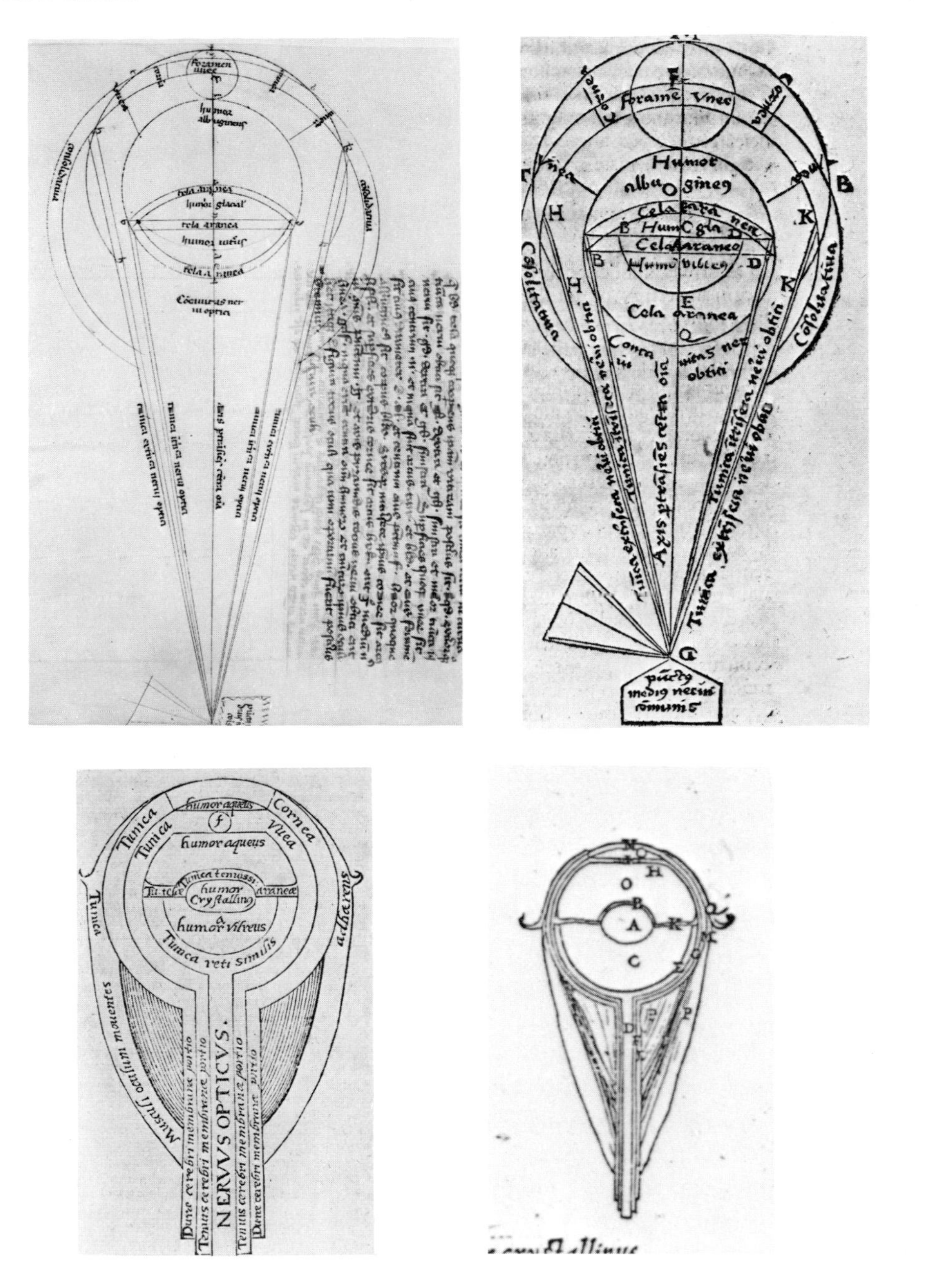

CHAPTER NINETEEN

The Population of the Celestial World

THE AVAILABLE SOURCES FOR this chapter constitute an embarrassment of riches that has necessitated a far more modest sampling of potential subjects than in other chapters. This consequent restriction, moreover, has had to be applied to kinds of illustrations of celestial objects as well as to individual pictures within a given kind. For example, representations of celestial objects and events within sculpture and architecture, on coins, medals, and gems, and on everyday artifacts of all sorts have been excluded even though a few token examples from this domain of pictorial material would have rendered the history of the illustrations of the heavens more complete. Nevertheless, the kinds of pictures selected for this chapter do reveal the salient features of this history, especially features involved in the visual presentation of the scientific and intellectual understanding of what one observes in the heavens as distinct, or at least distinguishable, from the aesthetic appreciation of such celestial experience.

Once again, the separation of object from theory that has been attempted in this chapter is an artificial one. Depictions of a celestial object, of what was seen, at times contained material explaining why the object was seen in the particular fashion that was represented. Why the phases of the moon are what they are, for instance, was almost always part of any picturing they received (see Illustration 234). In other cases, however, one kind of picture or diagram is used to present a given celestial phenomenon as seen, and another, quite different kind of picture, is used to set forth the theory behind what was seen. This is especially evident in the case of eclipses. Depictions of the visual interruption of the light of the sun or moon (see the last picture in Illustration 234) are easily separable from illustrative material providing the theory of eclipses (which has accordingly been relegated to Chapter Twenty-one).

Such a distinction of theory and object apart, we do know that celestial objects, shorn of most theory, were depicted frequently in antiquity. Petronius (first century A.D.), for example, speaks of a round plate or dish *(rotundum repositorium)* that displayed all twelve signs of the zodiac to the guests at Trimalchio's dinner (*Satyricon* 35), and Cicero (106–43 B.C.) tells us of a solid globe on which the stars and constellations were engraved (*Republic* I, 22)—an indication, one might say, of the kind of globe of which Ptolemy would later speak (see Illustration 222). Except for the Farnese Globe (Illustration 221), such artifacts are no longer extant. Indeed, the most important source for relevant illustrative material is—leaving aside ancient Egypt and Mesopotamia—once again to be found in manuscript books (although some extant astronomical instruments and anaphoric clock fragments are also pertinent). In books the two focal points for the subsequent history of the visual presentation of the celestial sphere are provided by treatises written by two authors, one Greek and one Arabic: Aratus and al-Ṣūfī.

Aratus of Soli, in Cilicia (ca. 310–240 B.C.), was the author of an astronomical poem in Greek entitled *Phaenomena*. In its more than 1,000 hexameters, he described the constellations, both zodiacal and north and south of the ecliptic. His major source was another work called *Phaenomena* by the fourth-century B.C. astronomer and mathematician, Eudoxus of Cnidus. Of greater significance for present purposes, however, is the fact that Aratus' poem was set into Latin a number of times: versions, rather than word-by-word translations, were made by Cicero, Germanicus Caesar (15 B.C.–A.D. 17), and Avienus (fourth century). These Latin versions, now usually termed *Aratea,* together with various scholia, furnished the raw material that was to nourish so much of the medieval Latin tradition in the artistic depiction of the constellations (see, in particular, Illustration 226). What mythologizing there is in Aratus' treatment of the constellations appears to have been already present in Eudoxus and when later sources contain mythological elements not present in Aratus (compare Hercules, for example, in Illustrations 221 and 224), the probability is that they ultimately derive from some later author like Eratosthenes (ca. 275–194 B.C.), who often added such elements of his own in his *Catasterismi.*

The translation of Aratus into Arabic in the ninth century appears to have had much less impact. Instead, in the Islamic realm, one central work had an influence all its own: *The Book on the Constellations of the Fixed Stars (Kitāb ṣuwar al-kawākib al-thabīta)* of ᶜAbd al-Raḥmān ibn ᶜUmar al-Ṣūfī (903–986). Al-Ṣūfī dedicated this work to Sultan ᶜAḍud al-Dawla, whose wish it was to have a more accurate knowledge of the fixed stars and their positions in the sky. In order to remedy what he saw as the inadequacies of earlier Arabic efforts in this regard, al-Ṣūfī used the catalog of stars in Ptolemy's *Almagest* (VII, 5–VIII, 1) as his point of departure. Correcting the values given there for stellar longitudes by 12°42′ (1° for every sixty-six years) to account for the precession of the equinoxes and critically revising Ptolemy wherever he deemed it necessary, he achieved, for completeness and accuracy, a singularly impressive result. In particular, the role of al-Ṣūfī's work in the visual representation of the fixed stars was of major significance and influence (see Illustrations 227 and 228).

Although the complete text of al-Ṣūfī's *Book of the Fixed Stars* was not translated into Latin, shortened versions or compilations of it, especially versions containing its pictorial material, did reach the Latin West. The result was an amalgamation of the Arabic tradition of stellar depiction represented by al-Ṣūfī with the classical tradition deriving from Aratus (see Illustration 228). Other ingredients occasionally worked their way into the mixture, but these were the two central elements in the visual representation of the constellations in the Latin Middle Ages.

The concluding illustrations of this chapter treat of matters beyond the representation of the fixed stars, initially so as astrological factors and imaginary constellations enter the field (Illustrations 229–230) and then, as a penultimate section, by a consideration of depictions of the planets, including the sun and the moon (Illustrations 231–234). The final Illustration 235, treating of comets, in one sense does not belong in the present chapter since, following Aristotle, comets were regarded as belonging to the sublunar realm. Nevertheless, their description and illustration were so often joined in the Middle Ages with the investigation of other celestial entities that their inclusion here is not totally anachronistic.

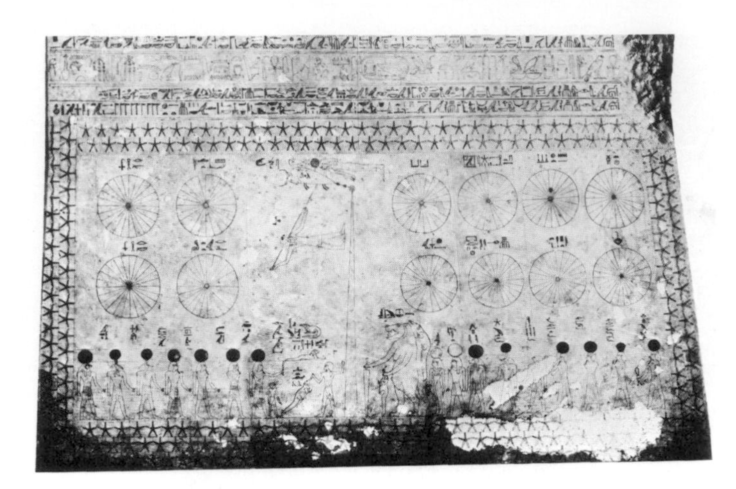

217. The Sky of the Egyptians I. Although the ancient Egyptians developed no mathematical astronomy like that of the Babylonians or the Greeks, their role in the early history of astronomical and calendrical conceptions and doctrines is fairly well documented in a number of extant monuments, portions of two of which are depicted here. The first, above, comes from the ceiling of the tomb of Senmut, "privy councillor" to Queen Hatshepsut (1504–1483 B.C.), at Deir el-Bahri, Luxor. The tomb itself dates from about 1473 B.C. Only the north half of the ceiling is given here (with north at the bottom, east at the right). The twelve "wheels" represent the twelve months of the Egyptian lunar calendar, grouped into three sets of four corresponding to the three Egyptian seasons. The division of each wheel into twenty-four sectors presumably represents twenty-four hours (a division of the day that is Egyptian in origin, the subsequent division of hours into minutes and seconds being Babylonian). The portion of this part of the ceiling of greatest astronomical interest, however, is that running from top to bottom between the four wheels on the left and the eight on the right. It contains, in an arrangement slightly deformed due to spatial considerations, representations of the northern constellations that are to be found in numerous Egyptian monuments of this sort. In fact, these representations can be summarized most conveniently by comparing them with the corresponding section, reproduced below, from the ceiling of a slightly later tomb: that of Seti I (Nineteenth Dynasty, 1303–1290 B.C.) in the Valley of Kings at Luxor. Two of the northern constellations were pivotal for the early Egyptian. The first is *Mes (Meskhetiu),* the Bull or Foreleg, appearing at the top in the Senmut ceiling as a lozenge-shaped bull with extremely tiny legs, but as an immediately recognizable bull in Seti I (this is Ursa Major or the Big Dipper, the only clearly identifiable Egyptian northern constellation). The second is the female *Hippo,* clearly recognizable at the bottom right of the central section in both Senmut and Seti I. A crocodile over her back, she rests her forelegs on a mooring post and a crocodile (Senmut) or simply on a mooring post (Seti I). The constellation *An* is the falcon-headed god in Senmut spearing *Mes* or the Bull; he has been transferred in Seti I to the role of a support of *Mes,* his feet resting on *Hippo*'s mooring post. The remaining northern constellations depicted in these two ceilings are *Lion* (easily identified with stars over his back at the left in Seti I, but much smaller and without stars in Senmut), *Sak* (the small crocodile with a curved tail), *Croc* (the crocodile with the straight tail at the lower left in both representations), *Man* (facing *Croc* in both, with arms positioned as if spearing him, though without spear), *Bird* (in Seti I only, as a falcon over *Lion's* head), and *Serket* (as a goddess at the upper left in Seti I only). In the top picture, the figures at the bottom to the left and right of all these constellations are mostly lunar day deities. Isis, however, heads the file at the right and can also be seen just behind *Hippo* in Seti I (the whole ceiling of which contains a full rank of such lunar deities).

218. The Sky of the Egyptians II. Representations of the zodiac, an astronomical notion that was Babylonian in origin, are not found on Egyptian monuments until well into the Ptolemaic period (304–30 B.C.), at which time Egypt was under Greek domination. This Babylonian-Greek conception was then often combined with "native" elements of Egyptian astronomy. What is perhaps the most famous monument presenting such a combination is reproduced here above. Originally part of the ceiling in the East Osiris Chapel of the temple of the goddess Hathor at Dendera (late Ptolemaic, before 30 B.C.), it is now in the Louvre. The central circular disk is supported by the four goddesses of the cardinal points and by four kneeling falcon-headed deities. The figures standing on the circumference of the central circle represent (one family of) the thirty-six Egyptian decans. Initially, these decans arose from the attempt to relate the thirty-six decades or ten-day periods of the 360-day civil year to the rising of certain stars or groups of stars; in turn, these decanal stars or constellations—all falling within a "decanal belt" approximately parallel to and south of the ecliptic—were used to mark out the hours of the night by means of their consecutive risings. When the zodiac was absorbed into Egyptian astronomy, the thirty-six consecutively rising decans were transformed into thirty-six 10° sections of the whole zodiacal belt, three for each sign (a relation that was later to play a significant role in ancient and medieval astrology, for which see Illustration 230). It is this latter "combination" of the decans with the zodiac that is represented by the Dendera ceiling. The zodiac itself occupies a circular band near the ceiling's center. To cite but three of the zodiacal figures depicted there, Cancer is most easily recognizable below the center and slightly to the left, Leo is below Cancer on the left and Gemini is below Cancer on the right. This positioning of the zodiac associates each of its figures with one or another of the decans. The planets depicted are naturally interspersed throughout the zodiacal circle, but all in exaltation, one of their most auspicious astrological positions. Thus, Jupiter appears as a crowned falcon-headed god directly below, and consequently in exaltation in, Cancer, while Saturn, as a bull-headed god, occupies the "nine o'clock position" together with Virgo, and so on. The very center of this ceiling disk is surrounded by the two pivotal northern constellations we have noted in the previous Illustration: *Hippo* stands in a horizontal position directly above center, while *Mes* is directly below her on the other side of the center, but now in the figuration of Foreleg and not of a proper bull. Other indigenous Egyptian constellations are crowded in and about the zodiacal belt. The two most notable (and at the same time the only two definitely identifiable) flank the bomb-shaped figure toward the bottom center of the ceiling: Orion to the right as Osiris, crowned and carrying a scepter and flagellum; and Sothis-Sirius, to the left, as a cow in a bark. Another amalgamation of the Babylonian-Greek zodiac with Egyptian elements is seen below on the inner face of the lid of the coffin of Soter, an administrative official in Thebes (first century of the Christian era). The central figure with arms outstretched is Nut, Egyptian goddess of the sky. The zodiac runs counterclockwise on her right side from Leo through Capricorn, and then, on her left, from Aquarius through Cancer. For more on Nut, see Illustration 272.

219. (Above) The Sky of the Babylonians. Very few of the numerous Babylonian tablets containing astronomical or astrological materials carry illustrations. One of this small minority is reproduced here. From Seleucid Uruk, this fragmentary tablet depicts three celestial "objects." The center position is occupied by the moon, its disk here filled on the left by a bearded, cudgel-bearing god (similar to other representations, it has been held, of the Babylonian god Marduk). Here, this god holds an animal—possibly a lion—by the tail, its curving body fitting into the right side of the lunar disk. Seven stars are clearly etched into the left portion of the tablet. They represent Pleiades or the Seven Sisters. On the right only part of a drawing of a leaping bull—that is, of Taurus—remains. The tablet reflects the frequent association with Taurus of the Pleiades, which are variously described in both Babylonian and Greek sources as stars belonging to the mane or back of the bull. The moon between the Pleiades and Taurus is presented in exaltation (Taurus being the traditional sign for that astrological position of the moon). The significance of the god and animal appearing on the lunar disk has been a matter of some controversy. It has been suggested, for example, that it is a representation of the new moon's victory over the old, or that the various bodily features of the god are meant to represent the different "seas" appearing on the moon's face. The fact that the text above the Pleiades-Moon-Taurus drawing presents an astrological omen having to do with an eclipse in the month of Aiaru (which corresponds to the zodiacal sign Taurus) raises the possibility that the figures within the lunar disk might in some way be intended to represent an eclipse. The fragmentary text at the bottom of the tablet is part of a list of the zodiacal signs, beginning with Taurus.

220. (Right) The Earliest Extant Illustrated Astronomical Treatise. The rather crude drawings reproduced opposite are taken from the earliest extant Greek papyrus (early second century B.C.; see also Illustration 68). Approximately two meters long, the recto side presents the text of an extremely elementary astronomical tract (the acrostic reproduced in Illustration 68 appears on its verso side). Thirty-two figures are interspersed throughout twenty-four columns of text, some of them having little or no bearing on the content of the text itself. The first figure, given here at the bottom left, presents a central zodiac circle, beginning with Aries (*κριός*) over the left part of the base of the left triangle at the bottom and proceeding counterclockwise through the twelve signs. The six triangles at top and bottom surmounted by disks or crescents carry the names of the planets (Venus occurring twice). The only relation this whole figure appears to have to the neighboring text is that the next column speaks of the times of the orbits of the five planets (Venus' "double name" again being noted); but there is no mention of the zodiacal signs, since they are irrelevant to the matter at hand. The set of drawings at the bottom right occurs several columns later in the papyrus. Here, deciphering the intent of the drawings, let alone relating them to the text, is exceedingly problematic. To begin with, the clearly identifiable figure

(Continued)

of a scorpion at the bottom of the column at the left may well be meant to refer to the zodiacal sign Scorpio, but the latter is nowhere mentioned in the text, either in the surrounding columns or elsewhere in the papyrus. It has been suggested that the crescent with an "arrow" running through it that appears in the center of the same column is a kind of ideogram (which occurs in a similar form elsewhere in the papyrus) referring to the necessity of the sun, moon, and earth to be in a straight line if eclipses are to occur (a contention that is mentioned only several columns later in the papyrus, the surrounding text speaking instead of simultaneously rising and setting stars). The ring in the central column enclosing the figure of an animal is inexplicable, save for the fact that the division of the ring into twelve parts may refer to the zodiac circle. The large circular figure at the top of the column at the right is labeled "sun" (*ἥλιος*) in its uppermost light portion and "moon" (*σελήνη*) in the lower, darker crescent. It has been suggested that this might well be interpreted as a representation of a partial solar eclipse, except that the shading is the reverse of what one would normally expect. But the immediately preceding text suggests another possibility: it tells that the moon derives its light from the sun and is not a self-luminous body. For if it did have light of its own, the part of it facing the sun would be dark (*ἀμαυρός*) and the remainder bright (*λαμπρός*), while just the reverse is true. Therefore, in terms of this text, the whole figure could be meant to depict the moon while the brighter upper part of the present circular figure bearing the label "sun" represents that part of the moon facing the sun, making it bright rather than dark or obscure. The final drawing at the bottom of the right-hand column with two smaller intersecting circles within a larger circle has been interpreted as referring to the familiar contention that the earth's shadow covers two lunar diameters (even though this contention is nowhere mentioned in our papyrus text). If this is a correct interpretation, it would seem proper to conclude that the illustrator of our papyrus either intentionally or unwittingly included more information in some of the figures accompanying the text than was to be found in the text itself. Alternatively, the figures may have been taken from a copy of a more complete version of the text, their relevance being occasionally obscured when portions of explanatory text were omitted.

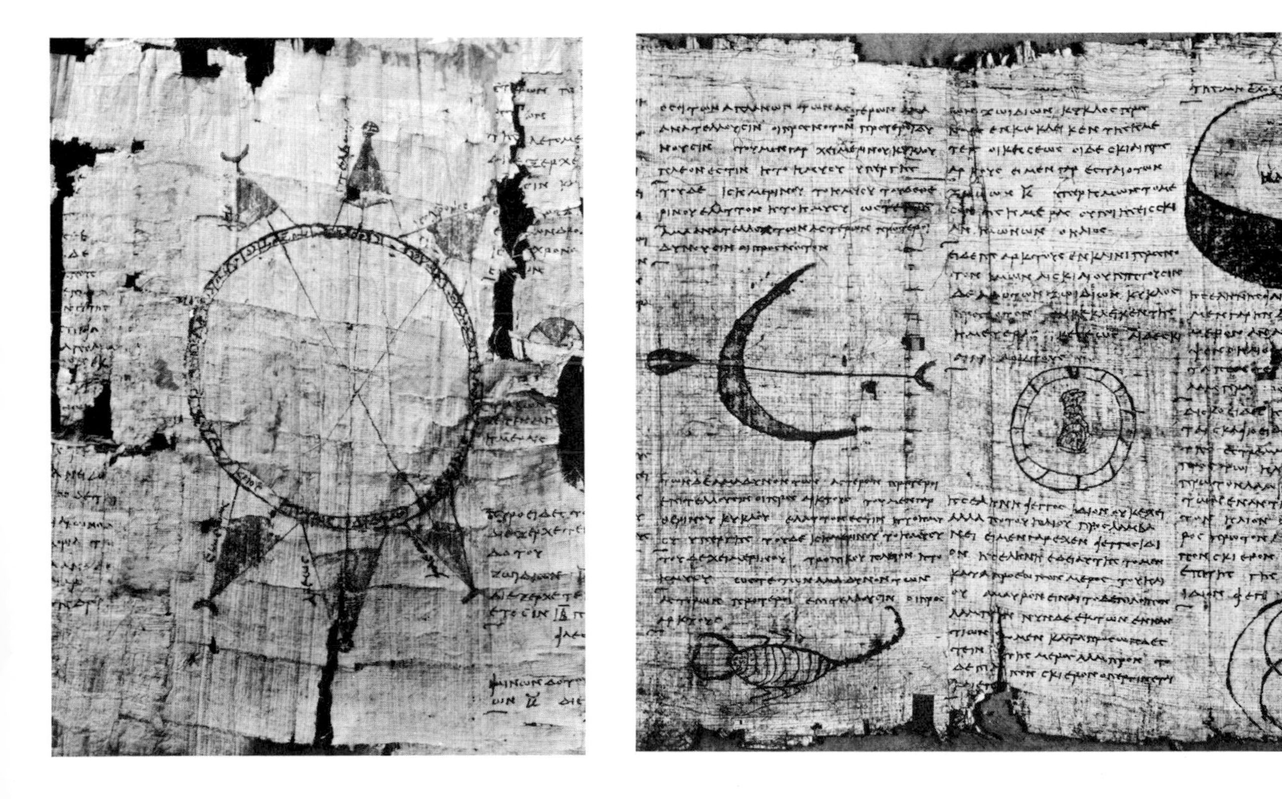

221. (Right) The Depiction of Constellations on a Sphere. Some of the earliest evidence we possess for the history of Greek representations of the fixed stars is the Farnese Globe. At present in the National Museum at Naples, this globe is a Roman copy (late first or early second century A.D.) of a Hellenistic original and is held overhead by a statue of Atlas (here cut away) that was added in the Renaissance. The equator, two tropics and polar circles, plus the equinoctial and solstitial colures, are given as "coordinates" in relief. The zodiacal belt is clearly depicted by three rings, that in the center being the ecliptic. The reproduction at the right reveals that segment of the zodiac bearing Virgo and Libra, part of Hydra's tail carrying Corvus appearing directly beneath. Below that one can observe Centaurus with Lupus and a barely discernible Altar before. Ophiuchus stands directly to the right of Atlas' hand, while above the Tropic of Cancer over Ophiuchus' head one finds Boötes, Corona Borealis, and Engonasin. Engonasin later became Hercules, but it would be wrong to assign him that more familiar name here since he is not depicted in his mythologized form with club and lion's skin (compare Illustration 224) but only as the Kneeling Man, which is precisely what his earlier name of Engonasin signifies. This unmythologized representation of Engonasin is quite noteworthy in that he is described in just such an earlier form in Aratus' *Phaenomena,* a fact that has allowed historians to conclude that the Farnese Globe goes back to an original that was closely associated with that Greek astronomical poem.

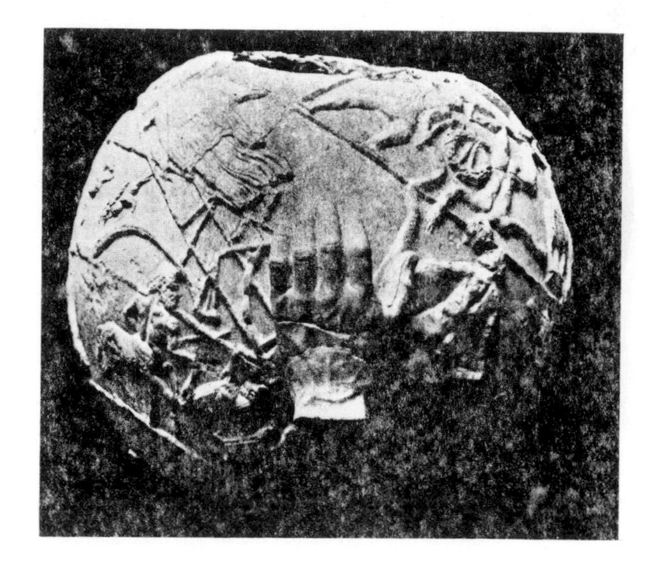

222. (Facing Page, Left and Center) The Heavens by Hemisphere. The two figures of constellations given here both present their subject in "hemisphere pictures," but not in the usually expected manner of hemispheres north and south of the equator. Instead, each hemisphere is determined by the colures running through the equinoctial points. Thus, the north pole is at the top of the circumference of each of these three hemispheres, the horizontal parallels representing (however inaccurately) the equator, tropics, and polar circles, while the verticals depict solstitial colures. The figure at the left on the facing page is taken from a Greek manuscript written ca. A.D. 813–820, though it has been shown that the figure itself derives from others dating from the latter half of the third century. Here, the hemisphere is in dark blue, the colure, parallel circles, and zodiac are in gold, and the constellations are sketched in slightly lighter and darker blues and thus do not appear in stark contrast to the background "night sky." It is interesting to note that the coloration closely follows directions given in Ptolemy's *Almagest* (VIII, 3) for the construction of a globe of the heavens. The globe should be of a deep color, Ptolemy tells us, corresponding to the night, not the day, and the constellations also should be depicted in colors that do not stand out too markedly from the background since a variety of colors would destroy the resemblance of the resultant images to what is truly seen. What is more, this matching of the present figure with Ptolemaic instructions fits nicely with the fact that the originator of such hemispherical figures most likely intended to depict what would have been represented on one-half of a globe, such as that described by Ptolemy, if it were viewed externally from the side. The zodiac in the present figure gives (none too clearly visible) images for the spring and summer signs running from Aries clockwise through Virgo. Five northern constellations are represented above the present zodiacal band, Auriga directly above Taurus being the most easily discernible. Similarly, southern constellations appear below the zodiacal band. Orion, for example, appears to the left between the equator and the Tropic of Capricorn, while Canis Minor and Hydra with Crater and Corvus on its back appear between the same parallels at the right. The other, barely visible, constellations can be picked out from the second figure reproduced here in the middle on the facing page. Clearly a descendant of our ninth-century Greek illustration, its "upright hemisphere" occurs in a ninth-century Latin *Aratea.* Corresponding to our Greek hemisphere, the northern constellations depicted here above the zodiac are (from left to right) Perseus, Auriga, Draco (crossing the vertical colure), Ursa Major, and, in the form of a vase, Coma Berenices. The southern constellations between the equator and the Tropic of Capricorn are the same noted above in the figure from the Greek codex. Those below the Tropic of Capricorn are (from left to right) Cetus, Eridanus, Canis Major, Argus, and, below the Antarctic Circle, part of Centaurus. Further evidence of the ancient ancestry of both hemispherical figures can be seen in the older form given to Aries (at the extreme left of the zodiacal band): he bears a girdle around his body.

223. (Far Right, Above) More than Meets the Eye. This fifteenth-century Italian manuscript contains elegant illuminations of the separate constellations in addition to the depiction reproduced here of the whole known heavens. One of the most significant factors in its composition is its attempt to represent the zodiac plus almost all the known northern and southern constellations in a single circular figure. Characterized by historians as one sort of planespheric illustration of the heavens (since it is a confused projection of the heavens onto a plane), it arranges its "complete map" of the fixed stars in concentric rings, the southern constellations occupying the outermost circular band, the zodiac roughly next, with the northern constellations taking up the innermost portions of the whole. Such a configuration presents the heavens in a manner in which they could never be seen, either in fact or as presented to view on a globe. Yet, following this presentation of more than could be observed from a single vantage point, one should note that the very center of the figure consists of Draco, winding S-like about Ursa Major and Ursa Minor. The next innermost ring carries various northern constellations; still others are crowded between these and the zodiac, which proceeds counterclockwise, beginning with Aries slightly to the left of bottom in the next to outermost ring. Although several zodiacal signs find themselves in the outermost ring (thus presumably representing the eccentricity of the zodiac in a planispheric figure such as the present one), this ring is the primary locus for the southern constellations. One can, for example, observe Lepus at bottom center with Orion overhead and Eridanus and Cetus to the left. Inasmuch as the codex from which this celestial illustration has been drawn is a presentation *Aratea* manuscript prepared in Naples for Ferdinand II and his court, it should occasion little surprise that its artistic merits outweigh its scientific ones. Although the positioning of the constellations is roughly correct, the inaccuracies of the figure are great enough to frustrate anyone attempting to use the figure to obtain a reasonably reliable comprehension of the celestial map. For example, its attempt to depict the eccentricity of the zodiac ecliptic through the location of several signs in the outermost ring is marred by Capricorn having a more inward place while Sagittarius and Aquarius maintain an outer position. The spacing of the zodiacal signs also veers radically from anything like equality, presumably in order to make room for a more striking display of various intervening northern and southern constellations. And Libra seems to have been omitted altogether. Finally, one should note that the arrangements of its concentric circles with the southern constellations occupying the outermost position and a slightly, if crudely, eccentric zodiac imply a planespheric projection from the South Pole, although the counterclockwise orientation of the zodiac means that the stars are not represented as seen but, as in the case of astrolabes (see Illustration 237), in the reverse order in which they would appear if seen from the North Pole. (For a further note on such projections, see the picture credit to this Illustration.)

224. From Manuscript to Woodcut. The picture on the left of the northern and zodiacal *imagines* of the heavens is a woodcut executed in 1515 by Albrecht Dürer. Together with its mate for the southern celestial hemisphere, this is among the very first printed celestial maps. There is, however, a manuscript prototype for Dürer's production. Its figure for the zodiac and northern constellations is reproduced at the right. Both it and Dürer's woodcut depict the stars as of 1424. Further, both view the northern hemisphere from outside, looking down on the north ecliptic pole, as the counterclockwise direction of the zodiac (beginning with Aries at the top) and the orientation of the constellations reveal. Both divide the zodiac into twelve equal parts and 360° and both use Ptolemaic numbers for individual stars. The manuscript figure—but not Dürer—occasionally includes Latinized versions of the Arabic names for certain stars (thus, *dubhe* over Ursa Major from the Arabic *Ẓahr al-dubb,* "the back of the bear," *alrucaba* for the pole star in the tail of Ursa Minor, and so on). The manuscript version is also alone in including a projection of the equatorial polar circle. Further, in addition to the names just mentioned, the manuscript map is Arabic in the particular figures it gives for various constellations. Thus, Hercules (to the right, just below center) appears clothed, wielding a scimitar in his right hand, while in Dürer he has been "westernized," appearing without clothes, a club replacing the scimitar, and with a lion's skin over his left arm (compare Luther as Hercules in Illustration 171). Similarly, Perseus appears in Arabic form at the upper left opposite Taurus in the manuscript map: he holds a scimitar in one hand, the severed head of an oriental demon (labeled *caput algol,* meaning demon's head) in the other. In Dürer, Perseus has regrown wings on his feet and, returning to proper classical mythology, holds the head of Medusa (appropriately labeled *caput meduse*). Dürer has also filled the corners of his celestial map with portraits of four scholars of astronomical importance, each of them perusing in one manner or another a celestial globe. Aratus Cilix (that is, of Soli in Cilicia; third century B.C.) appears at the upper left with Ptolemy at the upper right. The first-century Roman Marcus Manilius is at the lower left, while the remaining corner is occupied by the tenth-century al-Ṣūfī *(Azophi Arabus)*.

225. The Sky as Seen Then and There. Any number of buildings have ceilings that are decorated in what art historians are apt to call an "illusionistic" manner, the intent being to give the impression of an open air or sky. At times, the intent was not simply to present a starry heaven that had little or no regard for pattern, but to reproduce a sky that revealed the actual arrangement of the fixed stars. Thus, already in the eighth century an Arab prince had constructed a palace at Quṣayr ᶜAmra in which the constellations were depicted on the hemispherical dome of a cupola. Still extant, though damaged, it did not depict the heavens as actually seen, even though the relative positions of the constellations accurately reflect those of a proper celestial map; for it is set up so that the northern and southern constellations may be seen simultaneously. The depiction of the heavens on a cupola reproduced here below does, however, give a prospect of the sky as actually seen. Painted above the altar in the Old Sacristy (Sagrestia Vecchia) of San Lorenzo in Florence, it has a zodiac that is notably visible slightly below center. Divided into degrees in the narrow band at the center of this zodiacal belt, the signs begin with Aries at the right and proceed clockwise through Leo at the left (although portions—here not discernible—of Pisces and Virgo occur at the extremes). The solstitial colure appears as a central vertical line and the north polar circle is described about this colure toward the top of the present reproduction of these "cupola heavens." In addition to the constellation figures themselves, the constituent stars are depicted, but an even more interesting and informative feature is the representation of the sun in Cancer and the crescent of the moon in the Hyades (near Aldebaran) in Taurus. Given these solar and lunar positions as part of this particular "picture" of the heavens, one can determine that the celestial configuration in question is consistent with what would actually be seen at Florence around 6 July 1439, about noon. What we have here, then, is not an illusion of what might be seen, but a tolerably accurate representation of what was seen.

226. Scholia Figurata. This illustration of the constellation Eridanus, the River, comes from the earliest extant *Aratea* manuscript. Written in France in the ninth century, this particular codex migrated to England most likely in the tenth century and there served as the model for at least five extant English *Aratea* manuscripts made at various times through the middle of the twelfth century. There is also a possibility that, while still in France, this Carolingian manuscript was somehow associated with the ninth-century abbot, Lupus of Ferrières, whom we know to have been keenly interested in copies of Cicero's translation of Aratus, of which the present manuscript is a prime example. Representations of Eridanus vary throughout early *Aratea* codices. At times depicted as a serpent or as a goat-fish (as in Illustration 222), he is here "mythologized" into a river god, a form that was to become common in later representations. The most striking feature of the present figure of Eridanus, however, is the construction of the central part of his body out of a text written in a venerable (for the ninth century) *capitalis rustica* script. This text is not the Ciceronian translation; that appears in proper form below the figure. Instead, it is made up of "scholia" drawn from the *Astronomica* of the second-century Roman scholar Hyginus. Cicero's verse translation, like the Aratus original from which he worked, afforded a reader general information about location, form, and orientation, together with an occasional mythological legend relative to the constellation figure in question. Thus, we are told here that Eridanus, whose streams are besprinkled with tears from the sisters of Jupiter, courses snakelike from the sole of Orion's left foot to the spine of Cetus. The Hyginus scholia fleshing out the Eridanus figure supplements this Cicero-Aratus account with more specific information about the constellation's name, position, and individual stars. For example, we are told that it is also called Nilum or Oceanum, that it flows toward the feet of Lepus and the Antarctic Circle, dividing the winter heavens in two, and that it contains thirteen stars (of which only six appear below the present figure), the brightest of which is Canopus. As a whole, then, this picture and text of Eridanus combine the literary description of Cicero's Aratus with the more "scientific" information of Hyginus. One should also note the starkly classical style of this particular Carolingian depiction of Eridanus. Art historians have averred that his head could have been copied from some Pompeian fresco, and one scholar even found the overall appearance of the manuscript so classical that he dated it erroneously to the second or third century.

227. From Within and Without. Constellations depicted on the surface of a globe (as in Illustration 221) or in flat representations showing them as if they were on a globe (see Illustration 222) are naturally "mirror images" of their depictions as they appear in the sky to an observer. Each of these manners of representing the constellations can be found throughout the history of the "mapping" of the heavens, but *The Book of the Fixed Stars* of the tenth-century Arab astronomer ᶜAbd al-Raḥmān al-Ṣūfī provided both types of representation for each of the constellations. "In the sky," al-Ṣūfī tells us, "we see the stars in their true position, because we look upward from this center of the globe . . . and it is for this reason that we have included both positions; for otherwise the beholder might be confused if he saw the figure on the globe differing from what he sees in the sky. If we want to see the constellations in their true state, we must raise the page over our head and look at the second figure from underneath. We shall then see it conforming to what is found in the sky." An example of this "double depiction" is reproduced here from an Arabic manuscript executed in Samarkand in 1437. These two representations of Taurus appear on facing pages, that on the left headed by an inscription informing us that this picture of Taurus is "as seen on a sphere," while that on the right is announced as being the same as seen in the sky. If we restrict ourselves to this latter figure, we should begin by noting that the four cardinal directions are specified with North at the upper right, East at the upper left, South at the lower left, and West at the lower right. All the constituent stars are properly numbered. There are thirty-two internal ones (that is, falling within the body of Taurus), not counting the larger star in the northern horn (which is specifically labeled here as also belonging to the right foot of Auriga), plus eleven external stars, giving a total count of forty-four corresponding exactly with Ptolemy. This distinction between internal and external stars is also "color coded," the former being red with black numbers, the latter black with red numbers. In addition to the star common with Auriga, two others are provided with their names as well as numbers: first, *al-dabarān* (our Aldebaran) directly under the right eye of Taurus, and second, referring to the four stars (numbered 29 through 32) at the bottom of the "mane" of Taurus, *al-thurayyā,* the Pleiades. All of these annotations and numbers, including the cardinal directions, are naturally repeated in reverse in the figure "as seen on a sphere" at the left.

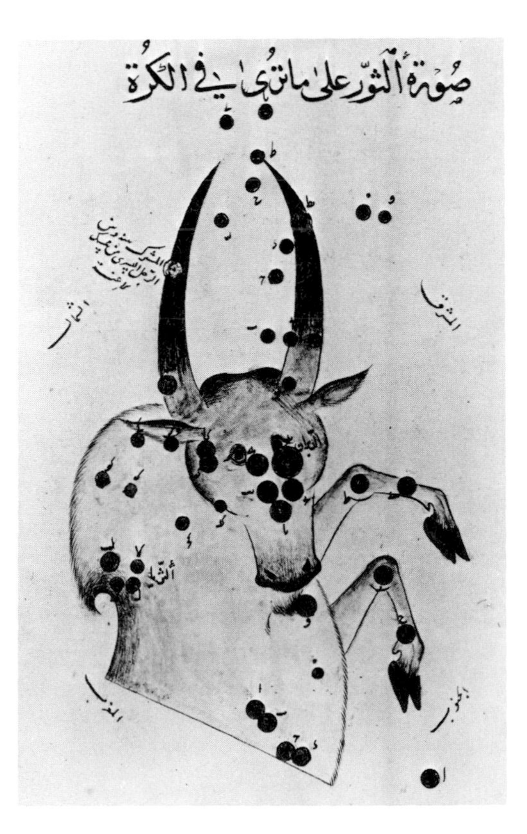

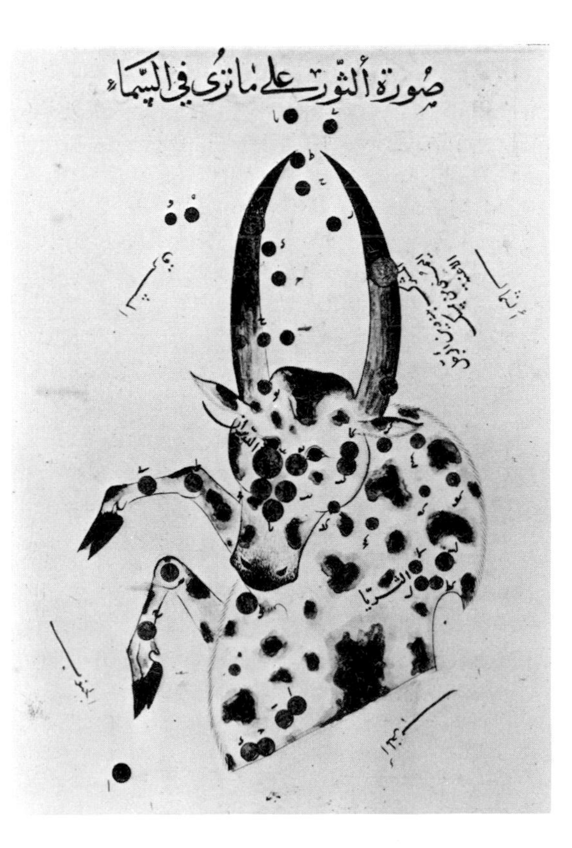

228. A Constellation in Classical, Arabic, and Arabic-Latin Dress. All three of the figures reproduced here are of the constellation Andromeda. The first at the left is from a ninth-century French codex of Germanicus' version of Aratus' *Phaenomena.* Classical in framing, in portraying its central figure as half-naked, as well as in other elements of style, the figure depicts Andromeda with arms outstretched, chained to two stone pillars. This is all part of the mythologizing set forth by Aratus and his Latinizers. Although mostly indiscernible here, twenty-five stars are placed over Andromeda's body without care for exactness of position, each painted in a flat gold. The second figure reproduced here is taken from an eleventh-century manuscript of ᶜAbd al-Raḥmān al-Ṣūfī's *Book of the Fixed Stars.* The colophon to this earliest extant codex of al-Ṣūfī's work tells us that it was written and illustrated in A.D. 1009–1010 (A.H. 400) by one al-Ḥusayn ibn ᶜAbd al-Raḥmān ibn ᶜUmar ibn Muḥammad, allegedly the son of al-Ṣūfī. Andromeda is one of the most interesting of the constellations treated by al-Ṣūfī since, in addition to the usual double "from within and from without" representation (see the previous illustration) given of one figure of Andromeda, there are two additional depictions of her, the final one of which is given here. Most notable are the two fish, one laid over the other, stretched across her breast. As the inscription overhead tells us, this is a picture of Andromeda (literally the "enchained one," *musalsala*) together with the northern fish described by Ptolemy. That is, in this figure al-Ṣūfī has combined Andromeda with Pisces, the zodiacal constellation with which she is closely associated in the sky. Thus, the large star in the lower segment of the two fish that is also located in what is approximately the center of Andromeda's body is labeled "the side of Andromeda and the heart of the great fish" (a reflection of an Arabic incorrect conflation of the modern Mirach or β of Andromeda with a star belonging to Pisces). One other star is labeled in our figure: *al-ᶜanāq* (the badger), just above Andromeda's left foot. The legend at the bottom of the picture informs us that the red dots belong to Andromeda while the black ones are part of the northern fish described by Ptolemy. The last illustration, on the facing page, combines the classical and Arabic traditions. Taken from an astronomical miscellany manuscript written in Salzburg in the fifteenth century, it puts the standard representation of Andromeda bound to two pillars together with two added depictions deriving from al-Ṣūfī. Yet another representation of Andromeda appears in the picture at the right as a small naked figure holding a cord tied about her waist. The legend at the top left opens by announcing that we have here the constellation Andromeda or the chained woman *(Mulier cathenata)*, also named the woman who has no husband (which is an appellation deriving from Arabic sources), and continues with astrological information about the zodiacal sign Pisces. The legend beneath the picture on the right merely gives instructions on how the constellation Equus (Pegasus) is to be positioned relative to Andromeda.

229. (Next Page) Stages of the Introduction of the Astrological. The illustration on the left on page 254 contains a considerable amount of information supplementary to the depiction of the constellation (Sagittarius) in question. The illustration is taken from a mid-fourteenth-century manuscript of the *Introduction to the Judgments of Astrology (Introductorius ad iudicia astrologie)* of the fourteenth-century Italian astrologer Andalò di Negro of Genoa. At the very top, as well as just above the framed figure, the sign of the zodiacal constellation is divided into thirty degrees, both of these divisions then serving as a "scale" against which to match most of the astrological information given. For example, beginning with the fourth line of text from the top, we are told just which segments *(termini)* of the sign of Sagittarius are ruled over by which planets. (Thus, matching the segments given here with the degree-scale above or below, we have Jupiter ruling over the first fourteen degrees of Sagittarius, Venus over the next five, Mercury over the next four, Saturn over the next five, and Mars over the final four.) The next line of text tells us which planets are associated with the three faces *(facies)* or equal third parts of Sagittarius' sign; in order, they are Mercury, the Moon, and Saturn. Similar divisions in succeeding lines tell us just which degrees or parts of the zodiacal sign are (1) feminine or masculine; (2) light, dark, smoky, or empty; (3) black; (4) unlucky *(azamena)*; and (5) red—the last three ascriptions being appropriately marked with blocks of matching color, yellow being reserved for *azamena.* The top three lines of our tabular text provide astrological information that is not related to the degree-scale of the sign. We are told that Sagittarius is the domicile or house of Jupiter; that the tail of the dragon (that is, the descendant lunar node) is in exaltation in 6°3′ of this sign; and that Sagittarius is the house for the triplicity of the Sun ruling by day, Jupiter by night, and Saturn as partner for both day and night. The figure of the constellation appearing below all of this information has its thirty-one constituent stars marked by red dots, the legend under the upper text specifying their magnitudes. The second figure on page 254 comes from a manuscript (ca. 1300) of a Spanish work that apparently belongs to the circle of Alfonso X. Since its interest is in the division of the zodiacal sign Leo and not in the constellation of that name, no stars are depicted on the central figure. Instead, it focuses on a visual presentation of the imaginary figures—invisible constellations one might say—that rise simultaneously with the successive degrees of Leo *(las figuras que suben en los grados del signo de leon)*. The simultaneous rising of stars and constellations was very much part of ancient and medieval astronomy and astrology (see Illustration 230) and this concern is in evidence in the present figure, albeit in a rather fantastic way. Thus, the thirty figures populating the outer ring of the wheel are the figures that rise

(Continued)

(Continued)

together with each of the thirty individual degrees of the sign Leo, the inscriptions on the relevant "spokes" specifying just what these curious figures are. For example, beginning with the first degree at the left (marked by the numeral 1) we learn that the figure represents lion's face *(cara de leon)*. Proceeding counterclockwise for the simultaneously rising figures for successive degrees, we have: a crane *(grua)* for degree 2, a hanging ship for degree 3, a happy man for 4, a large serpent for 5, a man with a sword for 6, and so on. Little more than a page of text accompanies this circular schema, only two sentences being devoted to each "degree." The first identifies the pertinent simultaneously rising figure in slightly more detail. The second broaches astrology through the addition of relevant nativity information. Thus, to cite the entries for the two final degrees appearing above the diagram, we are told that a bed with a lion on it rises together with the twenty-ninth degree and that whoever is born under that figure and degree will be a killer *(matador)*, a spiller of blood, and troublesome; while two lions in a single body rise simultaneously with the thirtieth degree and being born under it entails that one will be rich, strong in body, and deceptive.

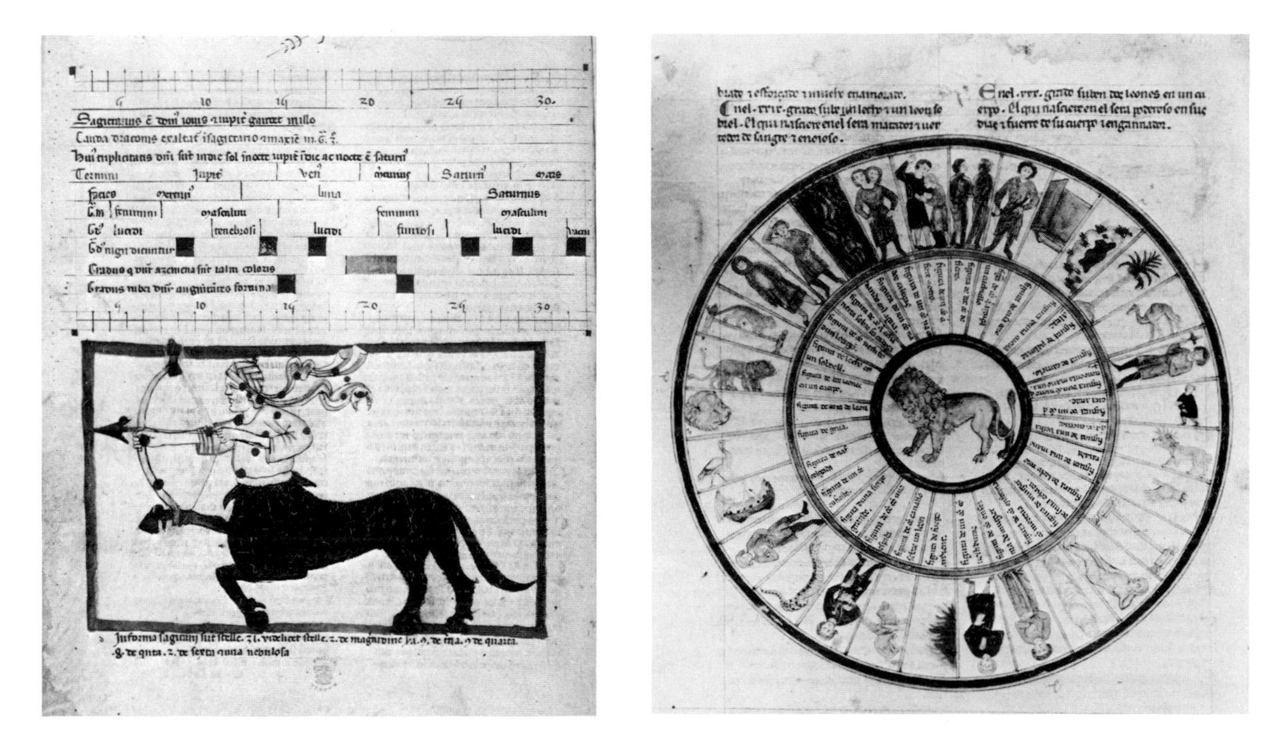

230. (Facing Page) Sphaera Barbarica. An understanding of this illustration will be facilitated by a brief, if necessarily oversimplified, indication of the doctrine behind it. The nucleus of the doctrine in question is expressed by the Greek word *παρανατέλλοντα*—literally, "those things which rise together"—the reference naturally being to those constellations, or parts thereof, north and south of the ecliptic, that rise together with certain degrees or segments of the ecliptic. (One aspect of such "simultaneous risings" has already been seen in the last figure of the previous illustration.) Viewed historically, the segments of the ecliptic or zodiac that were most frequently involved with such *παρανατέλλοντα* were (equal) third parts of each zodiacal sign called decans (initially an Egyptian conception—see Illustration 218—now assigned astrological duty). One could investigate, then, just which stars rose simultaneously with which decans (first, second, or third) of this or that zodiacal sign. However, it was not simply a question of such simultaneous rising of visible stars or parts of constellations, but also of imaginary figures associated with the various decans of the zodiac. The former constituted what became known as the *sphaera graecanica;* the latter, the *sphaera barbarica.* Omitting a great deal of intervening history, it should be noted that both of these "spheres" were carefully expounded in the *Great Introduction to the Science of Astrology (Kitāb al-madkhal al-Kabīr ᶜalā ᶜilm aḥkām al-nujūm)* of the ninth-century Arabic scholar Abū Maᶜshar. This work is the ultimate (textual) source for the pictures reproduced here. In it, Abū Maᶜshar divides the "forms" that rise with the zodiacal decans

(Continued)

into three groups: first, those of the "Persians, Babylonians, and Egyptians" (knowledge of which apparently reached him through a Persian translation of the formulation of the *sphaera barbarica* established by the first-century A.D. Greek Teukros of Babylon); second, those of the "Indians," of which he may have known in part in some way from the work of the sixth-century Indian astrologer Vâra-hamihira; and last, those of the Greek "philosophers, Aratus and Ptolemy." This threefold division is clearly reflected in our pictures. First, the circular diagram of Virgo on the left (taken from the same Alfonsine Spanish work cited in the previous illustration) has placed Abū Maᶜshar's "sphaera indica" in its outermost ring, next that *segund dizen los de babillonia e los de perssa e de egipto,* and, as innermost, that of "Ptolemy," Virgo herself, robed and winged, occupying the center. Three "spokes" divide Virgo's sign into three "faces" (*fazes,* equivalent to the decans). Turning to the outermost Indian *παρανατέλλοντα*, we note that a maiden standing between two trees (at the bottom of the circle) rises together with the first *faz* of Virgo. The second *faz* is accompanied by the figure of a black man, while in the third a maiden is about to enter a house of prayer. The middle ring is more complex, even if incompletely presented. Yet Abū Maᶜshar's specification of a star, part of the tail of Hydra, and heads of a raven and a lion (for the first *faz* or decan) is depicted, as are, to cite but a few, his half of a naked man, ploughshare, lion's tail, and two steers. The innermost ring brings us to the "Ptolemaic" *παρανατέλλοντα*, where the first decan witnesses the simultaneous rising of (parts of) the tail of Draco, the tail and rear foot of Ursa Major, the hind parts of Leo, and Crater on Hydra, and so on through the second and third decans, ending with the rear shank and legs of Centaurus. The *sphaera barbarica* received other, often more elaborate, representations in the Middle Ages and the early Renaissance (where we find its descendants decorating the walls, for example, of the Salone in the Palazzo della Ragione in Padua and the Palazzo Schifanoia in Ferrara). A more expansive manuscript representation can be found in a thirteenth-century codex of a version of Abū Maᶜshar prepared by Georgius Zothorus Zaparus Fendulus. The "spheres" for each zodiacal sign consume four pages, beginning with a full-page figure of the sign itself and followed by three pages representing the three *sphaerae* in matching horizontal strips. The last page giving the third decans for the sign Taurus is reproduced here at the right. The bottom strip gives the Greek *παρανατέλλοντα*: Perseus holding the reins of the Northern Horse (that is, Pegasus), the head of Taurus, and the end of Eridanus. The center strip gives the figures for the Indian sphere of the decans: a man whose body is half elephant, half lion, his son, the Northern Horse, and a reclining calf. Finally, the Persian-Babylonian-Egyptian sphere: the lower part of a dog-headed man, a man holding a serpent, two wagons, one pulled by horses, the other by cows, with a man holding a goat seated in the first. For other pictures from this Abū Maᶜshar manuscript, see Illustrations 11 and 232.

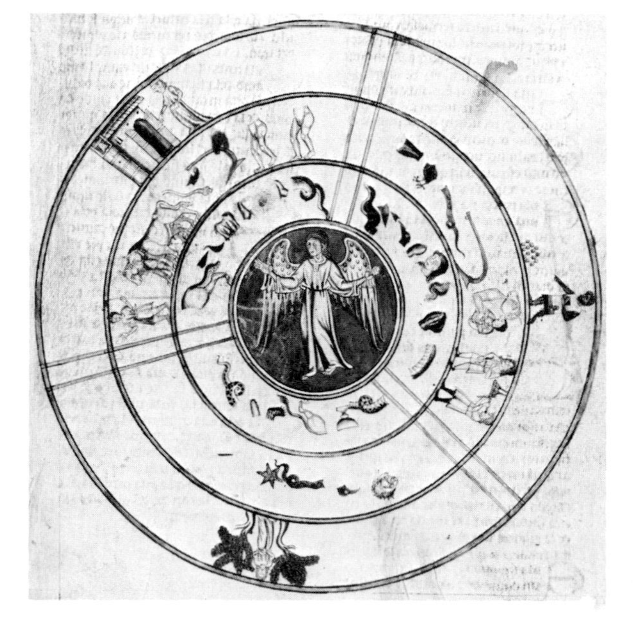

231. (Right) Planets Personified. The very name 'planet' means "wanderer"—from the Greek πλάνητες ἀστερές, "wandering stars"—a fact that excluded the planets from having their patterns mapped after the fashion of the constellations provided by the fixed stars. They could, however, be shown within one or another constellation belonging to the zodiacal belt (for astrological purposes—see the next illustration—or to represent the configuration of the heavens on a particular date—see Illustration 225), or they could be represented through personification. The latter type of representation is shown here from a fourteenth-century manuscript of Michael Scot's *Liber introductorius,* one of the central astrological handbooks in the Latin Middle Ages. A medieval, Christian guise, and not a classical depiction, has been given to each of the five planets represented on this manuscript page. There is not too much change in Mars (lower left) as a warrior and Venus (center) as a maiden, but, although Saturn (upper left) holds the sickle characteristic of his classical portrayal, he has been transformed into a medieval shield-bearing soldier. The changes in Jupiter (upper right) and Mercury (lower right) are much greater. The latter, for example, has lost his winged feet and hat, and appears in a bishop's robe with scepter, miter, and book. Jupiter, as a judge in the vestments of his office, is even more unusual. Each depiction follows the accompanying text with some care. In the description of Jupiter, for example, we are told that the *biretum* he wears signifies knowledge and good fortune and that his judge's rod and sheepskin bespeak dignity and judicial efficiency. The gloves he grasps in his left hand refer to a life of consolation, and the purse hanging below them signifies temporal riches. For other diagrams and figures from the same manuscript, see Illustrations 26, 80, 101, 251, and 277.

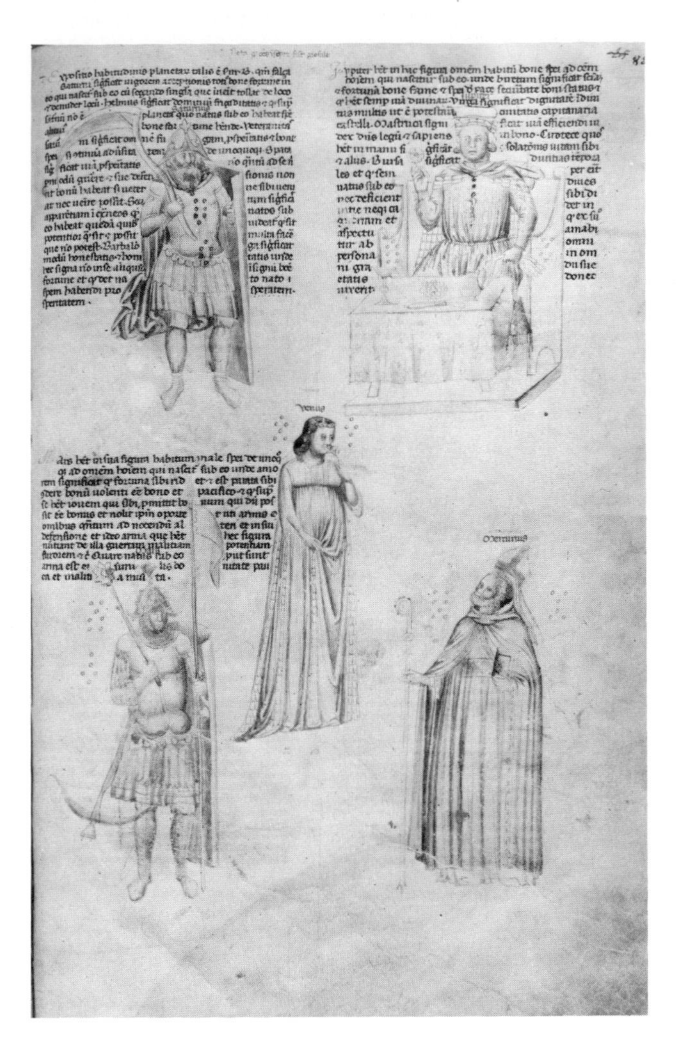

232. (Facing Page, Above) The Planets Within the Zodiac. Taken from the same fourteenth-century manuscript of Zothorus Zaparus' compilation of Abū Ma^{c}shar used in Illustration 230, the personification of a planet is here related to the signs of the zodiac. On rare occasions the reason for relating some planet to a zodiacal sign was to represent the configuration of the heavens as of a given moment or date (as was the case, for example, in Illustration 225), but in most instances, as in the present one, the purpose in establishing such a relation visually was astrological. The planet in question is Venus, depicted with a musical instrument (most likely a lute), something that was part of the Eastern tradition clearly presented by Abū Ma^{c}shar. The first picture at the top left represents Venus in exaltation in Pisces (properly represented at the bottom of the frame), while the second picture, top right, indicates that she is in depression (and hence appropriately depicted upside down in fall) in Virgo (who stands sternly at the left). At times, the exaltation and depression of a planet were more exactly located relative to the degree of the sign in question, as is true here for Venus' exaltation (but not for her depression) in the legend above the third picture: namely, 27° of Pisces. For other figures from the present manuscript, see Illustrations 11 and 230.

233. (Below) Black Balls on the Sun's Disk. During the third and fourth decades of the twelfth century, John of Worcester continued the composition of a chronicle apparently initiated by Bishop Wulfstan of the monastery in this English village. Among the most interesting facts and events recounted in this chronicle are several that reveal the monks of Worcester to have had more than a modicum of "scientific curiosity." We learn, for example, with what praise and comprehension they took note of the recent translation by Adelard of Bath of the astronomical tables of the ninth-century Arabic scholar al-Khwārizmī. And we can witness the care that went into their recording of unusual celestial phenomena: apparent eclipses, the aurora borealis, and sunspots. The latter are depicted on the twelfth-century manuscript page reproduced here below. The text surrounding this curious figure recounts that on 8 December 1128, two black balls *(due nigre pile)* appeared on the sun's disk, remaining there from morning till dusk. That located in the upper part of the disk was larger than that occupying the lower portion. Furthermore, the entry concludes, these two *pile* were directly opposite each other "in accordance with a figure of his sort" *(ad huiusmodi figuram)*, whereupon the reader's attention is drawn to the picture given here. There is little doubt that the chronicler intended the picture to fill out his account in an essential way. Indeed, one might even claim that he viewed its role as absolutely central. The fact that the script in which this account was written "stretches out" to fill the space as it proceeds indicates that the picture was drawn before the text was written and hence was of foremost importance in the chronicler's, or his scribe's, mind.

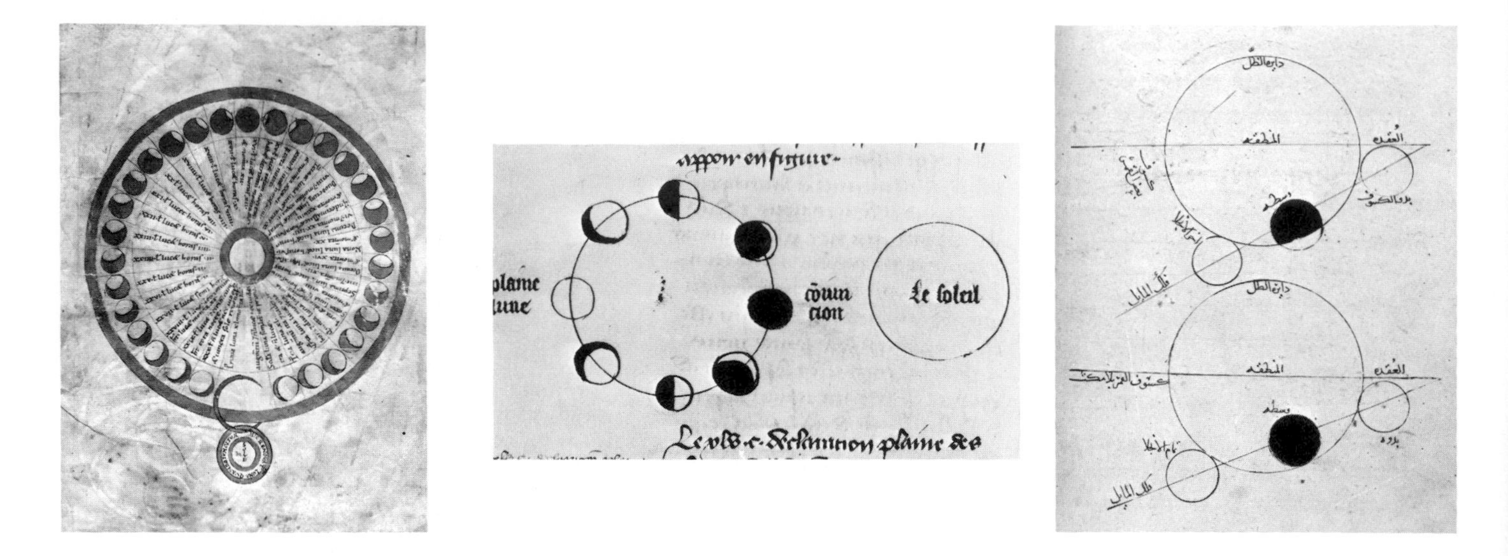

234. The Moon as Seen. The moon is surely the most notable nocturnal celestial object, and its different phases must have been the most easily observable change in the night sky. Depictions of this change, usually associated with diagrams explaining the reason for the different phases, therefore occur rather frequently in medieval manuscripts. That reproduced above at the left is attached to no text, but simply fills an otherwise empty page in a twelfth-century codex. The number of phases it illustrates is governed by the so-called thirty days of the moon, the time that elapses from new moon to new moon (the actual interval is approximately 29½ days). The spokes of the wheel formed by the diagram specify the days of the moon in question, give the total time (in hours and *momenta*—that is, fortieths of an hour) during which it will shine or be visible (compare Illustration 53), and occasionally state whether the moon is waxing *(in augmento)* or waning *(in detrimento)*. The small circle under the main diagram tells us that at this point the moon will undergo renewal since it is there in conjunction with the sun. The "colors" are reversed, making the full moon at the top of the diagram black, the new moon at the bottom white. Curiously, in spite of the effort spent to provide considerable information in writing within the diagram, the depiction of the phases is far from accurate. Thus, in addition to the new moon carrying an almost annular crescent, all the other phases are represented as increasing or decreasing crescents; no half or gibbous moon appears. The middle moon-phase diagram above is far more accurate, if less elaborate, than the one just examined. It is found in an early-fifteenth-century manuscript of Nicole Oresme's *Traitié de l'espere* that once belonged to Jean, Duke of Berry. In it, the orientation of "colors" correctly represents the lighted segments of the moon, and half and gibbous moons are appropriately shaped. The sun *(le soleil)* is at the right; the full moon, labeled *plaine lune,* is at the left; and the new moon, which receives the annotation *coniuncion,* is in between. Although far rarer both in occurrence and in manuscript depiction, the moon during a lunar eclipse is yet another kind of medieval illustration of the "moon as object." The present example of such an illustration, at the right, comes from a manuscript written at some point before the fifteenth century of *The Book of Instruction in the Principles of the Art of Astrology (Kitāb al-tafhīm li-awāʾil sināʿat al-tanjīm)* by the eleventh-century Persian astronomer and mathematician al-Bīrūnī. In the paragraphs preceding this illustration, Bīrūnī explains that a lunar eclipse may be partial or total, both of which cases are represented in his diagram. The top figure depicts what we see during a partial lunar eclipse, while that below refers to that species of total eclipse in which the moon begins to brighten on its eastern margin immediately after it reaches totality. The horizontal lines in both figures are properly inscribed as representing the ecliptic, the oblique straight lines as the moon's inclined orbit, and the intersections as the nodes. The large circles bear the label "circle of shadow" and both sets of three small circles depicting the moon in transit receive inscriptions (from right to left) pointing out that we have here representations of the beginning, middle, and end of the eclipse in question. The only point of major inaccuracy is the size of the earth's shadow. It should have had a diameter roughly double that of the moon. For other figures from the manuscripts utilized here, see Illustrations 49, 162, 189, and 279.

235. Bearded Stars, Torches, and Daggers. Comets were always considered to be among the most portentous of celestial phenomena, so it is not surprising that many of their depictions in antiquity and the Middle Ages reflect this concern with their ominous character. Alternatively, their description and portrayal often revealed the uncertainty felt about their nature and the fabulous variety of the forms they were reputedly observed to possess. Of the three depictions of comets given here, the first illustrates the former of these characteristics; the last two the latter. Above, the Bayeux Tapestry (possibly ca. 1066–1077) provides what is most likely the earliest extant representation of a comet. Portrayed in the upper border of this section of the tapestry, the comet is preceded by the words *isti mirant stellam,* "these men marvel at the star," a reference to the crowd below who gaze with terror at the celestial phenomenon. At the right, King Harold is informed of this evil omen, the empty ships in the border below being an apparition of the coming Norman invasion of England and thus of Harold's subsequent defeat and death at the Battle of Hastings (14 October 1066). Given these historical events, it is clear that the ominous celestial event depicted was Halley's Comet, which was visible in England from February through mid-May of 1066. The next picture (page 260, middle) comes from an English manuscript of the first half of the twelfth century. The depiction of the comet is unexpectedly sunlike, in spite of the fact that the surrounding text opens, with lines taken from Isidore of Seville's *Etymologiae,* by telling us that comets are called *crinite* by the Latins because they cast forth flames resembling hair *(ad modum crinium)*. The text goes on to say that the Stoics held there to be thirty kinds of comets, the names and effects of which have been set down by astrologers, and concludes by pointing out that certain scholars do not take comets to be stars (since they are neither fixed stars nor planets), but rather to be fire kindled by the Creator's will to signify some thing or event. The final pair of pictures at the bottom of page 260 comes from a late-fifteenth-century codex of a *Calendrier des Bergers* and explicitly addresses the problem of the various forms comets can assume. The full roster of these forms had already largely been filled in antiquity, and most of its members were duly transmitted to the Middle

(Continued)

(Continued)

Ages by Pliny (*Nat. hist.* II, 89–90), who offered most of the information in transliterated Greek. The Greek word *κομήτης* means "wearing long hair," so Isidore's reference to hair in the Latin (*crinite*) for comets is quite to the point, as are Aristotle's and Pliny's references to them as "bearded" (*πωγωνίας*; *pogonias*) stars. Although some of the most intriguing ancient forms (such as "daggers" and "darting serpents"—Pliny's transliterated *xiphias* and *acontiae*) are missing, many are still reflected in this fifteenth-century French *Calendrier.* Thus, the largest figure in the picture at the left is identified as a *lance de feu ardant,* clearly a descendant of the ancient description of a comet as "lance headed" (*λογχωτής*). Similarly, the first figure in the right picture bearing the inscription *colonne ardant* and further identified in the accompanying text as *comme ung pilier,* is kin to the ancient *κίων—columna/columen.* The identification of some of the other portrayals is much more problematic. At the left, the three burning objects labeled *chandelles ardantes* look less like candles than they do like the "jar shaped" *(pitheus)* comets of which Pliny speaks. When compared with the adjoining text, the four separate fires in the same frame seem to be intended to represent comets whose flames extend in different directions. The smallest of the four is identified as *feu qui est fol,* perhaps a reference to the *feu follet* of swamps as descriptive of one sort of cometary phenomenon. On the right, the star with angel's wings *(estoille volant)* is in all likelihood not a descendant of an ancient form, as two others that are mentioned just below the frame (but not pictured) clearly are: bearded and hairy stars *(estoille barbue, chevelue)*. The comet depicted to the right of the winged one is described as "tailed" *(couee)*; the text claims that this type lasts longer than all others. (The sun and stars in this section of the right picture refer to another topic—the planets or *estoilles erratiques*—and not comets.) For another figure from this *Calendrier,* see Illustration 266.

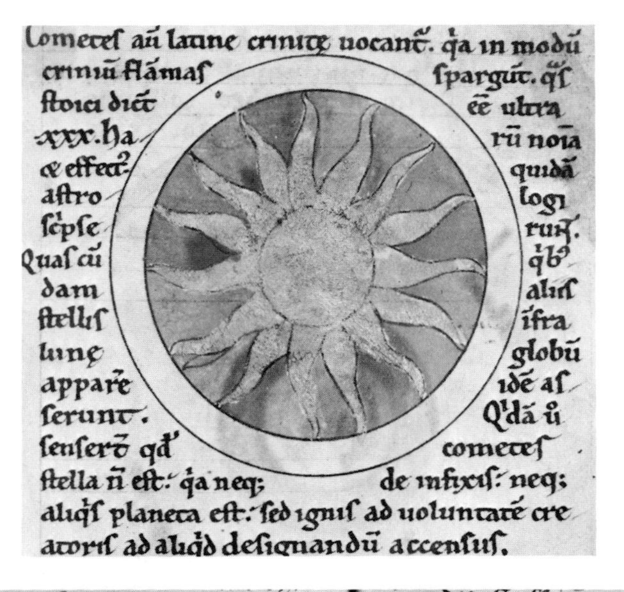

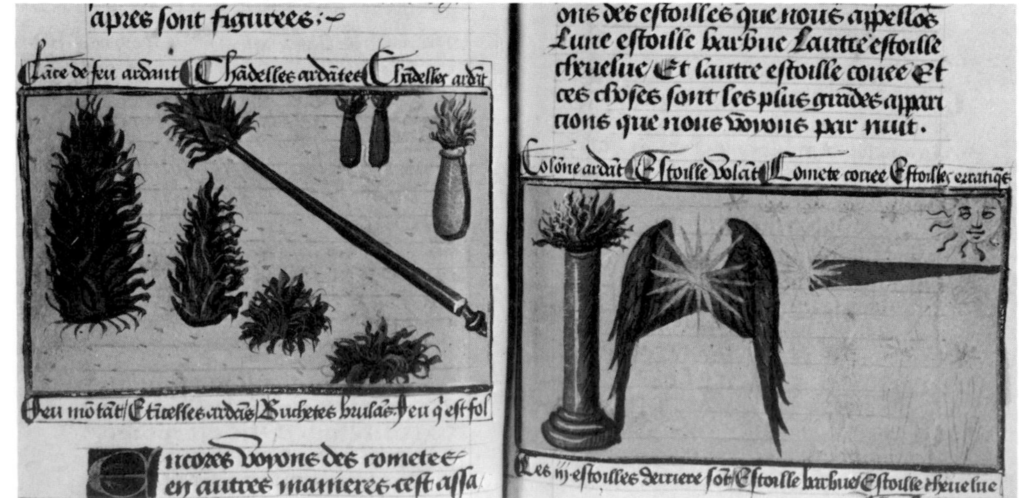

CHAPTER TWENTY

The Instruments of Science

IN THEIR CONCERN WITH the practitioner of science and with the locus in which the enterprise of science was carried out, Chapters Fifteen through Seventeen have already given some place to the depiction of scientific instruments. For the scientist was frequently portrayed, even when such a portrayal was used to personify the science itself, complete with the appropriate "tools of the trade." What better way to make it clear that one was depicting an astronomer than to put an armillary sphere into the picture, or that one was portraying a geometer than to equip him with a compass and a square?

In the present chapter, however, the central concern will be not with such supplemental representations of a scientific instrument, but with the depiction of the instrument itself. This will mean that on the one hand, the illustrations will be drawn from treatises on the construction and use of the instruments in question: quadrants, plane and spherical astrolabes, and other astronomical instruments, for example (Illustrations 236–240). Indeed, the fact that most medieval treatises on instruments deal with astronomical instruments explains why their illustration is dominant in this chapter.

On the other hand, diagrams and pictures of instruments will be taken from manuscripts of scientific texts that speak of the use (and sometimes the fabrication) of instruments relevant to the science being set forth—thus, texts on optics, alchemy, surgery, and materia medica (Illustrations 242–245). In one instance (the final picture in Illustration 243), the instruments portrayed are not attached to any specific scientific text, but rather serve as a kind of "pictorial preface" to a whole set of texts that follow in the same manuscript. Finally, two examples are given of the depiction of scientific instruments in use where that depiction is there for strictly decorative reasons (Illustration 241).

Considering the wealth of manuscript and other materials that can be used to illustrate the medieval visual history of instruments, this chapter could easily have been much longer than it is. Intentional restrictions have kept it to its present size. First, the decision to exclude illustrations that deal with technology from this volume has brought with it the exclusion of examples of the often striking pictures of gadgets and mechanical devices. (Such depictions could have been plentifully provided by manuscripts of the works of such figures as Philo of Byzantium, Hero of Alexandria, Pappus of Alexandria, and al-Jazarī.)

In the second place, the decision to devote this volume to material that pictures the objects, conceptions, procedures, and activity of science has meant that photographs of extant scientific instruments have been kept to a minimum. Those photographs that are included are there because they match and help explain a medieval drawing or picture of the instrument in question, or some part thereof. The same emphasis on pictorial material has even dictated the exclusion of those instruments, known as equatoria, for the calculation of planetary positions, since, though they often appear on or as the pages of manuscripts, they are actual instruments and not pictures or diagrams of instruments.

Finally, restrictions in space have entailed restrictions in how much could be said about the precise construction of each instrument, and about how it works or can be put to use for such and such a purpose. Yet the selected details of this sort that are given will provide some, if not totally adequate, information on such questions.

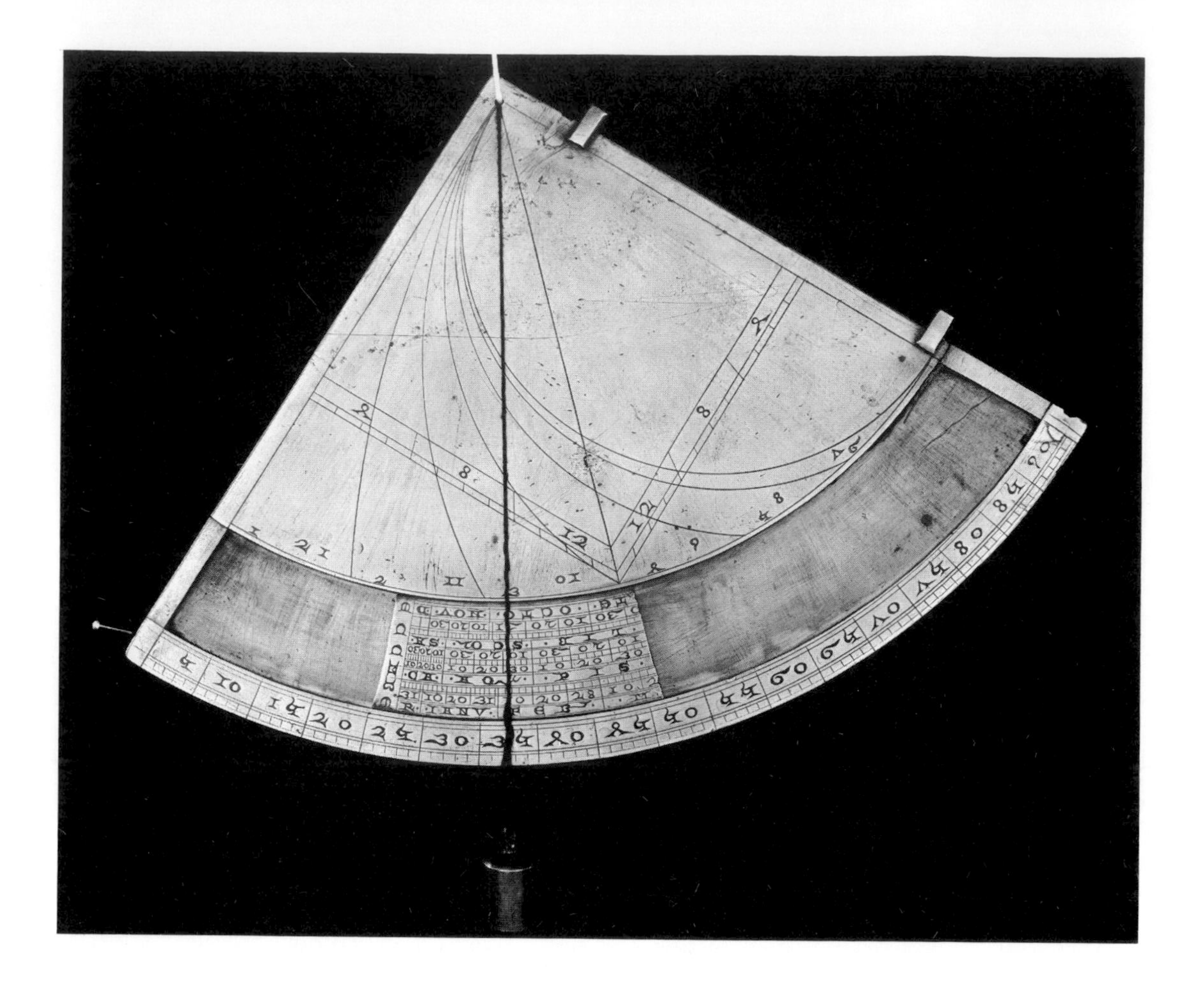

236. Measuring Terrestrial or Celestial. The subject of this illustration is the "old quadrant" *(quadrans vetus)*. We have already seen pictures of its application in the "practical geometry" of measuring the heights of inaccessible objects (Illustration 152) and will later meet a humorous representation of its use (Illustration 241). It received the qualification "old" at the end of the thirteenth century in order to distinguish it from the then recently invented "new quadrant" (a device that was the creation of the Montpellier Jew Profeit Tibbon [Profacius Judaeus in Latin] and that, in spite of certain features in common with its "old" predecessor, derived its essential features from the astrolabe). A photograph of the face of what appears to be the only extant *quadrans vetus* is given above, while a manuscript drawing of such a quadrant face is reproduced on the facing page. The latter is taken from a fourteenth-century English codex of the treatise on the *quadrans vetus* written (possibly in the twelfth century) by Johannes Anglicus of Montpellier. Constructed of brass, this quadrant has three basic features for its use as a sighting instrument: two pinnule-bearing sighting vanes fixed to one of its sides, a plumb bob or line (omitted in the manuscript figure) suspended from the intersection of its two sides, and the division of its quarter arc limb into 90°. One manner of employing these three features in measuring an inaccessible object such as a tower has been described in the caption to Illustration 152. Another involves using the calibrated square—called a shadow square—at the center of the instrument. Thus, given the (horizontal) distance of an observer from a tower, if one positions the quadrant so that the top of the tower appears in the pinnule sights and then notes the point at which the plumb line cuts one side of the shadow square, one has all the information necessary to calculate the height of the tower. For, to use the position of the plumb line in the photo at the left above as an example: (1) the triangle formed by the plumb line, the length it cuts off (here about 7.6) on the side of the shadow square, and the segment of the side of the quadrant equal to the other side of the shadow square (equal to 12), is similar to the triangle formed, respectively, by the line of sight, the height of the tower, and the distance of the observer from the tower; hence (2) by proportion 7.6/12 = unknown tower height divided by

(Continued)

the known distance. Of course, the same procedure will allow one to find one's distance from a tower if the height of the latter is known. In modern terms, the intersection of the plumb line with the sides of the shadow square determines the tangent of the angle of sight if the angle is less than 45°, its cotangent if it is greater than 45°. Other features of the *quadrans vetus* permitted its use to determine hours. This is effected by means of the unequal hour lines; that is, the arcs of circles (all of which have their centers on the pinnule-bearing side of the quadrant) dividing the quarter circle whose center is at the apex of the quadrant into six equal parts of 15° each, the diameter of the first of these hour lines (a semicircle) being equal to the radius of the divided quarter circle. The unequal hours determined by these lines are numbered 1 to 6 from left to right and 7 to 12 from right to left (in the photograph, the numerals in the latter series are reversed; thus 21 for 12 at the left). The unequal hours in question are those determined by the division of the day and night into twelve equal parts; this naturally means that they will vary in length in accordance with the season and that day and night hours will be equal only at the equinoxes. Given the scale determined by these hour lines, one can obtain the relevant hour by sighting the sun and noting the position of the plumb line in relation to the hour lines, relating the resulting reading to the known meridian altitude of the sun on the day in question. In order to be able to determine this meridian altitude without actual observation, a movable cursor was added to the quadrant. Shaped by the arcs of two circles with a length double the obliquity of the ecliptic, it was inserted into a groove bordering the limb of the quadrant (half of the cursor in the photograph is missing). Inscribed with a calendar and zodiac scale, when the cursor is moved so that its (equinoctial) center falls opposite the relevant geographic latitude marked on the quadrant's limb, the meridian altitude of the sun can be read off for any given date on the calendar scale. For other figures from the same manuscript, see Illustrations 152 and 238.

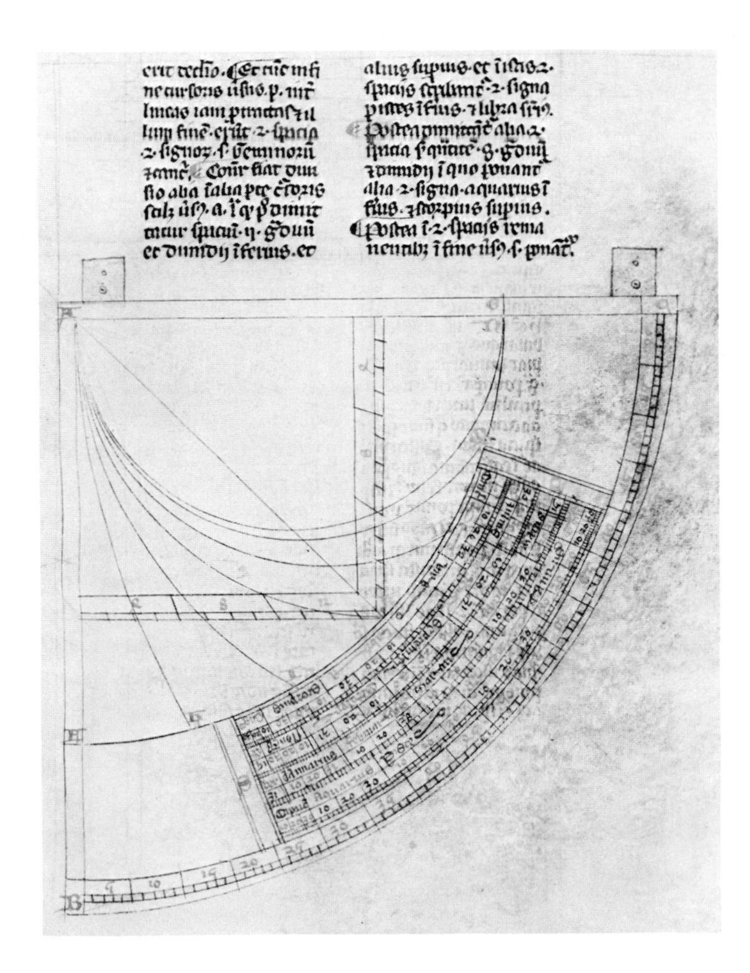

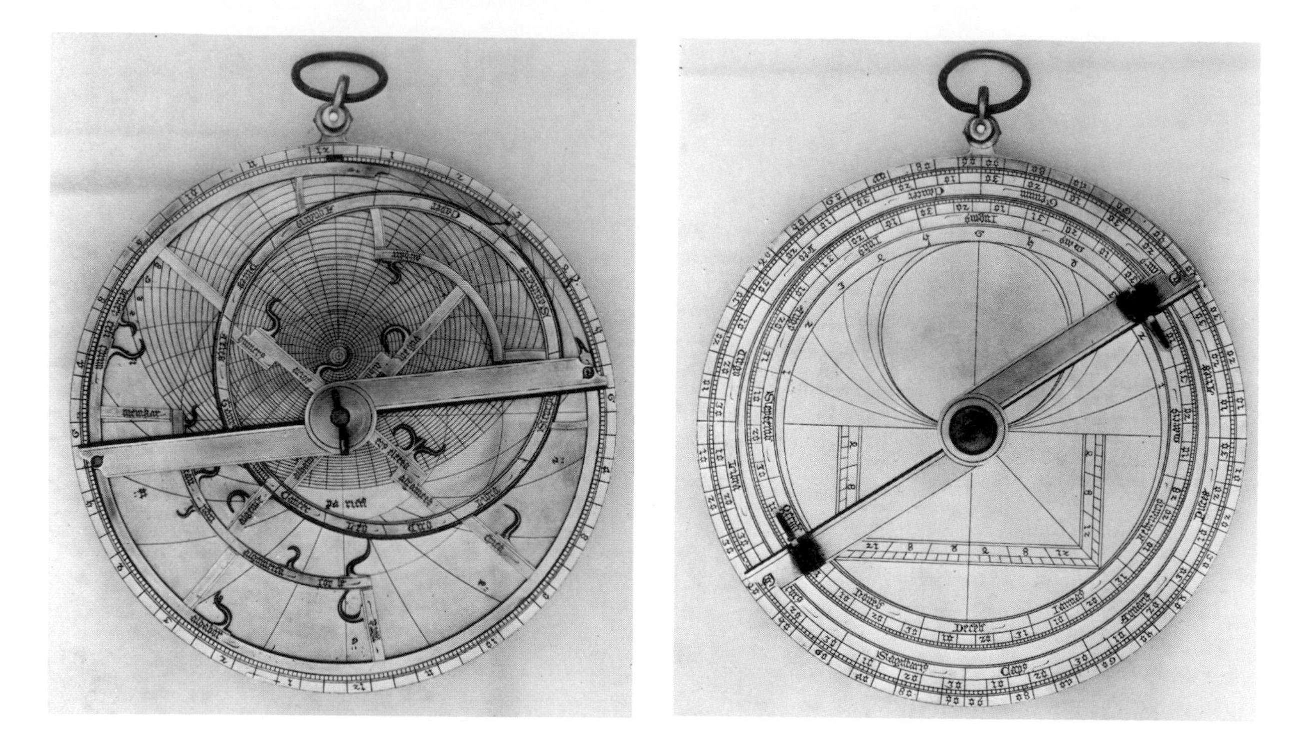

237. The Astrolabe in Fact. Of all medieval astronomical instruments, the planispheric astrolabe is surely the most familiar, a fact that is reflected in the relatively substantial number of actual instruments still extant, as well as in the number of medieval writings devoted to it. It was not, however, a medieval invention (either Arabic or Latin) since we have, or know of, Greek works on the instrument (by Theon of Alexandria [fourth century] or John Philoponus [sixth century], for example, the former being in part the source of the latter, as well as more completely of an early Syriac treatise on the subject by Severus Sebokht [seventh century]). More than that, we know that the mathematics of stereographic projection behind the planispheric astrolabe was a Greek creation. The most essential part of the astrolabe, its face, furnishes a two-dimensional "map" of the heavens. Just what this map consists of and how it functions are at least partly revealed by the modern "exploded view" of an astrolabe at the left, page 265. The whole instrument was usually made of brass, its main body being called the "mother" or mater. A number of different plates or tympana, each of them bearing various celestial coordinates specific to a given geographic latitude, could be inserted into the mater. The rete or "spider" *(aranea)* carrying pointers giving the relative positions of selected stars and a circle representing the ecliptic was placed on, and allowed to rotate over, the tympanum fixed in the mater below. The whole was held together by a pin or bolt that passed from the alidade on the back of the instrument through the mater, tympana, and rete, the pin itself being held in place by a wedge inserted into a slot or hole at its end. Since this wedge was often designed with its thicker end in the form of a horse's head, it was usually called the "horse" (Arabic *faras;* Latin *equus*). In almost all astrolabes, the relevant celestial coordinates are projected onto the plane of the equator (see the next illustration). The two tropics and the equator itself are fixed and hence the same on all tympana. The ecliptic on the rete is also a fixed coordinate. The other coordinates on each tympanum are local; that is, they are specific to the geographic latitude represented by the tympanum in question, since they define a celestial position in terms of angular altitude and direction in relation to the horizon and zenith of an observer at that latitude. Thus, the almucantars are projections of circles of constant altitude above and parallel to the horizon, while the lines of equal azimuth are projections of equally distant arcs of circles running from the zenith to the horizon. The horizon for the latitude in question is also projected onto the equatorial plane of the tympanum. Finally, lines of unequal hours—see previous illustration—are projected below the horizon line. All of these coordinates duly inscribed on a given tympanum will thus appear as in the second

(Continued)

modern figure at the right, p. 265. To turn to the function of the face of the astrolabe, it is clear that the rotation of the rete over a given tympanum will represent the daily rotation of the celestial sphere relative to an observer situated at the latitude represented by the horizon, almucantars, and lines of equal azimuth inscribed on that tympanum. The astrolabe, the face of which is pictured on page 264, top left, was constructed in the early years of the fifteenth century, a product of the atelier of Jean Fusoris (ca. 1365–1436), who not only directed the manufacture of astrolabes and other astronomical instruments, but also wrote a number of works dealing with these instruments and other aspects of astronomy. The limb of the mater pictured is divided into 360° as well as into twenty-four equal hours, and a rule or *ostensor* (enabling one to correlate readings on the limb with various "interior" positions) is fastened over the rete. The tympanum fixed in the mater is for the latitude of Paris (48°30′), as the inscription directly below center *(Parisius)* informs us. The azimuth lines are calibrated for every 10°, the almucantars every 2° (as the numerals 2, 4, 6, 8 at the left—east—circumference indicate). The picture at the right, p. 264, shows the back of the same Fusoris astrolabe. Its center is occupied by a shadow square in its lower semicircle, with unequal hour arcs above, both serving purposes equivalent to their role in the *quadrans vetus* (see previous illustration). The alidade fixed at, and capable of rotation about, the center carries the vanes with sighting holes more clearly pictured in the exploded view. A zodiac scale occupies the outer circumference, while a calendar scale is located eccentrically within it (the eccentricity accounting for the nonuniformity of the sun's motion through the ecliptic). The alidade can be used as a rule to correlate the position of the sun within the zodiac with a given date (ignoring inaccuracies due to leap years and the precession of the equinoxes). Of course, the back of the astrolabe could, like the *quadrans vetus,* be used to calculate the height of towers and the like as well as the altitude of celestial bodies. Moreover, used in conjunction with the celestial "map" on the face of the astrolabe, it could be employed to tell time, to determine geographic latitude, and to assist in the casting of horoscopes, to mention but a few of the astrolabe's uses.

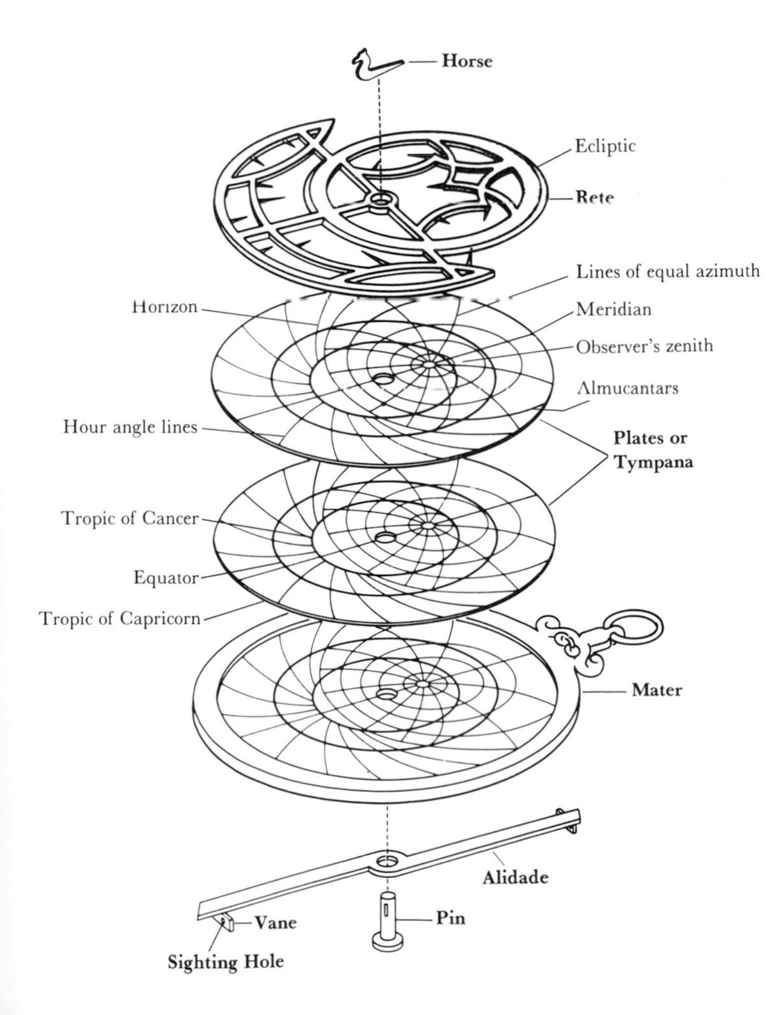

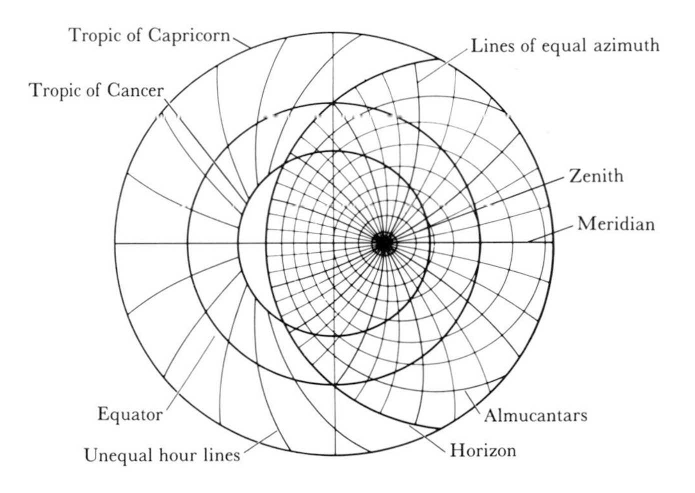

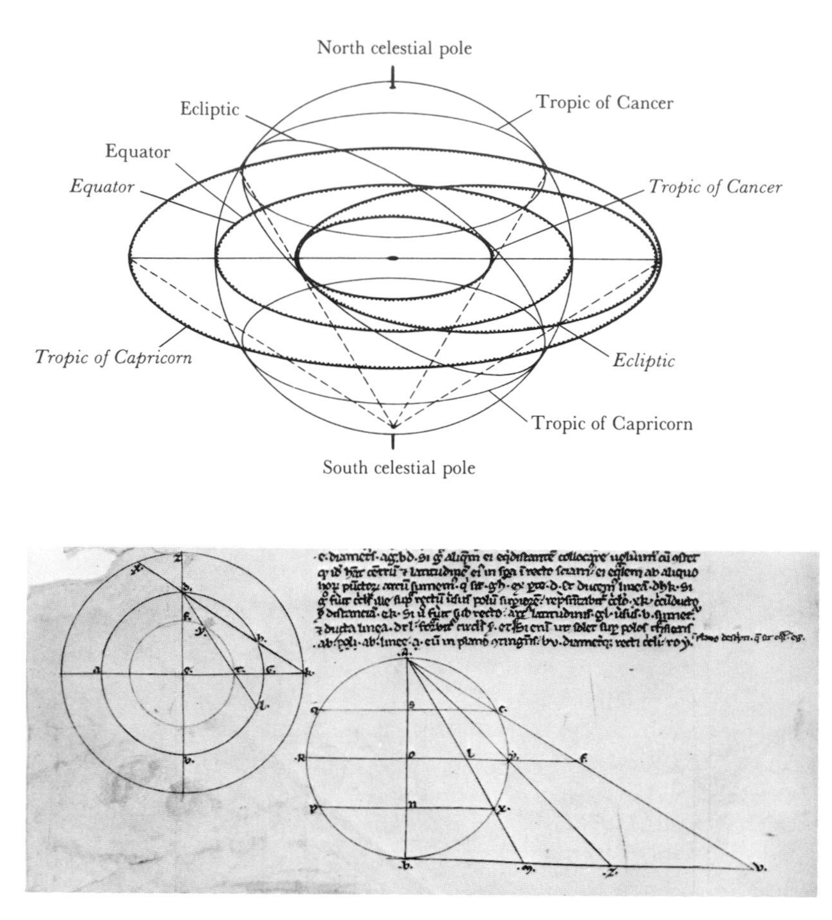

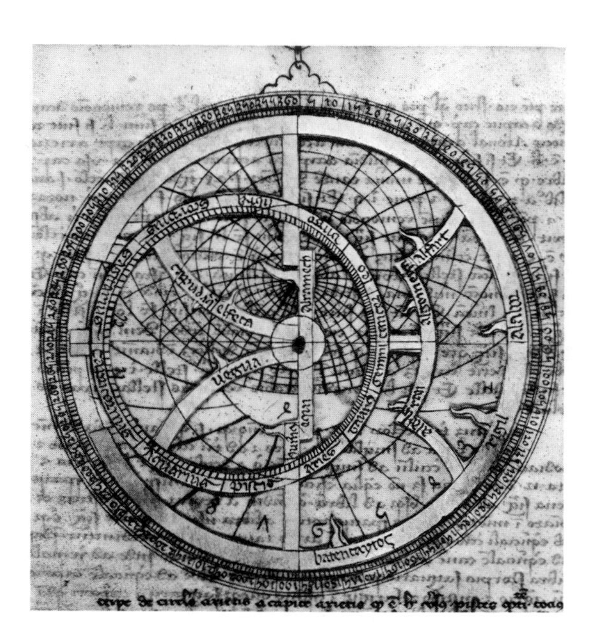

238. The Astrolabe in Manuscript. As mentioned in the previous illustration, the mathematics behind the astrolabe is that of stereographic projection. We possess a Latin translation entitled *Planisphaerium* of a work of Ptolemy (second century) that is lost in Greek which sets forth the geometry of such "mapping," but there is little doubt that the method of stereographic projection was known before Ptolemy wrote his treatise. In any case, the crux of this method is that the source or pole of projection is some point on the surface of a sphere, the plane of projection some plane perpendicular to the diameter through that pole. Thus, in the case of the stereographic projection involved in an astrolabe, the source of projection was usually the south celestial pole, the plane of projection that of the celestial equator. Further, the most important mathematical property of this kind of projection is that all circles on the sphere project as circles on the plane. Thus, in the modern drawing at the top left of page 266, the two tropics, the equator, and the ecliptic on the celestial sphere (in thinner lines with labels in roman letters) all appear as corresponding circles on the equatorial plane (in thicker lines with italicized labels), the equator naturally "representing" itself. Precisely the same kind of projection obtains for the horizon, almucantars, and lines of equal azimuth and unequal hours. In addition to the Latin translation of Ptolemy's *Planisphaerium,* medieval scholars had treatises of their own dealing with stereographic projection. At times, related material was added to works on the astrolabe (such as the Latin translation of the work of Māshāʾallāh [see below]), but the most extensively used treatise was the *De plana spera* of the thirteenth-century scholar Jordanus de Nemore (some of whose other works we have already met in Illustrations 96–98 and 139–141). The manuscript diagram at the lower left comes from a thirteenth-century codex (also seen in Illustrations 99, 120, 124, 133, 139, 141, and 149) of the same work. Setting aside the specific mathematics involved, as can be seen from these diagrams alone, they are both involved in establishing the appropriate correlation between circles on the sphere and their projections in the plane. The results of such a stereographic projection as that specified above can be

(Continued)

seen in the fourteenth-century manuscript figure of the face of an astrolabe reproduced at the right, top of page 266. Occurring in a copy of John of Seville's (twelfth-century) treatise on the construction of the astrolabe, it not only depicts the rete and tympanum of which the text speaks, but also presents the rete as a paper volvelle that can actually be rotated over the underlying tympanum. One has a diagram amounting, in effect, to at least part of a paper astrolabe. The next illustration, bottom left on page 267, of the back of an astrolabe (here turned upside down from its manuscript position in order to correspond to the usual figuration for such illustrations) is notable for two reasons. First, it comes from a very early (eleventh-century) manuscript of astrolabic material, belonging, as one historian has put it, to the first generation of translations from the Arabic of works dealing with this instrument. Secondly, this proximity to Arabic sources is succinctly revealed by the fact that the names of both the signs in its (outer) zodiac scale and the months in its (inner) calendar scale are given in Arabic and Latin, though it is notable that the latter are Arabic transliterations of the Roman names for the months. A shadow square appears clearly in the lower right quadrant. The Arabic inscription written diagonally within it tells us that the "maker of the astrolabe" was one named Khalaf. Presumably, the scribe who drew this manuscript figure (and others accompanying it in this eleventh-century codex) worked from an actual Arabic astrolabe that had been fashioned by this as yet unknown Arabic artisan. The final illustration at the bottom right comes from a fourteenth-century English manuscript containing the Latin translation of the Arabic treatise on the construction and use of the astrolabe written by the Abbasid astrologer, Māshāʾallāh (fl. ca. 762–815; Messahala in Latin). The best known and most used of such treatises in the Latin West, it provided the basis of Geoffrey Chaucer's English work on this astronomical instrument (ca. 1392) addressed to his son Lewis who wished "to lerne the tretys of the Astrelabie." The marginal pictures in the copy of Messahala reproduced here depict parts of an astrolabe. Above is a *novella;* that is, the rule that is placed on top of the rete. Below are the pin and wedge or horse used to hold everything together. Alternative names (some in abbreviated form) in Latin and transliterated Arabic are given for both. For the horse: *al-foraz* (from the Arabic *al-faras*), *cuneus* (meaning wedge-shaped), and *equus.* For the pin: *vectis* (literally, a bolt), *axis, clavus,* and *architob* (from the Arabic *al-quṭb*). For other figures from this last fourteenth-century English manuscript, see Illustrations 152 and 236.

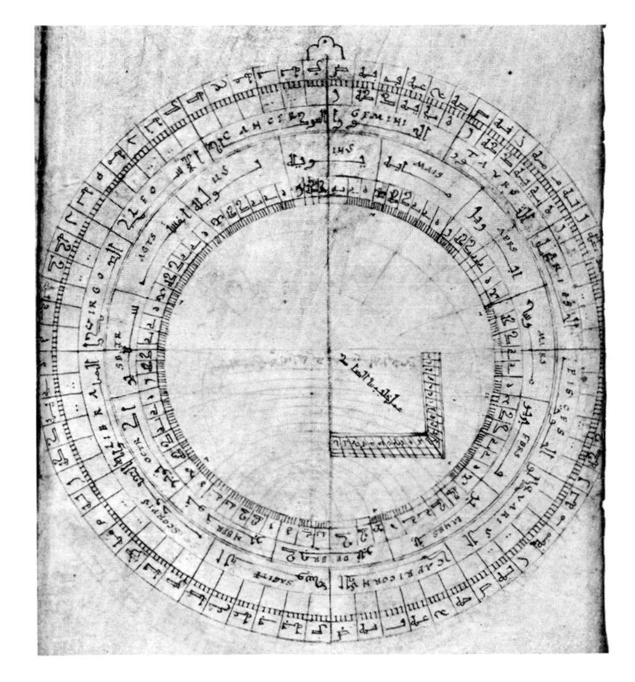

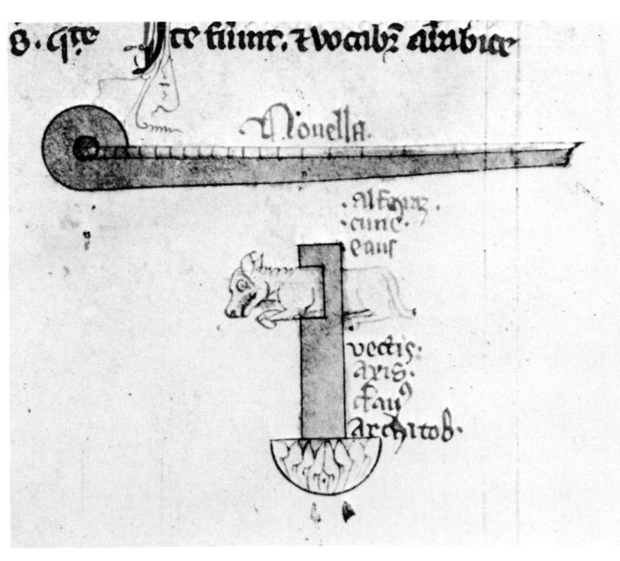

239. (Right) A Construction Detail for a Spherical Astrolabe. Although it served many of the same purposes as the planispheric astrolabe we have seen in Illustrations 237 and 238, by comparison the spherical astrolabe was exceedingly rare, both in terms of extant instruments—only two are presently known—and of treatises devoted to it. A page from one of these uncommon treatises is reproduced here. It is taken from a fifteenth-century manuscript of an Italian translation of the Castilian *Libros dell astrolabio redondo,* composed in the thirteenth century from as yet undetermined Arabic sources by Isaac ibn Sid, one of the most notable collaborators of Alfonso X, el Sabio (see Illustration 229), in his program of translating and preparing Arabic-based astronomical texts. The illustrations on this manuscript page are concerned with the construction of the two brass hemispheres that will, when joined, form the central sphere on which coordinates similar to those appearing on the plates or tympana of a planispheric astrolabe will be engraved. The figures are all but self-explanatory. Thus, in the left column, the top figure depicts the iron form *(la forma del ferro)* that is to fit into the stone mold *(la forma della pietra)* below it, the brass hemisphere *(la mezza spera)* whose casting they are to effect duly appearing between them. Form, hemisphere, and mold are appropriately positioned together *(tutto insieme)* at the bottom of the column. The circle in the right column represents the hemispheres thus formed joined together and hence bears the label "la spera congiunta." The whole of the first book of the present treatise is devoted to the construction of the spherical astrolabe; the second, to its use.

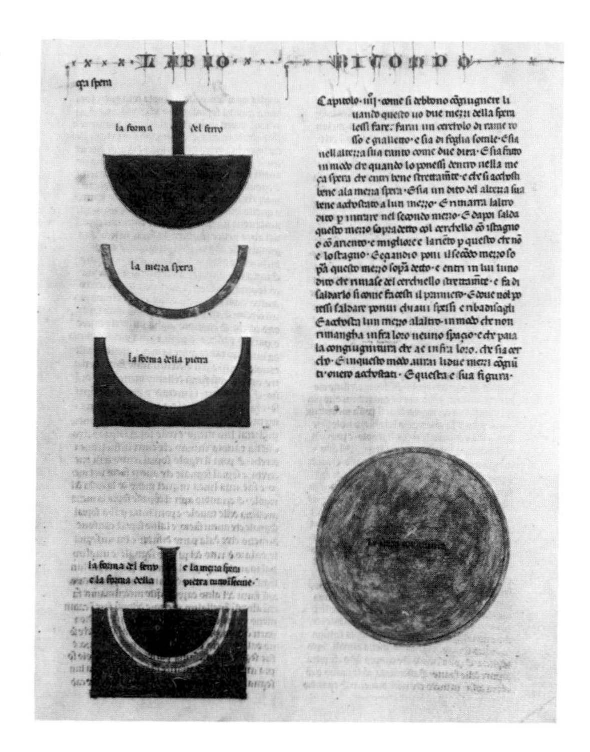

240. Construction Details for a Unique Astronomical Instrument. In 1326–1327, the English astronomer and mathematician Richard of Wallingford (see Illustration 158) composed his *Tractatus albionis* on an astronomical instrument of his own invention. During the same years, he wrote a treatise on another instrument of his creation: the *rectangulus.* Parts of this latter instrument are depicted here at the left from a fourteenth-century manuscript of Richard's work, the opening lines of which tell us that he had "designed the rectangulus to obviate the tedious and difficult work of making an armillary sphere." The advantage to which Richard refers with these words is most likely that the simple straight rods of his new instrument would allow one to avoid the difficult construction of the metal rings required by an armillary sphere. Nevertheless, his instrument could serve in place of an armillary sphere both in terms of making observations and of reducing one set of coordinates to another. When properly assembled, Richard's *rectangulus* would look like the modern reconstruction on the facing page, left. Its parts are as follows: (1) It has an alidade at the very top with a plumb line suspended from its extremity. (2) The alidade is fastened to a limb *(tibia)* directly below it in such a way that it may swing up and down, the limb itself bearing another plumb line at its end. (3) A rule *(regula)* is fixed beneath this uppermost limb in such a way that the two may be rotated right or left relative to one another. (4) A joint *(coxa)* directly beneath this uppermost rule is fitted above the "knob" of a second limb below it in such a way as to allow mutual vertical motion. (5) A second rule is positioned beneath this second limb, followed in turn by a second joint and a third limb and rule, vertical and lateral motions being correspondingly provided for as above. (6) The whole is mounted on a suitable column for support, again in such a fashion as to permit lateral rotation. Selected parts of this reconstruction are clearly identifiable in the marginal drawings reproduced on the facing page, at the right, from the fourteenth-century copy of Richard's treatise. The supporting column is depicted in both the left and

(Continued)

right margins. The first of these depictions gives a clearer representation of the lowermost rule as attached directly to the top of the column, the lowermost limb, labeled "B," in turn directly on top of that. The scalloped opening in the limb neatly reveals the peg and the pin—called a horse *(equus)* following the tradition of the astrolabe—through this peg used to fasten the rule and the limb to the column and to allow their required rotation. The truncated limbs over the supporting column at the right give adequate representations of the necessary joints, while the uppermost limb and the alidade above it have plumb lines depicted at their extremities. The marginal notes inserted in these figures refer to the fact that the rules there represented will have scales (not shown in the modern reconstruction) attached to them that will be divided as illustrated in the lower margin of the present page. It is noteworthy that the text of Richard's treatise makes explicit reference to these figures *(et apparebit talis figura, ut presens ostendit figura)* as essential to facilitating an understanding of the instructions it sets forth.

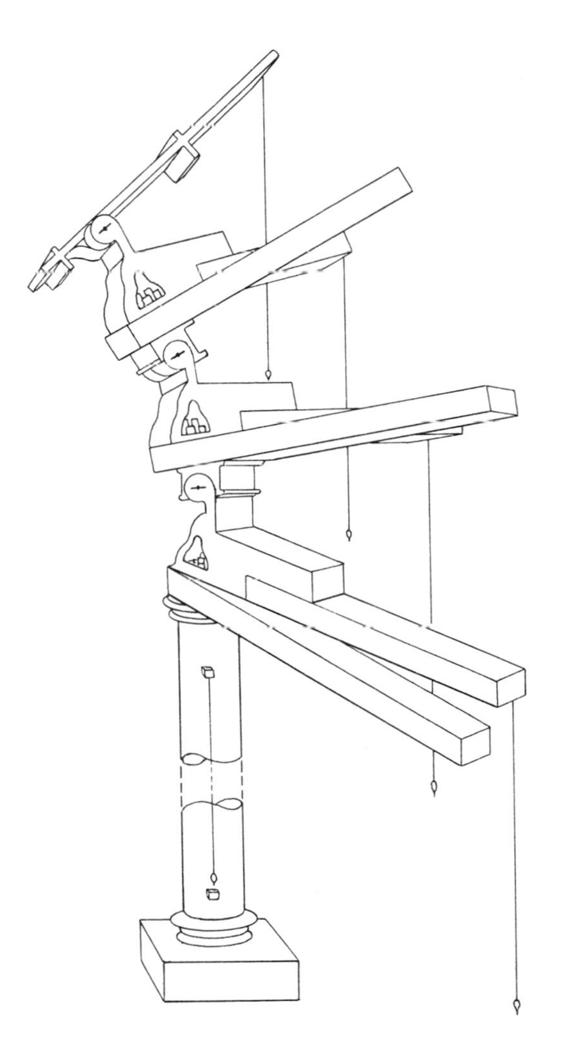

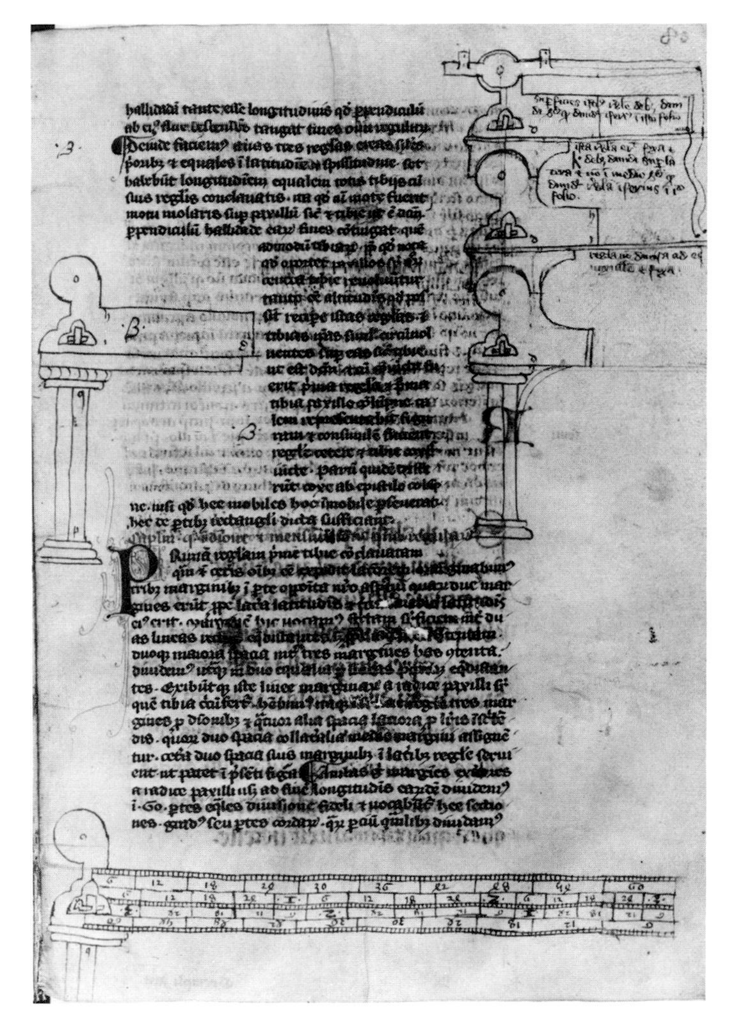

241. Astronomical Instruments in Use. The miniature at the left comes from a thirteenth-century French codex of the *Psalter of Saint Louis and Blanche of Castile.* The central figure holds what appears to be some kind of rule in his right hand while raising an astrolabe in his left. He is in the course of sighting some heavenly body through the instrument's alidade. If it is correct to say that an astronomer is being portrayed here, then, on the basis of what seem to be numbers in the book he is holding, the figure in front of this astronomer can be viewed as a depiction of a mathematician. The figure behind the astronomer is undoubtedly meant to portray some kind of scribe or clerk recording the information being provided by the celestial observation. The second picture, below, is quite different. Occurring on the initial page of a Latin translation of Ptolemy's *Almagest* as found in the fourteenth-century presentation manuscript we have referred to above a number of times (Illustrations 44, 59, 60, 62, 104, 178), it is an outstanding example of the animal grotesques or *drôleries* that so frequently decorate the margins of medieval manuscripts (compare Illustration 261). From right to left, this particular marginal picture shows a bear and a ram sighting through a quadrant, some kind of goat holding an astrolabe, a fox with an armillary sphere, and an ape holding what seems to be a cylindrical sundial in one hand, while probably pointing at the sun with the other. For other pictures of astronomical instruments in use, see Illustrations 155, 161, 162, and 177.

242. The Careful Construction of an Instrument to Measure Refraction. In his *Optics,* Ptolemy (second century) had already employed a calibrated copper disk in his investigation of refraction. But the description of such an instrument given by the eleventh-century Arabic scholar Ibn al-Haytham (in Latin, Alhazen) was much more complete than that provided by his Greek predecessor, not simply in terms of the details of description, but also of the instrument's construction. It is important to note further that Alhazen's account of this instrument was reproduced with very little variation by the thirteenth-century Polish scholar Witelo (on both Alhazen and Witelo see Illustrations 134 and 135). The

(Continued)

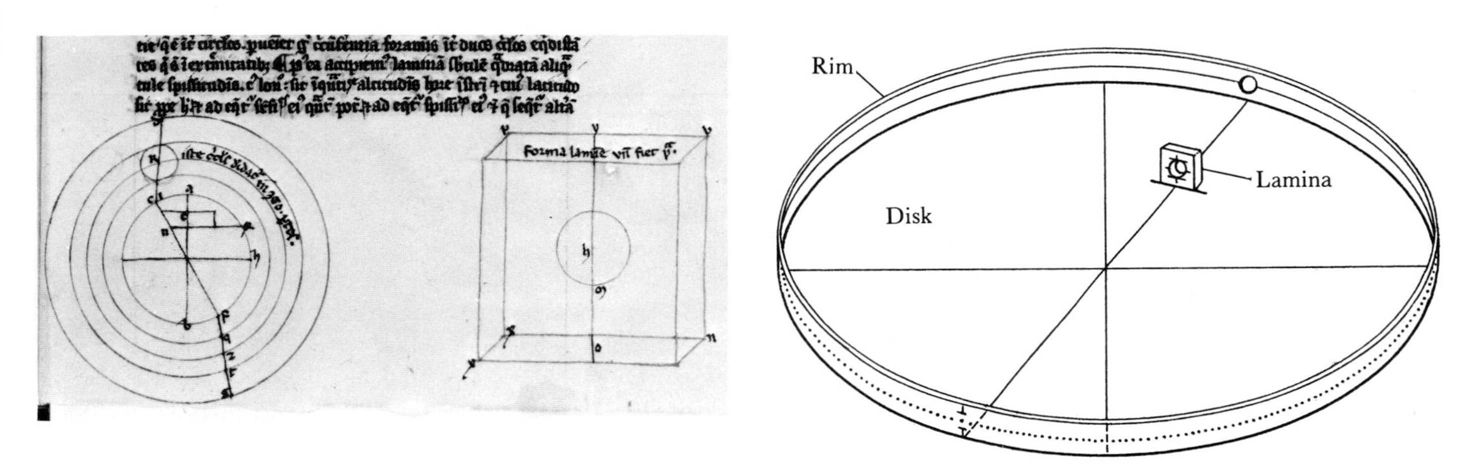

figures given above of the instrument in question come from a fourteenth-century manuscript of the Latin translation of the *Perspectiva* or *De aspectibus* of Alhazen. Since the figures themselves do not reveal the overall structure of the instrument adequately unless they are combined with a simultaneous reading of the accompanying text, a modern reconstruction of it has been reproduced at the right. Alhazen tells us that his instrument is constructed from a copper disk approximately one cubit in diameter, the perpendicular rim at its circumference being about two digits in height. Three parallel circles are inscribed on the inner surface of this rim. Only one of these circles is represented in the modern reconstruction, but all three are clearly shown in the manuscript figure at the left, for there the innermost circle depicts the disk itself, while the outer circles concentric to it all relate to the rim. Evidence of the care and exactness with which the disk, rim, and inscribed circles are to be fabricated can be seen from the instructions given to carry out their construction on a lathe *(tornatorium)*; by such a technique, for example, all can be rendered *perfectum per abrasionem.* Rim and disk, however, provide only the foundation of the instrument. One must next make a hole centered exactly halfway up the rim and an exactly matching hole in a lamina (clearly depicted in the manuscript figure at the right) whose center is the same height as that of the hole in the rim. As is clear from the modern reconstruction, the placement of the lamina relative to the hole in the rim will serve as a site through which an incident beam of light may pass. The manuscript detail of the lamina refers to this function by its inscription *forma lamine unde fiet visa,* although the placement of the lamina in the figure at the left of the whole instrument is somewhat confused. The next item to be noted in the construction of the instrument is its calibration. This consists of dividing the center circle on the rim into 360° (accordingly, the inscription in the manuscript figure reads: *iste circulus dividatur in 360 partes*), and, if possible, even into minutes *(et si possibile fuerit, in minuta)*. (It scarcely seems possible, however, that this latter division could have been carried out, since, with a disk approximately a cubit in diameter, this would require divisions in the neighborhood of 1/400 of an inch!) Finally, if we inscribe two diameters at right angles, plus a diameter corresponding to the line of sight, on the surface of the disk, the instrument is ready to be put to use. For example, in the case of refraction occurring as a ray of light passes from air to water, the disk is submerged vertically in such a way that its center falls on the refractive surface. Accurate positioning is facilitated by a round shaft attached to the center of the back of the disk, which fits into a support bar resting on the basin containing the water, an arrangement that allows smooth yet rigid rotation of the disk in order to effect different angles of incidence through the sighting holes. With the instrument so aligned, an incident ray of light passing through the rim hole and the lamina will not fall on the submerged rim at a point directly opposite the rim hole, but at some other point. The calibration of the rim allows one to locate that point exactly and hence determine the angle of refraction. Considering the exhaustiveness with which Alhazen describes the construction of this instrument and its use, it comes as some surprise that his *Perspectiva* does not include any measurements made by its means. Witelo's *Perspectiva,* by contrast, does. But when one compares the almost fifty measurements recorded by Witelo with those already given by Ptolemy, only one value is different. To return, however, to our manuscript diagrams, it is of some interest to note that the text of Alhazen contains no letters, although the figures accompanying the text do. This can most likely be explained by the importation of figures from Witelo whose corresponding text does contain the relevant lettering, if not precisely the same letters given here. For another instance of Witelo–Alhazen "figure migration" see Illustration 134.

243. Alchemy's Apparatus and Operations. We have already observed various alchemical paraphernalia as part of the adept's atelier (Illustration 182). But separate and clearer illustrations of the various vessels an alchemist was to employ were often produced for more strictly instructional purposes. Two examples of such a phenomenon are given here. The first consists of two marginal drawings intended to facilitate the comprehension of the text they accompany. The text is that of the *Summa perfectionis* of the as yet unidentified Geber, occurring in a fourteenth-century manuscript of Italian provenance. The figure at the left above shows an aludel (from the Arabic *al-uthāl*), a sublimation vessel, sitting in a furnace stoked through a door at its base with wood *(ligna)*. The hole *(foramen)* at the top of the aludel has some kind of wick *(licinium)* inserted through it, presumably for the purpose of testing the sublimation process going on within (if it collected powder, the process would not have reached completion). Several pages later in the same codex, the marginal figure reproduced at the right above represents another furnace. In this instance, an alembic (from the Arabic *al-anbīq*) sits atop a long tubelike vessel that descends into the furnace. The furnace itself is apparently vented by "windows" *(fenestra)*. The illustration below, right, serves as a kind of frontispiece to a sixteenth-century manuscript that once belonged to the French mathematician, astronomer, and cosmographer, Oronce Finé (1494–1555). Exhibiting the greater elaboration characteristic of so many sixteenth-century alchemical illustrations, the top of this page presents two different types of distillation furnaces: that on the left when the distillation is effected by placing the cucurbit capped by an alembic directly into the hot ashes; that on the right when the same arrangement is submerged in heated water. Of the two calcination spheres directly below the first of these distillation apparatuses, that on the right is the more interesting: it is identified as the philosopher's egg *(ovum philosophorum)* used for the calcination of spirits. Below that, a large furnace is depicted with lifted cover; it is said to be used for the calcination of copper and iron (represented at the base of the furnace by their alchemical symbols; see Illustration 79). On the right, a long *ampulla* carrying an inscription informing us that it is to be used for putrefaction and precipitation appears alongside a pot over a fire which, we are told, is to show us how a solution can be effected by submerging a vessel in boiling water. The lower right corner illustrates how the same is accomplished by steam. But perhaps the most interesting figure is that at bottom center. It is of a precipitation furnace constituted of various parts that are depicted separately above, dotted lines being added to show how these parts are to be put together. Thus, a long cylindrical glass precipitation vessel is to be inserted into an iron *precipitatorium*, which in turn is to be placed in the furnace. Presumably the long iron spatula just to the left of this *precipitatorium* is to be used for stirring the elements within the glass cylinder. The central figure of the furnace itself depicts all parts duly assembled (*furnus ad precipitandum completus cum suis instrumentis* is written at its base). For another picture from this second manuscript, see Illustration 257.

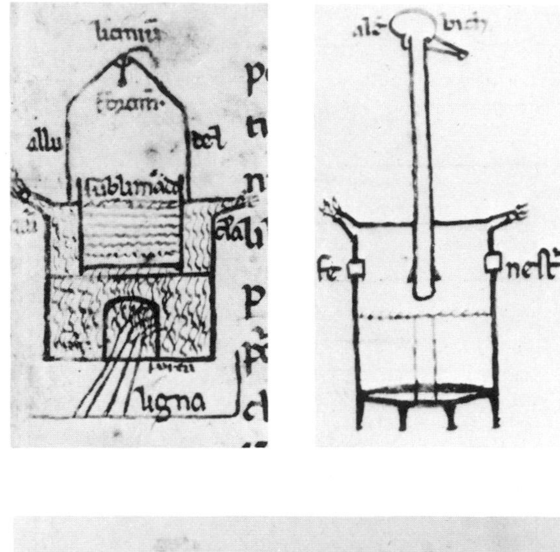

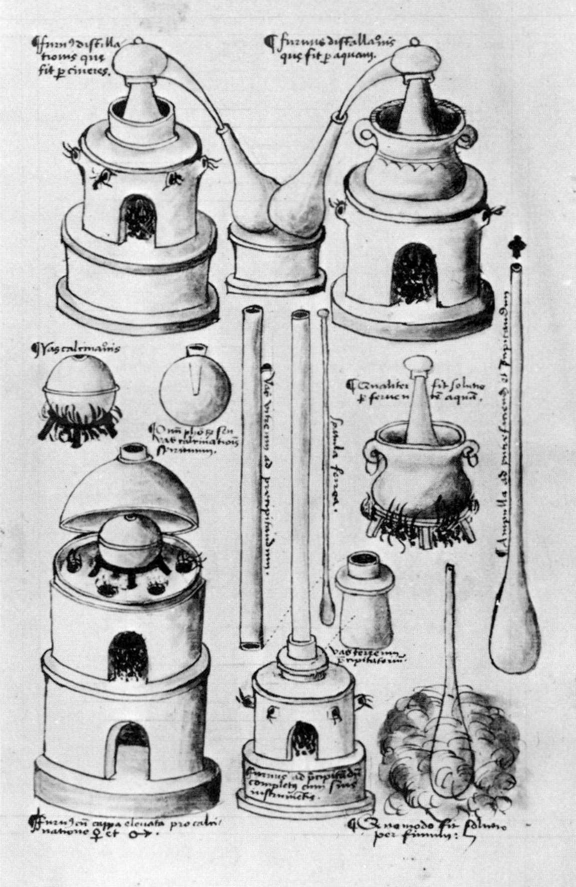

244. The Surgeon's Tools. One of the most important Arabic medical encyclopedias was that produced by the tenth-century scholar Abū l-Qāsim Khalaf ibn ᶜAbbās al-Zahrāwī, known in the Latin West as Alsaharavius or, most frequently, as Abulcasis. The extensive surgical part of his encyclopedic work was translated into Latin in the twelfth century by Gerard of Cremona and became a standard work on the subject for later medieval medicine. The illustration above on the right is taken from a fourteenth-century manuscript of that Latin translation, where it falls at the end of a chapter dealing with the treatment of fistulae and defluxions. Abulcasis concludes the text of this chapter by saying that he will depict various types of surgical instruments to be used in the kind of procedures he has just described as well as in other operations, adding that other instruments can be based on the types thus given. All of the sixteen instruments pictured in the two columns following his remarks are identified as saws *(sete)*, scrapers *(rasorii)*, or cutters *(incisorii)* of one sort or another. Most of them are described as being merely alternative or larger or smaller forms of the kind of instrument in question, but some receive further specification. For example, the scraper at the bottom of the left column is described as being stellate with the form of the head of a nail and as suitable for the scraping of large bones. That fourth from the top of the right column is noted as concave, that below it as fine, that third from the bottom as suited for cutting bone, and so on. The illustration below at the right comes from another fourteenth-century manuscript of the Latin version of Abulcasis' surgery. Now the subject is orthopedics and the instrument in question a more complex one: a traction machine for the reduction of vertebral dislocations. Twisting the poles at the right and left of the machine will apply the desired traction, while the other ropes horizontally crossing the "model body" will presumably anchor it properly during this procedure. (For another "orthopedic machine," see Illustration 168. Surgical instruments in use can be seen in Illustrations 167, 168, 265, and 269.)

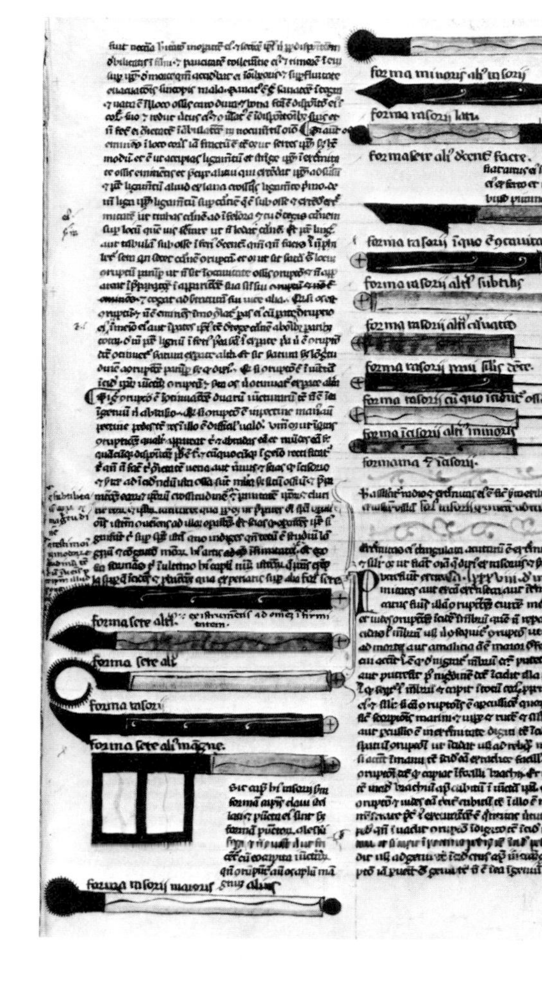

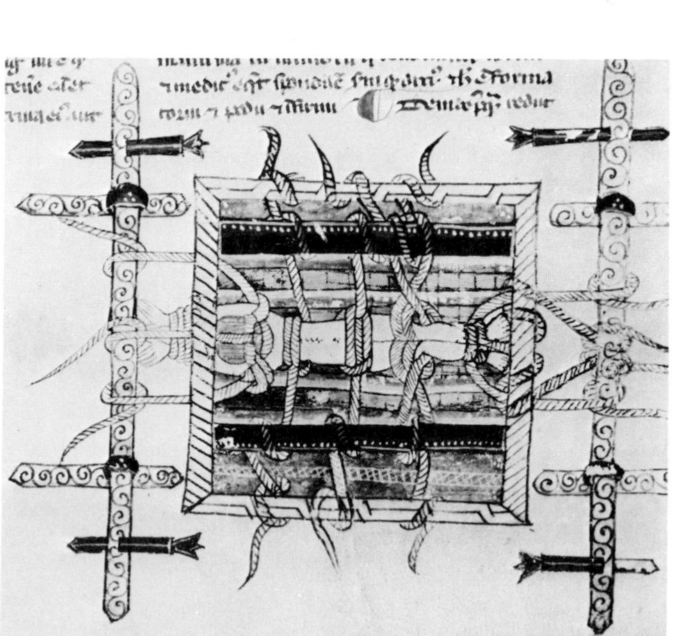

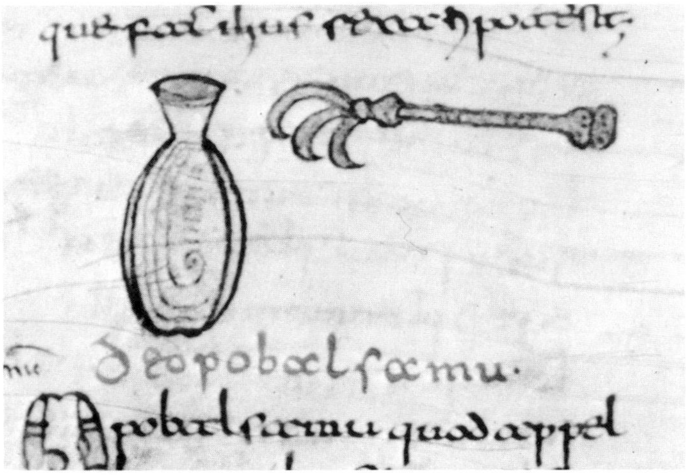

245. A Pharmacist's Means for Collection and Storage. As already noted in Illustrations 160 and 183, vials, flasks, and pots of numerous guises served the ancient and medieval pharmacist for the storage of his wares. The flask in the picture above at the left, however, is paired with an important supplement: a bark scratcher. Occurring in the same ninth- or tenth-century manuscript of the Latin translation of Dioscorides that provided the second picture of dracontea in Illustration 193, the subject is the balsam tree and the juice *(lacrimus)* it bears, called *opobalsamum* (or as we might say, the balm or even balsam of the balsam tree). This balm is quite costly, Dioscorides tells us, due to the exceedingly small yield given by each tree. It is best obtained on "dog days" by cutting into the tree with iron claws *(ungulis ferreis)*. Hence the depiction of the bark scratcher in the present manuscript. Our second picture below at the left is from a fifteenth-century Greek copy of Dioscorides (one of its frontispieces has been given above as Illustration 10). Here each flask is meant to contain the oil or balm (ἔλαιον; *oleum*) specified by its Greek and Latin labels (the latter being a direct transliteration of the former), the plant from which the oil in question is derived being duly depicted to one side. Thus, beginning at the left of the top row, we have oil of walnut (*caryinum,* to give only the transliteration), oil of quince *(melinum)*, and radish seed oil *(raphaninum)*. The second row presents us with oil of roses *(rosaceum)* and some kind of resin oil *(retininum)*, while the bottom row depicts oil of mastic *(schininum)*, mustard seed oil *(sinapinum)*, and oil of sesame *(sesaminum)*.

Part Six

REPRESENTATIONS OF THEORIES AND CONCEPTIONS

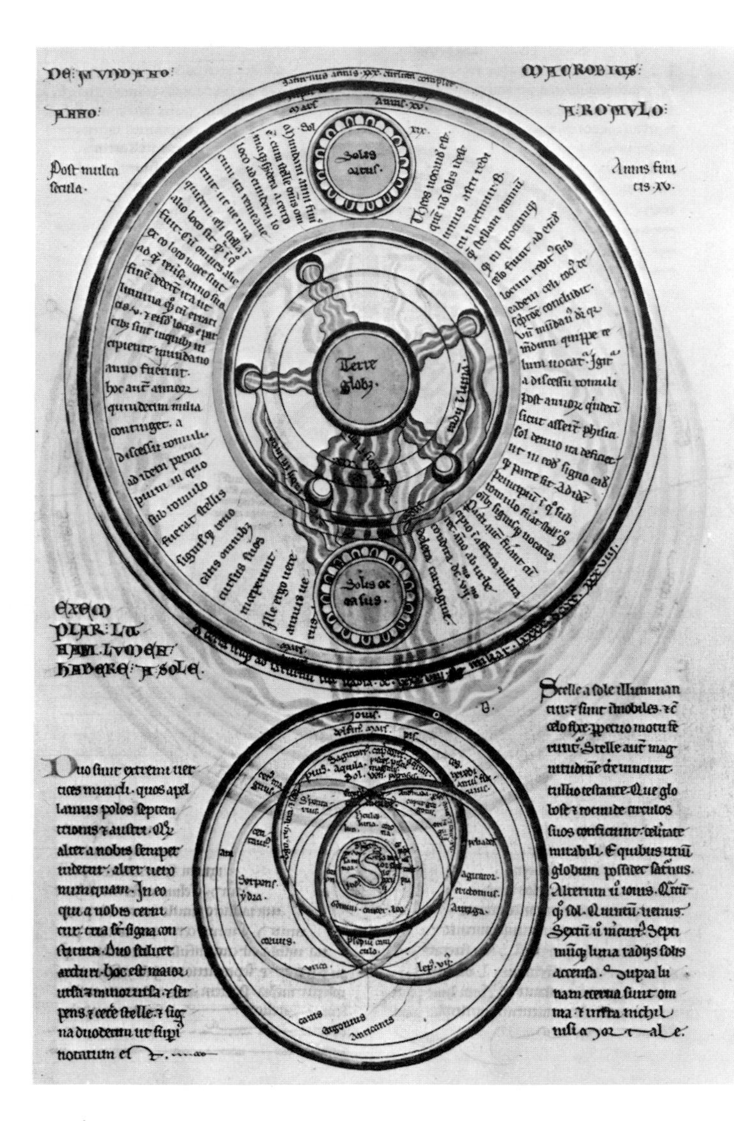

246. The Heavens and the World Year. The notion of the great or world year—a time when all heavenly bodies have returned to their initial positions—is alluded to as early as the fourth century B.C. by Plato (*Timaeus,* 39D), is discussed a number of times by Cicero, and occurs in Stoic sources. It appears in the upper circle of this illustration as the *annus mundanus* of Macrobius. Although the thirteenth-century manuscript from which this illustration is taken is not of Macrobius, but rather of the encyclopedic *Liber Floridus* of Lambert, Canon of St. Omer (twelfth century), the text it contains is Macrobean: "A world year will be complete," this text reads, "when all stars and constellations in the celestial sphere have gone from a definite place and returned to it, so that not a single star is out of the position it previously held at the beginning of the world-year, and when the sun and the moon and the five other planets are in the same positions and quarters that they held at the start of the world-year" (Macrobius, *Commentary on the Dream of Scipio,* William H. Stahl, trans. [1952]). The phenomenon described will occur every 15,000 years and, we are told, one such world-year cycle begins with the death of Romulus. In its outer rings, this upper circle also gives the orbital times of Saturn, Jupiter, and Mars as well as the distances between the earth and Saturn. The globe of the earth occupies the central position, the rising sun at the top, the setting sun at the bottom, while the riverlike lines from this lower sun to the moon depicted in different positions represent, as the legend at the left says, the fact that the moon derives its light from the sun. The lower circle confusingly pictures the fixed stars (given only by their names, save for the figure of Draco [*Serpens*] in dead center) and intertwined rings that represent the zodiac and, presumably, planetary orbits (which seem to have been appropriated from diagrams of another sort [see Illustration 249] and here are made to serve a quite mistaken role).

ALL OF THE SUBSEQUENT illustrations in this volume are visual representations of scientific ideas as distinct from scientific objects. As we will see, these "ideas" may be relatively simple scientific conceptions, more complex theories or doctrines, or even specific arguments set forth by specific texts. Concept and theory have also appeared in previous illustrations: among the various kinds of schemata that formed the subject of Chapters Five through Seven, as occasionally insinuated into the depiction of scientific objects in Chapters Eighteen and Nineteen, and especially in the mathematical illustrations of Chapters Twelve and Thirteen.

Some of the illustrations in the following chapters will also contain mathematical elements, most notably in the areas of natural philosophy and cosmology (largely covered in Chapter Twenty-two). Yet the diagraming of mathematics in these areas is quite different from what it was in Chapters Twelve and Thirteen. There, apart from the greater sophistication of the mathematics in question, the geometrical diagrams were an integral part of understanding the text. Consequently, such diagrams are found in almost all manuscript copies of these texts. Quite the opposite is the case in the mathematical diagrams of theory yet to come: their appearance in the relevant manuscripts is much more happenstance. The fact is that they are not necessary for the comprehension of the text to which they are attached; they are merely helpful. The doctrine or argument at hand could be, and often was, understood without their visual assistance, as the considerable number of diagramless copies of the works in question testify.

Any attempt to assemble a collection of illustrations that depicts the theories and concepts of ancient and medieval science is bound to leave much, indeed much of what is most important, unrepresented. For if one sets aside writings in exact science, manuscripts of the greater share by far of influential and significant works are completely without illustration. Codices of the works of Aristotle, for example, contain almost no pictorial material. And the same is true of Galen and of Avicenna's *Canon,* the translation of which was so important in the introduction of Galenic medicine into the Latin West.

The bearers of most of the visual representations of theory were not copies of central works such as these, but instead the encyclopedic and handbook works of Roman and early medieval heritage. The writings, for example, of Macrobius (fifth century) and Martianus Capella (ca. 365–440), of Isidore of Seville (seventh century) and the Venerable Bede (eighth century), of William of Conches (twelfth century), Hildegard of Bingen (twelfth century), Michael Scot (thirteenth century), and Thomas of Cantimpré (thirteenth century) are among the richest sources for this type of illustrative material. Nor is it merely the early manuscripts of such works that are plentifully covered with figures and diagrams, but also those that were written in the later Middle Ages, even though they then appeared side by side with the codices of the Aristotelian translations and other scholastic science that had, in terms of substance, all but totally replaced them.

This predominance of the encyclopedic in the domain of scientific illustration also helps to explain why so many, indeed almost all, of the illustrations in the final part of this volume are from medieval Latin works. There were far fewer works of the encyclopedic sort in Greek or Arabic. And where there were encyclopedic treatises in Arabic, for example, for every one bearing illustrations—like the *Wonders of Creation* of al-Qazwīnī (thirteenth century; see Illustrations 52, 126, 203, 208, and 253)—there were many more that were not illustrated, like the work of the Ikhwān al-Ṣafāʾ (tenth century; see Illustration 3).

As will be clear from the following illustrations of scientific theories and conceptions, they are to be found in medieval sources in all manners and sizes and in all degrees along the spectrum from crude simplicity to artistic elegance. One can, however, roughly classify them in terms of the scope of the conception or theory they were intended to illustrate and also in terms of the type of source from which the conception or theory derived.

Thus, to speak of scope, on the one hand the theory or concept being represented might be quite specific. In turn, such specific illustrations might be tied to an equally specific text (like the picturing of Aristotle's analogy of the powers of the soul and geometric figures in Illustration 254 or all the illustrations constituting Chapter Twenty-one) or connected with a rather specific argument that was found in a fair number of texts (such as the representations of the infinite spiral and "sphere wrapping" given in Illustration 255). On the other hand, the theory or concept being illustrated might be specific not in the sense of belonging to some particular text or argument, but rather because it related to a rather specific kind of event or fact in nature (like the diagrams of eclipses in Illustration 253 or of the form of the universe in Illustrations 276 and 277).

To look in the other direction, the theoretical or conceptual factor given visual representation could be quite general, not simply in the sense of its being explanatory of more than some single facet of nature or of not being tied to some specific text or kind of argument, but in a broader sense of its being involved in the explanation of a very great number of events and facts within nature. The prime example here, of course, is the theory of the elements (whether Aristotelian or Platonic does not matter), which will figure so largely in Chapter Twenty-five. Yet in another way one might cite as candidates for this kind of general illustration the macrocosm–microcosm concept implicit in Illustrations 266, 286, and 288, while final candidates could be found in the "world systems" pictured in the last two illustrations of this volume.

Alternatively, a different classification of theoretical–conceptual illustrations can be made by appealing to the kind of source behind the theory or concept being depicted. Here the division is somewhat more clear-cut. It is simply one of the source as a text versus the source as a tradition. Most of the illustrations in this final part are attached to some text as the source of whatever they are depicting. In fact, in some instances, even when an illustration appears in absolute isolation from any text, it is still a textual illustration because it has been detached from the specific text to which it belongs and without the knowledge of which it is scarcely understandable. (Some of the figures in Illustration 249 are clearly of this sort.) By contrast, an illustration standing alone might well have, indeed usually does have, a tradition as its source. To be sure, this does not mean that the tradition is not set forth in any number of texts. It merely means that the theory or concept being represented is traditional in the sense of being well enough known to be quite understandable without any reference to any specific text or even to any specific set of texts. The bloodletting and cautery men in Illustrations 264 and 265 or the urine diagrams of Illustration 260 belong to this category. So do, for example, the representation of music theory in Illustration 258, of man and the four elements in Illustration 288, and of the opening of the Book of Genesis in Illustration 273.

Finally, note should be taken of the fact that, for lack of space, maps and (for the most part) illustrations having to do with astrology have been excluded from Part VI. Note should also be made of the fact that the division of chapters in this part is not in any way meant to reflect a division in the kinds of theories or conceptions being represented.

CHAPTER TWENTY-ONE

Visualizing Specific Texts: Problems and Progeny

A FITTING INTRODUCTION TO the representation of scientific theories and conceptions that will form the subject of the concluding chapters of this volume can be provided by showing how many different ways there were to depict a given theory or concept as expressed in a particular text. Admittedly, the depiction of scientific objects also differed from one manuscript to another, but not with nearly the frequency or latitude of difference one can find when something theoretical or conceptual is being pictured.

Perhaps one of the reasons for this is that the depiction of an object usually bears, or at least is intended to bear, some resemblance to the relatively constant appearance of the object as seen. But another reason can be found in the histories of the manuscripts containing the illustrations in question. For although, like representations of objects, representations of theories or conceptions were often copied from some antecedent manuscript, this happened less frequently than in the case of the depiction of objects. In other words, illustrations of the theoretical or the conceptual were more frequently drawn or constructed from a reading of the text that presented the theory or concept. That such a reading could give rise to alternative illustrations will be especially evident from the first two illustrations of this chapter. Indeed, all of the illustrations in this chapter are the result of, or the offspring of, differing interpretations of a particular text.

Thus, Illustration 247 reveals variant depictions of the Platonic doctrine of the four elements as set forth in a single paragraph of the *De natura rerum* of Isidore of Seville. Similarly, Illustration 248 shows just how different could be the illustrations of a single argument dealing with rainfall. The argument comes from Macrobius' *Commentary on the Dream of Scipio.* Illustrations 249 and 250 are slightly different in the differences they reveal. The first provides an example of how variant diagrams were drawn in order to illustrate a given text, both when the text itself was present and when it was not. (The text in question was a fragment of the *Natural History* of Pliny.) Illustration 250, by contrast, shows how at least one of these variant diagrams was altered and then used to facilitate the understanding of quite different texts.

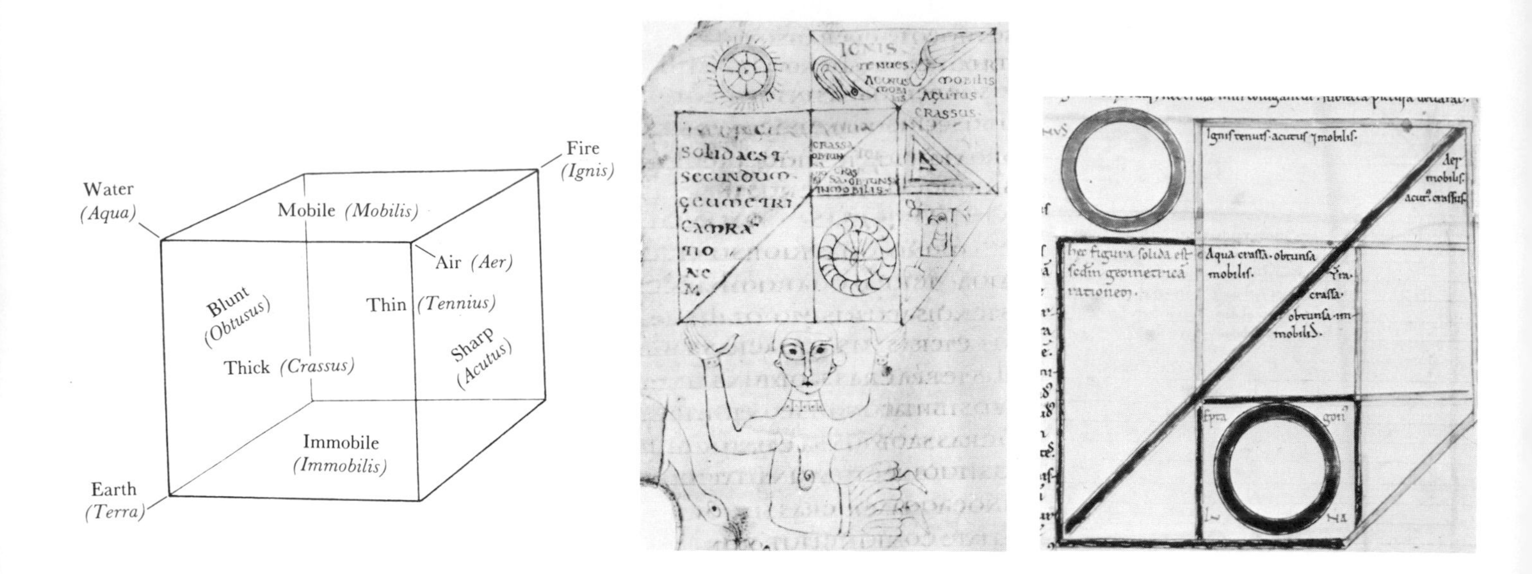

247. Cybus Elementorum. It has already been noted that the circular schemata appearing in manuscripts of Isidore of Seville's (seventh-century) *De natura rerum* were so central to this text that it often received the alternative title *Liber rotarum,* and some of these "wheel diagrams" (Illustrations 46–48, 279–80, 286) have been reproduced. But this work of Isidore's also contains one noncircular figure: it was—or was supposed to be—a cube, frequently labeled *cybus* in the text. The particular passage it accompanied is the following (*De natura rerum,* XI, 1): "There are four elements *(partes)* in the world: fire, air, water, and earth. Their nature is as follows: fire is thin *(tenuis)*, sharp *(acutus)*, and mobile *(mobilis)*; air is mobile, sharp, and thick *(crassus)*; water is thick, blunt *(obtusa)*, and mobile; earth is thick, blunt, and immobile *(immobilis)*. They mix with one another in the following way: earth as thick, blunt, and immobile is connected with the thickness and bluntness of water, while water joins with air in thickness and mobility. Further, air is tied to fire by having the sharp and the mobile in common. Earth and fire, however, are separated from one another, but are joined by water and air as two means. In order that these things be comprehended without undue confusion, I have expressed them in the following figure *(subiecta pictura)*. This figure is solid according to geometric ratio *(haec figura solida est secumdum geometricam rationem)*. Fire: thin, sharp, mobile; air, mobile, sharp, thick; earth, thick, blunt, immobile; water, thick, blunt, mobile." The theory presented in this passage is the Platonic doctrine of the characterization of the four elements in terms of triplets of shared primary qualities. (For more on this Platonic doctrine, including the relevance of the "geometric ratio" mentioned above, see Illustration 282.) One of the reasons the figure expressly called for by Isidore in his text was often termed a *cybus* of the elements is because it is specified as a *figura solida.* Ignoring the qualification "according to geometric ratio," the figure called for, but successfully constructed in no medieval codex of the *De natura rerum,* is given in the modern drawing on the left. Each face of the cube represents one of the six qualities—thin, thick, sharp, blunt, mobile, immobile—the contrary qualities among these six appearing on opposite faces. Given this arrangement, the four elements are represented by those four corners of the cube that are junctures of the various faces representing the relevant triplet of qualities characterizing each element. The extraordinary problems medieval scribes or illustrators had in reproducing this required cube can be seen from the following manuscript examples. First, in the middle, one of the earliest (eighth-century) codices (also seen in Illustrations 280, 286) of Isidore's *De natura rerum* gives what became relatively standard for most manuscripts of the work. The whole figure supported by a "folded" human, the somewhat larger inscription at the left in the "cube" asserts that the figure is "solid according to geometric ratio," while the remaining writing (some of which is scarcely visible) merely gives the required quality triplets for each element, in no

(Continued)

way indicating how the faces and corners of the cube are to reproduce diagramatically the shared triplets. More clearly drawn, the next figure, though considerably later in time, fares little better. It comes from a twelfth-century English (Peterborough) manuscript that contains the chapter of Isidore in the midst of other computist and cosmographic texts (see Illustration 57). The two circles decorating the eighth-century figure are now specified as the sun (*ennagonus sol,* above left) and the moon (*eptagonus luna,* lower right), but all other inscriptions are the same. A tenth-century codex (also seen in Illustrations 48 and 279) of the *De natura rerum* contains the even less cubelike figure given at the upper left on page 281. Once again, no more information is provided and there is no more success in explaining the doctrine. The more stylized attempt to depict the *cybus* given at the bottom left from an eleventh-century manuscript similarly fails to represent the idea behind Isidore's *subiecta pictura.* Material is furnished beyond the usual inscriptions, but this has nothing to do with the text or doctrine at hand. For example, the upper left is labeled the east, the lower right the west, while the circular area at the lower left is designated as the northern point *(cacumen),* connected by an "axis" running diagonally to the *cacumen australis* at the upper right. Finally, the figure reproduced at the extreme right from a late-eleventh-century English manuscript of Isidore lacks all resemblance to any cube or *figura solida.* Each element is tabulated together with its appropriate quality triplet, the whole "list" then simply enclosed in a circle.

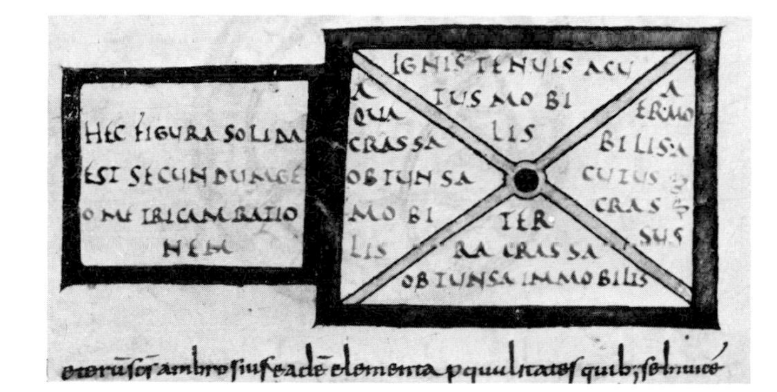

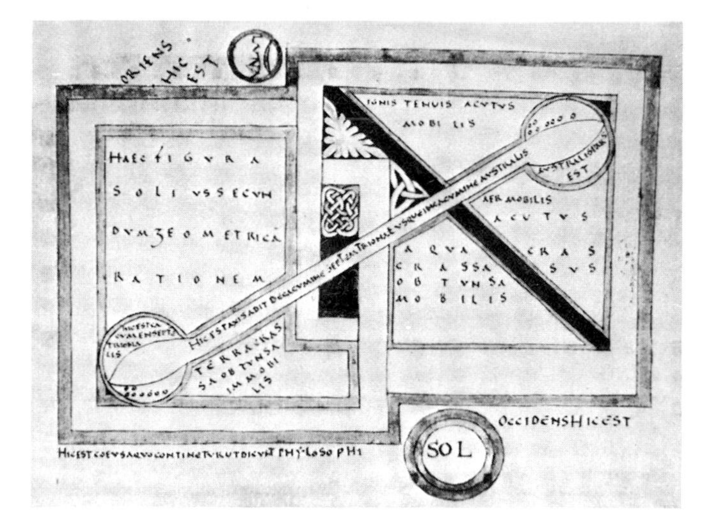

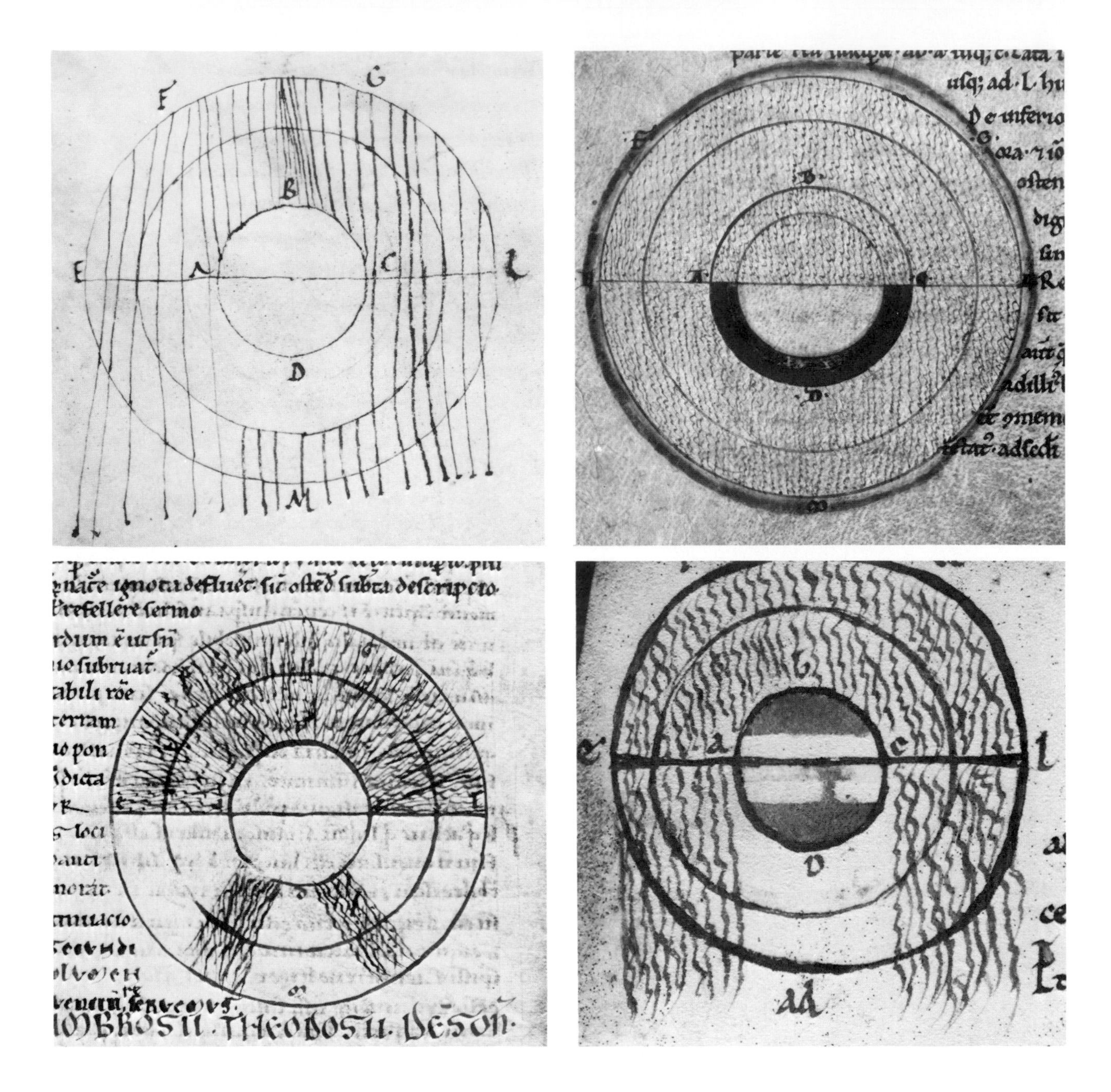

248. Diagraming Rainfall. The work is Macrobius' *Commentary on the Dream of Scipio,* the last chapter of Book I. The text to be illustrated is the following: "We might cite countless arguments as proof [of the fact that all weights tend toward the center of the universe], but particularly convincing is that of the rains that fall upon the earth from every region of the atmosphere. Not only do they fall upon this portion of the earth that we inhabit, but on the slopes that give the earth its sphericity; and, what is more, there is the same sort of rainfall in the region that we consider the underside. If air is condensed by the chill of the earth and forms a cloud which sends down a shower, and if, moreover, air surrounds the whole of the earth, then, assuredly, rain falls from all regions of the air except in those places parched by constant heat. From all quarters it drops upon the earth, which is the only resting place for objects possessing weight. Anyone who rejects this theory will be forced to admit that any precipitation of snow, rain, or hail falling outside the region inhabited by us would continue on down from the air into the celestial sphere. The sky is of course equidistant from the earth in every direction and is as far beyond the sloping regions and that portion which we regard the underside as it is beyond us. If all weights were not drawn to the earth, therefore, the rain that falls beyond the sides of the earth would not fall upon the earth but upon the celestial sphere, an assumption too ridiculous

(Continued)

to consider. Let a circle represent the earth; upon it inscribe the letters *ABCD*. About this draw another circle with the letters *EFGLM* inscribed, representing the belt of the atmosphere. Divide both circles by drawing a line from *E* to *L*. The upper section will be the one we inhabit, the lower the one beneath us. If it were not true that all weights are drawn towards the earth, then the earth would receive a very small portion of the rainfall, that which falls from *A* to *C*. The atmosphere between *F* and *E* and *G* and *L* would send its moisture into the air and sky. Furthermore, the rain from the lower half of the celestial sphere would have to continue on into the outer regions, unknown to our world, as we see from the diagram. To refute this notion would not benefit a sober treatise; it is so absurd that it collapses without discussion" (William H. Stahl, trans. [1952]). Part of the medieval history of this Macrobian text's illustration begins with the diagram at the top left of the facing page taken from an eleventh-century southern German manuscript. On balance, it seems to give a more adequate illustration of that required by Macrobius' text than any of the others reproduced here. The second figure, top right, occurs in a twelfth-century codex, but surely misses what Macrobius had in mind, as does the rather remarkable conception presented in the next figure (lower left) from another twelfth-century manuscript. The following two figures (lower right, and this page, upper left) are both found in the same thirteenth-century copy of Macrobius (also to be sampled in Illustration 250). Occurring on successive pages in this manuscript, they are clearly the result of the scribe's indecisiveness over just how the required "rainfall" figure should be drawn. Moving to the fourteenth century, the figure on the right below can be viewed as a relative—albeit in simplified form—of the first illustration. Finally, the figure at the bottom, found in a late-fifteenth-century manuscript, reveals that the later the codex need not in any way mean the more adequate the illustration.

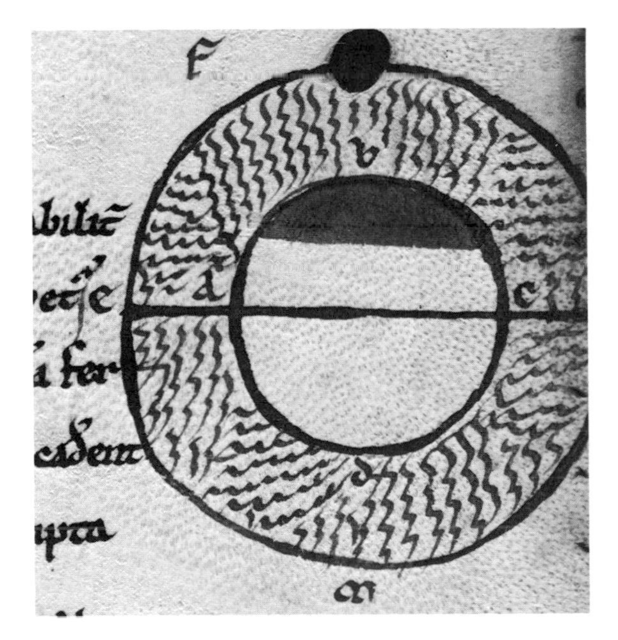

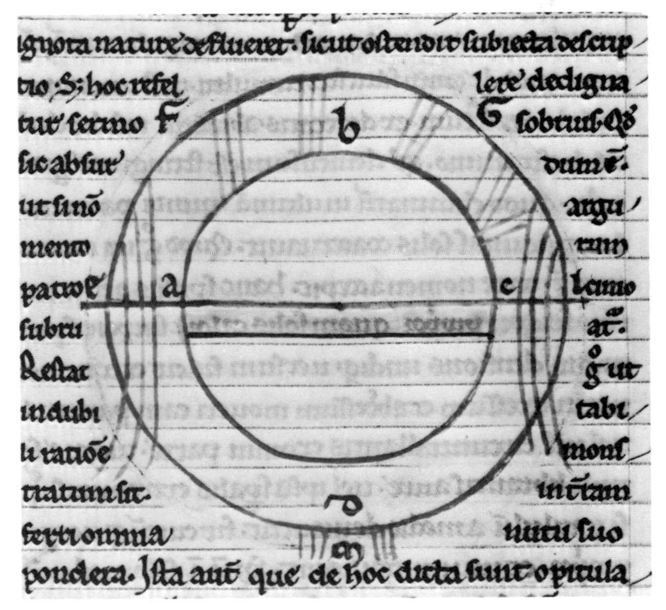

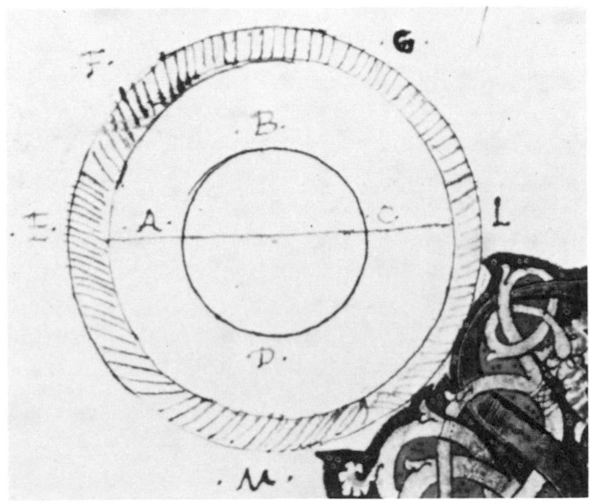

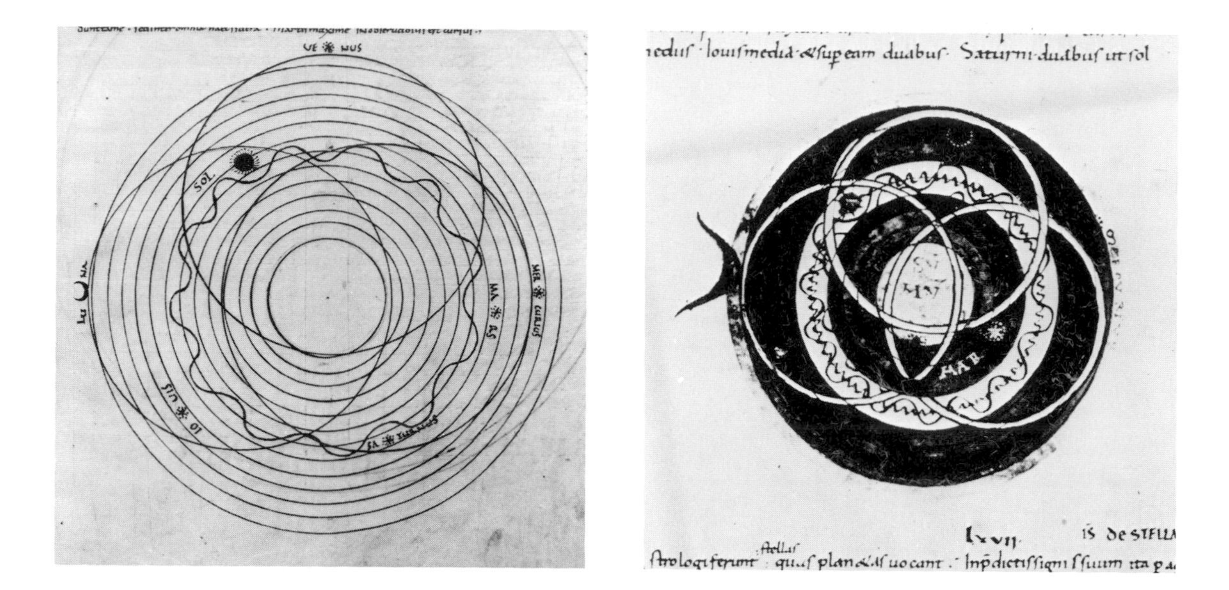

249. The Motion of the Planets in Latitude I. Pliny (ca. A.D. 23–79) was clearly one of the ultimate primary sources for a good deal of scientific writing in the earlier Latin Middle Ages, his *Natural History* often appearing in pieces as well as in codices containing the whole work. An especially notable example of the "fragmentary" Pliny was a collection of excerpts from Book II which, taken together, formed a mini-encyclopedia of astronomical and cosmographic knowledge. As a result, the collection was copied repeatedly from the eighth century on, a sign that it provided just the kind of information wanted on its subject by the early medieval scholar. One passage among these collected excerpts read as follows: "On the path of the [planets] through the zodiac circle. . . . They move toward the north and go toward the south. The latitude of the zodiac is oblique and they are moved through it. Nor are any lands habitable other than those that lie under it; those near the poles are wastelands. The planet Venus exceeds the [latitude of the zodiac] in both directions. The Moon wanders throughout its whole latitude, but does not go beyond it at all. The planet Mercury is the most irregular of all. Thus, while the latitude consists of twelve parts, it passes through no more than eight. Nor does it do so equally, but two in the middle, four above, and two below. Next, the Sun is carried through the middle, between two parts with the unequal, snakelike course of a dragon. The planet Mars [covers] four [parts] in the middle, while Jupiter [runs through] the middle and two above, Saturn, like the Sun, just two." To judge from the extant manuscripts, it seems that the earliest figures that diagram the information given in this text were circular ones. A typical example is given at the top left. Taken from a ninth-century manuscript, it accurately represents the division specified in the text of the latitude of the zodiac into twelve equal parts, a division that was conventional in medieval sources. Though also present in Geminus (fl. ca. 70 B.C.), for the Middle Ages this convention derived from Martianus Capella (ca. 365–440), who felt that the division of the longitude of the zodiac into twelve equal signs of thirty degrees each should entail the partition of its latitude into twelve equal parts as well (*De nuptiis*, 834). Note further that in this first diagram, the motions in latitude of some of the planets (Venus, the moon, and Mercury) are represented by circular orbits (a feature that can also be observed in Illustration 246). The motions of both the sun and Saturn are presumably given by the scalloped orbit passing through the center of the latitude. Jupiter *(Iovis)* and Mars seem to have no orbital line assigned to them. This particular diagram has erred in having both the moon and Venus exceed the latitude in their motion, since the text being visually explained clearly reserves this phenomenon for Venus (which, as a matter of fact, was true for the planets then known). The second circular figure, at the top right, comes from a manuscript written in Ripoll in A.D. 1056 (from which another figure appears in Illustration 283).

(Continued)

Its main features are the same as those of the ninth-century manuscript, though the whole is much obscured by the overlay of colors. There are now two snakelike lines, possibly one each for the sun and Saturn, while Mars has assumed a more interior position than in the first figure. Again, the required differentiation of the moon and Venus is missing and, to make matters worse, Mercury has now become "latitudinally equivalent" with them. As a whole, this second circular figure has sacrificed adequate representation of the text for presumed artistic gain. The next diagram, in the middle below, is a late-eleventh- or early-twelfth-century example of the simplest rectangular figure used to explain our text. Venus is now, correctly, the only planet broaching the boundaries of the zodiac belt. Mars—again correctly—occupies only "four in the middle" and Mercury's and Jupiter's wanderings also seem to reflect the text accurately. Saturn has no "orbital line" assigned to it unless the "course of the dragon" belonging to the sun is, once again, to do double duty. The longitudinal division in this figure into thirty-one parts is probably a miscount for thirty parts (of 10° each) for the length of the zodiac. The last diagram at the bottom comes from the twelfth-century astronomical–computist manuscript so frequently plundered before (see Illustrations 31, 51, 53, 69, 279, 285, and 290). The longitudinal division is now the familiar one of twelve equal signs, each with their names duly inscribed. The label in the frame at the lower left say's that, longitudinally, the zodiac consists of 365 parts (referring, of course, to "days" and not degrees), while the similar inscription in the frame at the right specifies the latitudinal division into twelve. The orbital lines here give a less adequate representation of the text than the previous, simpler, rectangular figure. A most substantial added feature in the present diagram derives from the importation of other Plinian material (*Natural History* II, 83–84). Appearing at the base of the whole figure, the first part of this addition appears at the left and tells us that the moon is 125,000 (Pliny has 126,000) stades from the earth, while the distance between the moon and the sun is twice that and from the sun to the zodiac three times as great. More elaborate is the stipulation of the so-called music of the spheres at the right, as the lowermost inscription "ARMONIA" indicates. The tones and semitones obtaining among the earth, each of the planets, and the zodiac are all specified in the appropriate intervals below. Taken together, they constitute an octave, as the (upside down) inscription "diapason" in the frame at the right appropriately reveals.

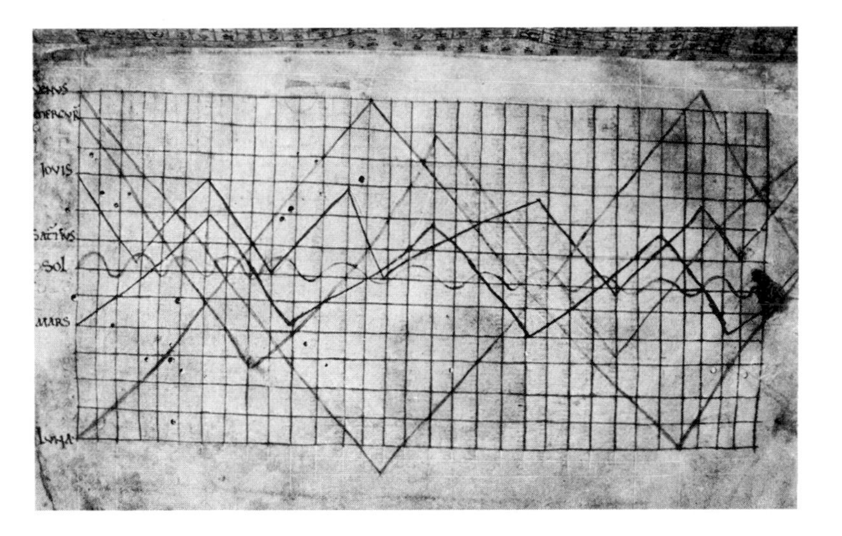

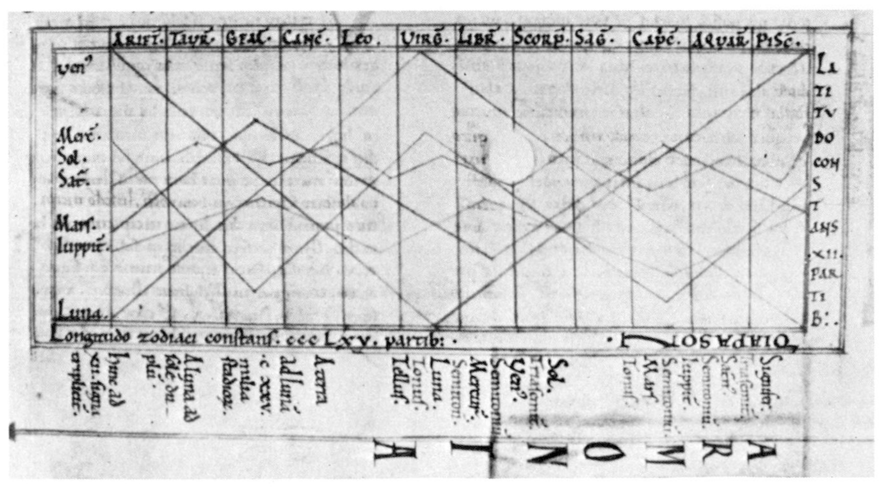

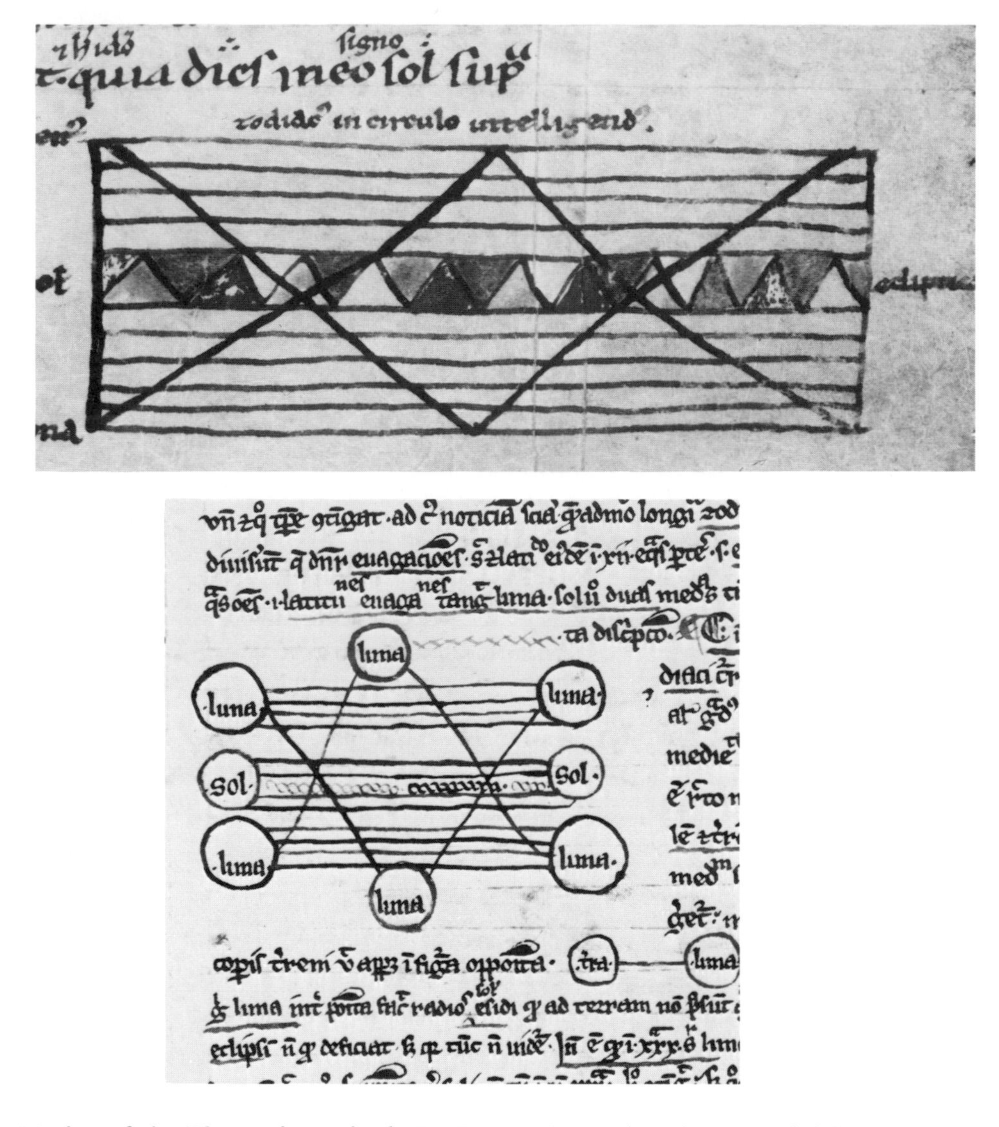

250. The Motion of the Planets in Latitude II. Far simpler, and explanatory of different matters, the three present "latitude figures" are in differing ways nevertheless related to those constituting the previous illustration. The first, at the top, comes from the same thirteenth-century manuscript of Macrobius seen in Illustration 248. The subject under discussion at the juncture in the text at which this figure occurs is the lunar sidereal period of, Macrobius says, approximately 28 days and the double crossing of the zodiac's latitude the moon makes during that period. The rectangle of the present diagram represents the zodiac, "to be understood as circular," the legend above it reads. The path of the moon begins at the lower left corner of the diagram, crosses the ecliptic line (at a point we would call a lunar node), continues to the other latitudinal extremity of the zodiac, where it "reverses course," crosses the ecliptic once again, and proceeds to the point (in longitude and latitude) in the zodiac whence it began, here at the lower right corner of the rectangle. This much clearly represents what is stated in Macrobius' text. Further, the labeling of the central parts of the figure as the path of the sun on the ecliptic (which Macrobius refers to as the *medietas latitudinis*) can be seen as related to the substance of the text. But the depiction of the frequently oscillating course of the sun through the ecliptic suggests some connection with the Plinian diagrams seen in the previous illustration. This connection is rendered

(Continued)

certain when one notes that the other path crossing the zodiac from the upper left and back to the upper right is expressly labeled as the path of Venus. For, although a planet of crucial significance in the Plinian figures, Venus is nowhere mentioned in the Macrobian text at hand. One is left with the conclusion, therefore, that part of some Plinian diagram was seen to be useful for explaining the present text of Macrobius and was thus imported to serve new duty there. Perhaps the unnecessary Venus was retained more for pictorial symmetry than by mere oversight. The other two figures reproduced here come from codices of the *Dragmaticon* of William of Conches (twelfth century). They occur with William's introductory words on solar eclipses, a subject that naturally involves a consideration of the moon's motion in latitude and the resulting lunar nodes. In order to know about solar eclipses, William says, "you should realize that, just as the longitude of the zodiac is divided into twelve equal parts called signs, so its latitude is divided into twelve equal parts called wanderings *(evagationes)*. The moon touches all twelve of these parts of the latitude; the sun, on the other hand, traverses [only] the middle of the latitude, all of which is made clear by the following illustration *(quod demonstrat subiecta descriptio)*." Whereupon the figures reproduced here appear. The first (middle of the facing page) comes from a fourteenth-century copy of William's work (also to be seen in Illustrations 276, 287). Quite similar in form to the Macrobius figure just examined, Venus has been replaced here by a second, alternative, path for the moon. The same is true in the second figure, below, taken from a highly stylized and decorative thirteenth-century manuscript (already seen in Illustrations 132 and 285). Yet in spite of its decoration, in addition to the double course of the moon and the central path of the sun given in the previous figure, it specifies the lunar nodes here by the inscription *"luna"* in the center of the figure as well as at top and bottom.

CHAPTER TWENTY-TWO

Natural Philosophy and Cosmology

INASMUCH AS THIS chapter will be a kind of catchall, containing material that falls outside of what can properly be considered as *philosophia naturalis,* it may serve as something of a second proemium, quite different from that provided by the preceding chapter, to this concluding part on the representation of theories.

As was noted in the introduction to Part VI, the central works in the Middle Ages on natural philosophy were for the most part bereft of illustrative material. At times, a specific doctrine required at least a modicum of diagraming for its comprehension. Such, for example, was the case with the doctrine of the intension and remission of forms, the figures of which were seen in Chapter Thirteen. It is also true that especially elegant copies of natural philosophical works were often furnished with illuminated initials (see Illustrations 44, 158, 178, and 210 for examples of this relating to other disciplines) or, though more a Renaissance than a medieval phenomenon, with elaborate frontispieces depicting their concern (Illustrations 159 and 162). Yet most of the illustrations or diagrams one finds in manuscripts of medieval works on natural philosophy do not relate to the subject of the work as a whole or even to some general doctrine expounded by the work in question. They relate instead to specific arguments set forth by some segment—perhaps of only a few lines—of text. This is especially true when the argument involves mathematical considerations of one sort or another. Illustrations 254 to 256 provide examples of that.

The case is somewhat different in the "handbooks" dealing with natural philosophical and cosmological matters. There one does find illustrative material attached to more general theories or doctrines. The doctrine may be—in fact, most often was—explained by an accompanying text, but, unlike the figures that on occasion appear, for example, in copies of commentaries on Aristotle, the illustrations usually were not tied to specific arguments. The figures in Illustrations 251 and 253 setting forth theories of the tides and eclipses are intended to provide examples of this kind of "visual aid" (Illustration 252 furnishing, for comparative purposes, material relevant to some of the mathematical factors involved in eclipses).

A final note should be made of the fact that the visual representation of theories within natural philosophy, cosmology, and related areas has already received appreciable attention in the dichotomy diagrams, wheels, squares of opposition, and mnemonic figures and symbols that were the subject of Chapters Five through Eight.

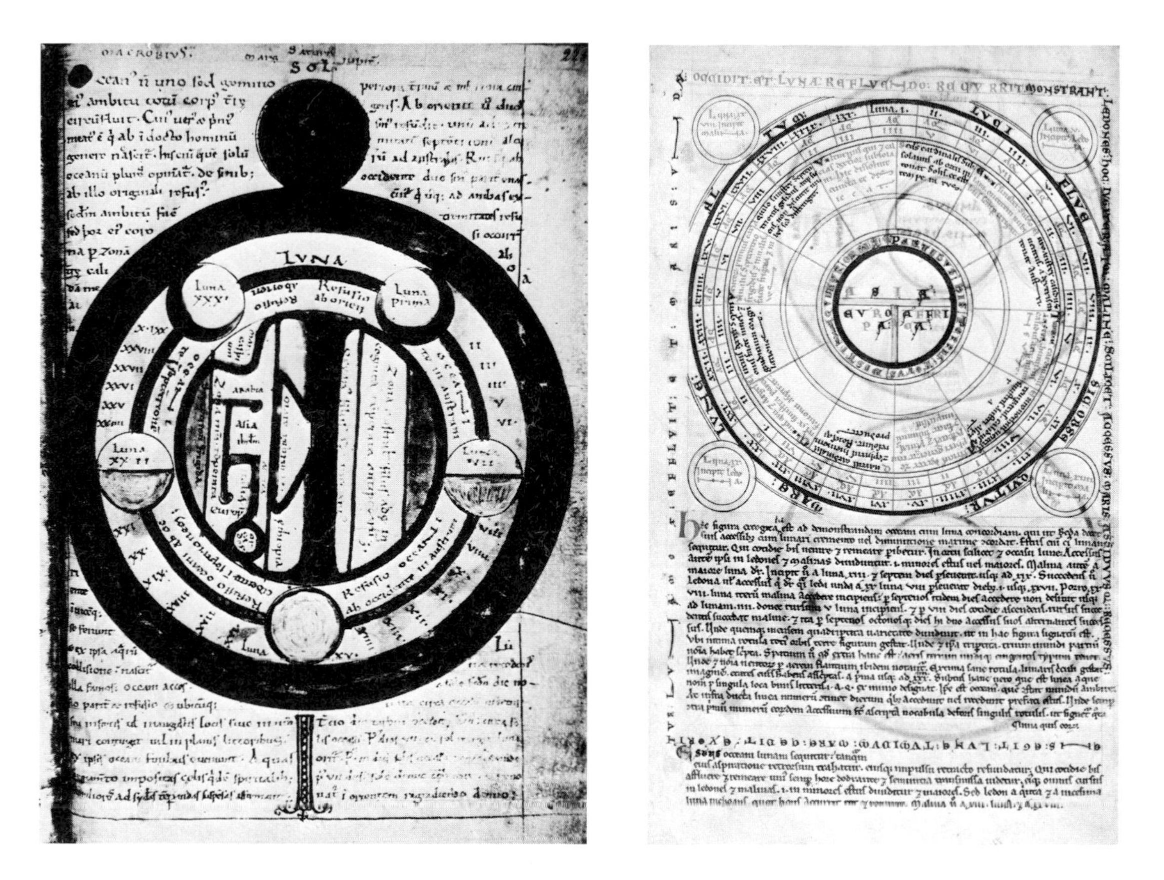

251. Tide Wheels. We observed wheels or *rotae* being put to use in the visual representation of all manner of notions in Chapter Six. These were appealed to particularly in setting forth doctrines having to do with phenomena related to the globe of the earth or the vault of the heavens, where spheres, circular orbits, and cycles made such an appeal extremely natural. The three tidal *rotae* pictured here are good examples of this kind of use. The first at the left, above, is taken from Lambert of St. Omer's own manuscript (written in 1120) of his encyclopedic *Liber floridus* (see also Illustration 246). The theory of tides this *rota* is intended to represent is that of Macrobius (fifth century), as is clearly indicated by the inscription of his name at the upper left of the page and appropriately confirmed by the appearance of the relevant segments of Macrobius' text *(Comm. Scip.*, II, 9, 1–3) in all the "quadrants" bordering the *rota*, save that at the lower right. The operative notion in Macrobius' tidal theory was his belief (deriving ultimately, it seems, from Crates of Mallus [second century B.C.]) that the main course of the ocean was equatorial, but that its flow was in turn divided, in both the East and the West, into two streams that headed north and south. These streams of the Eastern and Western "meridional oceans" meet at the poles, where "they rush together with great violence and impetus and buffet each other" and, by their collision, give rise to that "remarkable ebb and flow of the ocean" *(illa famosa Oceani accessio pariter et recessio)* constituting the tides. These details are given by the text surrounding the present *rota,* the *rota* itself then reminding the reader of this Macrobian theory by means of the four inscriptions on its innermost circular band. They draw attention to the "back-flow of the ocean" *(refusio oceani)* toward the north and the south from both the east (in the top half of the *rota*) and the west (in the bottom half). The outer ring labeled 'luna' at first glance makes it appear as if the moon were somehow related to the tidal theory in question; but this outer ring is merely an added feature giving the days and phases of the moon (compare Illustration 234), a matching explanatory text being appropriately provided at the lower right. The center of the *rota* is occupied by a map of the world. To call attention to only some of its features, one can begin by noting that, in perfect accord with the direction of the *refusiones oceani* given for the Macrobian tidal theory, east is at the top, north at the left. Thus, the dark vertical area in the center of the map is Macrobius' equatorial ocean, while the two darker segments at the extreme left and right are the frigid and uninhabitable regions of the earth at the poles. The best-known part of the ancient and medieval world occupies the left half of this circular map. Here we can easily see the inscriptions for *arabia, asia* (directly below which is *Iheru-*

(Continued)

(Continued)

salem), *europa,* and (vertically, just to the left of bottom center) *ethiopia.* The small circle carrying the label 'orcades' represents the present-day Orkney Islands. In contrast to what one finds in Macrobius, the second *rota*, reproduced at the right, top of page 289, represents a lunar theory of the tides. The source is Bede (eighth century), though the present figure appears in a collective manuscript of the late twelfth century (also seen in Illustrations 50, 131, and 280), where it accompanies a text carrying incorporated glosses and other material added to the Bedean original (from his *De natura rerum,* ch. 39). The tides *(aestus oceani)* follow the moon, Bede felt, being alternately drawn to and repelled by it, the flux and reflux occurring twice each day. Moreover, there are lesser or weaker and greater or stronger tides, called, respectively, *laedones* and *malinae* (words of uncertain etymology, though glossators have tried to derive *laedon* from *laesa* ["injured"] and *malina* from *maior luna*). The text below the *rota* tells us that *malinae* begin on the thirteenth day of the moon and last through seven days, whereupon *laedones* commence on *luna* XX and last eight days, *malinae* begin again on *luna* XXVIII, and *laedones* again on *luna* V (the commencement of *malinae* therefore being roughly at full and new moon). All of this information is also contained in the *rota.* The four small circles outside the *rota* announce the inceptions of *malinae* and *laedones* and are further positioned at or near the pertinent days of the moon inscribed in the outermost ring of the *rota.* The next ring repeats the word 'aqua' throughout its circumference (presumably in reference to the tides being explained), while the ring directly within it carries Roman numerals setting forth the relevant durations of *laedones* and *malinae.* The remaining information presented in the *rota* is nontidal. Thus, the wide circular inner band contains information about the twelve winds (compare Illustration 280), and the very center gives an *O-T* map of the world, Asia and the East, as traditional, at the top. Yet another tidal theory is represented by the last *rota,* below. It is taken from a fourteenth-century manuscript (seen also in Illustrations 26, 80, 101, 231, and 277) of Michael Scot's *Liber introductorius.* The accompanying text tells us that four angles *(anguli)* of the earth are known through four angles of the heavens, which are in turn determined by four of the zodiacal signs regarded as fixed; namely, Taurus, Leo, Scorpio, and Aquarius. As the inscribed division of the small circle at the center of the *rota* indicates, the "angles" of Taurus and Leo are above the earth and those of Scorpio and Aquarius are below it. This arrangement in hand, when the moon enters an angle—at Leo, for example—the tides begin to increase *(aqua maris incipit crescere)*, reaching high tide when the moon is in the middle of the arc between Leo and Taurus, whereupon they recede until the moon reaches Taurus, and so on with respect to the remaining angles. This means that the tides increase twice during the day and twice during the night, a fact depicted on the *rota* by the positioning of two of the fixed signs determining the relevant angles in the upper hemisphere and two in the lower. The arcs of the moon's path related to the increase and decrease of the tides are properly labeled *accessus aque* and *recessus aque* just inside the wheel's circumference (though the labels in the lower hemisphere are incorrect). The function of the *rota* in explaining this tidal theory is clearly expressed by Michael Scot at the conclusion of the text. All that he has just said, he claims, will be made known to both eye and intellect *in presenti rota.*

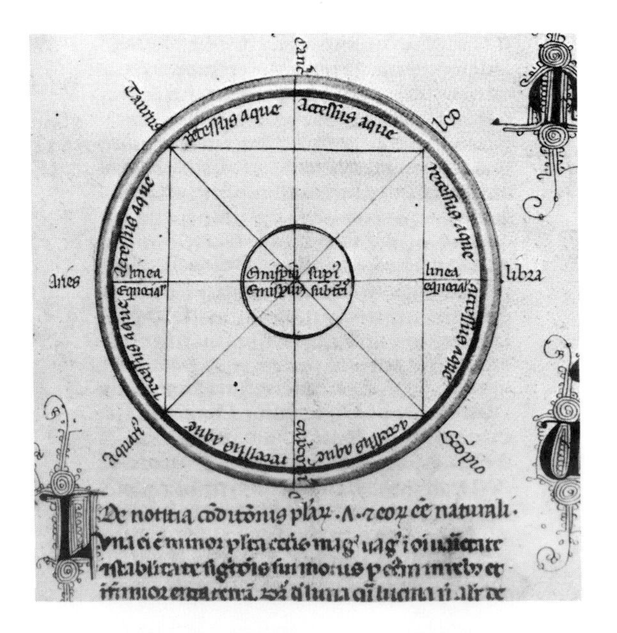

252. Shadow Cones. Since either in the accompanying text or within the diagram itself the circles in both of the figures reproduced here are identified as the sun, earth, and moon, one might surmise that they represented some part of a discussion of eclipses. Such, however, is not the case (though the figure given at the right, above, does appear in a manuscript of Gerard of Cremona's translation [see Illustration 89] of Ptolemy's *Almagest* shortly before he does embark on an investigation of eclipses). This first figure occurs in the fourteenth-century presentation manuscript we have frequently seen before (Illustrations 44, 59, 60, 62, 104, 178, and 241), and sets forth the geometry involved in Ptolemy's determination of the sun's distance from the earth. Elegant but somewhat lacking in overall accuracy, the upper circle with the slightly misplaced diameter *ADG* represents a great circle of the sun. The second largest circle, with the diameter *KNM,* represents the earth, the small circle *ETH* in gold between them depicts the moon, and the lowest and smallest circle marks the vertex of the earth's shadow cone. The mathematics accompanying this diagram allows Ptolemy to conclude that the mean distance of the sun is 1,210 times the earth's radius (actually, it is closer to 24,000 times). He also infers that the vertex of the shadow cone is 268 times the earth's radius and the mean distance of the moon at syzygies 59 times the earth's radius. The second diagram on the right comes from a thirteenth-century manuscript (also seen in Illustration 130) of an anonymous *De speculis,* translated from the Arabic and erroneously ascribed to Euclid. Again, though the sun *(sol)*, earth *(terra)*, and shadow *(umbra)* are clearly labeled, the topic at hand is not a discussion of eclipses. It is, rather, the establishment of several facts: that a ray of light is emitted from every point on the sun's surface, that these rays (examples of which are appropriately lettered in the figure) illumine the half of the earth facing the sun, and that, so covering the earth, they progressively narrow, forming a shadow with the earth "like a pine cone" *(sicut pinea)*. It is notable that the author of this brief treatise specifically calls for a figure such as that reproduced here *(in hac nostra forma figuravimus)*.

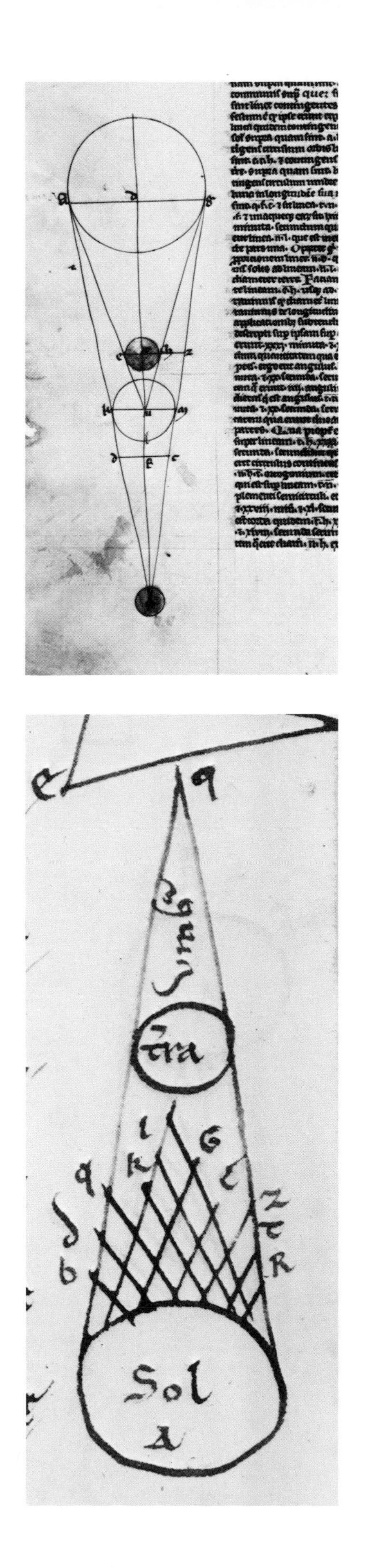

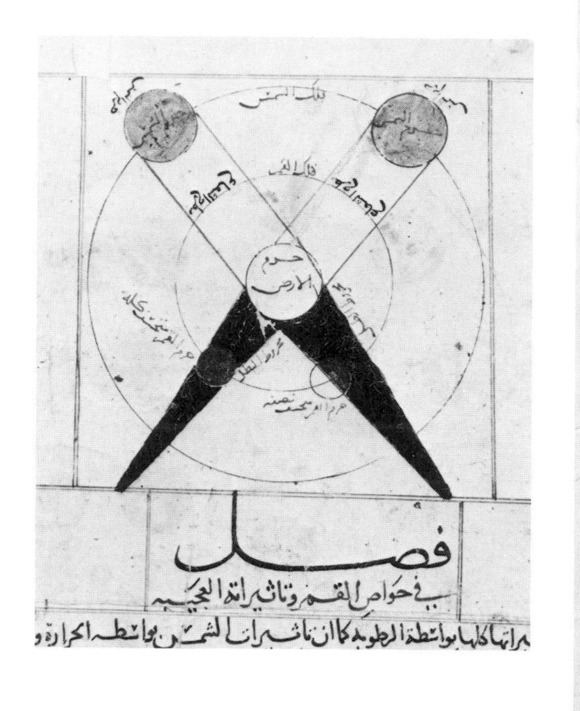

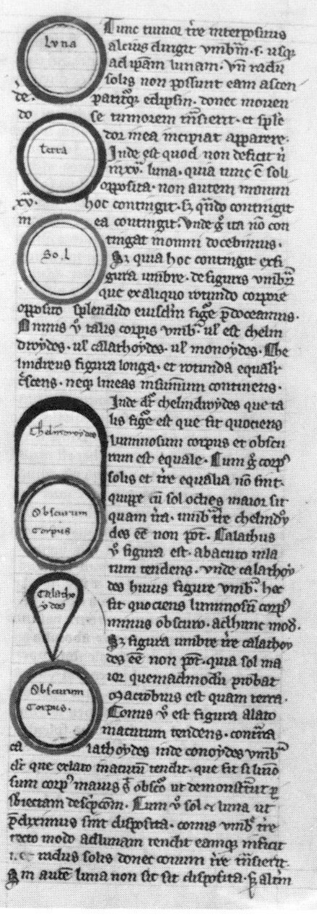

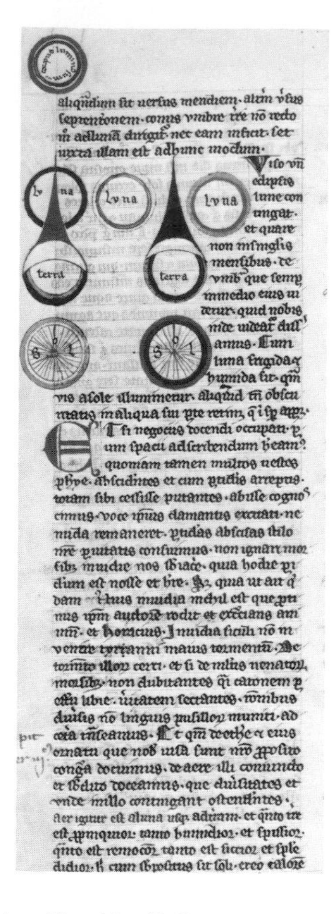

253. How Eclipses Come About. Illustration 234 has already provided a "scientific object" diagram of how the moon looks during an eclipse. Here we have several figures representing the theory of its—or the sun's—eclipse, even though these figures are not taken from some fundamental work in astronomy. None of them comes from an astronomical work in which the relevant mathematics is presented. Instead, the sources are cosmographic, poetic, or, in one case, even theological. Thus, the first figure at the extreme left, above, occurs in the thirteenth-century manuscript of al-Qazwīnī that has been used before (Illustrations 52, 126, 203, and 208). The treatment of eclipses in his *Wonders of Creation* occasions no surprise: the form of the present diagram is one that is found among lunar eclipse figures in any number of other cosmographic works. Two positions of the sun are depicted on an outer circle, the earth occupying the center. Light rays extend from both suns to the earth, whose conical shadows end in vertices that usually fall, as here, beyond the circular "orbit" of the sun. Two corresponding positions are given for the moon. In this Qazwīnī figure, one of the moons is depicted in totality, while the other undergoes only partial eclipse, though in many of the other illustrations of this sort, both moons are pictured in total eclipse phase. The second cosmographic source for eclipse figures is a thirteenth-century copy of the *De philosophia mundi* of William of Conches (twelfth century). The second set of figures given here from two pages of this work belongs to a chapter that addresses itself to the cause of lunar eclipses, to the shape of the shadows involved, and to why the moon does not undergo an eclipse every month. The three circular figures at the top, left column, give a simplistic representation of the cause of the lunar eclipse; namely, the interposition of the mass of the earth *(tumor terre interpositus)* between the sun and the moon, no shadow yet in sight in the diagram. William goes on to explain that since the sun is not the same size as the earth but is indeed eight times larger, the shadow of the earth cannot be cylindrical *(chelindroydes)*. Further, the shadow cannot be basket-shaped *(calathoydes; calathus)*, since then the sun would have to be smaller than the earth. These two different kinds of "rejected" shadows are depicted in the lower half of the left column. Only the shadow-casting body *(obscurum corpus)* and the shape of the shadow are pictured, though the circle at the very top of the column at the right is labeled *corpus luminosum* and may well have originally been placed below the two figures at the lower left in order to represent the light source in question. If this was so, then the migration of this *corpus luminosum* to the top of the next column surely indicates that the

(Continued)

illustrator of this manuscript did not have a very penetrating understanding of what he was illustrating. The upper half of the right column depicts the remaining possibility, that the shadow of the earth in a lunar eclipse is conical *(conoydes; conus)*, as indicated by the teardrop-shaped shadow cones. William points out that the moon does not suffer eclipse every fifteen days when it is opposite the sun, but only when it falls within the shadow cone and not to the north or south of it. Thus, the double figure with teardrop shadow cones is meant to represent, on the one hand, the moon falling within the cone and therefore undergoing eclipse and, on the other hand, the moon positioned north and south of the cone and therefore not suffering eclipse. The figures in this copy of William of Conches may seem crude and inexact, but they are far more adequate than that reproduced below, left, from a fourteenth-century manuscript of a Provençal poem *(Le breviari damors)*. Since the moon is positioned between the sun and the earth, a solar eclipse is in question here, but more attention has been paid to the decorative character of the background than to anything connected with explanatory effectiveness. The final, almost full-page, set of figures given at the right below comes from a fourteenth-century manuscript of the *Sentence Commentary* (written shortly before 1350) of the Cistercian Pierre Ceffons. The figures—and there are more on other pages of this work—are intended to represent not only eclipses of the sun and moon, but also the illumination of the moon and the planets by the sun. For example, the figures in the right column with two illuminated bodies between the sun and the central earth are meant to depict various situations in which the inferior planets Venus and Mercury can be found. Ceffons included a great deal of mathematics, logic, and natural philosophy in his *Sentence Commentary,* but just how he managed to work the present material into his major theological work is as intriguing as the figures accompanying this material. The standard initial topic in Book II of the *Sentences* was the creation of the world. Ceffons skews this slightly to concentrate on the creation of the heavens and then chooses as his lead question whether one can, by natural reason, prove that there are nine celestial spheres. This question allows him to launch into astronomical matters—many of them poorly understood—with a vengeance, frequently citing Ptolemy and Arabic works in support of his show of encyclopedic knowledge. He is then led on to questions concerning the sphere of the fixed stars; the continuity or contiguity of the heavenly spheres; the character of the motion of the heavens; and finally to the problem depicted in the present illustrations, of whether, knowing the number, order, and disposition of the celestial spheres, one may infer that some planets can suffer eclipse.

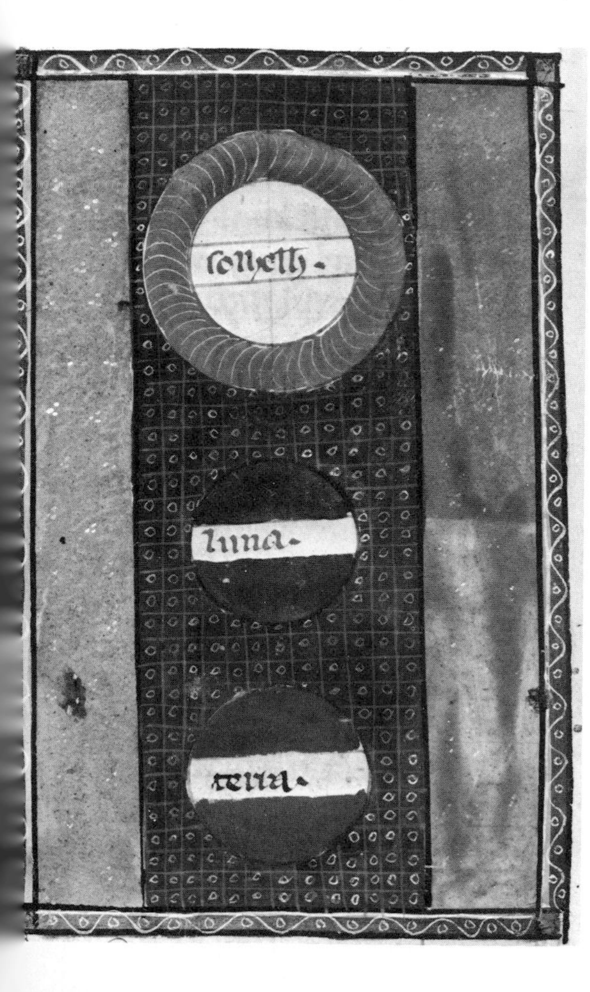

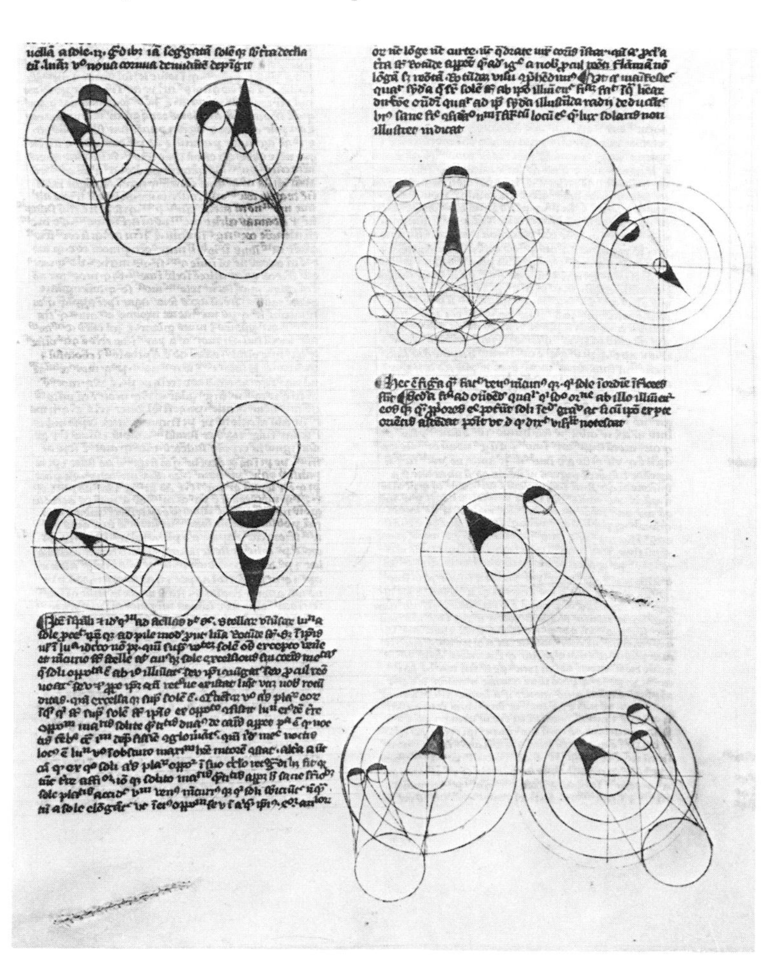

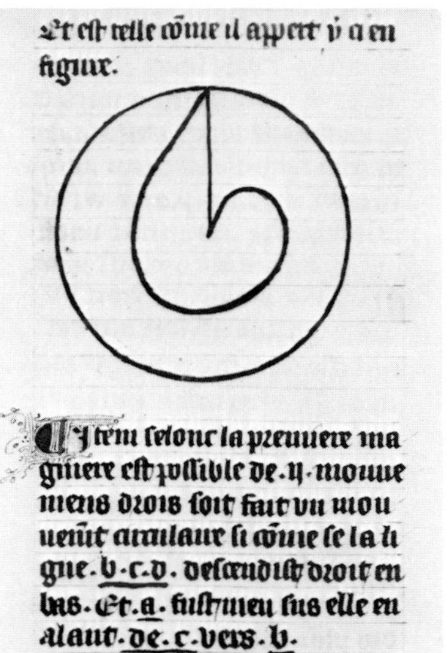

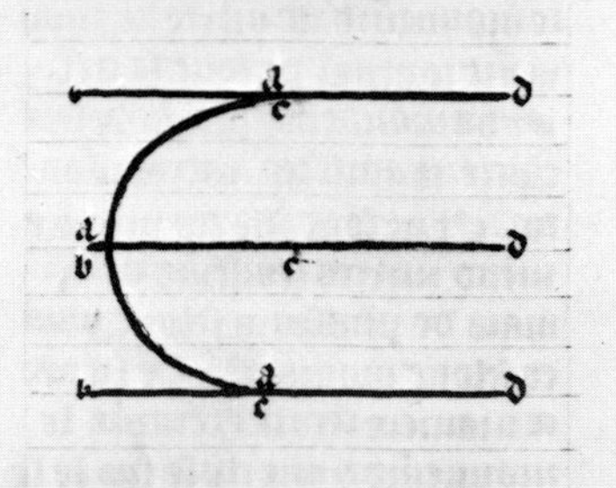

254. Appealing to Geometry in Natural Philosophy. Although Aristotle was quite critical of the mathematization of philosophy, especially as he witnessed it in Plato's successors at the Academy, he did not hesitate within his own natural philosophy to utilize a mathematical argument or appeal to some mathematical point. Three instances of this are given here, together with the late medieval diagrams to which they gave rise. The first two on the left are both from a fourteenth-century manuscript (see also Illustrations 64, 255, and 278) of Nicole Oresme's *Livre du ciel et du monde,* a commentary on Aristotle's *De caelo* that treats almost every line of text (translated into Old French). The circular diagram at the top left relates to Aristotle's claim (Book II, ch. 14) that the earth's position in the center of the universe is supported by the assertion of mathematicians relative to astronomy, in particular by the changing figures they used to represent the position of the stars. Oresme glosses this by pointing out that if the earth were not at the center of the universe, this would have to occur in one of three ways: either on the axis of the universe or outside this axis, and, if the latter, either equidistant from the two poles or not so equidistant. This is what the present circular figure represents (though the second alternative position specified for the earth is not depicted very accurately). Given any of these alternative positions, Oresme continued, we would not be able to see—as we do—half of the heavens from any place on earth, we would not see the stars in the way we do, we would not have days and nights as they presently are, and so on. All of these things are clearly apparent to the imagination *(par ymaginacion)*, he concludes, either with or without a figure such as the present one. The pair of diagrams given below at the left are part of Oresme's attempt to explain Aristotle's reference in the second chapter of Book I of *De caelo* to mixed motions. A motion of this sort might combine rectilinear and circular motions, such as the motion of an object on a rotating radius, the result being described by a line called a helix *(elycen)* which, Oresme notes, is used by Archimedes in his *Quadrature of the Circle.* "And this is such as appears here in a figure" *(Et est telle comme il appert y ci en figure)*, reads the inscription directly above the first of the two diagrams. The lower diagram relates to Oresme's example of the derivation of a circular

(Continued)

motion from two rectilinear motions. We are to imagine an object moving to the left from point *A* on the uppermost straight line while that line drops straight down to the position shown by the middle straight line. Given the proper velocities of these two rectilinear motions, the motion of the object will describe a quarter of a circle. Similarly, if the object then reverses course and moves to the right on the straight line while the line in turn drops to the lower position, another quarter circle will be described. The two squares in the next figure, top right, come from an anonymous dialogue *de naturalibus scientiis* found in a sixteenth-century manuscript (see also Illustrations 35, 169, and 281). The subject at hand is a passage in the *De anima* (Book II, ch. 3) in which Aristotle sets up an analogy between the parts of the soul and geometrical figures. In both cases, he claims, the prior always exists potentially in that which follows; for just as a triangle exists potentially in the square, so does the nutritive faculty of the soul exist in the sensitive faculty. Precisely how triangles exist potentially in squares is presented visually by our anonymous sixteenth-century author in the two squares reproduced here. The final figure at the bottom right pictures an argument that is found in both geometrical and natural philosophical works in the Middle Ages. It is alluded to, for example, in Roger Bacon in the thirteenth century, in Thomas Bradwardine and Nicole Oresme in the fourteenth, and in the margins of various medieval manuscripts of Euclid's *Elements,* as well as in the present fourteenth-century codex of a geometrical work ascribed to one Gordanus (not to be identified with Jordanus de Nemore). The argument itself relates to the meniscus of water or any other liquid contained in a vessel (since any point on such a surface is equally distant from the center of the universe). But this means that the closer the vessel is to this center, the more liquid it can contain when "full." This is so because the circular arc determining the surface of the liquid is "more curved" when the vessel is closer to the center of the universe; that is, the meniscus then "bulges" higher over the rim of the vessel. Indeed, it was even maintained that if such a vessel were absolutely full of liquid, moving the vessel further from the center of the universe would cause some of the liquid to overflow, since the surface of the liquid would become less curved.

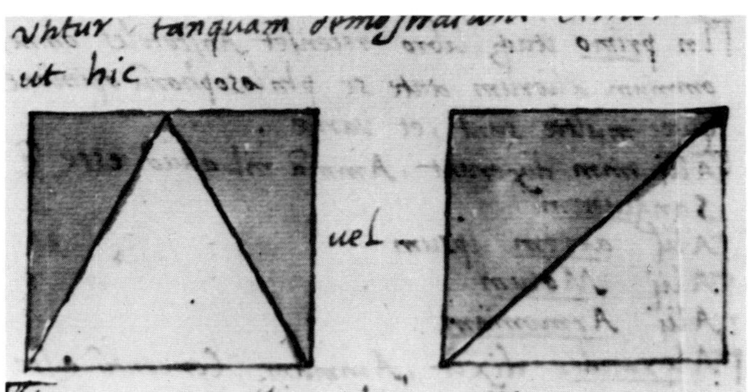

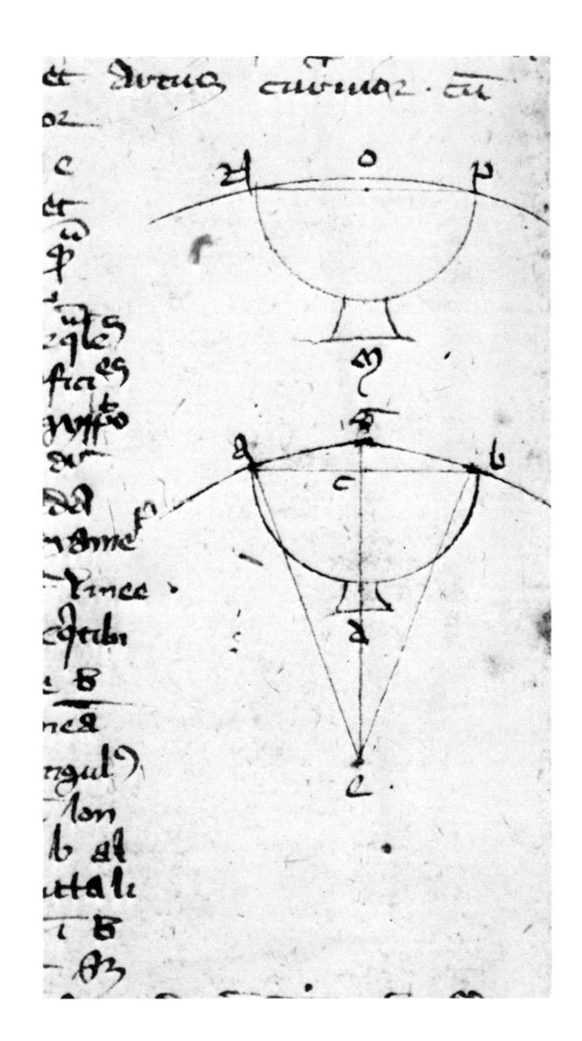

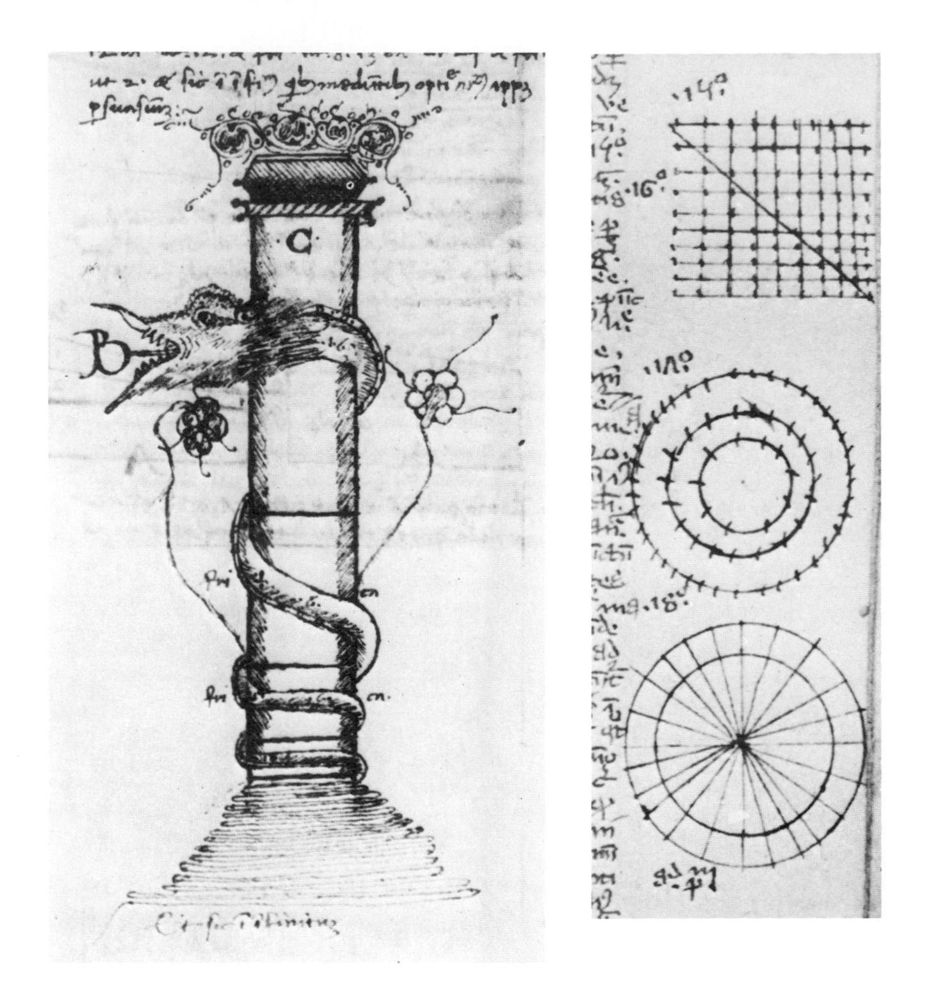

255. The Infinite and the Continuous. It was Aristotle's view that there could be no completed or "actual" infinites and that the only allowable infinite was that involved in the *potentially* infinite divisibility of a continuous magnitude (the infinite multitude of number being in turn grounded on that infinite divisibility). Further, this potential infinite only obtains with respect to the divisibility of a magnitude, *not* its size. But some medievals believed that they could establish an inconsistency in these contentions. For if one permits the potentially infinite divisibility of a magnitude, let that be represented by a cylinder divided according to its proportional parts $\frac{1}{2}$, $\frac{1}{4}$, $\frac{1}{8}$, . . . , $\frac{1}{2n}$, . . . , *ad infinitum.* Next, allow a spiral to be described on the surface of the cylinder so divided in such a way that the spiral makes one complete revolution over each and every proportional part of the cylinder. Is not the resulting spiral line potentially infinite in length? Such an infinite *linea girativa* appears at the left above, drawn rather fancifully as a snake in a sixteenth-century manuscript (already seen in Illustrations 13, 143, and 144) of a commentary on Richard Swineshead's *Liber calculationum* (see next illustration). Another side of the medieval elaboration of Aristotle's views about the infinite had to do with his contention that all continuous magnitudes were (potentially) infinitely divisible into always further divisible parts. Therefore, no continuum could be composed of indivisibles; Aristotle himself constructed a number of arguments proving that this must be so. In support of Aristotle, his medieval followers constructed even more such arguments, many of them mathematical in nature. Diagrams relating to two of the most popular such *rationes mathematice* are given at the right on page 296 from a fourteenth-century manuscript of the *Sentence Commentary* (see Illustration 72) of a fourteenth-century English scholar named Crathorn. (It is notable that arguments about the infinite divisibility of continua and other scientific

(Continued)

matters often appeared in late medieval theological *Sentence Commentaries;* compare Illustration 253.) These particular diagrams depict mathematical arguments using parallel and radial projection to overturn indivisibilism. Thus, if parallels are drawn from all of the indivisibles comprising one side of a square to all of the corresponding indivisibles in the opposite side, these parallels will cut the diagonal of the square in the same number of indivisibles as in the sides. It follows, therefore, that the diagonal is equal to the side (a conclusion drawn both when the component indivisibles were held to be finite in number and, erroneously, when they were allowed to be infinite in number). But since the equality of the diagonal and the side of a square is absurd, the indivisibilism that entails it is also absurd. And a similar argument, here depicted in two different figures, was formulated by constructing all of the radii from the center of a circle to all of the component indivisibles in the circumference. These radii would cut all inner concentric circles in the same number of indivisibles, thereby implying their equality with the outer circle. Yet another example of medieval deliberations about the infinite is pictured below. These figures are taken from a late-fourteenth-century manuscript (also seen in Illustrations 64, 254, and 278) of Nicole Oresme's *Le livre du ciel et du monde* (fourteenth century). The topic at hand is Averroës' (twelfth-century) comment on Aristotle's *De caelo* that an infinite in all directions is greater than an infinite in only some directions. Oresme opposes this by showing that one can transform a "single-directional" infinite into an "all-directional" infinite. Let the all-directional infinite sphere be *A*, represented at the top of our diagram by the straight lines diverging from the central point *A*. Let the single directional infinite be "a body *B* one foot in width and one foot in depth and infinitely long in only one direction—namely toward the right"; further, divide this foot-square infinite column into all of its cubic-foot constituent parts. This much is clearly depicted in the center, the constituent parts labeled *C, D, E, F, G,* . . . The next and final "sphere wrapping" move in the argument is illustrated by the concentric circles below this divided infinite column. We are to take the first cubic foot of the column and form it into a sphere, take its second cubic foot and wrap it around the sphere so formed, the third cubic foot in turn around that, and so on *ad infinitum*. The result will be that the single directional infinite which is the column *B* will be transformed into an infinite sphere equal to *A*. Q.E.D.

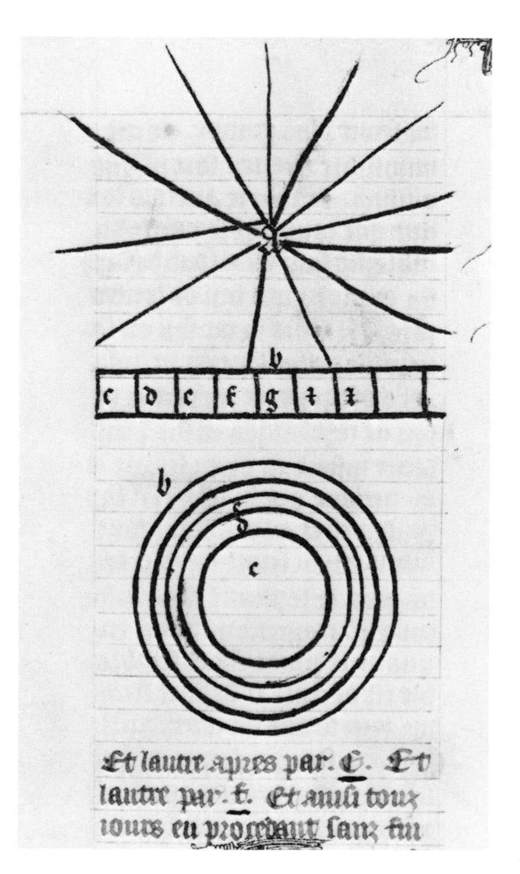

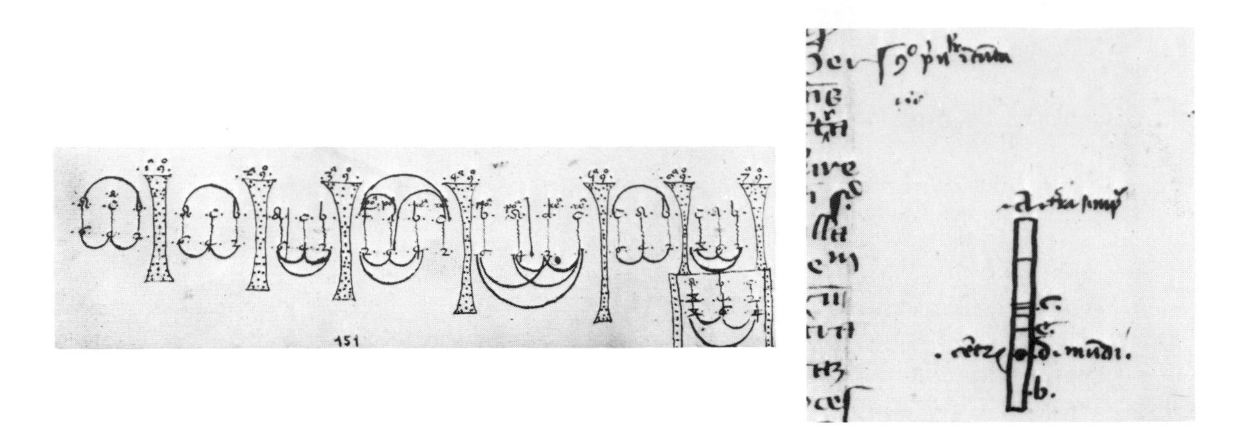

256. The Calculator: Richard Swineshead. In the year 1328, the Oxford scholar Thomas Bradwardine wrote a work the major burden of which was to determine what changes in speeds or velocities corresponded to the changes in force–resistance ratios determining those velocities. In our terms, his answer was that while velocities increased or decreased arithmetically, the force–resistance ratios increased or decreased geometrically. (For example, if V results from or corresponds to F/R, then $2V$ results from $(F/R)^2$, $3V$ from $(F/R)^3$, $\frac{1}{2}V$ from $\sqrt{F/R}$, etc.) Bradwardine's answer was supported and developed by a number of fourteenth-century scholars, most brilliantly by his Oxonian successor, Richard Swineshead, in his *Liber calculationum,* for the authoring of which he became known as the "Calculator." In one of the sixteen tractates constituting this work, Swineshead set down forty-nine "rules of local motion" that, assuming Bradwardine's view of the relations obtaining between changing V's and their corresponding F/R ratios, "catalog" just which *kinds* of changes in velocity correspond to which *kinds* of changes in force and resistance and vice versa. The marginal illustration at the left occurs at the beginning of these rules in a fourteenth-century manuscript of the *Liber calculationum.* Using the "arc diagrams" already observed in Chapter Ten as quite standard in the illustration of the mathematics of musical ratios, the scribe of this manuscript has tried visually to represent Swineshead's rules or *conclusiones.* To explain but two of the resulting marginal diagrams, that at the extreme left depicts the first rule setting forth the mathematics of force–resistance changes (no velocities yet in question). We are to assume some resistance B (at the right in this arc diagram, given the sample value 2 below) on which a force C (in the center, given the value 4) is acting, but which increases to A (at the left, given the value 6), the resistance remaining constant. Given this, Swineshead concludes that the ratio of the final (increased) force to the resistance is "composed of" the ratio of the initial force to the resistance "plus" the ratio of the final force to the initial force [that is, in our terms, $F_2/R = (F_2/F_1) \times (F_1/R)$]. The diagram illustrates this by showing that the upper arc (representing F_2/R) spans just as much as the two lower arcs (representing F_2/F_1 and F_1/R). The diagram fourth from the left belongs to Swineshead's fourth rule, the first in which velocities are taken into consideration. The rule stipulates that when, for instance, a single increasing force operates on two equal or unequal, but constant, resistances, the increments of velocity (as we would say) relative to these resistances are equal. Thus, in the corresponding diagram, the final and initial values of the increasing force are given at the left and the two resistances are given at the right. (Sample values of 12 [erroneously written 2], 6, 4, and 2 are given underneath.) This in hand, the fact that the larger arc above shows a "gain" over the smaller arc at the right above that is the same as the "gain" of the larger arc below over the smaller arc at the right below represents the equal increments of velocity claimed by the rule. Another tractate of Swineshead's *Liber calculationum* applied Bradwardine to a quite specific problem. Briefly put, the problem was whether a thin rod in free fall near the center of the universe will ever reach that center in the sense that the center of the rod will eventually coincide with the center of the universe. The problematic part of the question derived from the fact that as soon as any part of the rod passes the center of the universe, that part may be considered a resistance against the rod's continued motion. Assuming that the rod acts as the sum of its parts and that the relevant forces and resistances determined by these parts follow Bradwardine's "law," Swineshead concludes that the center of the rod will never reach the center of the universe (which is correct, under the assumptions made, since the time intervals for each increment of distance will increase *ad infinitum*). The marginal sketch at the right accompanies this particular text of Swineshead in a fourteenth-century manuscript of his work. Possibly drawn by a reader trying to puzzle his way through this segment of the "Calculator," the rod (here termed *terra simplex* to indicate that it is a heavy body) is appropriately divided into parts, one of them depicted as already having passed the center of the universe, which is duly labeled *centrum mundi.*

257. Theories of the Art. Of all the various symbols belonging to the history of alchemy, that of a serpent biting or devouring its own tail—the Ouroboros, to use its Greek name—is undoubtedly one of the oldest. Such tail-eaters are referred to, for example, in Egyptian sources around 2300 B.C. and depicted in papyri as early as the sixteenth century B.C., though their function was then naturally not an alchemical one. Indeed, the symbolic role in cosmology of some kind of encircling serpent continues to be found in astrological, Gnostic, and even apocryphal biblical texts contemporary with the alchemical treatises that put it to use for their own purposes. The earliest appearance of the Ouroboros in a Greek alchemical text is of the eleventh century (in the manuscript utilized in Illustration 78), where it occurs as one among other figures (many of them of apparatus) on a single page entitled *Chrysopoeia* ("goldmaking") of Cleopatra. There, the encircling serpent symbolizes the unity of matter, as indicated by the single inscription in its center: *ἕν τὸ πᾶν*; "One is the All." The drawing of the Ouroboros reproduced here at the left comes from a fifteenth-century Greek codex. The accompanying text assigns the tail-eating serpent not the role of unifying matter but of representing that operation of the art that consists in the dissolution or levigation (λείωσις) of bodies. The four feet of the serpent stand for the *τετρασωμία*, an alloy of the four metals lead, copper, tin, and iron, while the three ears represent three vapors and twelve compounds. We are also told that the green underbelly of the serpent symbolizes putrefaction or fermentation, while its venom denotes vinegar, an important ingredient in alchemy. The picture at the right, taken from the same sixteenth-century manuscript that has already provided a full page of alchemical apparatus (Illustration 243) presents an image that is very much later than the Ouroboros. The section of this manuscript in which this picture occurs carries the title *Duodecim figurae scientiae maioris,* a collection of texts (from alchemical works ascribed to Hermes, Arnald of Villanova, Geber, Hortolanus, and numerous others) put together with these twelve figures by the fifteenth-century adept Johannes Andreae. Copulating figures were frequently used to symbolize the conjunction or unification of opposites. As the legend for this figure indicates, their function here is to represent the solution of bodies into quicksilver *(argentum vivum)* and thus into a fixed and white *aqua permanens* (an alternative name for quicksilver) that is like unto tears *(fit aqua permanens fixa et alba, ut lachrima)*. The white color of the resulting liquid is properly labeled *(albus color)* at the bottom of the flask. Finally, as the second marginal annotation to the left of the flask tells us, the four heads above the copulating couple represent the four elements *(ex quatuor elementis iste lapis*—that is, the Philosopher's stone or elixir—*compositus est: igne, aere, aqua, terra)*.

258. The Discovery and Confirmation of Musical Ratios. We have already had ample opportunity (see Illustrations 63, 95, 100–105) to examine the diagraming of the mathematics of musical ratios. The present pictures relate to the same topic, but in a quite different way. For the most part, the diagrams in the previous illustrations employed arcs in providing visual representations of the composition of ratios and of how those ratios constituted various musical scales. Here, however, the pictorial burden is the portrayal of how, according to legend, the basic musical ratios were discovered and confirmed. The legend in question derives from Nicomachus of Gerasa (fl. ca. A.D. 100). It is preserved in his *Enchiridion* or *Manual of Harmony* (ch. VI), but it was almost certainly also included in his lost *Isagoge* or *Introduction to Music,* since this work was the primary source for the first two books of Boethius' *De institutione musica* (early sixth century), where the story also appears (I, chs. 10–11) and whence it was transmitted to the Middle Ages. The legend recounts how, chancing to pass the workshop of blacksmiths, Pythagoras noticed that the different sounds resounding from the blows of their hammers upon the anvil somehow produced a single consonance. On further investigation, he was able to determine that this was not due to the strength of the blows or the shape of the hammers, but to their different weights. Thus, if one hammer weighed twice that of another, the two resounded in a diapason consonance, or an interval of an octave. Similarly, the intervals of the fifth, the fourth, and a whole

(*Continued*)

tone could be produced by different hammers of appropriate weight. In sum, if the weights of four hammers were 12, 9, 8, and 6, one could account for all of these basic musical consonances or intervals. (Of course, differently weighted hammers will not in fact produce the phenomenon Pythagoras is reputed to have heard, but this was not clearly shown until the seventeenth century.) The legend goes on to relate how Pythagoras sought to confirm what the weighted hammers had revealed to him. He suspended weights equal to those of the hammers from strings, substituted pipes whose lengths stood in the same ratios as these weights, and placed similarly measured weights of liquid in glasses; plucking the strings, blowing the pipes, and striking the glasses all confirmed the musical intervals he had "discovered" by the smithy's hammers. The four drawings filling the manuscript page reproduced on the facing page succinctly present all of the important features of this story. Appearing in a fifteenth-century codex of the *Theorica musicae* of Franchinus Gafurius (1451–1522), the pictures served as the source of an identical plate of illustrations in the 1492 printed version of this early Renaissance musical treatise. In the depiction of the hammer legend at the upper left, Pythagoras has ceded his position to another protagonist: Jubal, according to the *Book of Genesis* (4:21), the father of song, cithara, and organ. Each of the hammers in this miniature carries an indication of its weight (4, 8, and 12 on the left and 6, 9, and 16 on the right). The same series—4, 6, 8, 9, 12, 16—is found in the depiction of the weighted strings at the lower left, the pipes at the lower right, and the water-filled glasses at the upper right. This last picture includes a supplement to the legend: the striking of bells that bear the same ratios to one another as the hammer weights and other "experimental instruments." Pythagoras is the actor in all three of these miniatures, though in that at the lower right he is accompanied by the Pythagorean Philolaus of Croton (b. ca. 450 B.C.) who, according to Nicomachus, was also a contributor to early music theory. The top half of another full-page miniature in a much earlier (twelfth-century) manuscript is reproduced below. Here Pythagoras is depicted in the course of striking bells with a hammer, his left hand holding a balance on which other hammers are resting, an obvious reference to the initial part of the legend we have just recounted. He is framed by couplets, one of which tells us that he discerns the weights by means of a balance, thus spurning dissonance *(Pondera discernit, trutinans et dissona spernit)*. Boethius is seated at the left, plucking a monochord, and, as several of his framing couplets inform us, aurally judging sounds by running his index finger up and down its single string *(Ut videat vocum, discrimina per monochordum; Iudicat aure sonum, percurrens indice nervum)*.

CHAPTER TWENTY-THREE

Medical Theory: Diagnosis, Prognosis, and Therapy

THE RICHNESS OF available sources for material illustrative of constellations and other celestial objects required a very limited selection of pictures for Chapter Nineteen. The same is true for this chapter. Illustrations in medical manuscripts are, if anything, even more plentiful than those representing the population of the heavens. There is, however, a notable difference in these two areas of scientific illustration. In the case of constellations, a great share of the vast number of available pictures is, as was noted in the introduction to Chapter Nineteen, to be found in copies—Greek, Arabic, and Latin—of certain particular works. In the case of medical illustrations, the same is decidedly not true. The occurrence of relevant visual material runs the gamut of medieval medical writings: from illustrations accompanying almost any sort of medical text to medical miscellanies containing figures and pictures belonging to no definite text or writing.

This chapter's very restricted sample of such illustrations has been chosen from manuscripts randomly located throughout the whole spectrum of medical works. The only guiding principle has been that the illustrations represent some aspect of medical theory. It has already been noted that a fair amount of theory is present in the pictorial material in Chapter Eighteen, dealing with human anatomy. And the same thing is true of some of the illustrations of medical practitioners and their environs in Chapters Fifteen and Seventeen. Here, however, unlike most of the previous illustrations, the emphasis is not on the depiction of the "scientific object"—that is, on the visual representation of patients, physicians, and tools of the trade—but rather on one or another aspect of medical theory, even though in many instances the picturing of theory necessarily brought with it the incidental picturing of the medical "objects" as well.

The particular illustrations selected for this chapter can roughly be divided according to three aspects of medical theory: diagnosis, prognosis, and therapy. We begin with a visual presentation of a catalog of diseases and their likely locations (Illustration 259). The next two illustrations treat of the two diagnostic procedures that were of greatest importance in ancient and medieval medicine: the evaluation of a patient's urine and of his pulse. The obstetrical information given in Illustration 262 is also diagnostic in its visualization of the various presentations of the fetus, though it includes elements of therapy as well in the form of advice on how to remedy undesirable presentations.

Prognosis is the central concern in Illustration 263; it presents (rather fanciful) material that bears directly on the question of the fatality or nonfatality of the illness in question. Prognostic considerations are also involved in whether the wounds shown in Illustration 268 are curable, and they occasionally also form part of the diagnostic judgments made on the basis of a patient's urine or pulse (though such elements are not present in the relevant illustrations reproduced here).

Almost all of the remaining illustrations in this

chapter relate to therapy—to cautery, bloodletting, surgery, the administration of drugs—and to astrological considerations that were deemed important in governing the application of such therapeutic measures.

The final illustration of the chapter stands rather by itself. Its subject is the localization of mental processes within the brain or heart and, as such, it relates to anatomical contentions and not to diagnosis, prognosis, or therapy. Yet because this subject is so highly theoretical, it has been given a place here rather than among those more observational, object-based, anatomical illustrations in Chapter Eighteen.

An extremely important, if also unusually difficult, question is precisely what role illustrations such as those in the present chapter played in actual medical practice or teaching. Clearly they do not represent diagnosis or therapy actually being carried out, nor do they present a record of something actually seen during diagnostic or therapeutic procedures—even when the features of the illustration in question could, charitably interpreted, be taken as representing such things. They are, instead, schematic; their purpose is to call to mind this or that particular element of evaluation or procedure, to allow the physician or surgeon or midwife to double-check what he or she was about. That medieval medical practitioners in fact did at times appeal to visual representations of elements of theory, such as those reproduced in this chapter, is evident from the persistent history these representations have in manuscripts of medical works and especially from their appearance in the form of girdle books (Illustration 267) where "information at the ready" was the intent. In these, the collection of visual material was surely an instance of production for use.

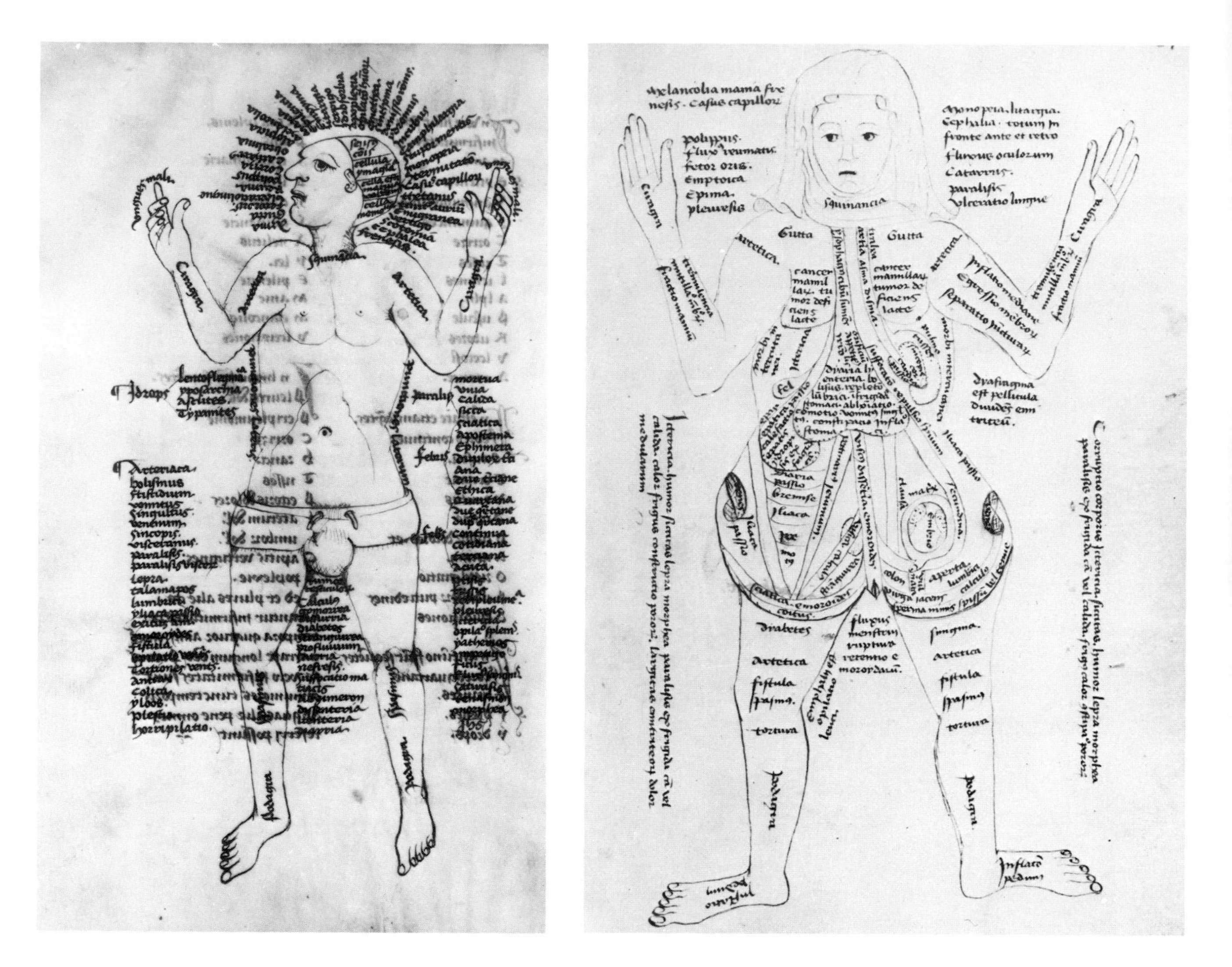

259. Figurae de infirmitatibus. Both of the pictures reproduced here are taken from a late-fourteenth-century miscellany containing what appears to be the earliest manuscript version of a group of texts that were later collected by an all-but-unknown Johannes de Ketham (fl. ca. 1460, Vienna?) and then printed in Venice in 1491 under the title *Fasciculus medicine.* This particular brief set of texts was traditionally accompanied by illustrative material, including urine circles and bloodletting, wound, and zodiac men (see Illustrations 260 and 265–268), in addition to the two figures given here. The function of these figures is to present a catalog of diseases that would simultaneously be more informative and more visually arresting than a listing of the contents of such a catalog. This is effected by employing eye-catching drawings of human figures on which, or about which, diseases could be listed, the name of each disease appearing near that part of the body it commonly afflicted. The "disease man" on the left, page 304, is less successful than one might hope in setting forth such a local pathology. The names of very few diseases are written directly on his body. Discounting the four faculties of the soul whose labels *(sensus communis, cellula ymaginativa, cella estimativa rationis, cella memorativa)* appear on the posterior part of the head (and obviously also ignoring those names showing through from the reverse side of this vellum page), one finds only the following: quinsy *(squinantia)* on the throat, ar-

(Continued)

thritic gout *(artetica)* on the upper arms, gout of the hand and the foot *(ciragra, podagra)* on the lower arms and ankles, spasm (*spasmus*) near the knees, and injured or diseased nails *(ungues mali)* over the fingers. The disease names radiating from the head as well as those inscribed between the legs are also "locally indicative," since almost all of the diseases so listed in some way affect, respectively, the head or the genitals and urinary or lower intestinal tracts. The lists to the left and right of the body do not serve as indicators of location, though some of the maladies listed are grouped as types of dropsy *(idrops)*, paralysis, or fever *(febris)*. As a whole, however, this manuscript disease man is more satisfactory than some of his printed "offspring," where one finds much more external listing, but in mere alphabetical order. By contrast, the disease woman at the right is more successful in revealing the location of diseases than her male counterpart. Not only are many more labels placed where they belong on her body, on either side of her head, and between her legs, but there is a sketchy depiction of various features of internal anatomy that provides an even more specific location for some diseases. A stomach in the form of an upside-down heart appears at the center, the heart and lungs to the right and slightly above, the gall bladder *(fel)* and liver to the left and slightly below. The two kidneys *(renes)* are located in extreme lateral position near the hips. She is also depicted as pregnant, two small faces labeled *embrio* peering from her womb *(matrix)* on her left side. For another picture from this manuscript see Illustration 266.

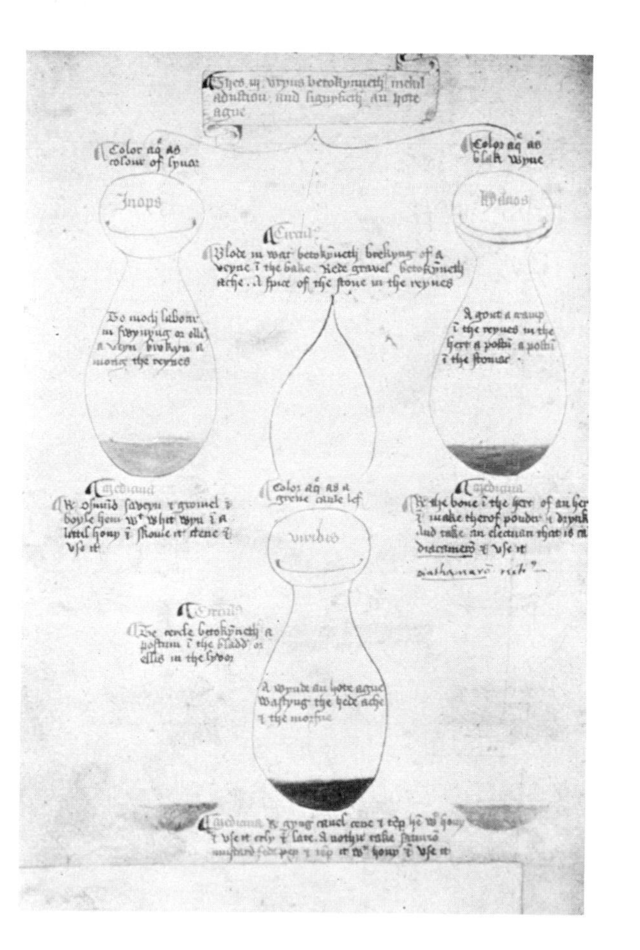

260. Uroscopy Diagrams. One of the major diagnostic procedures of ancient, and especially medieval Arabic and Latin, medicine was the examination of a patient's urine. The physician inspecting a flask of urine has already been seen in Illustration 183, but here our concern is with some of the details of uroscopic diagnosis. Although there is no genuine Galenic treatise dealing specifically with such diagnoses, a Latin translation of the Greek work on urines written by Theophilos Protospatharios (seventh century) became a central text on the topic in the Middle Ages while a Latin translation of an Arabic tract by the tenth-century Jewish philosopher and physician Isaac Israeli also became standard fare. These two works, together with the Latin poem *De urinis* by the twelfth-century French physician Gilles de Corbeil, were the most popular writings on uroscopy. The theory behind these writings was that the urine was connected with the blood and other humors (red and black bile and phlegm) and that, consequently, its examination would reveal pathological conditions deriving from humoral anomalies in addition to urinary disturbances. Attention was paid to three factors in examining the urine: color, consistency, and deposits (though quantity, smell, and even taste were at times also considered). The three flasks at the left, taken from a fifteenth-century Middle English

(Continued)

(Continued)

medical miscellany, reveal one form of illustration that served as an *aide-mémoire* in carrying out uroscopy. The legend at the top announces that "thes III uryns betokynneth miche [the text mistakenly has 'nichil'] adustion [that is, drying out] and signyfieth an hote ague." The colors of the urines in question are given as *inops,* which is explained in English as the "colour of lyvor," *kyanos* or "blak wyne," and *viridis,* "as a grene caule [that is, cabbage] lef." Another observable feature is tabulated between the top two flasks and just to the left of that at the bottom: frequently referred to as the *corona,* but here given as *circulus,* this feature is what one might observe at the circumference of the surface of the urine where it touches the flask. (Thus the *circulus* at the bottom left tells us that the "cercle betokynneth a postum—that is, an aposteme or abscess—in the bladder or ellis in the lyvor.") Indication of the disease involved is written within each flask (thus, the "black wyne" urine at the upper right may signify gout, a cramp in the kidneys [*reynes*], or an abscess in the heart or stomach), while appropriate treatments announced by the label 'medicina' are listed below each flask. (Thus one of the things prescribed for a "blak wyne" sufferer is to drink a potion containing powder made from the bone in a stag's heart.) "Urine circles," such as that reproduced below from a late-fourteenth-century English astrological-medical miscellany, were more prevalent as uroscopic schemata, but, because of their greater scope, contained much less information. The illustration is labeled at its center as a *Tabula de iudiciis urinarum per colores* and twenty different "specimens" in varying shades of red, black, gold, white, yellow, green, and so forth populate its circumference. The three urines we have seen in our Middle English illustration appear here at right center and are correctly identified as denoting excessive dryness *(nimiam adustionem)*. The three flasks below them containing exceedingly dark or black urines exemplify terminal illnesses *(significant mortificationem)*, while the two urines just to the left of top center, in different shades of reddish gold *(refus, subrufus)*, illustrate the other end of the spectrum, since they are signs of perfect digestion. For another figure from the codex containing this urine circle, see Illustration 266.

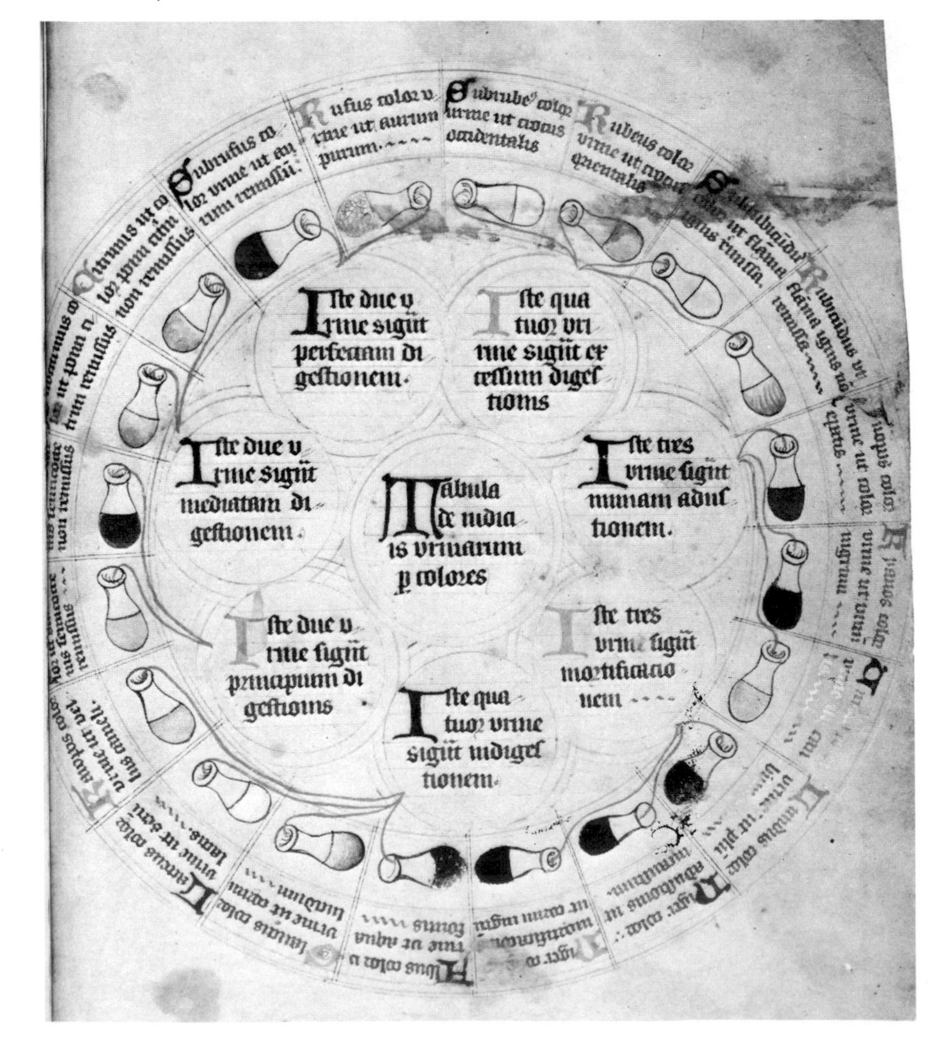

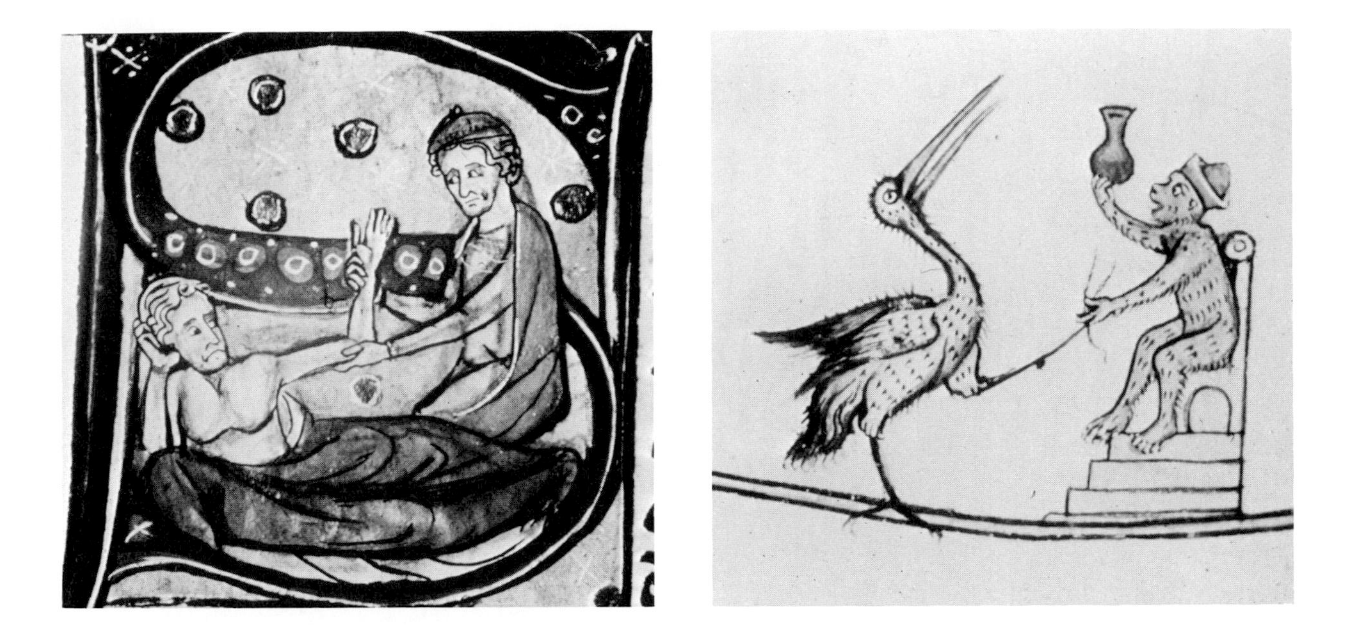

261. De pulsibus. Within early medicine, evaluation of the pulse held an importance equal to that of uroscopy as a diagnostic procedure. Indeed, since the pulse relayed information relative to the heart and not, like urines, merely to the liver, there were some who saw fit to champion the knowledge that could be derived from the pulse over that which could be had through the inspection of urine. Further, properly interpreted, the pulse was more definitive than uroscopy. As the fourteenth-century physician Bernard de Gordon put it, "the pulse is a messenger that does not lie" *(pulsus est nuncius qui non mentitur)*. By contrast to uroscopy, there were a number of genuine Galenic works dealing with the pulse, but they do not seem to have been those most frequently used in the Middle Ages. Instead, we find repeated reference to a Latin version of a treatise on the topic by the seventh-century Byzantine physician Philaretos and, once again, to a poem, *De pulsibus* by Gilles de Corbeil. In addition, most medieval medical "handbooks" contained a section on the pulse and, though the relevant details naturally varied from treatise to treatise, all of them noted that attention should be paid not simply to the frequency and duration of heartbeats, but also to the condition of the artery in question, the duration of diastole and systole, the strengthening or weakening of pulsations, and the regularity or irregularity of the beat. Some idea of the consideration paid to factors of regularity and irregularity can be seen from the importance ascribed by many authors to the relation between heartbeats and various musical tempi. It can also be seen from some of the varieties of irregular pulse that were cataloged by some medieval physicians: the goat-leap pulse *(gazellans)*, the vermicular *(vermiculosus)*, the antlike *(formicans)*, the shrew-mousetail *(cauda soricina)*, the spasmodic, tremulous, and twisted *(spasmosus, tremulus, retortus)*, and so on. Pulse treatises also contained instructions for taking the pulse—which hand to use, how to place the fingers, where one should stand or sit relative to the patient, etc. Pictures of the procedure often ignored or glossed over such details, though the general features were depicted with tolerable accuracy. A case in point is the illuminated initial at the left, taken from a thirteenth-century manuscript of Gerard of Cremona's translation of the *Breviarium* of Serapion the Elder (Yaḥyā ibn Sarāfyūn; ninth century). The fingers of the physician's right hand are placed in the right position to record the pulse of the radial artery, while his left hand holds the patient's arm at the elbow. Some historians have thought that in miniatures of this sort the thumb of the left hand is positioned in such a way as to indicate the taking of the pulse of the brachial artery, but this is doubtful, even though instructions concerning the brachial pulse were part of the relevant literature. The picture at the right simultaneously portrays uroscopy and pulse taking, a combination found in a fair number of medieval miniatures. In this case, however, the double diagnosis is being carried out on a stork or crane by an ape-physician. Apes were medieval favorites when it came to illustrated animal parodies of humans and human endeavors, and the physician or surgeon and the medical profession were not excluded from such *drôleries*. In almost all instances of such parodies, the figures were marginal ones. In the present case, the marginal illustration comes from a 1316 French manuscript of the Metz Pontifical.

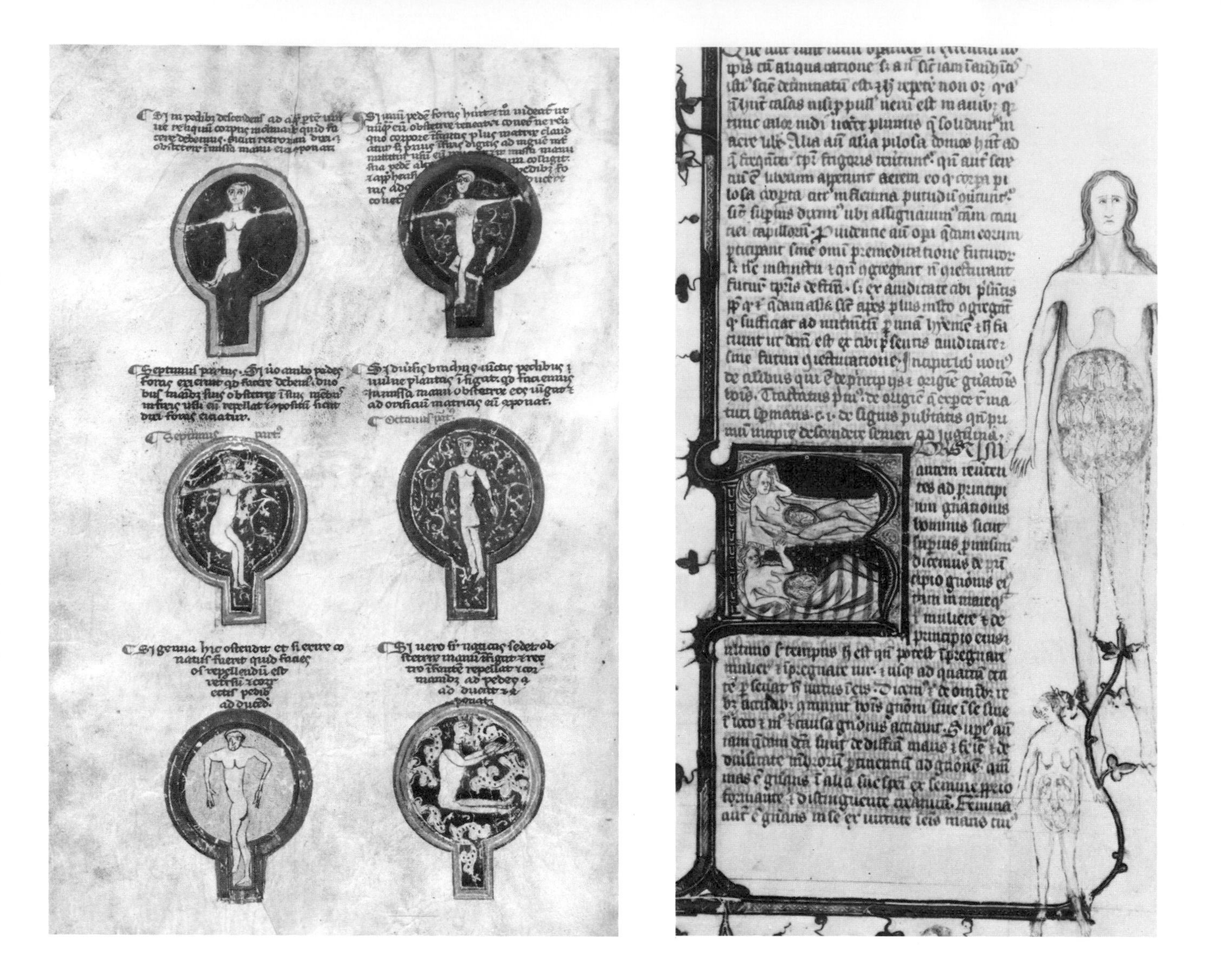

262. The Fetus in Situ and Different Presentations at Birth. The central ancient work on gynecology and obstetrics was written by Soranus of Ephesus (fl. A.D. 98–138), the most important figure of the methodist sect of medicine. Entitled *Gynaecia* (Περὶ γυναικείων), the four books of this work covered the qualities and qualifications needed to become or to be a good midwife; obstetrics in general, including natal care as well as knowledge required for delivery; and women's diseases. Soranus also composed an epitome of parts of his larger work that was presumably intended to serve as a kind of catechism for midwives. Its text is lost, but some parts of it are most likely preserved in a Latin translation made in the sixth century by Muscio. (It was retranslated into Greek in the fifteenth century—turning Muscio into Moschion—from which the resulting "new Greek" was turned back into Latin!) Many of the medieval manuscripts of Muscio's work contain illustrations of the various presentations *(partus)* described in the text. One page from a thirteenth-century example is given above, at the left. Although other manuscript illustrations of Muscio at times depict the umbilical cord, if not the placenta, those reproduced here fill their keyhole-shaped uteri with no more than the fetus in differing presentations plus various elements of decoration. It is difficult to imagine that these particular drawings would be of much use to a prospective midwife, especially since the texts accompanying each presentation are full of errors. The texts begin by describing the problematic presentation pictured below them and

(Continued)

conclude with an indication of what procedure should be followed to ensure a proper delivery. Thus, to cite but three of the examples given on this page, the text (appropriately corrected) at the upper right tells us that we have here a presentation in which one foot is outside the uterus and the other still inside; the text at the center left describes a presentation with both feet exiting from the uterus, while that at the bottom right deals with the situation in which the buttocks appear first. In each instance the "therapy" recommended amounts to turning or pushing the fetus back and then drawing it out with feet and hands in a proper position. The second set of in situ illustrations reproduced at the right, page 308, is of a quite different sort. Taken from a fourteenth-century manuscript, not of an obstetric, let alone medical, work, but of the *De animalibus* of the thirteenth-century scholar Albertus Magnus, the illustrations occur at the beginning of Book IX on generation. Without shedding any light on the accompanying text, these drawings are artistic rather than instructive in intent. Beginning with in situ depictions for the two women (one with twins) within the illuminated initial, the artist was somehow inspired to give us what seemed to be triplets in a smaller marginal figure and then no less than a "litter" of some twenty-six in the second, larger, and rather doleful-looking marginal woman. Yet whatever the number, the "presentations" are all the same: knees drawn up with hands covering the face. For another illuminated initial together with "marginal overflow" from this manuscript, see Illustration 210.

263. To Live or to Die. In most instances in ancient and medieval medicine, prognosis proceeded from observed symptoms of a disease and the outcome was inferred by applying one or another aspect of medical theory. In some cases, however, prognostication assumed an astrological base or was grounded in some kind of magical or occult contention. The two present illustrations address the problem of this latter sort of medical prediction. The first, above, depicts the Caladrius bird, whose representation, as well as medical relevance, was most frequently found in illustrated medieval bestiaries. The picture reproduced here comes from a thirteenth-century manuscript of that genre. The bestiary account opens by telling us that the Caladrius is completely white without a speck of black and that its dung can serve to treat dimness of sight. The account continues with the specifics of its prognostic value: "Now this

(Continued)

(Continued)

bird is generally to be found in the halls of kings, owing to its peculiar properties. For if anybody is very ill indeed, you can tell from a Caladrius whether the patient is going to live or die. When the sickness is mortal, as soon as the Caladrius sees the patient he turns his back on him, and then everybody knows that the fellow is doomed. If on the other hand it is not a mortal illness, the Caladrius faces the patient. It takes the whole infirmity of the man upon itself, flies up toward the sun, sicks up the man's infirmity and disperses it into the air. Then the patient is cured" (quotation from T. H. White, *The Book of Beasts* [1954]). The king in the present illustration clearly will enjoy the latter good fortune. In addition, one might infer that uroscopy was in some way involved in the diagnosis of his illness since a urine flask rests on a shelf over his bed. Note might also be made of the fact that this bestiary account goes on to nonmedical matters by likening the Caladrius to Christ, its total whiteness relating to His sinlessness; its turning away, His turning away from the Jews because of their lack of belief; its turning toward the patient, His turning toward Christians, thus bearing their infirmities and removing their sins. The second illustration presents a more complex sort of magical prognostication. Taken from the same twelfth-century English manuscript already seen in Illustrations 57 and 247, the figure is a Latin version of the Greek "Sphere of Democritus" (of which original the earliest example occurs in a fourth-century papyrus). The Latin sphere probably dates from the sixth century, though extant manuscripts containing it only occur from the ninth century on. In these manuscripts it is variously ascribed to Pythagoras, Plato, Apuleius, and others as well as to Democritus (to no one of whom was it in fact related). The "mechanism" of the sphere was at once numerological and astrological. One began by determining the numerical value of the patient's name. This was simpler in the case of a Greek sphere, since the letters of the Greek alphabet also served as numerals. In Latin, this initial move required establishing one or another numerical value for each Roman letter, a requirement that is carried out in our illustration in the columns of paired Roman letters and numerals to the left and right of both spheres. The next step in prognosis was to add the day of the moon on which the patient fell ill to the numerical value of his or her name. The total was then divided by thirty. If the result was to be found among the numbers in the upper segment of the sphere, one concluded that the patient would live, but if it appeared in the lower segment, the prognosis was that the patient would die.

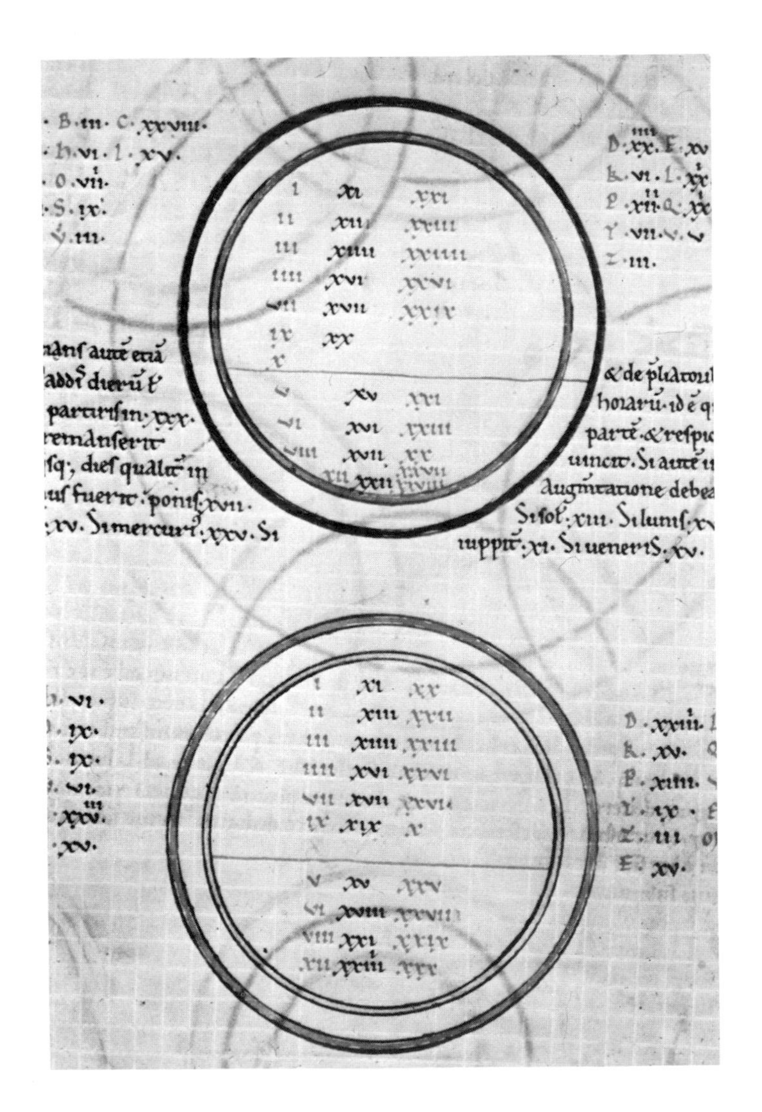

264. Incenditur Sic: Where to Cauterize. The Hippocratic *Aphorisms* close with the following admonition: "Those diseases that medicines do not cure are cured by the knife [σίδηρος]. Those that the knife does not cure are cured by fire [πῦρ]. Those that fire does not cure must be considered incurable" (VII, 87). This is but one indication of the importance of cautery in early medicine. Not only other Greek physicians, but Romans like Celsus (fl. ca. A.D. 25) and Caelius Aurelianus (fifth century) fre-

(Continued)

(Continued)
quently mentioned its therapeutic value, though not without criticism and indication that its use was occasionally controversial. Nor was its role any less prominent in Arabic medical literature and practice. Thus, Avicenna (eleventh century), who devotes less space to it than many others, praised cautery as exceedingly useful in preventing the spread of disease, in comforting an afflicted organ, in doing away with diseased matter within an organ, and in restricting the unwanted flow of blood. It may well be true that the use of cautery began to wane in the High and Late Middle Ages, but one still finds it strongly recommended in this period of incipient decline (a decline, however, that never ruled out its occasional use, even to the present day). Thus, an anonymous illustrated thirteenth-century cautery tract opens by pointing out how beneficial cauterization is to the body: it dries up excess humors and draws them within, while at the same time weakening and purging their sources. We have already seen cauterization and cautery points represented in several pictures showing the medieval surgeon at work (Illustrations 167–168). The set of pictures on page 311 (taken from a twelfth-century manuscript) omits the practitioner and devotes itself exclusively to the depiction of just which locations are appropriate for cauterization for various given diseases. The accompanying legends specify the disease or illness in question and then merely "point to" the figure bearing the relevant locations for cautery by the phrase "burn in this manner" *(incenditur sic)* or even by the single word "thus" *(sic, ita)*. Twenty-five different ailments are covered in three pages of illustrations like those reproduced here. Beginning at the top left, we learn that the appropriate cautery points for a toothache are between the ears and the temples (though the indication of these points on the pertinent figure is rather hard to perceive). The central figure in this quadrant illustrates where to cauterize a patient with a tertian fever (that is, one recurring every forty-eight hours), while the final figure in the same frame gives similar information for someone suffering from dropsy. The smaller figure in the upper right frame specifies the point for kidney or hip illness and the larger figure in frontal view gives the points appropriate for both cautery and bloodletting in the case of a patient with quartan fever. Turning to the bottom left, the two naked figures are both concerned with the cautery points for sciatica sufferers, while the remaining clothed figure represents a patient afflicted with quotidian fever and displays the relevant locations for both cauterization and phlebotomy. The last frame at the lower right contains a seated figure ill with podagra, the barely visible points now said to indicate where such a person should be burned and cut. The figure below holding the bleeding cup for his infirm partner displays the proper cautery points for one with tumors or some other illness of the knees, while the remaining figure at the far right gives the same for hernia. Other, similar schematic illustrations of cautery points contain somewhat more text in their legends, most notably texts specifying the type of cautery iron to be used (for example, round, flat, or sharp). A few such cauterization schemata even include the names of famous physicians and philosophers who are imaginatively paired with each cautery procedure (for example, Hippocrates for the eyes, Galen for the kidneys, Aristotle for the head, and Chrysippus for the stomach). Other pictures from the same manuscript can be seen in Illustrations 167 and 191.

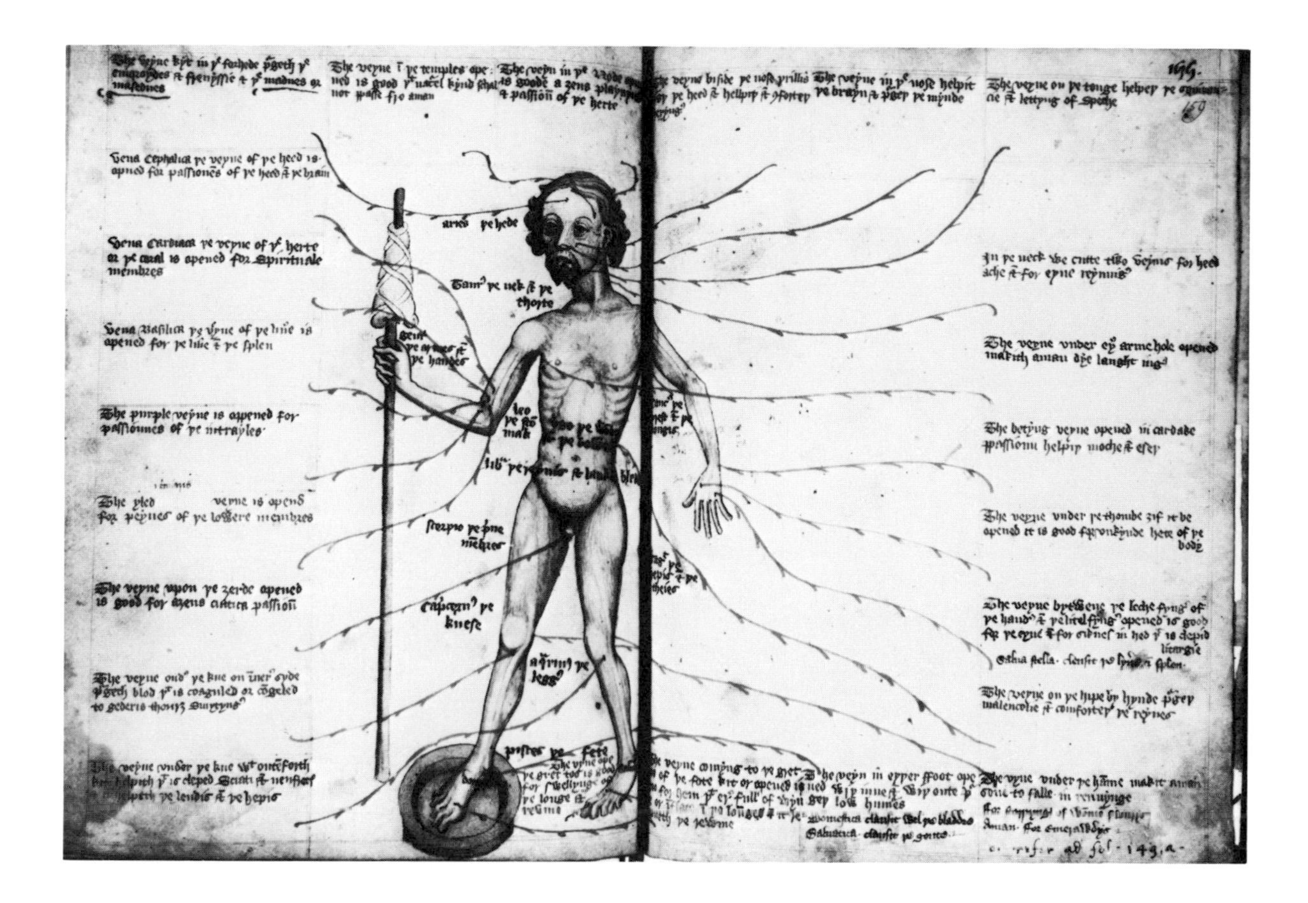

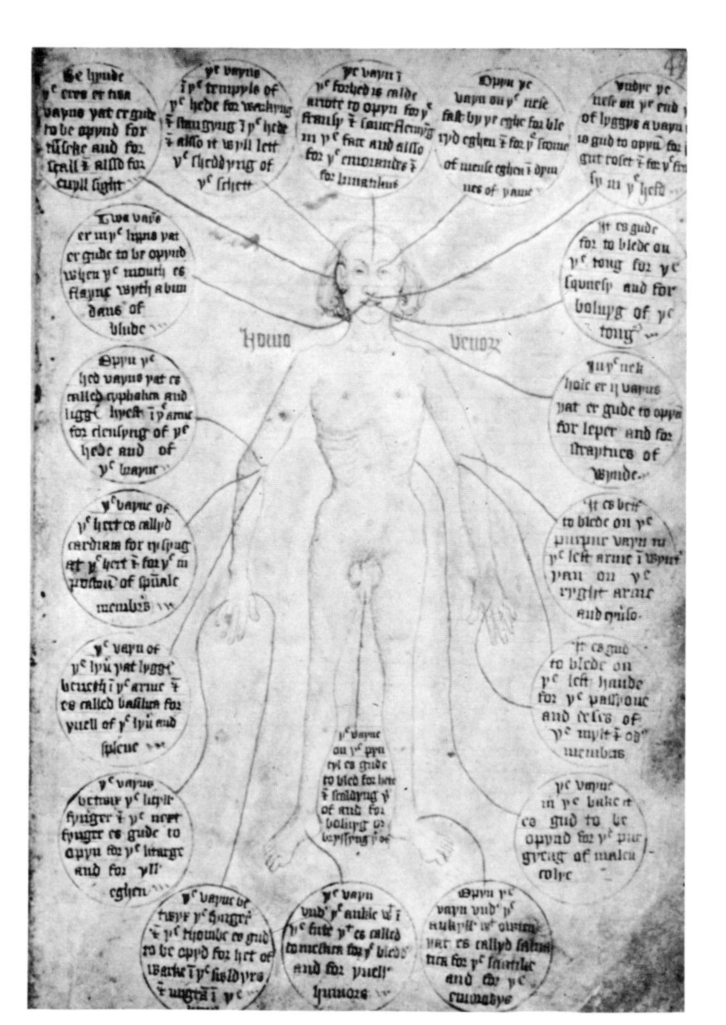

265. It es gude for to blede. An anonymous work in a twelfth-century manuscript recounts the beneficial character of bloodletting by claiming that "it contains the beginning of health, it makes the mind sincere, it aids the memory, it purges the brain, it reforms the bladder, it warms the marrow, it opens the hearing, it checks tears, it removes nausea, it benefits the stomach, it invites digestion, it evokes the voice, it builds up the sense, it moves the bowels, it enriches sleep, it removes anxiety, it nourishes good health" (Lynn Thorndike, *A History of Magic and Experimental Science* [1923], 1:728). Not all ancient and medieval physicians would have waxed so eulogistic about the values of bloodletting, but it was clearly one of the most recommended and used of therapeutic procedures. Mentioned in the Hippocratic corpus and in Aristotle and Galen, bloodletting raised a central question—where to bleed for what. Galen had established correspondences between particular bloodletting points and the diseases that were cured or relieved by bleeding there, but one must turn to medieval manuscripts to find visual representations of these correspondences, such as the vein man reproduced at the lower left, page 313, from a fifteenth-century medical miscellany. Circles surrounding a central figure of a man contain legends in Middle English that describe from which veins blood is to be let and the ailments thus treated, lines being drawn from each circle to the vein in question. Thus, the cir-

(Continued)

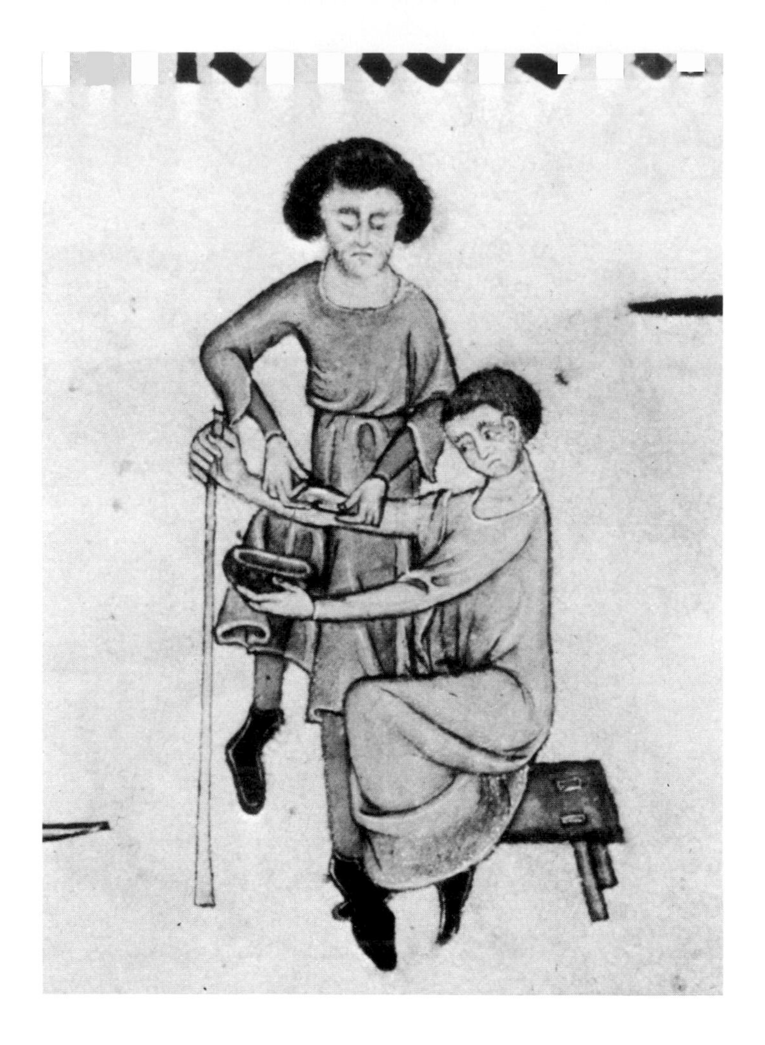

(Continued)

cle second from the top at the right announces that "it es gude for to blede on the tong for the squnesy [that is, quinsy] and for bolnyg [a swelling or tumor] of the tong." In this case, the point of venesection is close to the seat of the illness being treated. In more instances, however, the vein incised was quite remote from the location of the disease. Hence (second circle from the bottom left), one should let blood from the vein between the little and ring fingers in order to relieve lethargy and eye ailments or (third from the top left) one should incise the "vayne that es called cephalica" located near the elbow to effect "clensyng" of the head and brain. The number of bloodletting points located on the inner arm at the elbow is most likely the result of easier access to veins in that locale. Knowing which points of venesection relate to which illnesses was only a part, albeit the most important part, of the art of bleeding. Knowing when to bleed was also most important. It was often held that it was better, for example, to use the left side of the patient in autumn and winter, the right during spring and summer. Even more significant, however, was knowing when *not* to bleed from this or that point. Here the relevant knowledge was astrological: one should not bleed from a vein in a part of the body governed by this or that zodiacal sign when the moon was in that sign. This connection between bloodletting and the astrological allotment of bodily parts is found as early as the *Carmen astrologicum* of Dorotheus of Sidon (first or second century) and was frequently depicted in medieval manuscripts in the form of any number of "zodiac men" (for which see the next illustration). Such information was also occasionally included in bloodletting figures themselves. An example is reproduced at the top of page 313 from a fourteenth-century manuscript. The legends at the top, bottom, left, and right give (again in Middle English) the kind of information seen in the first vein man. Inscriptions on or near the central figure specify which parts of the body are assigned to which zodiacal signs. One should also note that the schematic patient in this illustration stands with his right foot in a bloodletting basin, his right arm holding a staff surmounted by a bloodletting bandage. The latter is an indication of the support usually provided for a limb undergoing venesection, allowing proper relaxation of the limb and furnishing a prop upon which it could rest. This feature is also part of the next illustration (top left), taken from the same manuscript as the first vein man. Here the seated patient holds the bloodletting basin in his free hand, while the physician applies pressure at appropriate points around the incision. Finally, at the lower left, is a marginal depiction of an instrument of incision taken from a fifteenth-century manuscript of John of Arderne's *Practica chirurgia* (for more of which see Illustration 269).

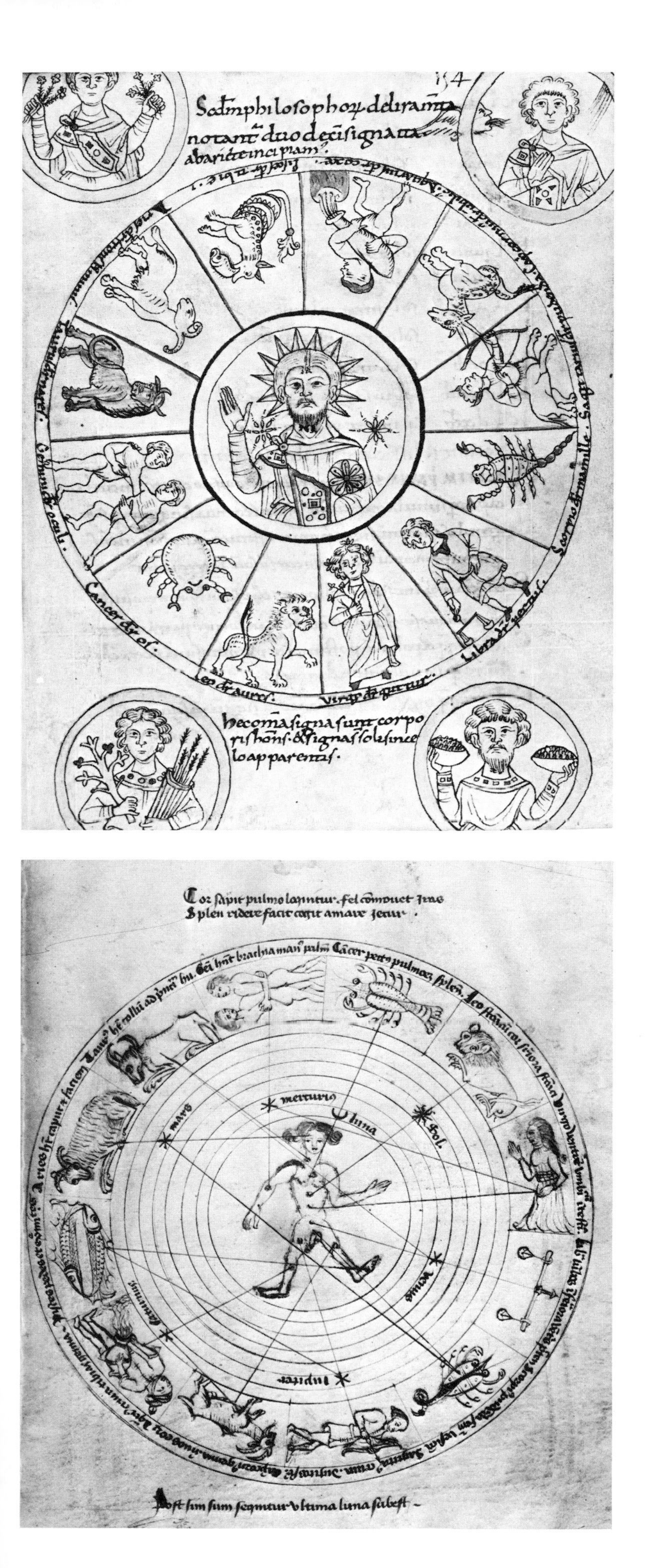

266. Medicine and the Heavens. Surely the aspect of astrological medicine that was most frequently pictured was the doctrine of *melothesia:* the assignment of various limbs or parts of the human body to different zodiacal signs or planets. Although the extant pictorial material is medieval (a sample of which was already seen in the previous illustration), the doctrine itself is ancient. For example, in his *Tetrabiblos,* Ptolemy (second century) drew attention to the association of the planets with various bodily parts and also with various ailments. The assignment of parts of the human frame to the zodiacal signs, beginning with Aries at the head and finishing with Pisces at the feet, is already found in finished form in such sources as the Latin astrological poem, *Astronomicon,* of Manilius (first century) and the astrological handbook of Firmicus Maternus (fourth century). Sextus Empiricus (fl. ca. A.D. 200) mentions (*Adv. math.* V, 21–22) the same body-zodiac allotment, but adds that the reason behind it is that the presence of a maleficent planet in a zodiacal sign at the time of birth will entail the imperfection of the bodily part paired with that sign. (The usual pairing of signs and bodily parts is as follows: Aries the head, Taurus the neck, Gemini the shoulders and arms, Cancer the breast or heart, Leo the sides, Virgo the belly, Libra the loins, Scorpio the sexual organs and groin, Sagittarius the thighs, Capricorn the knees, Aquarius the shins, and Pisces the feet.) The earliest known *melothesia* figure is reproduced at the top left. Taken from an eleventh-century manuscript, it was most likely crafted by a Christian scholar distrustful of pagan learning, since the whole scheme is announced as being based on the foolishness of philosophers *(secundum philosophorum deliramenta)*. A Christlike sun dominates the center of the main circle: its appropriateness is apparent when we realize that the depicted zodiacal signs are not only allotted to various parts of the body, but also specified as signs through which the sun moves. Reading the inscriptions at the circumference of the circle, we learn that Aries (at the upper left) is assigned to

(Continued)

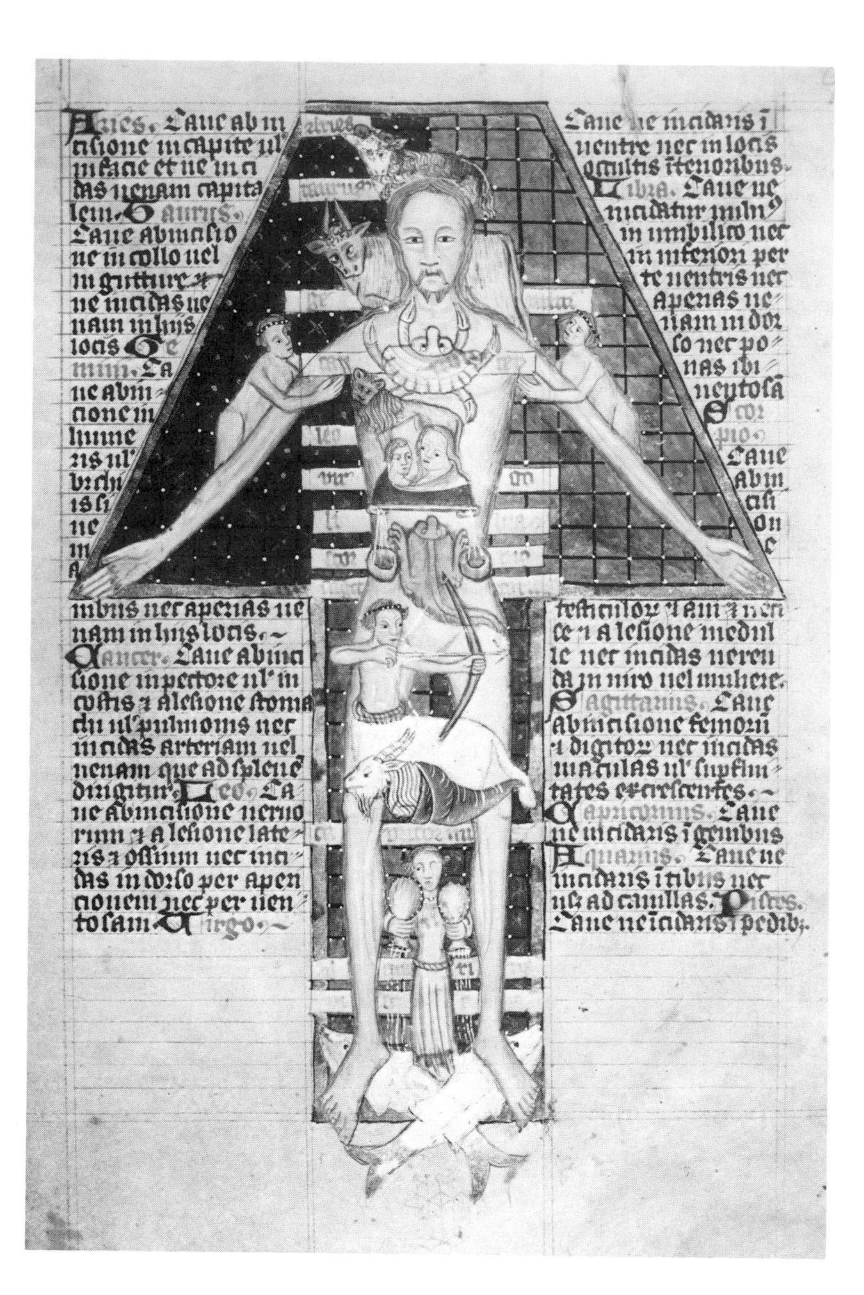

(Continued)

the forehead, Taurus the nose, Gemini the eyes, Cancer the mouth, Leo the ears, Virgo the throat, Libra the chest, Scorpio the breasts, Sagittarius the pudenda, Capricorn the navel, Aquarius the hips, and Pisces the shins. Although this order progresses generally from top to bottom of the body, for the most part it is clearly not the pairing of signs and bodily parts that was traditional long before the execution of this drawing. By contrast, the standard pairing is given, albeit with some of the minor variations and supplements it was always subject to, in each of the next two illustrations (bottom of page 315 and on page 316). The first, circular figure is taken from the late-fourteenth-century manuscript just seen in Illustration 259, while the second, Christlike zodiac man comes from the late-fourteenth-century codex that provided the urine circle in Illustration 260. The author of the circular schema has indicated the zodiac-body correspondence in two ways: first, by drawing lines joining each zodiacal sign with the relevant bodily part; second, by writing the names of the specific parts on the circumference of the circle directly outside the depiction of each zodiacal sign. Further note should be taken of the planet-bearing celestial spheres surrounding the central zodiac man. Beyond depicting the planetary orbs themselves, these intermediate spheres have provided a way to furnish additional information: the lines joining each planet (save the sun and the moon) with a sign indicate one of the astrological houses (either solar or lunar) of these planets. The second zodiac man is of a much more prevalent type. Depictions of the zodiacal signs are placed directly on his body. In this particular figure the names of the signs are also given, but in similar instances such literal identification was often omitted, the placement of pictures of the signs—or merely of their symbols, as in Illustration 81, or sometimes of the names of the signs only—being considered suffi-

(Continued)

cient. Note that in the present example the text surrounding the central figure cautions against carrying out certain specific medical procedures when the moon is in one or another sign. For example, when it is in Taurus, one should refrain from making any incision in the neck or throat or from cutting any vein in that vicinity. Such information is, of course, a more elaborate exposition of the kind of astrological considerations we have seen for bloodletting in the previous illustration. The final picture, at the extreme right, is of a planet man. Occupying a full page in a *Calendrier de Bergers* (also seen in Illustration 235), the illustration shows the star-bearing circles of each planet—with the sun and the moon having their own distinctive depictions—connected by lines to the organs with which these heavenly bodies are associated: the sun, the stomach; the moon, the head; Saturn, the lungs; both Mercury and Venus, the kidneys; both Mars and Jupiter, the liver. This is not the traditional allotment of planets and bodily parts, in which only one planet was assigned to an organ: the sun, the heart; the moon, the head; Saturn, the spleen; Jupiter, the liver; Mars, the gall bladder; Venus, the kidneys (and, at times, also the testicles); Mercury, the lungs. The four corners of this shepherds'-calendar plate are occupied by representations of the four temperaments. In addition to specifying the psychological characteristics displayed by bearers of these temperaments, each is associated (by both word and picture) with a particular element and animal. Thus, clockwise from the upper left we have the choleric temperament, fire and a lion; the sanguine, air and an ape; the melancholic, earth and a pig or boar; the phlegmatic, water and a sheep. Legends above or below these corner pictures indicate the most favorable times at which to bleed a person of the temperament in question—for example, it is best to bleed a choleric patient when the moon is in Aries, Leo, or Sagittarius.

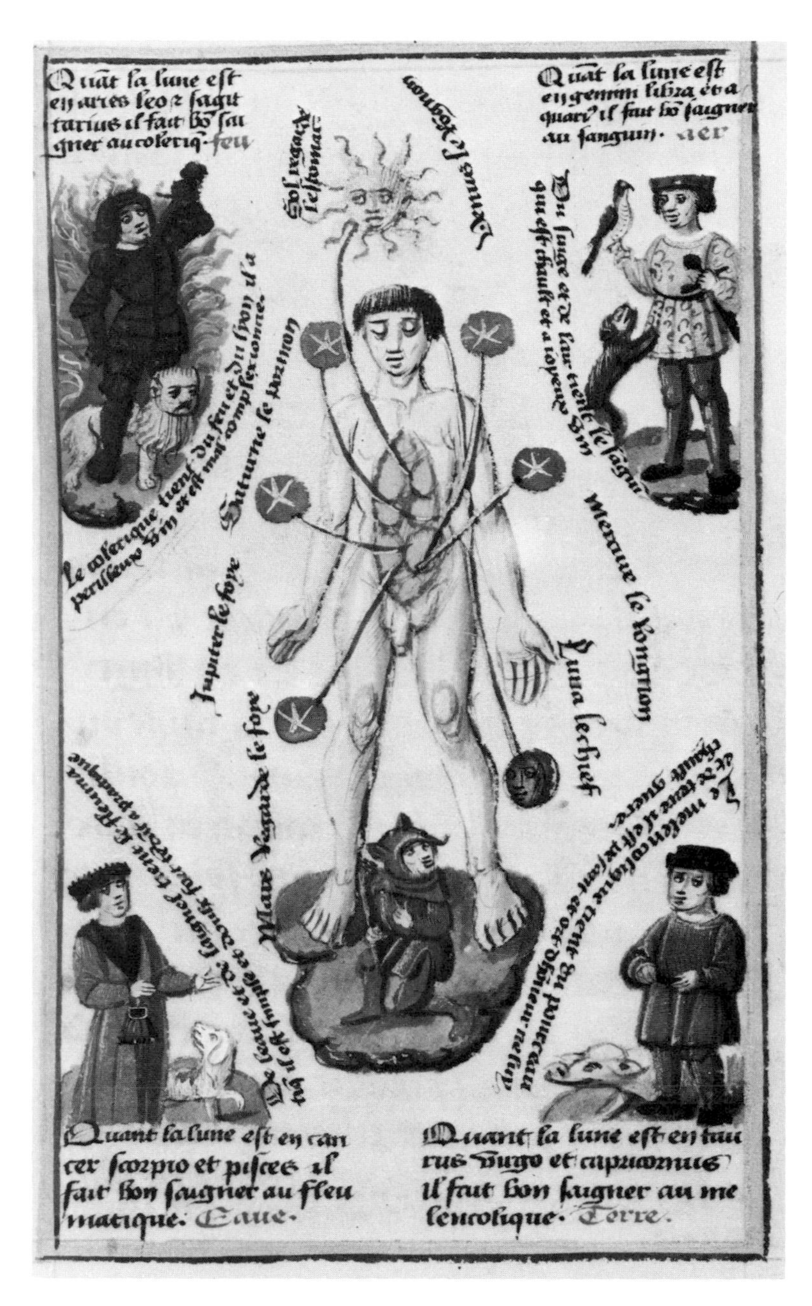

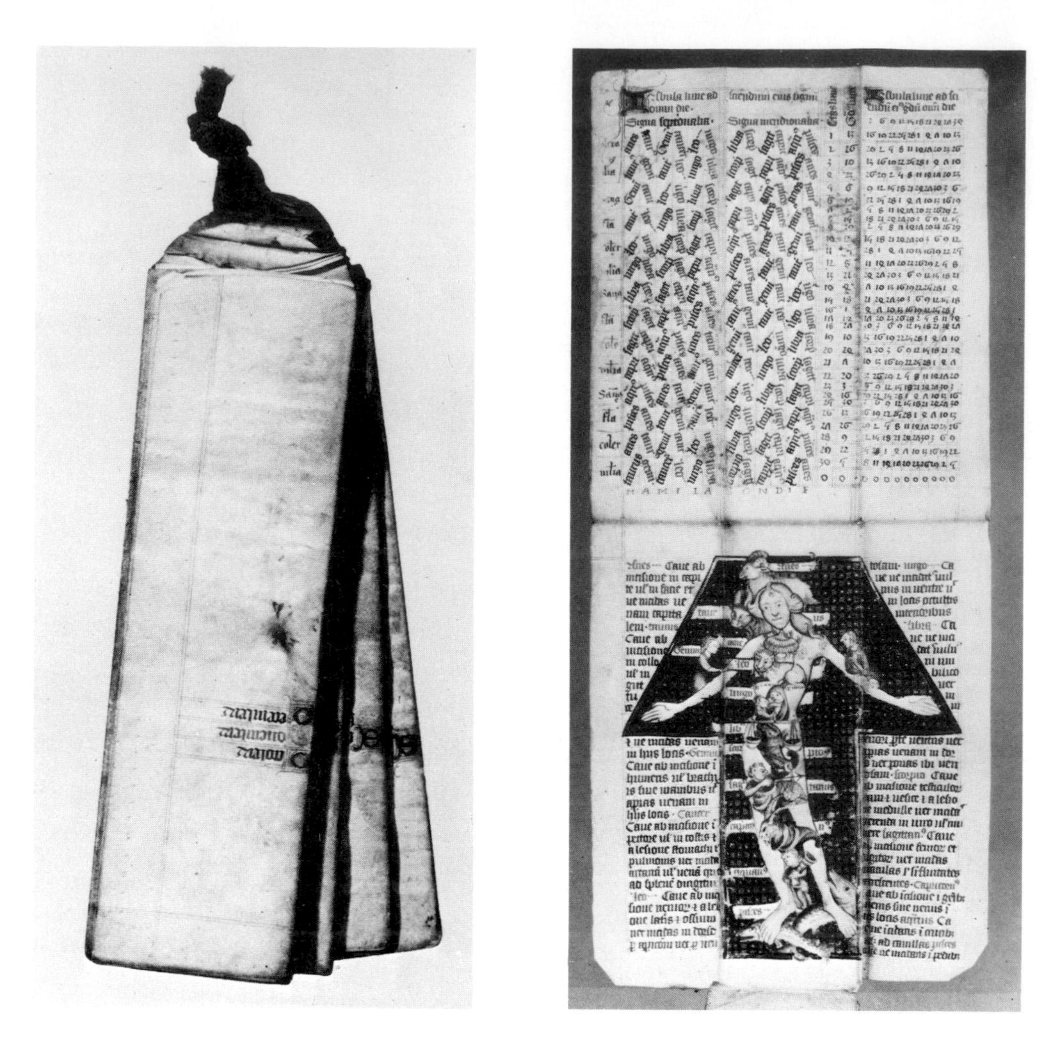

267. Portable Basics for Diagnosis and Therapy. A fifteenth-century astronomical girdle book has already been shown in Illustration 93. The present set of pictures is from a book of the same type that was intended for the use of medieval physicians. Composed at the end of the fourteenth or the beginning of the fifteenth century, this manuscript girdle book consists of six leaves of vellum approximately fourteen by six inches in size, each leaf folded once lengthwise and then folded again into thirds, the resultant form being the seven-by-two-inch folded leaves shown at the extreme left. The whole was meant to be suspended from the belt, each leaf identified on the outside by an appropriate indication of its contents. In addition to this external view, the inside faces of four of the six leaves of this girdle book are reproduced here. As will be apparent, a number of the *aides-mémoire* we have seen in the immediately preceding illustrations can also be found among the present pictures. First, second from the left, there is the kind of zodiac man already met in Illustration 266. Directly above him on the same unfolded sheet appear two lunar tables: the first enables one to determine the zodiacal sign in which the moon appears on a given day; the second, the degree of the moon for each day. Both of these tables provided information of importance for the medieval physician since, as we have seen (Illustrations 265 and 266), the position of the moon in the heavens at a particular time was often crucial to what medical procedures should, or should not, be carried out. The next leaf pictured (third from the right) gives eclipse data. Together with similar illustrations on other leaves, the dates and times of lunar eclipses are listed from 24 October 1398 to 11 June 1462 and of solar eclipses from 18 October 1408 to 20

(Continued)

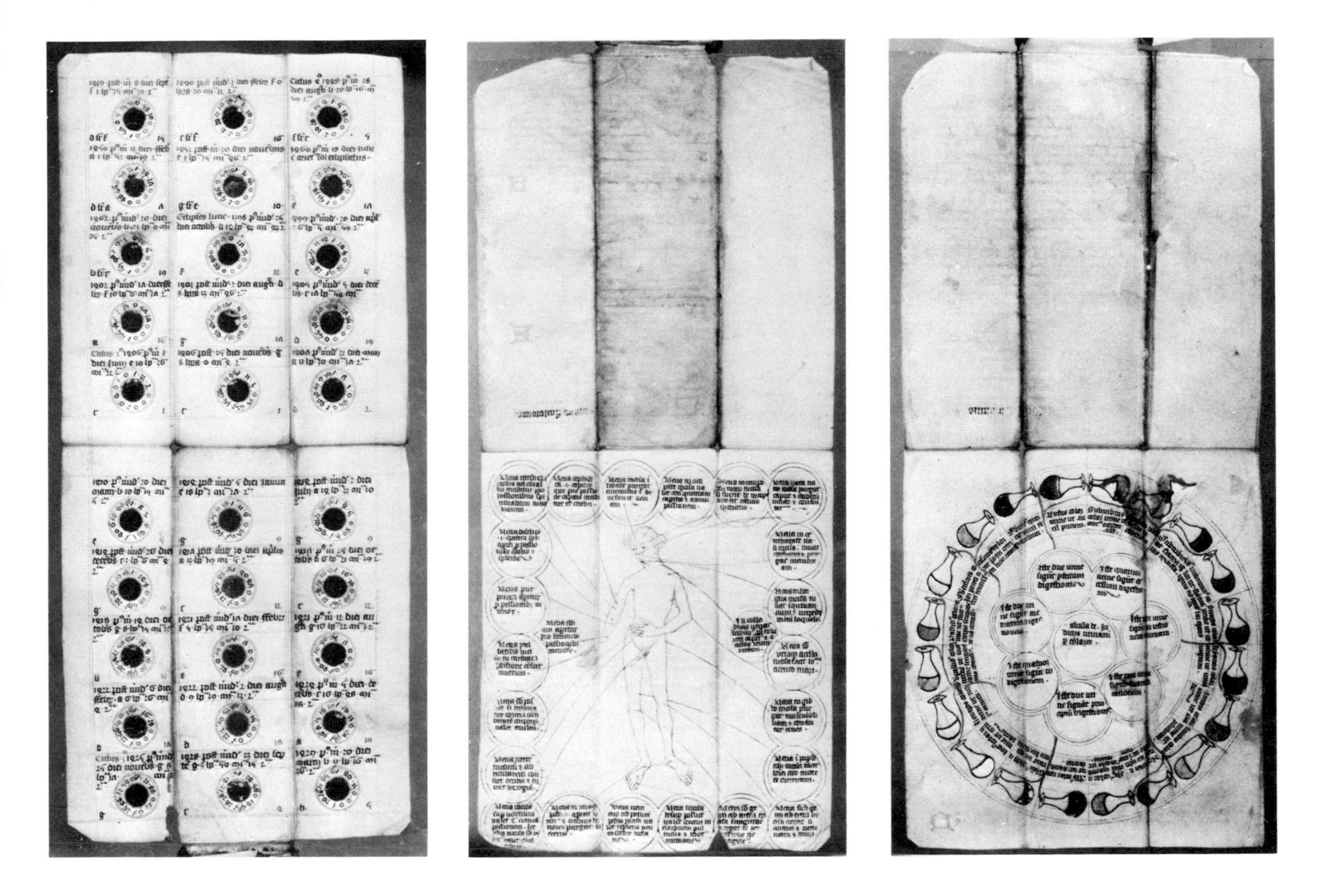

November 1462. As in the girdle almanac pictured in Illustration 93, in addition to the times of eclipses, the center of each circular figure on this leaf depicts which part and how much of the sun or moon will suffer eclipse (although these details are all but invisible in the present reproduction). On the other hand, unlike the earlier girdle almanac, some of the kinds of supplementary eclipse data seen in Illustration 92 are here furnished in the rings surrounding each circular "picture"; namely, how much (in *puncta*) of the body in question will be eclipsed and data enabling one to determine the middle, end, and total duration of the eclipse. Finally, the dominical letter and golden number of each year in which the eclipse occurs are given at the bottom left and right of each circular figure. (The "golden number" of each year indicates where it will fall in the nineteen-year lunar cycle traditionally used to correlate lunar with solar calendars [see Illustration 51], while the "dominical letter" of each year automatically enables one to determine on which day of the week any date in that year will fall; these numbers and letters were both crucial to predicting the date of Easter and other ecclesiastical feast days.) Unfortunately, it is nowhere explained in this girdle book just why or how all of this information was relevant to the cares of a medieval physician. On the other hand, the significance of the figures on the last two leaves appearing at the right is apparent without any explanatory text: the first is a bloodletting man such as that given in Illustration 265; the second is a urine circle similar to that depicted in Illustration 260. The leaves of the girdle book not shown here provide a calendar, a table of planetary positions, and texts that briefly explain the various schemata contained in the book.

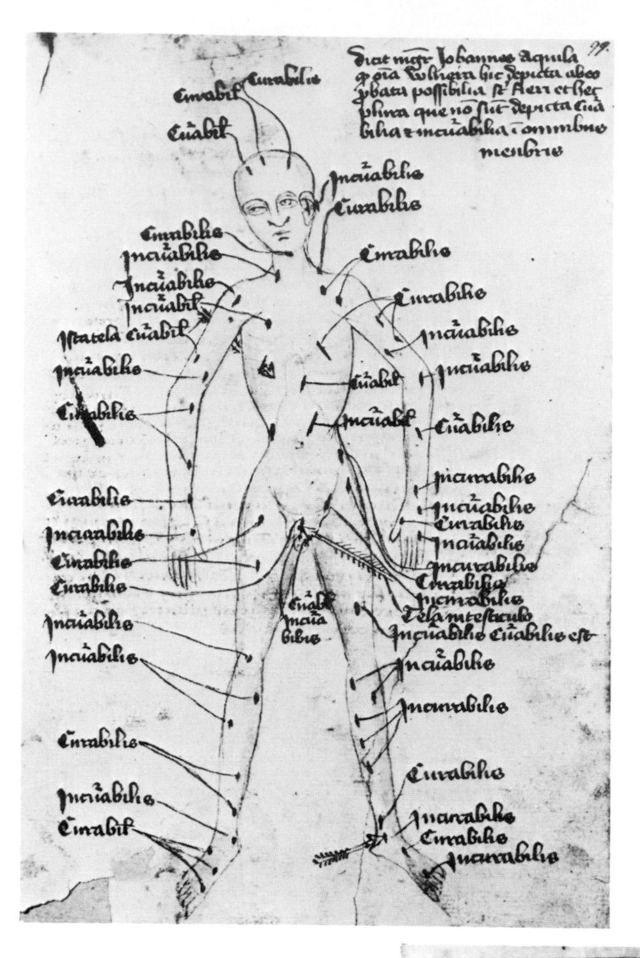

268. Wound Men. This chapter began with reproductions of a so-called disease man and disease woman. Close relations of these earlier illustrations are the "wound men" depicted here, the most important difference being that, unlike the disease figures, the emphasis here is not on a mnemonic picturing related to diagnosis, but rather on prognosis and therapy. These schematic wound men can be found in varying degrees of complexity, two extremes of which are represented here. The first, on the left, comes from a late-fourteenth-century manuscript. A legend in the upper right-hand corner tells us that a certain Magister Johannes Aquila maintained that all the wounds depicted on the accompanying figure—as well as many others that are not depicted—are either curable or incurable. Although at least one historian has raised the possibility of identifying this Johannes as a little-known thirteenth-century French physician named Johannes Aquilanus who wrote a poem on phlebotomy, such an identification seems highly unlikely. Whoever this Johannes may have been, his wound man is only minimally informative. Almost all of the nearly sixty wounds on the body are merely labeled curable *(curabilis)* or incurable *(incurabilis)*, there being almost no indication of how the wounds were caused, let alone of how those designated as curable

(Continued)

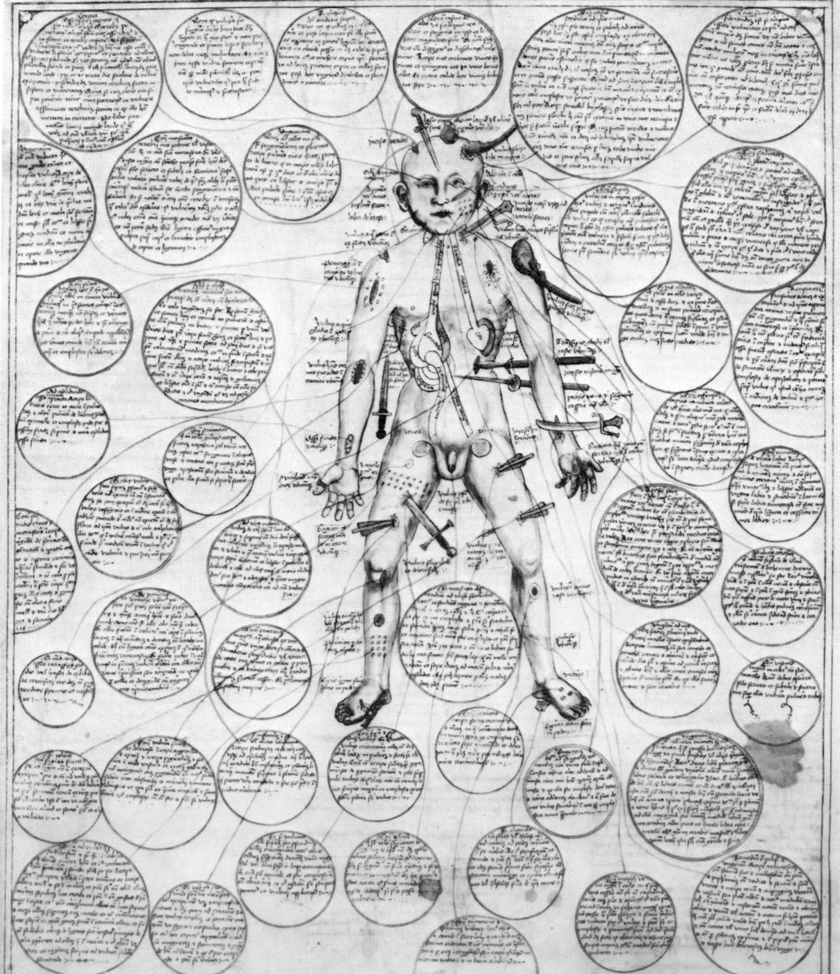

were to be healed. The only exceptions to showing causes are for the three wounds caused by arrows: one in the right thorax said to be curable, another in the left ankle labeled incurable, and a third in the testicles held to be curable *(tela in testiculo curabilis est)*. By contrast, the second wound man at the bottom of page 320, taken from a sixteenth-century manuscript, fairly bristles with information. Knives, clubs, daggers, splinters, and nails, as well as arrows, graphically announce the cause of many of the wounds in question. Moreover, in addition to injuries caused by external violence, such things as tumors, verucas, inflammations, and even bloodshot eyes and facial blemishes are included in the roster of wounds presented by the figure. Brief legends surrounding the body describe all such conditions and wounds in words. Though unrelated to this catalog of wounds, additional anatomical data are given through the depiction of the tracheal artery and the heart and lungs on the right, and the esophagus, stomach, liver, gall bladder, intestines, etc., on the left. But most informative of all are the forty-five text-bearing circles filling out the remainder of this illustration. They tell us how to treat almost all of the infirmities depicted. Thus, the text in the circle second from the bottom at the extreme left joined by a line to the lower arrow in the left thigh contends that, in the case of such a wound, the surgeon should withdraw the arrow by using two probes as guides, one inserted from each side of the wound. Alternatively, the circle third from the top left associated with the gash in the right shoulder refers to a case of white pustules on the circumference of the wound for which one should employ a plaster made from centipedes crushed in a nonsalty fat or grease. As a final example, one might note that the circle next to the left foot gives the prescribed remedy for facial blemishes *(macule faciei)*: thoroughly burn a mixture of unshelled snails, alum, and other ingredients, mixing the resulting ashes with a soap made from berries, and bathe the afflicted face with the resulting product each morning and night.

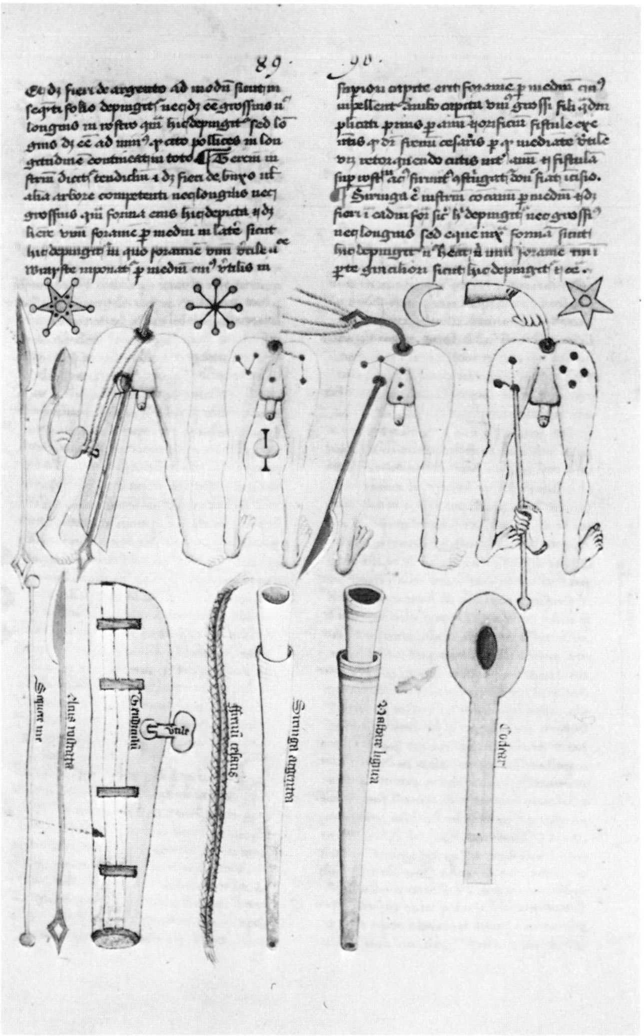

269. Surgical Cures. The procedures of bloodletting and cautery that have been noted in some of the foregoing illustrations were part of the charge of the ancient and medieval surgeon. So were the procedures—more recognizably surgical to modern eyes—depicted in the present illustrations. The first, at the left, is from an early-fifteenth-century manuscript of the *Practica de fistula in ano* written in 1376 by the English surgeon, John Arderne (see Illustration 265 for a picture from another of his works). John's treatise on anal fistulae was evidently quite popular, so much so that it merited translation into English in the fifteenth century. Modifying appreciably the treatment of fistulae given by Abulcasis (al-Zahrāwī; see Illustration 244), he designed original instruments for his surgical procedure. These instruments are depicted here together with a representation of how they are to be employed. The first instrument to be used is shown at the lower left: it is a probe appropriately, and rather quaintly, named a *sequere me* (follow me). Its use is depicted at the upper right: the surgeon is to insert a finger into the rectum, at the same time using his other hand to insert the probe into the external opening of the fistula (that closest to the rectum if, as in the case of the present illustration, there are multiple orifices), pushing the probe through the fistula until it is felt in the rectum. Having thus determined the track of the fistula, the probe is to be removed and replaced by a curved needle *(acus rostrata)* having an eye at one end. This instrument (second from the left) is to be passed through the fistula eye end first and drawn out of the rectum, whereupon it is threaded with a four-stranded ligature (called the *frenum Cesaris*). This whole procedure is shown in the schematic figure of the patient second from the right. The next stage of the operation is depicted at the top left. Here an additional instru-

(Continued)

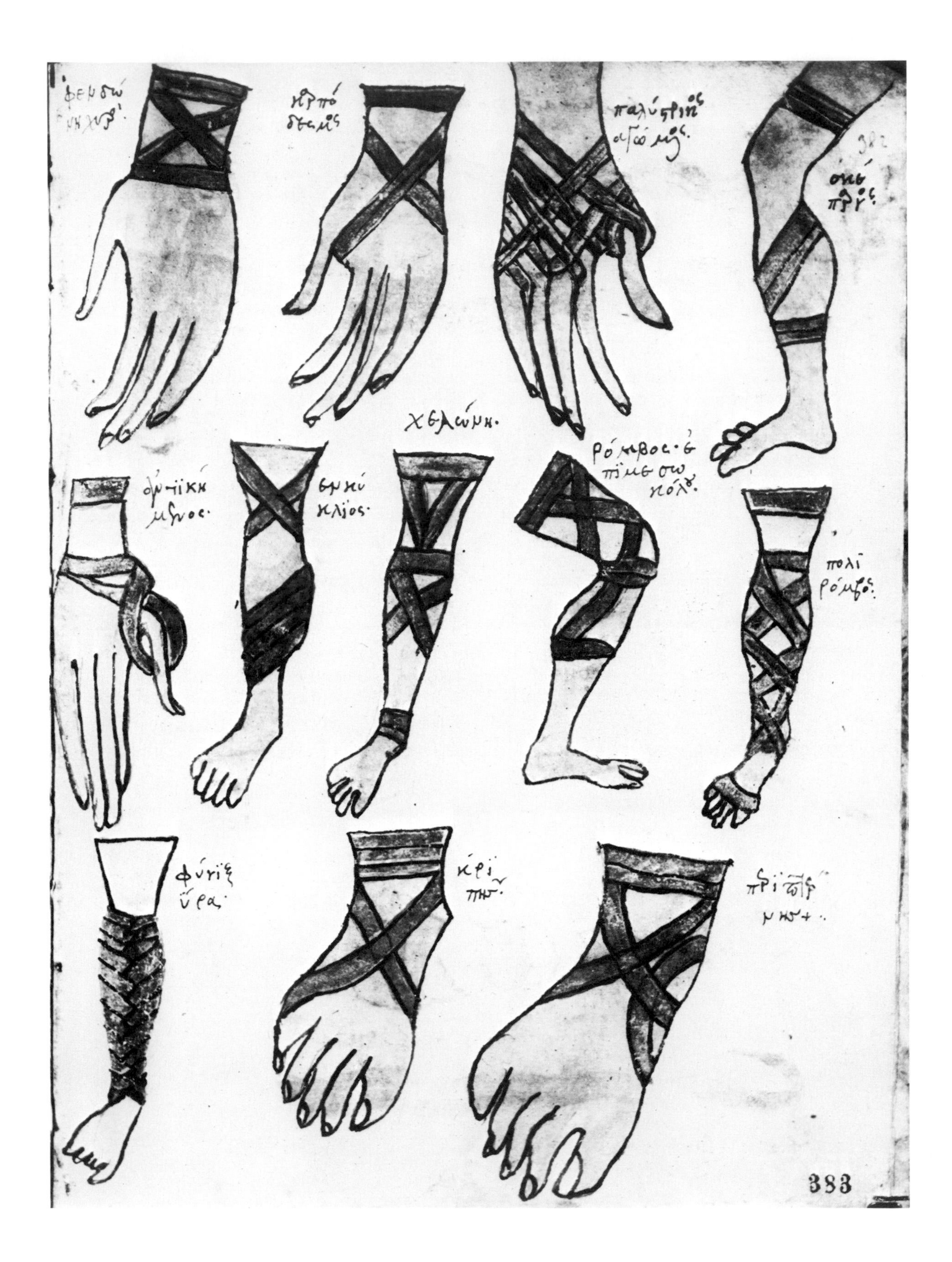
383

(Continued)
ment is employed; called the *tendiculum,* it is pictured third from the left below. A peg *(vertile)* inserted into a hole in this instrument affords a means of tightening the ligatures when they are wound around it. Finally, the spoon-shaped implement (called a *coclear*) pictured at the lower right is inserted into the rectum, the end of the *acus rostrata* fitting into its concave end (to protect the opposite wall of the rectum while cutting). A knife or scalpel may now be introduced into the fistula along the *acus rostrata.* Drawing the scalpel, *coclear, acus,* and ligatures out through the rectum will then cleanly divide the fistula. (The two other instruments depicted below are both syringes; they are not directly involved in the central details of the operation.) The second illustration, facing page, is reproduced from a fifteenth-century Greek manuscript (also seen in Illustration 160) of the **Περὶ ἐπιδέσμων** *(On Bandages)* of the second-century physician Soranus of Ephesus (see also Illustration 262). The techniques of bandaging described in this work draw heavily on the work of Soranus' contemporary Heliodorus as well as on a pseudo-Galenic treatise. The particular bandages depicted in the present illustration all concern the hand, leg, or foot, but there are almost fifty other pictures dealing with the proper bandaging of all other parts of the body. The text to which these pictures belong names each bandage and describes briefly how each bandage is to be wound about the ailing part and held in place. The names alternatively refer to the part being bandaged—such as the "hand sling," *σφενδόνη χειρός*, and the "wrist bandage," *καρπόδεσμος* (the first two at the upper left); to the form of the bandage itself—"circular," *ἐνκύκλιος* (second from center left), or the "boot," *κρηπίς* (bottom center); or even to the function of the bandage—"opposed," *ἀντικείμενος* (for the bandage pulling in the thumb at center left). Even more intriguing are those names presumably given because of some resemblance to an otherwise irrelevant object: "axe" or "adze," *σκέπαρνος* (at the upper right); or "tortoise," *χελώνη* (in dead center). These particular illustrations have a long and interesting history. They also appear in the ninth-century Greek codex seen in Illustration 168, and there is some evidence that they go back to the time of Soranus himself. Looking in the other direction, they were also reproduced in the sixteenth century by il Primaticcio along with others that were in turn used in the same century by Ambroise Paré and Conrad Gesner (on whom see Illustration 206).

270. Theory Behind Medication. Many of the herbal pictures reproduced at the beginning of Chapter Eighteen contained some indication, either visually or in the text, of the medicinal uses of the plants shown. To do so was to present at least a modicum of theory. On occasion, however, herbals and related medical works provided somewhat more of the theoretical underpinings of the discipline they represented. For the most part, this increased presentation of theory occurred in the accompanying text and not in the illustrations themselves, as in the two examples given here. The first, above, is taken from a mid-fourteenth-century Provençal herbal. The plant depicted is specified as *caparus sive caparis,* that is, caper *(capparis spinosa).* We are told that it grows beyond the sea (information that might account for the lion depicted to the left of the plant). Therapeutically, it comforts an ailing stomach (a comfort that may, as the illustration reveals, include the inducing of vomiting) and is useful in com-
(Continued)

(Continued)

bating illnesses of the spleen and scrofula, among other things. More unusual, however, is the inclusion of a fragment of theory that does not relate to the diseases against which the drug is effective—namely, the strength of caper as a medicament expressed in terms of its primary qualities. This theoretical information is presented immediately after the name of the drug: *es ca e seca al ii* g^{a}, that is, expanding the abbreviations and translating, "it is hot and dry in the second degree." Galen had already characterized simple medicines in terms of pairs of primary qualities and, more importantly for present purposes, had "measured" the intensity of these qualities by a scale of degrees that was limited by temperateness at its lower end (in which the medicine would have no perceptible effect) and progressed from qualities intense in the first degree (where the effect was minimal) through the fourth degree (in which the effect was immediate and dramatic, if not catastrophic). It is this tradition that is reflected in the Provençal herbal. The second illustration, above, is from a fourteenth-century Italian manuscript of the *Tacuinum sanitatis.* The fairly numerous illustrated copies of the *Tacuinum* derive from a Latin translation of the (unillustrated) *Taqwīm al-ṣiḥḥa (Tables of Health)* of the eleventh-century Christian physician, Ibn Buṭlan. The illustrations themselves are notable for the manner in which the medicinal plants at hand are placed in a relatively realistic setting, a phenomenon naturally of much importance to the history of art. Our interest, however, is primarily in the captions to these illustrations, for it is there that the theory behind the object depicted is revealed. The subject of the present illustration is, as announced above the framed picture, the medicinal use of sweet melons *(melones dulce)*. Seven kinds of relevant data are given, some of them more theoretical than others. (1) Quality or *complexio* of such melons: cold in the second degree, wet in the third degree. (2) Selection *(electio)*: melons from Samarkand that are ripe and of good substance, color, and odor. (3) Use *(iuvamentum)*: they break up (internal) stones and clear up the skin. (4) Detriment *(nocumentum)* to health: they cause looseness of the bowels. (5) Means of avoiding this detriment *(remotio nocumenti)*: administer in an acetic syrup. (6) What they produce in the body *(quid generant)*: a moderate amount of blood. (7) Patients to whom suited *(conveniunt)* and when and where: those of a dry and choleric temperament, both old and young, in the autumn or at the end of the summer in temperate regions.

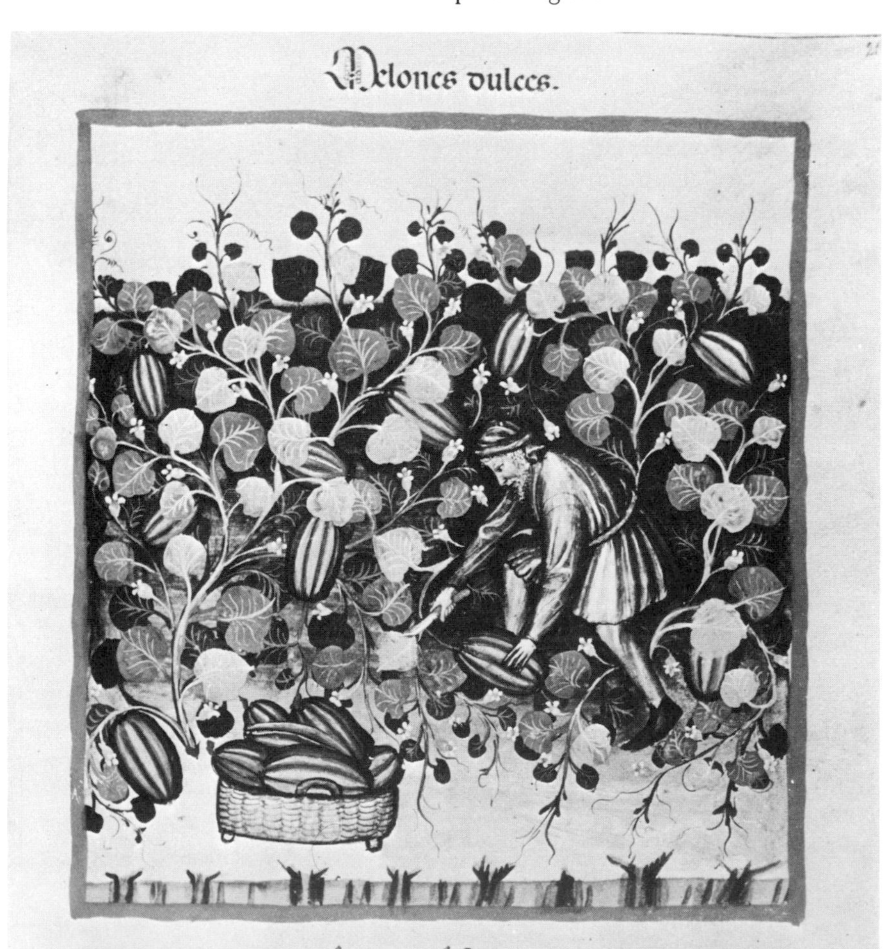

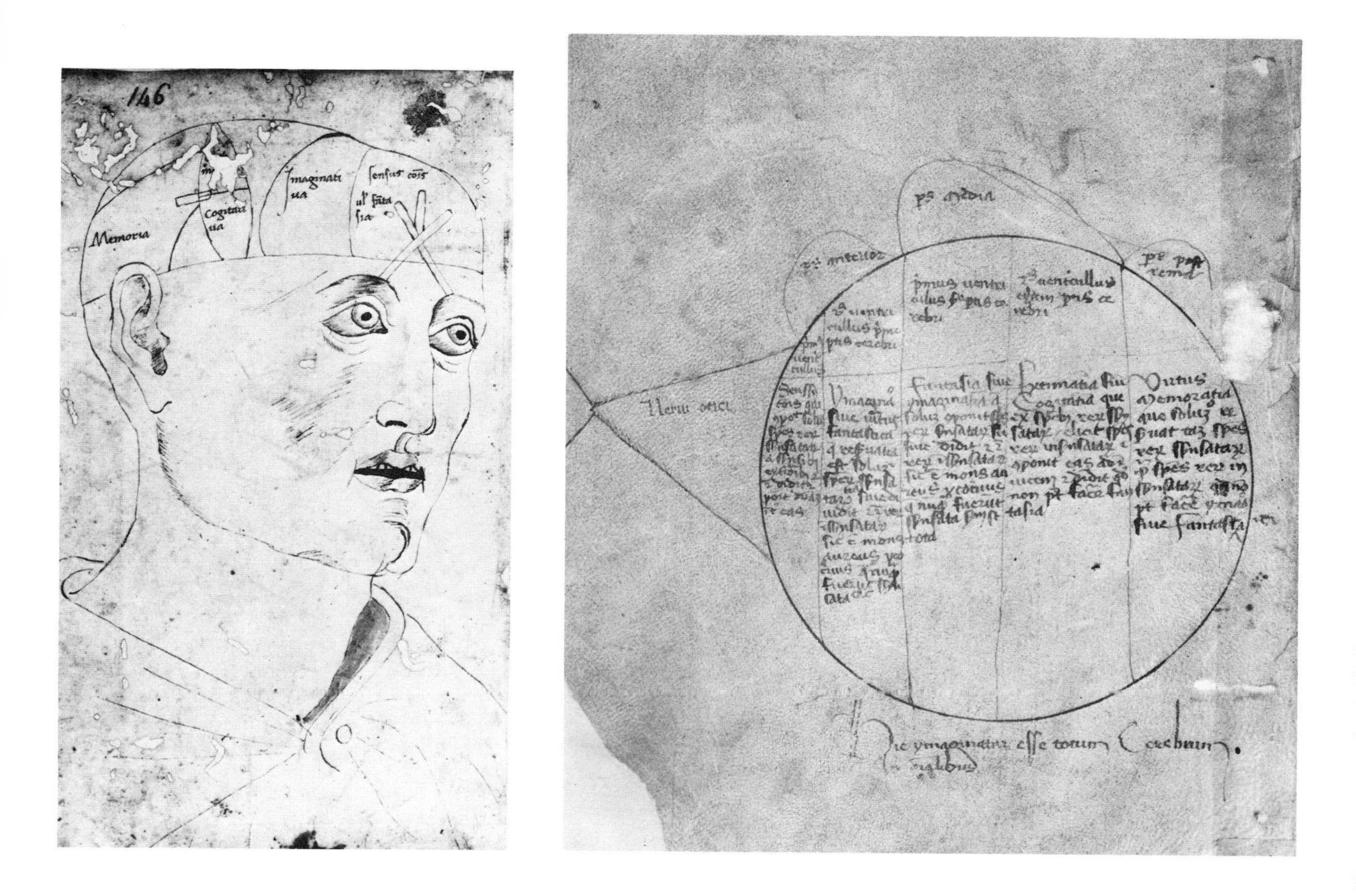

271. The Anatomy of the Brain and the Functions of the Soul. "Tell me where is fancy bred, Or in the heart or in the head?" (*Merchant of Venice,* 3.2.63–64). Although Shakespeare's "fancy" refers to love and not properly to one of the mental processes whose bodily locale had been the subject of so much controversy in antiquity and the Middle Ages, the alternative proposed in his question clearly goes back to the two opposing views at the center of that controversy. One camp, most notably represented by Aristotle, maintained that the seat of the soul was the heart; while the opposition, including Galen and other medical writers, claimed that the brain served this function. No ancient diagrams illustrating this debate are extant. For that we must look to medieval sources where, though the Aristotelian heart party does receive representation, by far the greater share of pictorial material is devoted to showing where the functions of the soul reside in the brain. The assignment of these functions in the disease man has already been shown in Illustration 259, where the rear of the head carried labels that revealed the location of each faculty of the soul. An almost identical distribution of these faculties is seen in the drawing at the left above, taken from a fifteenth-century medical miscellany. Here the assignment of the faculties does not occur as a mere supplement to other medical information, but is presented in and for itself. The theory being presented is what historians have come to call the "medieval cell doctrine." Deriving from elements within Galen, this doctrine of localizing mental processes in the cells or ventricles of the brain received an important medieval stamp of approval when it was set forth by Christian scholars such as St. Augustine (354–430) and Nemesius of Emesa (fl. ca. 400). Although there were slight differences in the allotment of mental functions to the various cells of the brain, this illustration gives a relatively standard view of the matter when four cells were involved. The first cell or ventricle at the forehead carries the label *sensus communis vel fantasia.* It was almost uni-

(Continued)

(Continued)

versal to locate the "common sense" in this initial position, but *fantasia* was much more frequently ascribed to the second ventricle, which it shared with the imaginative faculty (here alone in this part of the brain). The third ventricle is occupied here by that mental process termed *estimativa* (mostly missing due to damage to the manuscript) or *cogitativa*—a faculty that, in human beings, enabled one to recognize the benefit or harm in physical things not comprehended by the exterior senses and to come to know the imaginary representations derived from these things. Finally, the fourth ventricle is revealed as the seat of *memoria.* The circular schematic illustration at the right, page 325, is taken from a late-fourteenth-century manuscript and is explicitly announced to be a diagram of "the whole brain in animals." The triangle at the left is said to represent the optic nerves, while the fivefold division of the brain apparently follows a variant arrangement deriving from Avicenna (eleventh century). The upper segment of the circle carries labels specifying the various ventricles; the last division (giving a fifth ventricle) remains unlabeled. The scallops above indicate the three parts of the brain as anterior, medial, and posterior. The text in the first division informs us that this first ventricle is the seat of the *sensus communis,* which composes species from things sensed by the exterior senses and also differentiates among them. The next two ventricles both deal with *imaginatio* or *fantasia* (they are labeled, respectively, *ymaginativa sive virtus fantastica* and *fantasia sive ymaginativa*), a doubling that was characteristic of the five-cell Avicennian doctrine. The text in this figure does not make the reason for this apparent duplication clear. It is simply that the imaginative faculty in the second ventricle is representative (the text mistakenly has *reservativa* for *representativa*) of the species of both sensible and insensible things (examples of golden mountains and goat stags being given for the latter), while the imaginative faculty located in the third ventricle is preservative of the same. The fourth ventricle is the home of the *extimativa* or *cogitativa* faculty met before, while the fifth ventricle harbors the *virtus memorativa.* The third illustration, page 326, comes from a manuscript written in 1449–1450 of the *Philosophia naturalis* of one Henricus Plattenberger. The first two illustrations clearly fall in the Galenic medical camp in locating mental functions in the brain. This illustration is equally clearly Aristotelian, as is obvious from the location of a great number of mental processes in the large heart depicted in the center of the body. In addition to such expected functions as *cogitatio, intellectio, sensatio, ymaginatio, fantasia,* and so on, we find essentially logical operations such as *abstractio, discursus,* and *refutatio* among the more than twenty processes populating the heart. Four of the five exterior senses are appropriately located at the nose, mouth, hands, and eyes, a visual pyramid proceeding outward from the last. The optic nerve is indicated in a lozenge in the dead center of the head, while the very top of the head has received the label *congregatio specierum,* the brain itself appearing as another lozenge directly below. Since no mental processes find their seat in the brain, it simply carries the label *cerebrum.* Similarly, although the rear of the head bears the inscription *organum reservativum,* this is meant to indicate only the organ involved

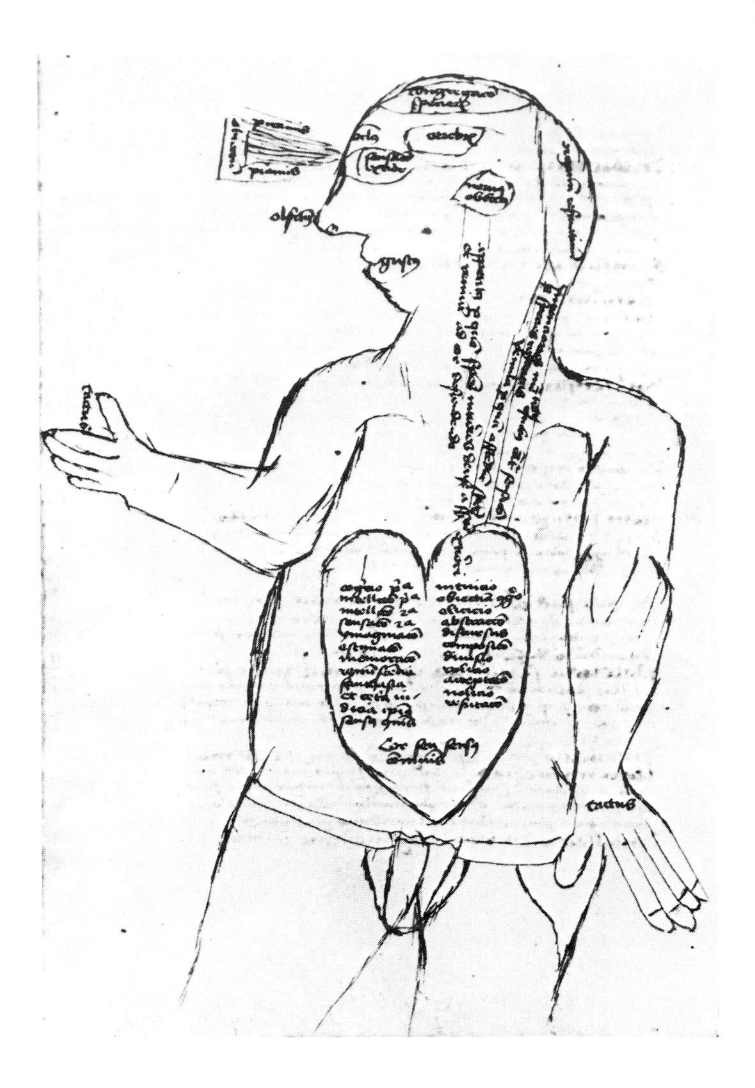

(Continued)

in the retention of images and not the mental process of memory. The latter is located (as *memoratio*) in the heart. Finally, note should be made of the channels between the heart and the head. They are the paths through which information derived from the exterior senses descends to the heart, a second channel at the rear of the body providing the means by which the *spiritus* or vital spirit acquired through respiration ascends to the head. The final illustration, on page 327, comes from the same thirteenth-century manuscript seen in Illustrations 209, 211, and 212. Indeed, its extremely schematic character is similar to that of the bodily organs in the second picture of Illustration 211. Inasmuch as the surrounding text has nothing to do with this curious figure, its excessive schematism makes it difficult to determine just what is being represented. Were it not for the rather obvious nose at the bottom of the figure, one might not realize that the two circles are the eyes from which the optical nerves proceed upward. This in turn indicates that the remainder of the figure at the top of the page is intended as a diagram of the brain. By appealing to similar figures in other manuscripts, one can establish with some likelihood that the four diamonds (two of them split into triangles) are meant to represent the four cells. Beyond that, nothing can be said with any certainty at all of the significance of the remaining shapes. For another diagram of the brain, see Illustration 287.

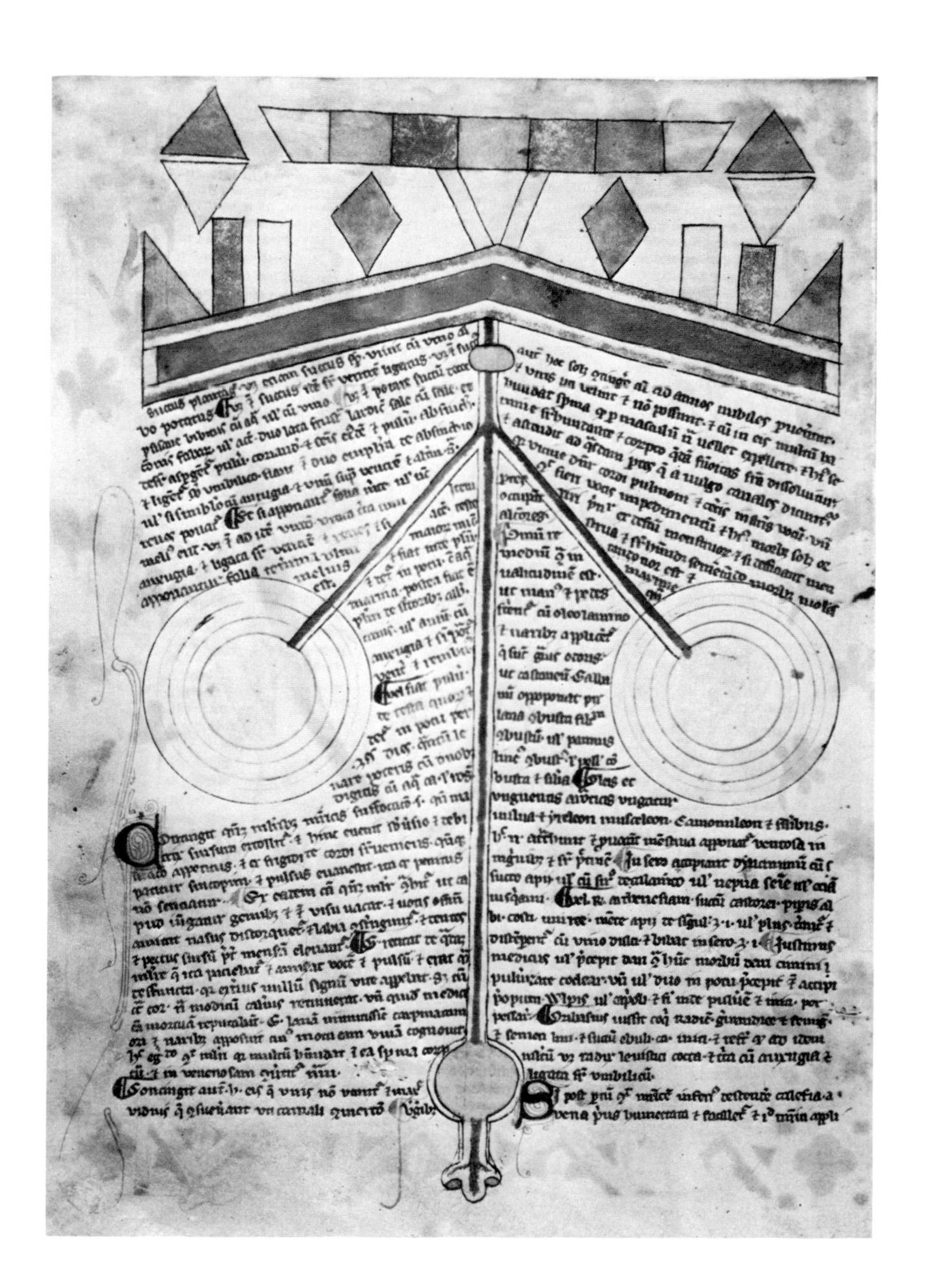

CHAPTER TWENTY-FOUR

The Creation, the Universe, and Its Parts

THE SPHERICAL ASTRONOMY of the universe was dealt with in Chapter Twelve and the objects populating it were the subject of Chapters Eighteen, Nineteen, and Twenty-Two. This chapter will treat visual material concerned with how that universe came to be and with the various theories of its form, of the relations obtaining among the creatures within it, and of such earthly features as climata and winds; consideration of the elemental constituents of the universe will be reserved for the next chapter.

With the exception of the opening illustration of this chapter, which provides a glimpse of the ancient Egyptian view of the cosmos, all the pictorial material is, once again, medieval. At that time, the predominant "world system" was Aristotelian, a factor that caused no small amount of intellectual commotion, since Aristotle maintained the existence of an eternal world, while medieval scholars—Christian, Jewish, or Islamic, no matter—firmly believed in its temporal creation. The challenge, then, was to overturn the Aristotelian contention of the eternity of the world—though some saw fit to allow its possibility while denying it in fact—and at the same time to embrace the remaining elements of the Aristotelian system, all neatly worked into the context of a world created in time and dominated by a supreme creator.

This medieval perspective is naturally at the very heart of depictions of the creation, almost all of them belonging, as in Illustration 273, to the account of the creation set forth in the first chapter of the Book of Genesis. But it is also very much part of most of the other illustrations in this chapter. God and the divine are dominant elements—for example, in the Christian-Platonic "mechanism" of creation depicted in Illustration 274. And the diagrams provided in Illustration 275 of the range of beings within the universe also contain God, as the supreme limit of all, together with various of His ministering angels. Even depictions of the shape of the cosmos (Illustrations 276 and 277) necessarily include reference to the Creator and His work, most frequently in terms of an empyrean heaven or some other outermost celestial sphere that is His abode. In the other direction, one occasionally finds a depiction of hell (Illustration 276) or, symbolically, of the devil (Illustration 277).

The visualization of terrestrial features, such as the zones of the earth (Illustration 279) or its winds (Illustration 280) did not require depiction of the divine element in the world system. It was also absent in most instances from diagrams setting forth the doctrine of the harmony of the celestial spheres (Illustration 278). Though discussed in medieval works that may have treated the divine origin of the world in other passages, this ancient doctrine received its most detailed examination in treatises on musical theory and in works reporting the elaboration it was given in such treatises. It was part of *musica mundana,* as distinct from *musica humana* and *musica instrumentis* (Boethius, *De musica,* I, ch. 2), and included the harmony to be found in the diversity of the seasons *(temporum varietate)* and the agreement of the elements *(compage elementorum)* as well as in the heavenly spheres, even though for the most part only the last-mentioned was to become the subject of medieval diagraming.

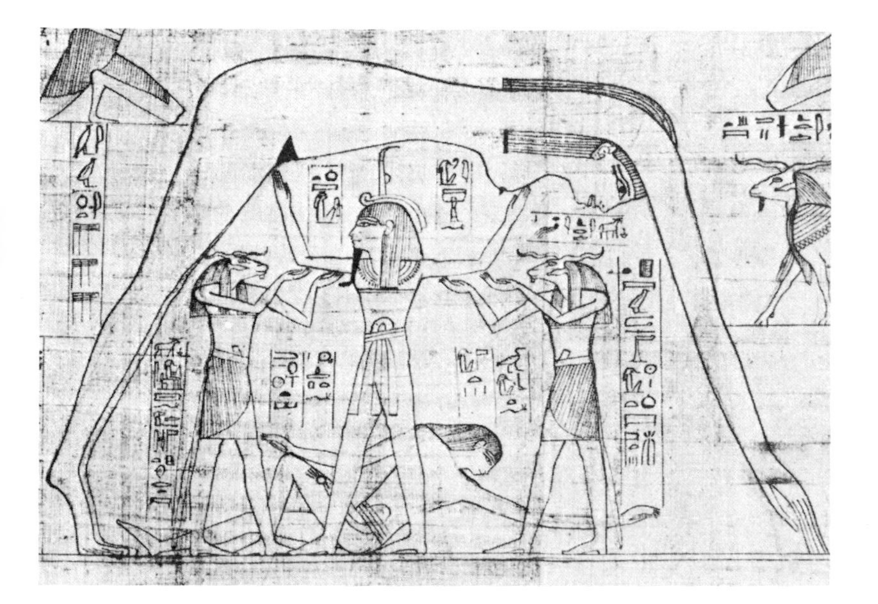

272. The Egyptian Cosmos. In the Egyptian creation story associated with the city and temple of Heliopolis, the god Atum, high upon a hillock in the midst of the primordial watery waste of Nūn, creates two cosmic-entity gods: Shū, the god of air, and Tefnūt, the goddess of (perhaps) moisture. They in turn bear Geb, god of the earth, and Nūt, goddess of the heavens or sky. This Old Kingdom (2800–2250 B.C.) story provided the divine personages needed for the visualization of the cosmos according to Egyptian eyes. It has already been noted (Illustration 218) that, as sky goddess, Nūt often furnished appropriate decor for the inner face of coffin lids, thus providing the deceased pharaoh with an eternal view of the heavens overhead. Here, however, Nūt appears as part of the more complete representation of the cosmos given at the upper left. Taken from a tenth-century B.C. papyrus from Deir el-Bahri of the *Book of the Dead,* she is arched over a prostrate earth god Geb beneath. Shū, as god of the air, supports Nūt with his upraised arms (which are in turn supported by ram-headed wind spirits on either side). As a whole, the drawing can be interpreted as a depiction of the separation of heaven from earth, a theme that was traditional in many ancient creation myths. It is also worthy of note that the Egyptian conception of the female Nūt as sky goddess and the male Geb as earth god is just the reverse of what is usually encountered in cosmic mythology. Nūt was, however, not the only figure Egyptians used to represent the heavens. The Divine Cow often served as an alternative. Shown here at the lower left as represented in the tomb of Seti I (1303–1290 B.C.) in the Valley of Kings at Luxor (compare Illustration 217) the divine celestial cow has a belly covered with stars, with barks of the sun poised between her front and back legs. Once again, the air god Shū supports the celestial vault represented by the divine cow, while eight spirits symbolizing the pillars of heaven lend similar support to her legs.

273. In principio creavit Deus celum et terram. This opening verse of the first chapter of the Book of Genesis was often interpreted as God's creation "in the beginning," not simply of the heavens and the earth, but of the four elements. Based on the contention that heaven (as fire) and earth mentioned as extremes imply the inclusion of the intermediate elements of air and water, this interpretation naturally related the Genesis account of creation to natural philosophy. Moreover, the explanation of other verses in this account brought forth still other fragments of science and philosophy. Although in most instances this interpretation occurred only in writing, some received the added attention of visual representation. An especially interesting case in point involving a fair amount of theory will be treated as part of the next illustration, but the present pictures also provide a visual interpretation of parts of the first chapter of Genesis, albeit at a more elementary level. The first example at the left is taken from a late-thirteenth-century French *Bible moralisée*. A compass appeared frequently in medieval creation miniatures such as this one. The theme behind this motif most likely was that of the Creator as designer or architect of the universe; its ultimate source probably was Plato's *Timaeus,* in which the notion of a craftsman god who impressed forms on a preexistent unformed matter or chaos was central to the world's creation. Indeed, in this miniature, one might imagine

(Continued)

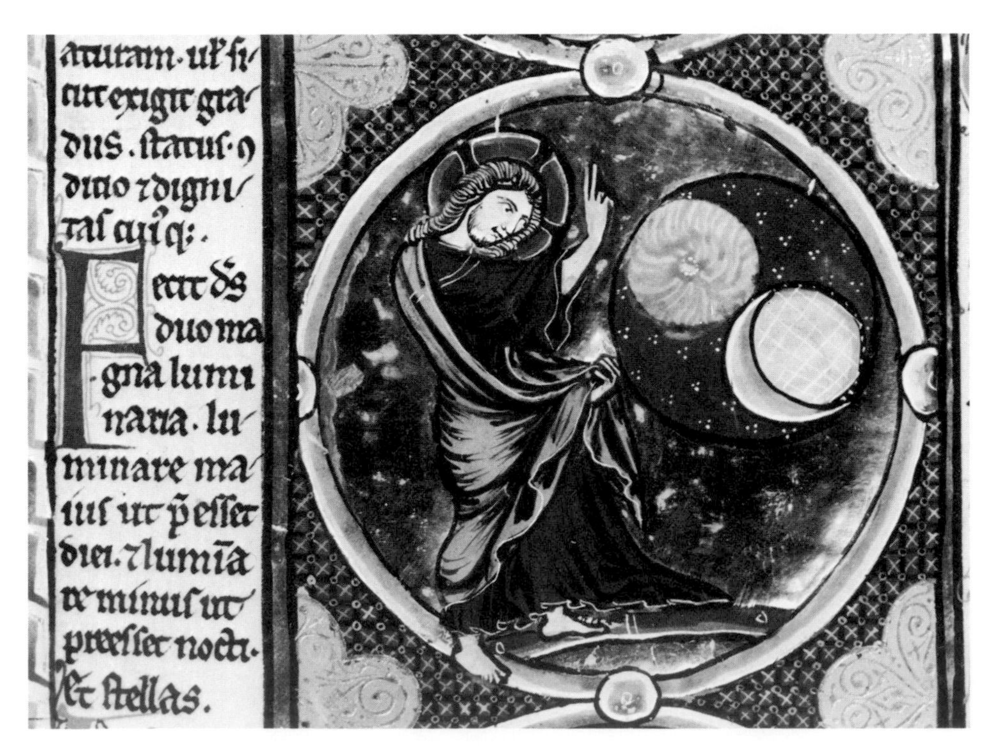

that the amorphous mass flanked by the sun and the moon at the center of the universe, of which God is taking measure with his compass, in some way reflects this Platonic idea of unformed matter, even though there is no textual basis for such a suggestion. The second illustration, bottom of page 330, taken from another thirteenth-century French *Bible moralisée,* depicts the fourth day of the six days of creation in the Genesis account. It is then that God created the sun, as the greater light ruling the day; the moon, as the lesser light ruling the night; and the stars. All of these heavenly luminaries are shown in a circle in front of God, who is gesturing in a manner intended to depict an act of creation. The verse (I, 16) represented appears at the left: *Fecit Deus duo magna luminaria, luminare maius ut preesset diei et luminare minus ut preesset nocti; et stellas.* The third illustration, above, also depicts a verse from the first chapter of Genesis, but unlike the first two pictures it is not taken from an illustrated biblical text. It comes instead from an early-thirteenth-century English bestiary. Such a phenomenon was not unusual, since a fair number of bestiaries contained illustrations of selections from Genesis 1 and 2 as prefaces to their proper charge. The present miniature depicts the first stages of the sixth day, when God brought forth the "living creature after his kind, animals of burden, those that crawl, and beasts of the earth." The verse in question (I, 24) frames the miniature on two sides: *Dixit quoque Deus: producat terra animam viventem in genere suo, iumenta et reptilia et bestias terre secumdum suas species. Factumque est ita.* The pictures of animals are very much in the bestiary style, as can easily be seen by comparing them with other figures reproduced from the same manuscript (the last picture in Illustration 201 and the manticore in Illustration 206).

274. A Christian–Platonic World. Both of the pictures reproduced here are taken from a twelfth-century manuscript of the *Clavis physice* of the early twelfth-century scholar Honorius Augustodunensis. The *Clavis* was to a great extent an adaptation of the *Periphyseon* or *De divisione naturae* of Johannes Scotus Eriugena (ca. 810–877). Some of the features of the present illustrations depict elements set forth by the *Clavis* while others represent notions (largely twelfth-century Platonic) not explicitly found in that work. In the miniature on the left above, the uppermost rank is dominated by a kingly *Bonitas* at the center, flanked on the left by four female figures representing *Justicia, Virtus, Ratio,* and *Veritas,* and by three others on the right standing for *Essentia, Vita,* and *Sapientia.* This is likely a variant of the theme of Wisdom or Philosophy, with the Seven Liberal Arts as seven daughters (compare Illustration 173), here serving to depict—as the legend directly underneath announces—the "primordial causes" of the world. The causes are divine names, having come to Honorius' *Clavis* from Pseudo-Dionysius the Areopagite (fifth century) via Scotus Eriugena. At least part of their effect on the world can be seen in the next rank of pictures, labeled *effectus causarum.* Time *(tempus)* and space *(locus)* as two initial elements or categories needed for an organized world appear at the left and right, while a rather fascinating depiction of "unformed matter" *(materia informis)* occupies the center of the second rank. Shaped as four-faced, this *materia informis* is that "earth without form and void" of *Genesis* 1:2 *(inanis et vacua)* in the Latin Vulgate (ἀόρατος [invisible] and ἀκατασκεύαστος [unformed] in the Greek Septuagint), given a twelfth-century interpretation as the chaos or primordial matter of Plato's *Timaeus.* The third rank, entitled *natura creata, non creans,* provides a visual representation of the four elements (from left to right: fire accompanied by three angels, a connection not uncommon in

(Continued)

Christian sources; air; water; and earth). The very bottom of the miniature pictures God as *finis,* for God is both the beginning and the end, the alpha and the omega, of all—an idea once again appropriated by the author of the *Clavis* from Scotus Eriugena. At the right, a second schema from the same manuscript is substantially more involved with twelfth-century Platonism than the first. The central figure of a woman represents the *anima mundi* (world soul) that was so often a feature of twelfth-century natural philosophy. Thus, William of Conches maintained that the *anima mundi* gave rise to the motion of the stars, vegetation among plants, sensibility in animals, and rationality in man, a contention that is reflected in the present schema by medallions of the sun and the moon joined by the legend *sydera* (stars) overhead, and the inscription on the scroll held open by the personified world soul: *vegetabilis in arboribus, sensibilis in pecoribus, rationabilis in hominibus.* Yet another Platonic element appears in the three faculties of the human soul, the names of which appear at the feet of the central figure: *racionabilitas, concupiscibilis, irascibilitas.* The four elements occupy the four corners of the schema, beginning with air at the upper left, proceeding clockwise through fire, earth, and water. They are characterized by both Aristotelian pairs and Platonic triplets of primary qualities. These two manners of characterizing the four elements were already seen in Illustrations 66 and 247, and they will be examined in greater detail in the next chapter. There is, however, a notable feature of their depiction in the present schema that deserves comment now: namely, the arms joining the corners occupied by the elements and framing the whole schema in a rather striking fashion. Each of these arms bears the name of a single (Aristotelian or Platonic) primary quality, a quality shared by each of the elements joined by the arm in question. The arms thus indicate visually which elements share which primary qualities. At the same time, they provide information about the transformation of these elements into one another, since it is on the basis of their shared primary qualities that such transformations occur. Another schema from the same twelfth-century manuscript is given in the next illustration.

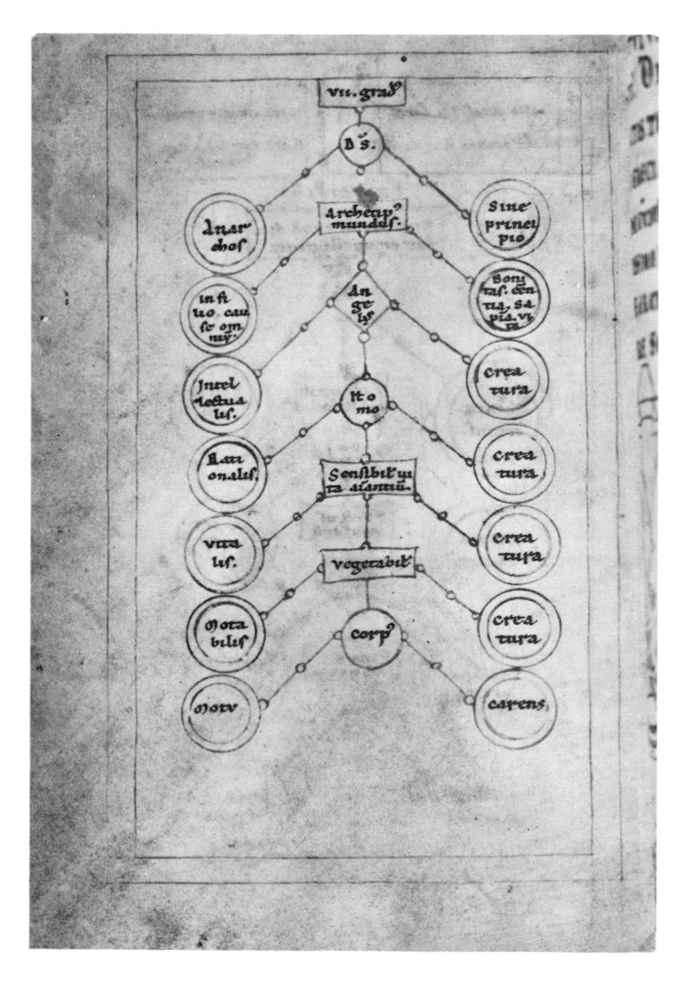

275. The Scale of Being. The idea of a hierarchy of being extending from prime matter, or even from nonbeing, through all manner of existents to god or some other supreme being at its other limit was professed by any number of authors from antiquity on. Often viewed as a scale of being at any level of which God could, as an expression of the plenitude of this scale, create beings of a perfection matching that level, the whole idea has been christened by one historian the "Great Chain of Being." The illustrations reproduced here are three variant medieval expressions of that chain. The first, at the left, comes from the twelfth-century codex that appeared in the previous illustration. The central column of the "tree" forming this schema announces the various degrees of being, the circles on either side indicating properties descriptive of each degree. Thus, it begins with mere body *(corpus)* totally lacking in motion at the base of the tree. The next stage is occupied by *vegetabilis* being, which does harbor a principle of growth or motion within it *(motabilis creatura)*. Sensitive living being *(sensibilis vita animantium; vitalis creatura)* is next, followed by man as *rationalis creatura.* Angelic being as *intellectualis creatura* occupies the next stage. Above the angelic realm is the *archetipus mundus,* characterized (in the circle at the right) by several of the primordial causes found in the miniature in the immediately preceding illustration: *bonitas, essentia, sapientia, vita.* The cor-

(Continued)

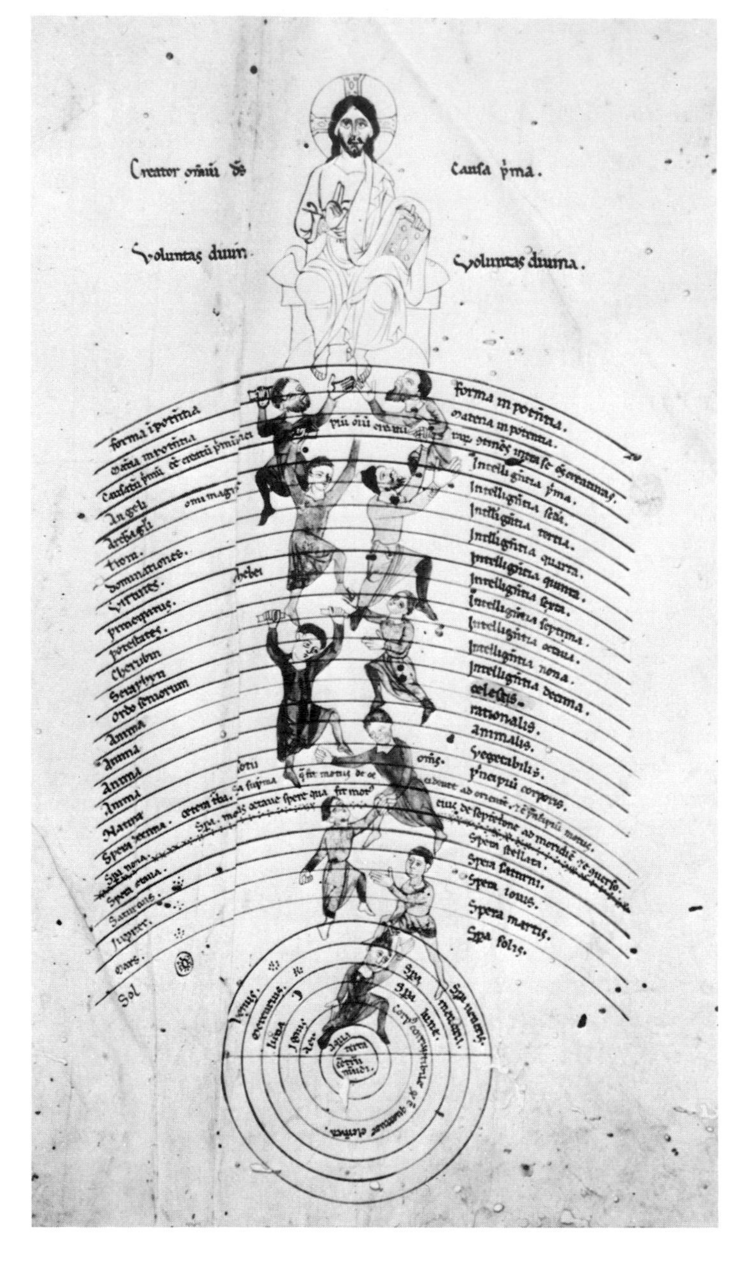

(Continued)

responding circle at the left reminds us that these causes dwell in Christ the Word *(in Filio cause omnium)*. The apex of the diagram is God, identified as "degree seven" *(VII gradus)* and as without beginning *(anarchos, sine principio)*. The second illustration, top left, taken from another twelfth-century manuscript where it accompanies a brief text on the nature of man and the destiny of his soul, is a quite different and much more complex chain. It proceeds from the small innermost circle at the bottom (labeled *terra, centrum mundi*) to the domineering figure of Christ in majesty at the top, surrounded by the inscriptions *Cre-*

(Continued)

ator omnium Deus, Causa prima, and (twice) *Voluntas divina.* The three element circles of water, air, and fire enclose the central earth below. They are followed by circles representing the spheres of the moon, Mercury, and Venus. Then, beginning with arcs in place of complete circles, we have the spheres of the sun, Mars, Jupiter, and Saturn. These spheres are followed by three others: the eighth sphere of the fixed stars; a tenth accounting for motion from west to east (presumably of the sun, moon, and planets against the "background" of the fixed stars), also said to be the beginning of motion; and between them a ninth sphere, on which the inscription speaks of a north-to-south motion—perhaps a confused reference to the motion of the planets in latitude (see Illustration 249). The next arc carries the legend *natura principium corporis,* showing that it is meant to account for the material aspects of the world—both celestial and sublunar—beneath it, just as the next four arcs representing the world soul are meant to account for its nonmaterial aspects. The next ten arcs cover the Avicennian doctrine (*Metaphysica* IX, 4) of the ten celestial intelligences, here correlated not with heavenly spheres but with nine choirs of angels and the twenty-four elders of the *Apocalypse* (here referred to as the *ordo seniorum*). Inexplicably, the artist has inverted the order of these intelligences, seraphim and cherubim thus erroneously falling lower

(Continued)

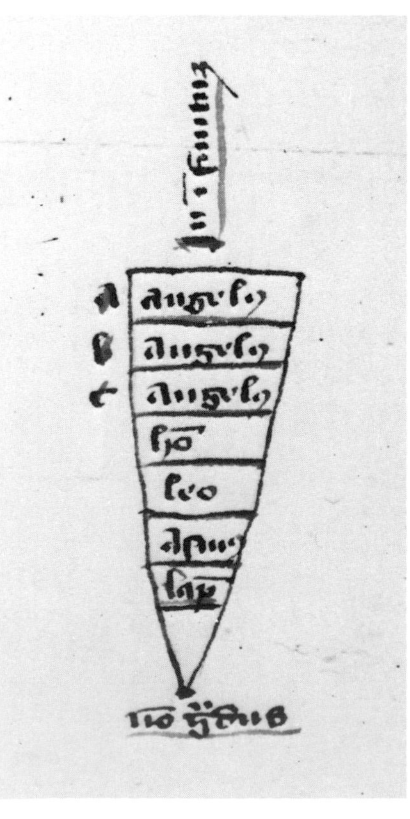

than mere angels and archangels. Three final circles remain between this series and the enthroned Christ. The first, it has been suggested, corresponds to the Plotinian νοῦς (its inscription is: *causatum primum esse creatum primum principium omnium creaturarum continens intra se omnes creaturas*), while the final two represent, respectively, *materia in potentia* and *forma in potentia.* The ten human figures stretching from the earth below toward Christ above, increasing in age over this upward path, apparently symbolize the ascension of the human soul. The final chain of being at the bottom of page 334 is reproduced from an early fifteenth-century manuscript (also seen in Illustrations 67 and 142) of the *De perfectione specierum* of Jacobus de Sancto Martino (fourteenth century). Here the chain or scale is expressed through a diagram belonging to the doctrine of the latitude of forms (see introduction to Chapter Thirteen and Illustrations 142–145). Within this doctrine, the scale of being was often termed the *latitudo entium,* occasionally, as here, pictured geometrically as the inverted isosceles triangle. Beginning from a point representing no being at all—*non gradus,* as labeled here—the increasingly greater parallels to the triangle's base correspond to greater degrees of being or perfection. The first of these parallels here represents the species stone *(lapis),* followed by *asinus, leo, homo,* and three angelic species. The vertical inscription over the top of the inverted triangle reads *in infinitum* and is meant to convey the fact that the triangle continues upward to an infinitely distant God. Indeed, the notion of God as an infinitely distant extrinsic limit to the whole *latitudo entium* found frequent expression in fourteenth-century theological works. The natural philosophical doctrine of the latitude of forms was put to work, as it were, in a remarkable number of contexts.

276. The Shape and Strata of the Universe, I. An example of the most frequent medieval depiction of the form of the universe is seen at the right, taken from the very decorative fourteenth-century manuscript already seen in Illustrations 21, 45, and 58. Surrounded by excerpts from the introductory chapters of a commentary on the *De sphaera* of John of Sacrobosco (for whom see Illustration 128) written by the thirteenth-century Franciscan John Pecham, three outer spheres enclose the shell belonging to the fixed stars. The first is that of the empyrean heaven, the motionless abode of god, spirits, and the throne of Solomon. Next there is the *primum mobile* or tenth sphere, described here as a heaven between the empyrean and the crystalline and said to be moved with the simplest motion. Directly within this there is a *celum cristallinum et applanes,* which, according to Pecham's text, undergoes a slow west-to-east motion, the daily rotation from east to west being provided by the *primum mobile.* The next eight spheres are those of the fixed stars and the seven planets, followed by the four elemental spheres constituting the sublunar realm. The excerpts from Pecham conclude with the reflection that the total number of spheres entails that one ascend through fifteen degrees in proceeding from the center of the world to the throne of Solomon in the empyrean. This count ig-

(Continued)

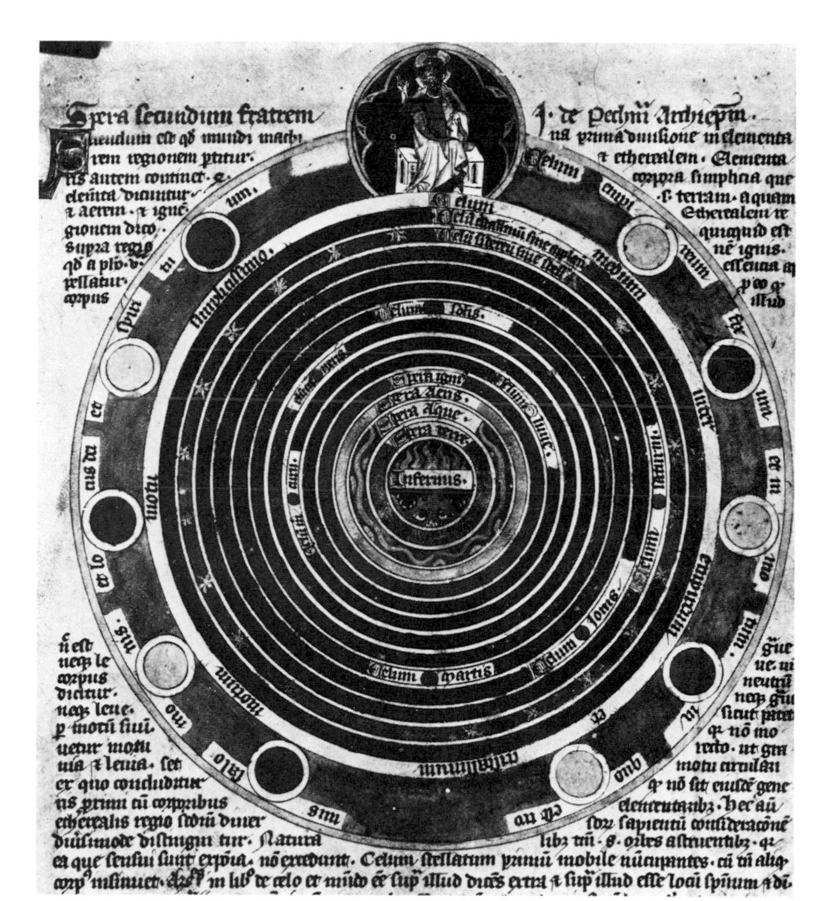

(Continued)

nores the very center in the present picture: it is occupied by hell *(infernus)*, shown here complete with flames and a rather charming devil peeking out from its depths. The next circular diagram of the world, at the left, comes from a fourteenth-century manuscript (also seen in Illustration 250), where it accompanies the *Dragmaticon* of William of Conches (twelfth century). Its division into four *regiones* is Platonic. In the *Timaeus* (39*E*), Plato spoke of four types of living creature corresponding to fire, air, water, and earth; but, like all medievals, William read the *Timaeus* in the translation and commentary provided by Chalcidius (fourth–fifth century). There, other parts of the Platonic corpus were mined, resulting in a fivefold division of the universe, which provided the background for the present diagram. The two outermost regions are both fiery, the first being a divine celestial, rational, immortal, and impassible living creature *(animal)* providing the abode of the stars, while the second is also a rational and immortal living creature, but now passible and hence less noble than that of the first region. The third and fourth regions have good and bad angels as their inhabitants, while the fifth region is that of man, a mortal, rational, and passible living creature. As a whole, this schema can be viewed as a kind of

(Continued)

conflation of the *anima mundi* with the notion of the universe as composed of a celestial realm together with the sublunar spheres of the four elements. The final circular figure, bottom of facing page, comes from the same fourteenth-century copy of the Provençal *Breviari d'amor* of Matfre Ermengaud that furnished the extraordinarily crude eclipse picture in Illustration 253. The numbers written on the various spokes of this earth-centered circle serve to illustrate the preceding lines of the poem, which address themselves to the problem of the distance between the sublunar and celestial regions. But the most interesting features of the figure are the two angels. They are visualizations of the intelligences that move the heavenly spheres, and they move these spheres by turning an Arctic and an Antarctic crank.

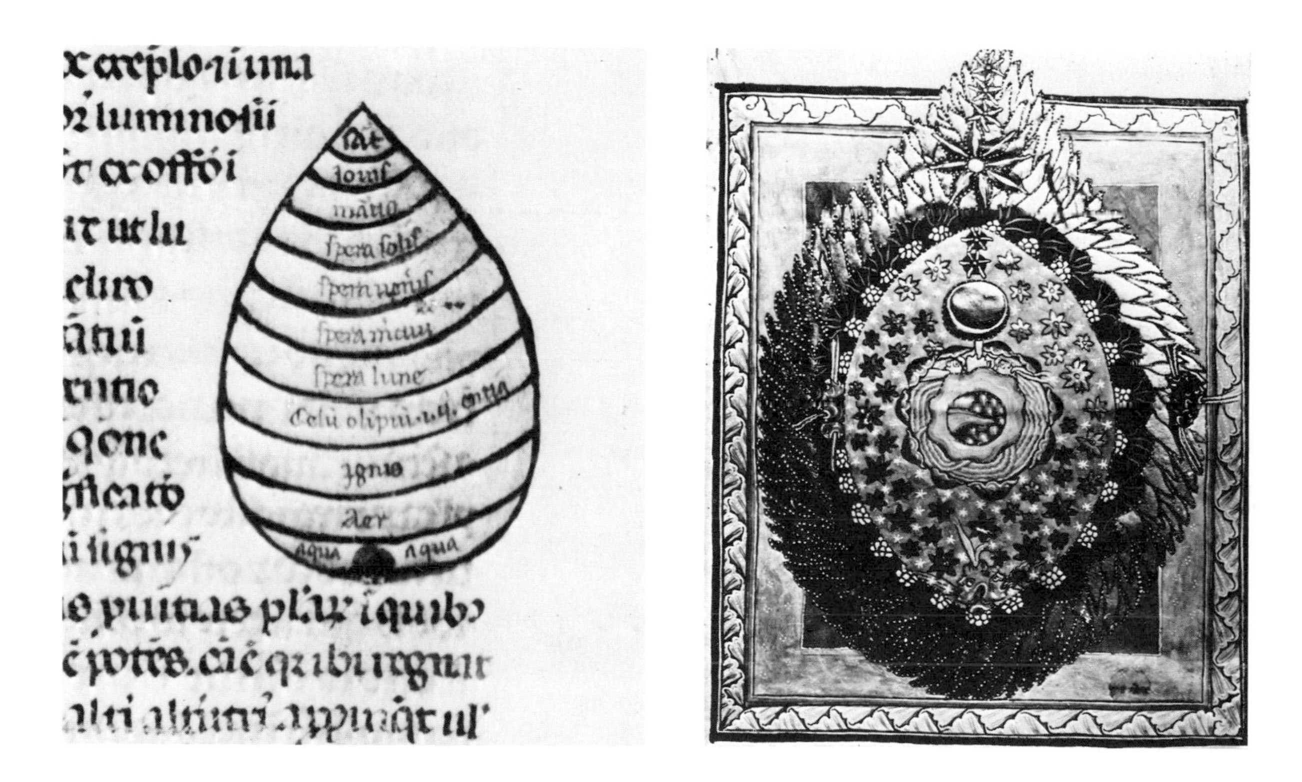

277. The Shape and Strata of the Universe, II. The unusual "beehive" universe depicted at the left, above, is one of many figures and diagrams populating the fourteenth-century manuscript of Michael Scot's *Liber introductorius,* already frequently seen (Illustrations 26, 80, 101, 231, and 251). The text that the picture reproduced here accompanies tells us that God, as Father of all visible and invisible creatures, posited a fifth essence above the four elements and that He distinguished seven regions *(provincias)* in that essence, each of a certain size and like unto the rings of an onion *(ad instar circulorum cepe)*, in which he placed a single luminous body. All this is depicted in our figure, beginning with the three lowest "slices of the onion" as water, air, and fire, earth presumably represented by the unlabeled black circle at the bottom "pole" of the onion. These layers are followed above, first by a stratum signifying the *celum olimpium* or the fifth essence alone, and then by layers representing the spheres

(Continued)

(Continued)

of the seven planets, from that of the moon through that of Saturn at the very top of the figure. The second picture, at the right on page 337, of an egg-shaped universe, comes from a manuscript of the *Liber Scivias* of the twelfth-century visionary Hildegard of Bingen that may well have been written and illustrated within the lifetime of its author. It reproduces with tolerable accuracy the vision of the universe experienced by Hildegard: "I saw a gigantic image, round and shadowy; like an egg, it was less large at the top, wider in the middle and narrower again at the base." With east at the top of this egg and north on the left, the small circle at the very center reveals the terrestrial world, the various objects within it representing the four elements. Two wavy circular layers surround this central circle, the first being that of bright air or *alba pellis* (white skin), which furnishes moisture for the earth below, the second that of aqueous air containing three small faces representing the east winds at the top (faces for the other winds appearing further out within the egg at the north, west, and south). The innermost ovoid area, composed of *purus aether,* is sprinkled with the fixed stars as well as the moon, clearly depicted directly above center followed by two stars representing Mercury and Venus. The largest star above these is the sun, topped in turn by Mars, Jupiter, and Saturn. The outermost region occupied by the sun and the three superior planets is that of bright fire *(ignis lucidus)*. Between it and the central ovoid region of pure ether there is another layer of fire, this time black or dark *(ignis niger)* like a shadowy skin. It is the abode of hail and lightning, shown as pineapplelike figures throughout its circumference, as well as of the north winds (depicted as the three faces at the left). Each element of this curious universe also harbored a symbolic meaning. Bright and dark fire, for example, symbolize God and the Devil; the sun, Christ; the three superior planets, the Trinity; the *alba pellis,* the water of baptism; and so on.

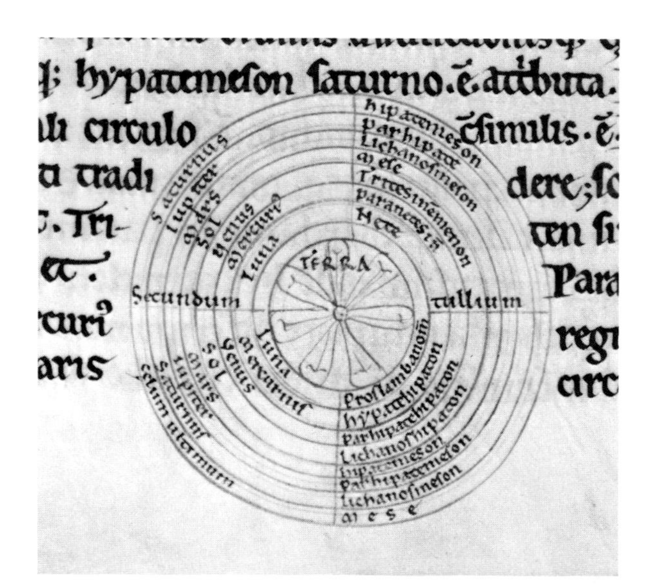

278. The Music of the Spheres. The belief that the stars and planets produce harmony was much discussed—both pro and con—in antiquity. Aristotle, for example, claimed (*De caelo,* I, 9, 290b15) that those who, like the Pythagoreans, held this belief were in error. The fact that his criticism did not erase the belief from ancient and medieval thinking is adequately presented by the first two figures reproduced here. That on the left is taken from a twelfth-century manuscript (also seen in Illustrations 95, 100, and 104) of Boethius' (ca. 480–524/525) *De institutione musica.* Visually epitomizing the contents of bk. I, ch. 27, this circular diagram presents both alternative views of the music of the spheres discussed by Boethius. Its upper semicircle presents the minority opinion that Boethius took from Nicomachus (fl. ca. A.D. 100), which maintained that the highest tone was sounded by the moon, the lowest by Saturn. This assertion is recorded by the appearance of the planet's name and the name of the tone, or string giving that tone, within the same circular band: thus, *Saturnus* in the outermost band is associated with *hipatemeson, luna* in the innermost with *nete.* Translating this information from the Greek scale (see Illustration 77) into modern symbols, we find that Saturn and the

(Continued)

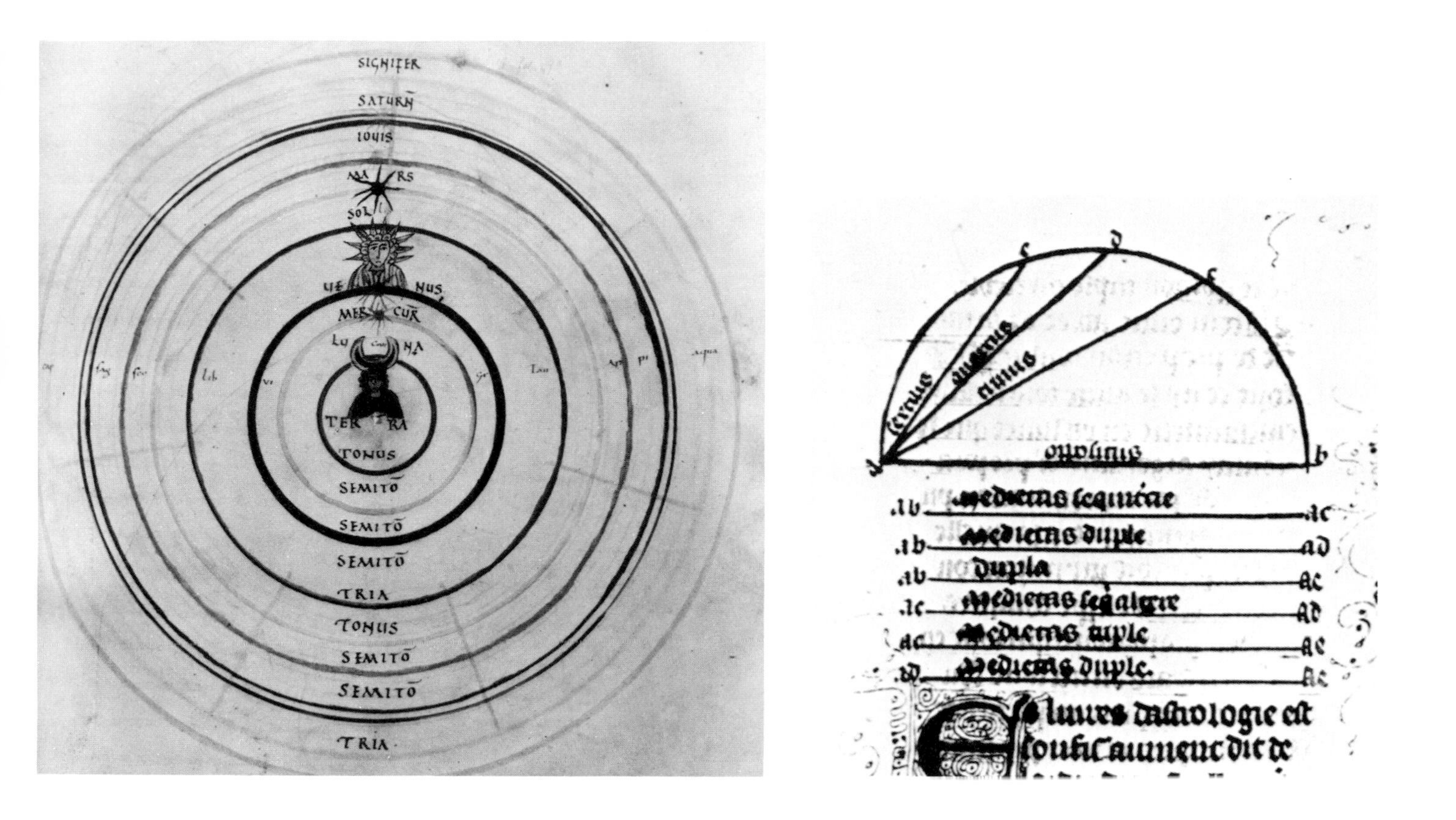

moon harmonically bridge the octave *e*——*e′*. The words *secundum Tullium* inscribed on the horizontal diameter of this diagram indicate that its lower half presents the second view: that of Cicero or Tully (106–43 B.C.; as in his *Somnium Scipionis*). Here the moon sounds the lowest tone *(proslambanomenos)*, the sphere of the fixed stars the highest *(mese)*. Since this system avoids any interval of a tone and a half between planets and poses all in terms of single tones and single semitones, the addition of the sphere of the fixed stars is needed to account for an octave *A*——*a*. The second circular diagram, above left, is drawn from a ninth-century codex composed at Salzburg. It occurs in the midst of extracts from Pliny (ca. A.D. 23–79) and duly represents his description of the music of the spheres (*Hist. nat.*, II, 84). Only the intervals between the tones sounded by the spheres are given, not the tones themselves. The name of a tone is written in full *(tonus)*, while those for the intervals of a semitone or three semitones are abbreviated. Since the Plinian system stipulates an interval of three semitones between the zodiac *(signifer)* and Saturn instead of a single semitone (as was the case, for example, in the otherwise identical system set forth by Censorinus [A.D. 238]), the intervals between the earth and the zodiac cover two fifths, or an octave plus a tone. The Plinian position that appeared as a supplement in the last diagram in Illustration 249 corrected this excess by omitting any interval between the earth and the moon. In the present figure, the scribe has erroneously increased the excess by inserting an extra semitone between the sun and Venus. The inscriptions in a very small later hand running horizontally across the circular diagram give the first few letters of the names of the zodiacal signs that are the solar (on the left) and lunar (on the right) houses of the five planets and indicate the single houses for the moon and the sun directly over their personifications. The final diagram at the right comes from that fourteenth-century manuscript of Nicole Oresme's *Livre du ciel et du monde* already used a number of times (Illustrations 64, 254, and 255). The burden of the diagram is not an explanation of the music of the spheres, but a related notion: that the ratios between the astrological aspects of the heavens are all harmonic, or parts of harmonic, ratios, a doctrine that Oresme claims he "did not learn from the teaching of others." These different aspects are represented by the inscription in a circle of the side of a hexagon *(AC)*, a square *(AD)*, and an equilateral triangle *(AE)*, plus the circle's diameter *(AB)*, drawn here in half of the circle in question. The resulting lines yield, respectively, the sextile, quartile, and trine aspects, plus opposition. Oresme's discovery is tabulated below the semicircle. Thus, the ratio *AB:AC,* or that of opposition to trine, is a "half [part] of a sesquitertian ratio." Since the half or half part of any ratio is, in our terms, the square root of that ratio, Oresme is saying that opposition stands to trine as the square root of 4:3, where 4:3 is the "harmonic" ratio of the fourth. The remainder of the legends under the semicircle give similar information for the ratios obtaining between other aspects. Another diagram of the harmony of the spheres is given on the first page reproduced in Illustration 289. Diagrams from the ninth-century Salzburg codex are also found in Illustrations 280 and 284.

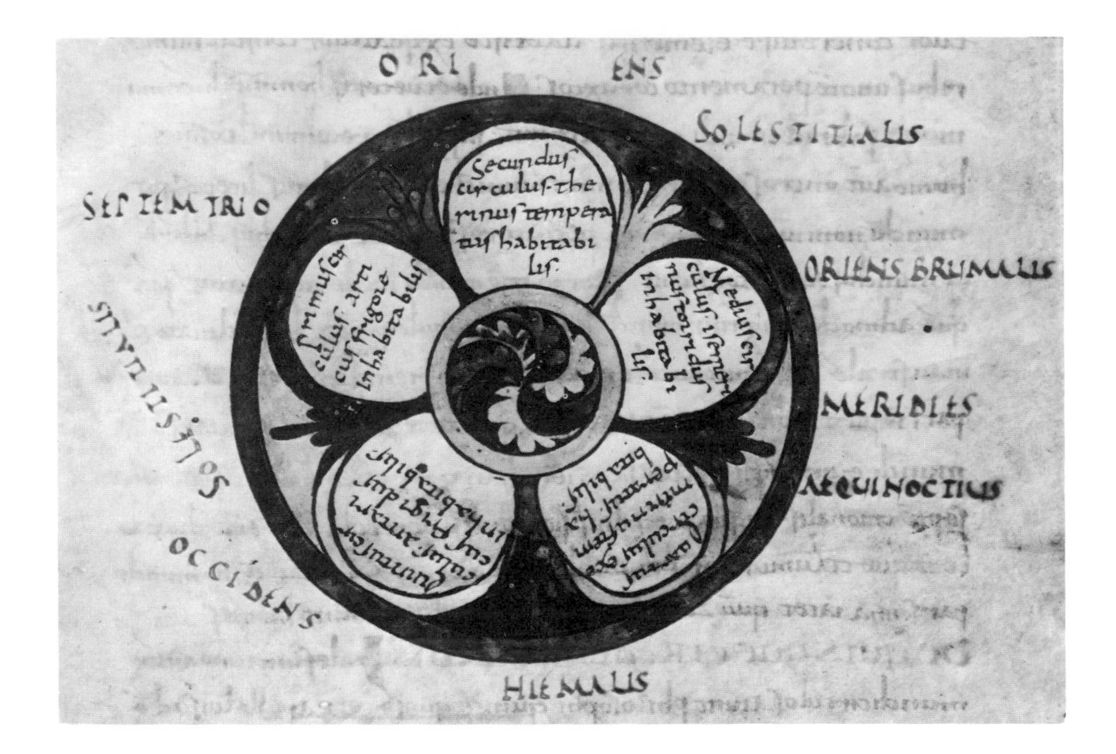

279. Habitable or Uninhabitable: Zones of the Earth. As part of ancient astronomy, seven terrestrial climata were distinguished, each clima being a parallel or narrow zone determined by the value of the longest daylight for that parallel or zone. (These seven climata have already been indicated as part of the circular diagram in Illustration 54.) It was not long before these climata were put to geographic use, since they provided terrestrial boundaries between which kinds of inhabitants, cities, and other geographical features could be cataloged (in Pliny, for example, *Hist. nat.,* VI, 39). Most medieval discussion of terrestrial zones was quite different, however. If one sets aside excerpts copied from Pliny, this discussion dropped almost all reference to the mathematics involved in the Greek climata and directed its attention to five zones associated with the two tropics and the Arctic and Antarctic circles. The most frequently copied diagram of these zones was that belonging to the *De natura rerum* of Isidore of Seville (seventh century). Shown here above as it appears in a tenth-century manuscript (also seen in Illustrations 48 and 247) of that work, the function of this diagram is purely mnemonic, its circular form having no relation to the spherical form of the earth itself. In fact, it has been suggested that the form of this diagram may have been derived from the mnemonic location of the five zones on the open palm of a hand. Each interior circle representing a zone is numbered, beginning in the ten o'clock position, and carries an indication of its name (in transliterated Greek), character, and habitability or uninhabitability. Thus, the first zone is *articus,* uninhabitable due to extreme cold; the second is *therinus* (summer), temperate and habitable; the third or middle zone is *isemerinus* (equinoctial), uninhabitable because it is *torridus;* the fourth is *exemerinus* (for Χειμερινός: winter), temperate and habitable; and the fifth is *antarticus,* uninhabitable because it is *frigidus.* Isidore also gives the Latin names for these zones in his text and the scribe of the present manuscript has attempted—with considerable confusion—to enter them outside the "wheel," with the names of the cardinal directions between which, according to Isidore, these zones fall. The second picture, p. 341, comes from a manuscript (ca. 1200; also seen in Illustrations 281 and 288) of Bede's eighth-century computistic work, the *De temporum ratione* (compare Illustrations 51 and 75). These two zone diagrams follow Bede's chapter on climata

(Continued)

and have been imported to supplement—since they do not match—the text. The figure at the right is clearly the Isidorean one just examined. That at the left most likely derives from Macrobius (early fifth century), who expressly recommended a diagram of the zones as providing better comprehension than that afforded by mere verbal explanation (*Comm. Somn. Scip.*, II, 5, 13). Several versions of this Macrobean diagram differing from that reproduced here can be seen on page 362. A most notable feature of the present diagram is that the fragments of text it contains do not come from Macrobius or from the work of Bede it accompanies, but rather from Bede's *De natura rerum.* The zones delineated in the second figure are the same five set forth by the diagram from Isidore. Now, however, the central torrid zone and the two temperate zones are all said to be bounded in one way or another by the zodiac (that is to say, by the two tropics). Unfortunately, the zodiac itself is mistakenly drawn perpendicular to the parallels, thus obscuring its relation to the tropics and hence to the zones they define. The circular diagram at the middle left, page 342 (taken from the twelfth-century computistic–cosmographic miscellany also seen in Illustrations 31, 51, 53, 69, 249, 285, 290) gives yet another medieval manner of diagraming the five terrestrial zones. The view is one that looks down on the earth from the north (at the bottom of the circular figure). The two uninhabitable frigid zones are thus represented between the two uppermost arcs and within the small circle below. The area between that circle and the next arc is the northern temperate zone, followed by the uninhabitable torrid zone and the southern habitable temperate zone. The second circle within the figure in its lower half mentions the mythical Riphaean mountains *(Riphei montes)*, traditionally held to be located at the northern boundary of Europe. The top half refers to Ethiopia *(Ethiopum terra)*, an indication of the traditional belief that habitation was possible in that part of the torrid zone bordering the temperate region. The next figure, at the bottom of page 342, is taken from a fifteenth-century French manuscript of the *Dialogi cum Moyse Judaeo,* written by the twelfth-century physician and scholar Petrus Alphonsus as a vindication of Christianity against Judaism. Though extremely stylized, Petrus' zonal diagram clearly reports the ancient doctrine of seven habitable climata. The only uninhabitable area indicated is the extreme northern zone shown at the bottom of his diagram. The upper half of the figure, decorated by three towers, carries the

(Continued)

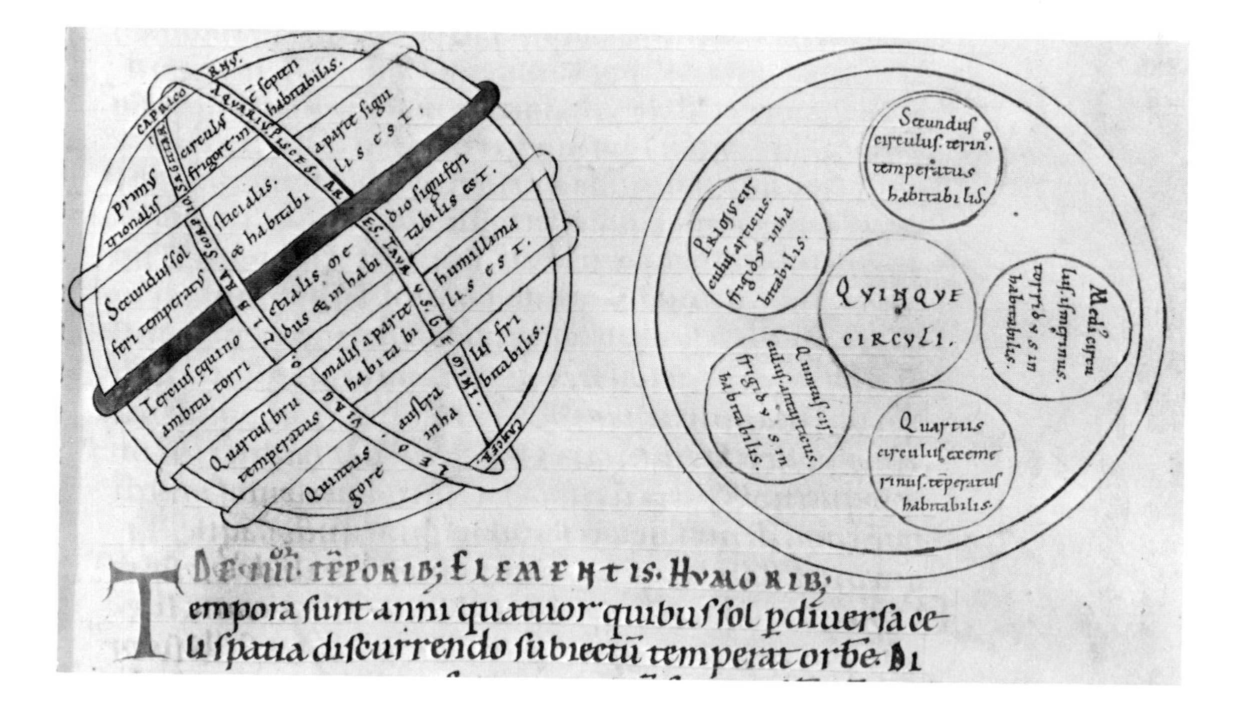

(Continued)

inscription *Aren civitas,* a reference to the city of Arim or Arin as the navel of the world (compare Illustration 130), testifying to Arabic influence on Petrus. The final set of diagrams at the right, below, comes from the same codex of al-Bīrūnī's (eleventh-century) *Art of Astrology* that furnished the eclipse diagram in Illustration 234. After discussing the traditional Greek climata, Bīrūnī turns to other ways to divide the world. The Isidore-like diagram of seven circles reproduced here represents the view of the Persians, who, Bīrūnī claims, "divided the world by the kingdoms into seven regions [*kishwarāt*] and attributed this division to Hermes." India is assigned the first region in the ten o'clock position, followed clockwise by Morocco, Syria, Rūm and Slavonia, Khazaria and Turkestan, and China and Tibet, Persia itself being the fourth region at the center. The square diagram below represents the system of the Hindus, who "divide the world into nine portions called *nūkand (navakhanda)* to eight of which they give names in their own language corresponding to the points of the compass, while the ninth is the central part." The Hindu names for the cardinal directions are written in the inner set of squares, those in Arabic occupying the frame of the whole diagram. South is at the top, east at the left.

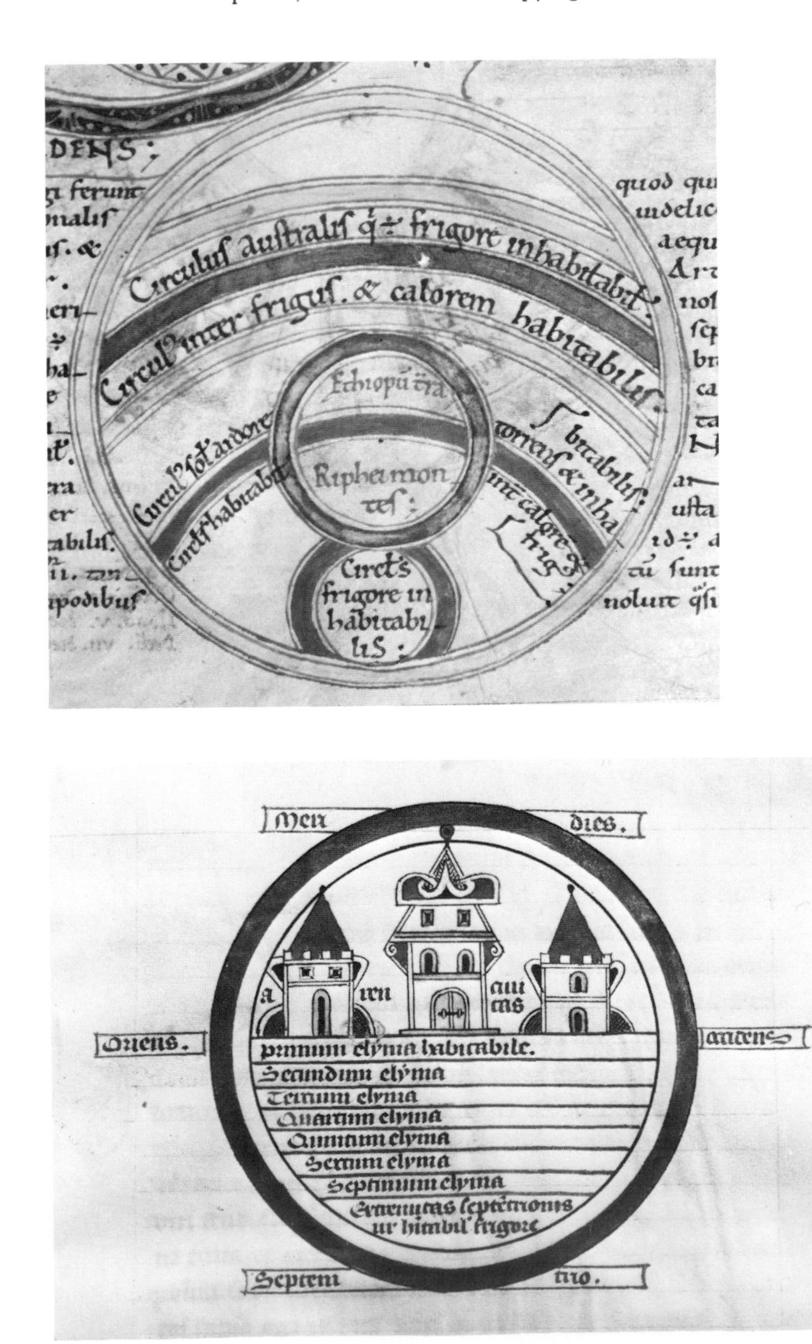

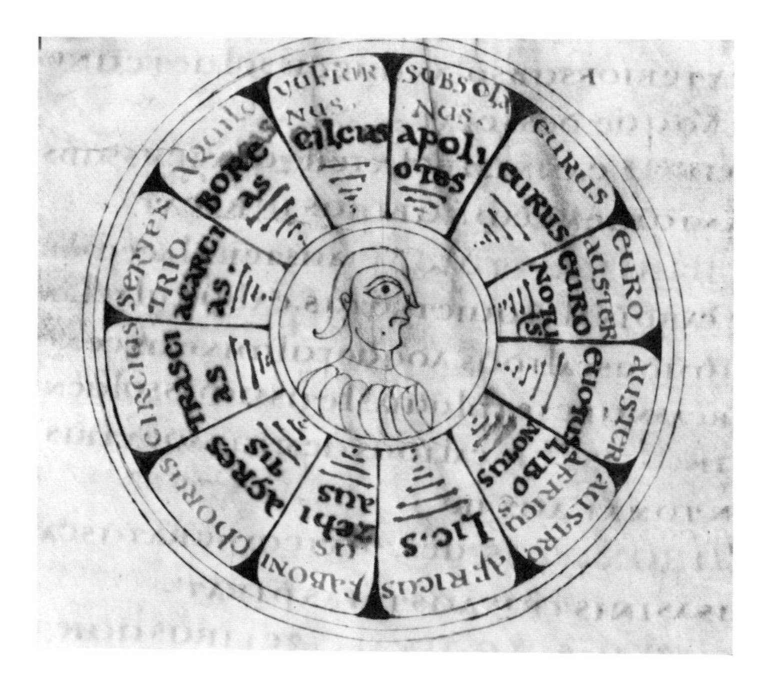

280. The Winds. It was relatively standard procedure in antiquity to consider two factors as most important in treating winds: their causes and their positions. It is the latter that is of concern here. Any number of ancient authors may be consulted in order to discover the "system" of wind positions, but it is fair to say that the basic elements of such systems were devised by Aristotle. For present purposes it is important to note that Aristotle introduces his discussion of the winds (*Meteorology,* II, ch. 6) by saying that he must use a diagram (ὑπογραφή) as well as reason in setting forth his contentions. The basis of Aristotle's diagram is the circle of the horizon. On it he determines eight points by (1) the equinoctial sunrise and sunset together with a diameter perpendicular to these equinoctial

(Continued)

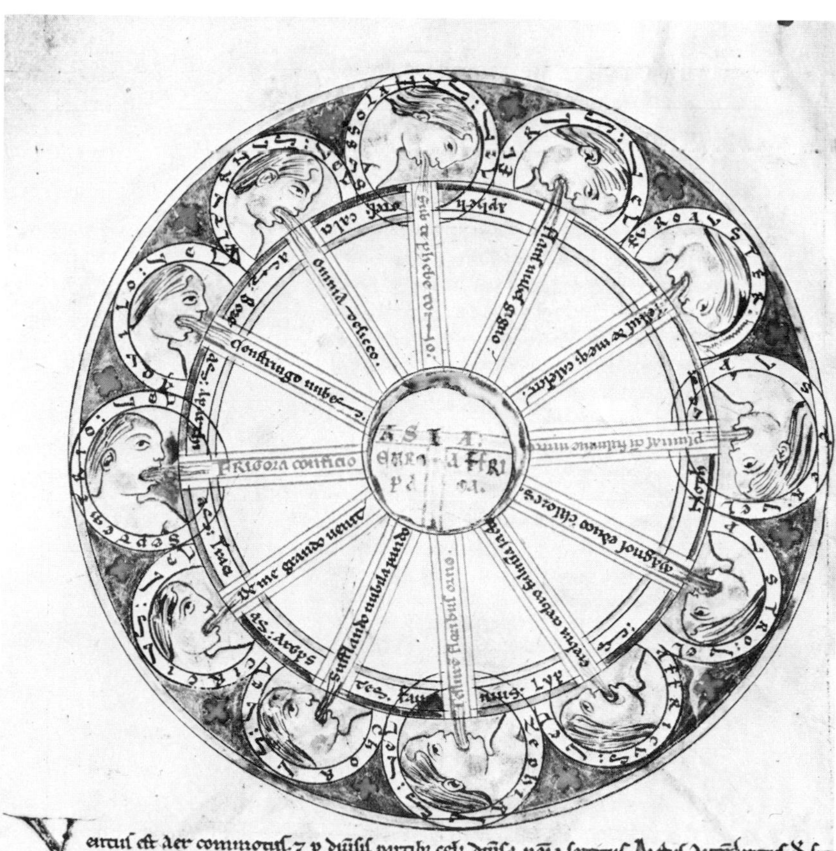

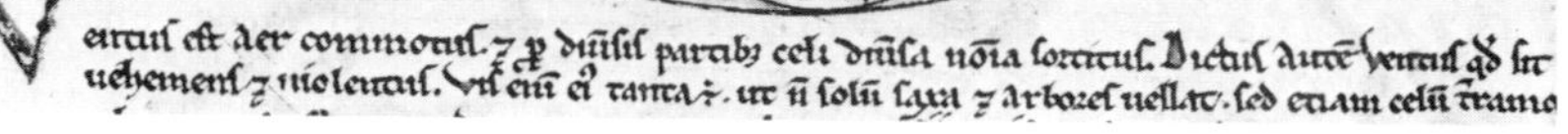

(Continued)

points and (2) sunrise and sunset at the summer and winter solstices. These points provide him with eight winds. Two others are added at the ends of a chord roughly corresponding, Aristotle claims, to the ever visible circle, while a possible additional wind might be posited opposite one of these latter two. In sum, then, Aristotle has delineated positions for ten or eleven of what were to become the twelve standard winds of antiquity and, hence, of the Middle Ages. (Note that the circular diagram from Ptolemy in Illustration 54 adds four other winds, for a total of sixteen.) The first medieval diagram of the winds, at the left on page 343, is another of Isidore's wheels from his *De natura rerum,* taken this time from an eighth-century manuscript of the work (also seen in Illustrations 247 and 286). North (Septentrio) is at the left; south (Auster) at the right. The Latin names of all twelve winds are given just inside the wheel's circumference, while the Greek names occupy (with much misspelling) an inner circle. A development of the Isidorean diagram is given at the bottom of page 343. It appears in a cosmographic miscellany (ca. 1200; also in Illustrations 50, 131, and 251) where it precedes part of the chapter on the winds from Isidore's *Etymologies* (XIII, xi, 1–14). This figure has one notable difference from the traditional Isidorean wind wheel: human faces are portrayed as blowing each of the winds, a feature that was to become somewhat common in visualizing this subject (compare the winds in Illustration 291). East is at the top and north at the left in this particular diagram, an orientation matched by the *O-T* map of the world (compare Illustration 251) with Asia on top of the *T,* Europe at the left, and Africa on the right below. Once again, in most instances both the Latin (on the circle enclosing each face) and the Greek names (on the larger inner circle) are given for the winds (save at the very bottom, where the Latin Favonius has changed places with the Greek Zephirus). The spokes of the diagram carry inscriptions (most likely derived from Isidore's *De natura rerum*) that specify some of the effects of the various winds.Thus, Aquilo or Boreas (just above Septentrio or Aparthias [ἀπαρκτι-άς] at the left of the horizontal diameter) restrains clouds, while Vulturnus or Calcias (καικίας) just

(Continued)

above it dessicates everything. A quite different depiction of the winds occurs in a ninth-century manuscript of Bede's *De natura rerum,* seen on page 344. The names (again both Greek and Latin) and selected properties of the winds are given on the various inner arcs and above the two central figures as well as (here barely visible) on the rectangular frame of the whole picture. The major disadvantage of this scheme is the confusing way it represents the directions of the winds. A reader who did not already know of these directions from a text or from some diagram like the first two reproduced here would be unlikely to be able to fathom them on the basis of the present figure alone. The final, extremely decorative, diagram below comes from a sixteenth-century manuscript of a French translation of Ovid's (43 B.C.–A.D. 17) *Metamorphoses.* With west at the top and north at the left, the spokes of the wheel and its circumference give, when space permits, the relevant Latin and Greek names—albeit confusedly—for the twelve winds. Additional vertical strips point out the equator, tropics, and Arctic and Antarctic circles, and the hub of the whole provides an indication of the five terrestrial zones, though without mention of their habitability or uninhabitability. (The ninth-century manuscript of *De natura rerum* is also seen in Illustrations 278 and 284.)

CHAPTER TWENTY-FIVE

The Basic Constituents of the Sublunar Realm: The Elements and Man and the World as Quadripartite

ALTHOUGH MYTHIC COSMOGONIES and earlier, pre-Socratic natural philosophers viewed various of the four elements or basic opposites (such as the hot and the cold) as fundamental ingredients within the world, the first clear and explicit statement that the four elements are the primary cosmic constituents is to be found in Empedocles (fl. ca. 450 B.C.). He described their fundamental role in the world by claiming that "besides these nothing else comes into being nor ceases to be. . . . Nay, there are these things alone, and running through one another they become now this and now that and yet remain ever as they are." That is to say, each element is incapable of change, becoming "now this and now that" natural substance by being brought together by Love into proportionally different mixtures and then, separated by Strife, perishing or changing into another natural substance. Thus, as one later doxographer tells us, "Flesh is the product of equal parts of the four elements mixed together and sinews double portions of fire and earth mixed together . . . and bones of two equal parts of water and of earth and four parts of fire mingled together."

Empedocles' view of fire, air, water, and earth as the basic building blocks of the world was assumed by almost all later Greek practitioners of natural philosophy and medicine, most notably by Plato and Aristotle, who in turn transmitted the conviction of the central role of the four elements to the Middle Ages, both Islamic and Latin.

In the *Timaeus,* Plato mathematicized the doctrine of the four elements, stipulating that air and water must fall as two geometric means between fire and earth as extremes (see Illustration 282). He also assigned four of the five regular solids to the elements, while at the same time claiming the construction of these solids—and hence of their corresponding elements—out of two kinds of triangles that compose the regular polygons forming the faces of the regular solids. Fire was assigned the pyramid, earth the cube, air the octahedron, and water the icosahedron. This would explain, for example, how one particle of water could be transformed into one particle of fire and two of air, since the twenty equilateral triangular faces of an icosahedron of water could be seen as yielding both the four equilateral triangular faces of a pyramid of fire and the sixteen equilateral triangular faces of two octahedra of air.

Ancient and medieval commentators eagerly filled in much of the specific mathematics Plato had left out and even went so far as to supplement what they could glean from the *Timaeus* with additional doctrines about the four elements that could be found nowhere in Plato himself (see Illustrations 282 and 283).

Aristotle would have none of this quasi-mathematics in his theory of elements. As the very simplest of sensible bodies, the four elements were, he believed, logically but not physically analyzable into a substratum and various contrary primary qualities. In Aristotle's eyes, the contrary qualities needed must have the power to act upon or to be acted upon by one another, a requirement only satisfied by the contrary pairs hot-cold and dry-moist. The elements are qualified, then, in the following fashion: fire, hot and dry; air, dry and moist; water, moist and cold; earth; cold and dry (see Illustration 284). Furthermore, the whole scheme would explain the action and reaction and the combination and transformation of the elements, since such physical phenomena could be accounted for in terms of the primary-quality pairs belonging to the elements.

As will be evident from the illustrations provided in this chapter, medieval natural philosophers extended and developed both the Platonic and the Aristotelian theories of the elements. It was already part of ancient medicine to characterize the four bodily humors—and hence also the temperaments based on these humors—by the same contrary pairs of qualities Aristotle had applied to the elements (see Illustration 36), and the medievals did not hesitate to emphasize this correspondence (see, for example, Illustration 286). But they went further and saw "fourness" everywhere in the universe. Indeed, this tendency can be found as early as Victorinus, bishop of Pettau (d. ca. 304). In his brief *Tractatus de fabrica mundi*—which as a whole celebrates seven as involved in the six days of creation plus the seventh day of rest—Victorinus pauses to note the importance of the tetrad not simply in the four elements, but also in the four seasons, the four animals before the throne of God, the four gospels, the four rivers in Paradise, and the four generations or ages of man between Adam and Christ. Although later medievals might not have repeated all of Victorinus' "fours," they certainly did not hesitate to add to the list and thus to view the world as thoroughly quadripartite (see Illustrations 287 and 288).

Other diagrams and pictures involving the four elements can be seen in Illustrations 23, 66, 247, 266, 273–277, and the illustrations in the next chapter.

281. The Elements Alone. In most of the following diagrams and pictures in this chapter, it will be seen that the depictions of the elements indicate something of their primary qualities. Such information is absent from the two pictures reproduced here. The first, left, is taken from a twelfth-century manuscript written at the monastery of Prüfening in Bavaria (also seen in Illustrations 279 and 288) containing extracts from Pliny and Bede and others dealing with computistic, astronomical, and cosmographic matters. It personifies the four elements. *Terra* as Mother Earth appears at the lower left, astride a centaur who feeds from her breast. *Aqua,* mounted on a griffin at the lower right, is pictured after the fashion of the zodiacal sign Aquarius. Above, *Aer* rides an eagle and carries an inflated bladder with a human head at its neck, while *Ignis* is seated on a lion (the face of which might be compared with those of the other lions shown in Illustration 203), holding a torch in his hands. The second illustration, page 348, comes from an anonymous dialogue, *De naturalibus scientiis,* found in a sixteenth-century manuscript (also in Illustrations 35, 169, and 254). Its purpose is to show the form and order of the strata occupied by the four elements. It might be considered a magnified detail of the relevant central parts of a dia-

(Continued)

(Continued)
gram of the whole universe, such as that given in the first picture in Illustration 276. Here, however, not only are the spherical layers of the elements portrayed, but also their occasional penetration into one another's regions. The sun and the moon at the top of the illustration mark the outermost limit of the sphere of fire and of the whole sublunar realm and hence the beginning of the celestial divisions of the universe.

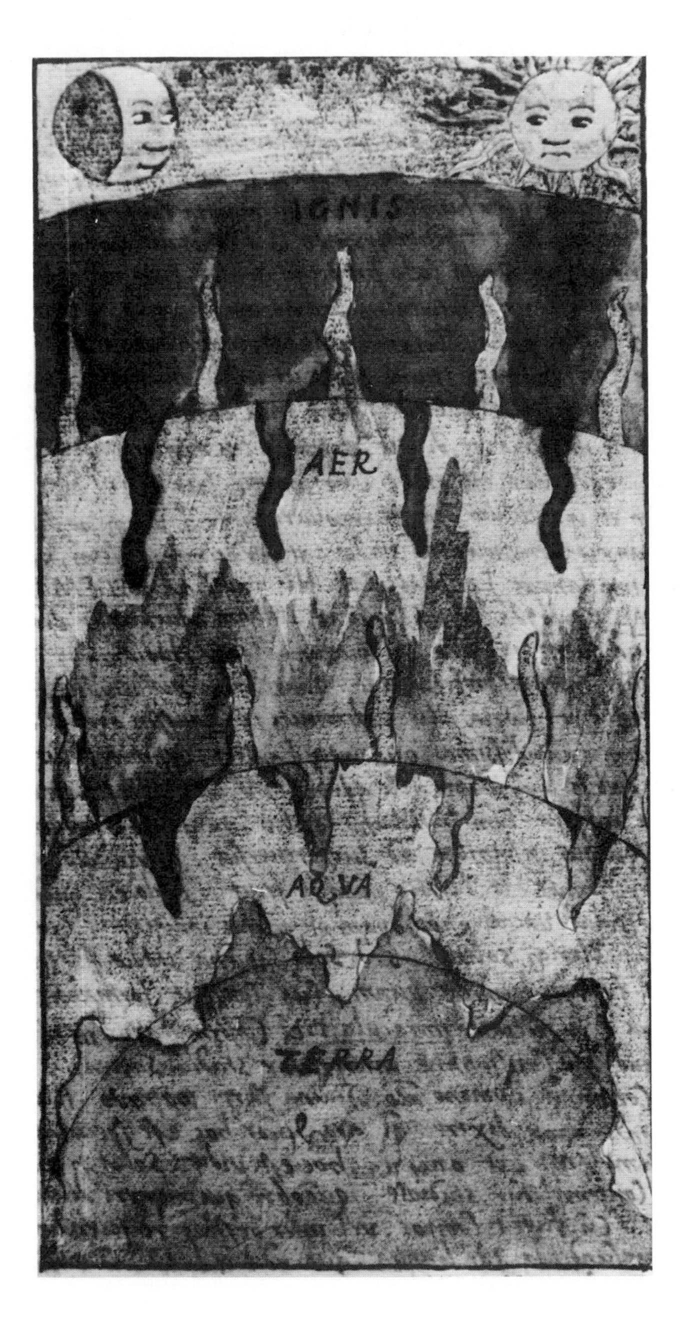

282. The Platonic Theory of the Elements. Although the four elements clearly play a role in the natural philosophy Plato set forth in the *Timaeus,* medieval writings and, especially, diagrams treating the Platonic doctrine of the elements equally clearly are a development and considerable expansion of anything maintained by Plato himself. Thus, in the diagram on p. 349, left—taken from a twelfth-century manuscript of Adalbold of Utrecht's (ca. 970–1026) commentary on Boethius' *Consolation of Philosophy,* III, *metrum* 9—the central feature is the characterization of each of the elements (in the circles at the left) in terms of triplets of primary qualities (in the right-hand circles) that we have already noted as the subject of the *cybus elementorum* in Illustration 247. The six qualities (*acutus, subtilis* or in other texts *tenuis, mobilis, obtunsus, corpulentus* or *crassus,* and *immobilis*) providing the material for these triplets are nowhere to be found in Plato. Adalbold probably derived them from what seems to be their earliest extant occurrence in the commentary of Chalcidius (fourth to fifth century) on the *Timaeus.* Chalcidius is rather notorious for taking material for his commentary from others, but in this case his source remains unknown. At the same time, some elements of the present diagram can be traced to Plato's *Timaeus,* not surprisingly so, since the particular verse of Boethius on which Adalbold was commenting
(Continued)

is a kind of poetic summary of contentions found in that Platonic dialogue. One line of the verse claims that God binds the elements together by numbers *(Tu numeris elementa ligas)*, as clear a reference as one could expect to Plato's craftsman creator binding together the visible and the tangible. In the lines of the *Timaeus* (31*B*–32*B*) directly preceding this claim, Plato points out that fire is necessary to render the universe visible and earth to render it solid and, hence, tangible. Yet these two extremes, he continues, are best bound together by a continued geometric proportion. Further, since fire and earth—and the world of which they are elements—are solid in form (στερεοειδῆ), two means are necessary to join them. These two means are, of course, the elements air and water. This theory is elaborated by Adalbold in his commentary and indicated in the diagram by the numbers and annotations at the left-hand side. Adalbold knew enough of Boethius' *Arithmetica* to see that what was relevant to the matter at hand is the fact that two means fall between two cube numbers standing in geometric proportion. Therefore, with fire as the cube number 27 (XXVII) and earth as the cube number 8 (VIII), air and water fall between them as, respectively, 18 (XVIII) and 12 (XII), thus satisfying the continued geometric proportion: 27:18 = 18:12 = 12:8. The abbreviated annotations at the left of the diagram tell us not only that

(Continued)

(Continued)

air and water (and hence their numbers) function as means *(medietates)* between fire and earth and their numbers, but also that each of the ratios constituting the resulting geometric proportion is sesquialtern *(sesqualtera)*—that is, is equal to the ratio 3:2. Furthermore, Adalbold notes that the differences between 12 and 8 and between 27 and 18 are the square numbers *(tetragoni)* 4 and 9, and that these in turn stand to 6 (the difference between 18 and 12) in sesquialtern ratio. With the information provided by this Adalboldian diagram in hand, if one turns back to the *cybus elementorum* in Illustration 247 it is possible to see the relevance of the claim made there that the figure in question was "solid according to geometric ratio," a claim that was in no way explained or understood in the text accompanying that figure. Finally, note should be made of the fact that the arcs at the right-hand side of the diagram duly point out which of the qualities characterizing the elements are contraries. Arc diagrams such as this one, explaining the so-called Platonic theory of the elements, are found in numerous manuscript copies of many medieval works (see, for example, Illustration 289). But the picture reproduced at the right on page 349 is unique. Not from a manuscript, but from a thirteenth-century fresco in the crypt of the Cathedral of Anagni, it clearly shows the same arrangement of elements and Platonic triplets, numbers and all.

283. More Platonic Numbers. The illuminated initial at the left is from the beginning of the Book of Genesis as found in a bible from St. Hubert (ca. 1070). The intertwined capital *I* and capital *N* provide the first word of the opening verse that continues below: *In principio creavit Deus celum et terram.* Christ in Majesty occupies the initial's central circle, an alpha and an omega inscribed at his right and left, and the letters of the word *lux*—denoting Christ as light, as in the opening of the Gospel of John—written on the rays of his halo. Of present interest, however, are the four circles surrounding this central figure of Christ. They depict personifications of the four elements (fire above, earth below, air at the left, water at the right), often seen as implied by the creation of heaven and earth that is announced by the first verse of Genesis (cf. Illustration 273). Each figure holds attributes of his element, the name of the element (here barely visible) inscribed on the outer ring of each circle. These frames contain even more: numbers taken from the Platonic theory of elements, a feature that would seem appropriate to those who, like St. Augustine (*City of God,* VIII, ch. 11), set the opening of Genesis side by side with the *Timaeus.* The relevant numbers are given in both Roman numerals and the Latin of their factors: *bis bina bis* (2 × 2 × 2) = VIII, *ignis; bis bina ter* (2 × 2 × 3) = XII, *aer; ter tria bis* (3 × 3 × 2) = XVIII, *aqua; ter tria ter* (3 × 3 × 3) = XXVII, *terra.* The whole, of course, furnishes the Platonic numerical system of elements just seen in the preceding illustration. Still more numbers are found in the next diagram of the

(Continued)

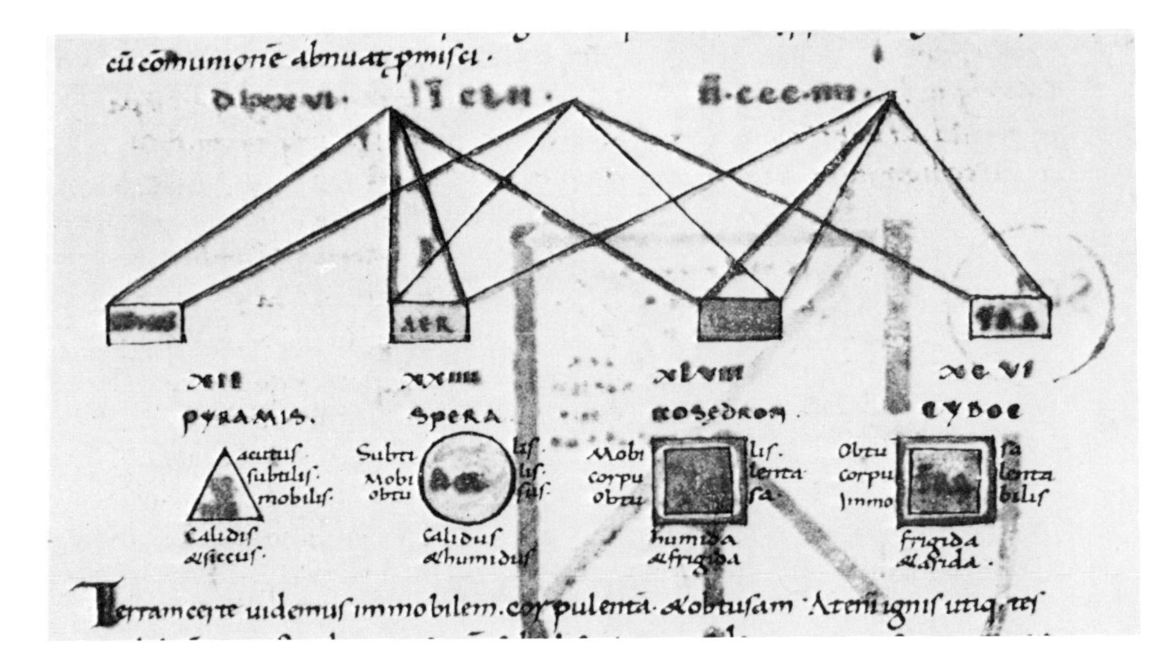

Platonic theory, above, taken from a manuscript (ca. 1065) of an anonymous treatise on the four elements. The elements themselves appear twice, first in a lower row, in accordance with Plato's identification of the elements with four regular solids (though air now is a sphere instead of an octahedron and the icosahedron for water is drawn the same way as the cube for earth). The standard Platonic triplets of qualities are written next to each of these solids, with the Aristotelian pairs inscribed below. A new numerical series—12, 24, 48, 96—appears above the name for each solid. Its extremes are no longer cube numbers, as in the traditional Platonic series, but they do constitute a geometric progression, with the numbers for air and water serving as two means between the extremes for fire and earth. The rectangles in the upper half of the diagram also carry the labels of the elements, lines ascending from these rectangles joining above in three further numbers related to the lower series in the following way: 576 = 48 × 12; 1,152 = 12 × 96 = 24 × 48; 2,304 = 24 × 96. The final diagram at the right comes from a twelfth-century codex (also in Illustration 289) of Chalcidius' translation of Plato's *Timaeus,* where it marginally illustrates a passage (35*B*) that occurs shortly after the passage establishing the proportionality of the elements mentioned in the previous illustration. In this later passage Plato discusses the division of the World Soul—metaphorically considered as some kind of malleable material—that he has placed within the perfect body of his cosmos. The first step of the division amounts to the two series of square numbers 1, 2, 4, 8, and 1, 3, 9, 27. In later authors these series were "joined" by the initial unit they have in common and visually presented in a "lambda" diagram similar to that reproduced here. Chalcidius, for

(Continued)

(Continued)
example, includes such a diagram—as well as the present one—in his commentary (ch. 32), taken, as was his habit, from an earlier writer. It is also worthy of note that he later applies an intercalation of the two series of the first lambda diagram (as 1, 2, 3, 4, 8, 9, 27) in his explanation of the harmony of the spheres (ch. 96), something nowhere found in the *Timaeus* but very much part, for example, of Macrobius (see the first diagram in Illustration 289). By contrast, the present lambda diagram represents the second step in Plato's division of the World Soul. He tells us (36*A*) that we should insert harmonic and arithmetic means (see Illustration 95) within each of the intervals of the initial two series of square numbers. Taken from Chalcidius' commentary (where it was naturally derived from other authors), it has been transported here to the relevant passage of the text of the *Timaeus* itself. In order that the required insertion of harmonic and arithmetic means be expressible in whole numbers, the two series of squares that are to suffer the insertion have been multiplied by 6, resulting in the two series 6, 12, 24, 48 and 6, 18, 54, 162 the members of which appear at the apex and on the protruding shelves of the diagram. Then the demanded insertion of means completes the lambda, yielding the two series 6, 8, 9, 12, 16, 18, 24, 32, 36, 48 and 6, 9, 12, 18, 27, 36, 54, 81, 108, 162. Q.E.D. Plato himself goes on to specify a process for the insertion of even further means, but that is not part of the diagram.

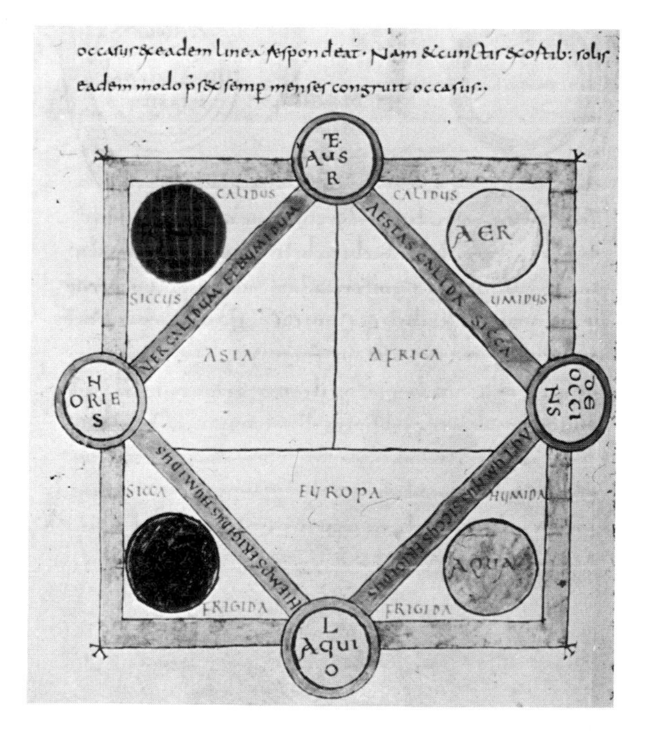

284. The Aristotelian Theory of the Elements. The complications, fascinating though they might be, of the mathematics of proportion involved in the Platonic theory of the elements are absent in the diagrams illustrating the theory deriving from Aristotle. There, the heart of the matter was simply a visual representation of the paired primary qualities shared by the elements (fire, dry and hot; air, hot and moist; water, moist and cold; earth, cold and dry) in such a way as to render obvious the opposition of these elements as expressed by these primary qualities. Compared to diagraming the Platonic system, this was usually a simple pictorial task—so simple, in fact, that additional information was frequently put into such diagrams. The first diagram at the left, taken from the ninth-century miscellany also seen in Illustrations 278 and 280, is one that is so supplemented. (For an Aristotelian diagram of the elements without any added information, see the circular figure on page 4 of Ill. 289.) The top of the square of this diagram is hot *(calidus)*, the bottom cold *(frigida)*, the left side dry *(sicca)*, and the right moist *(humida)*. Given this visual distinction of primary qualities, placing the elements in the four corners of the square (fire, air, water, earth, clockwise from the upper left) automatically as-
(Continued)

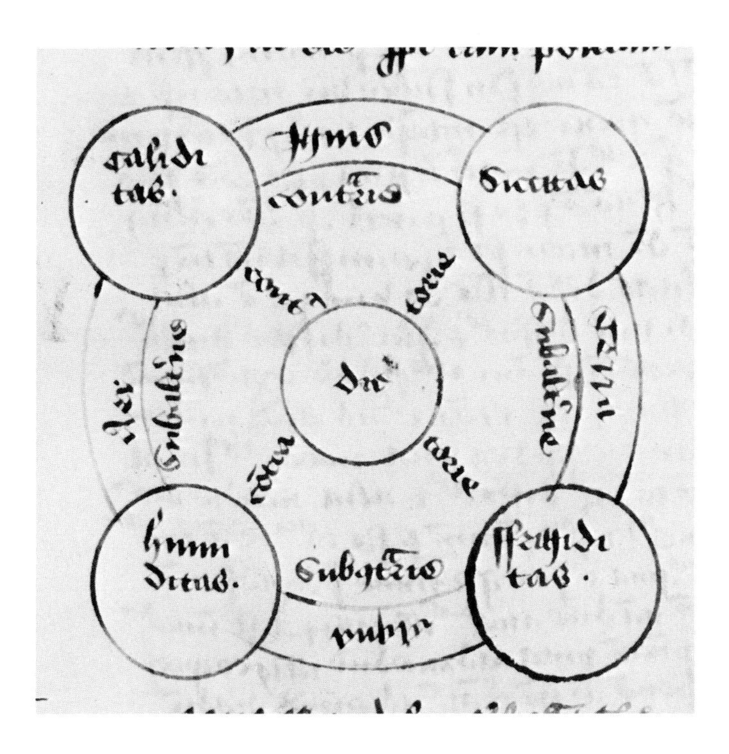

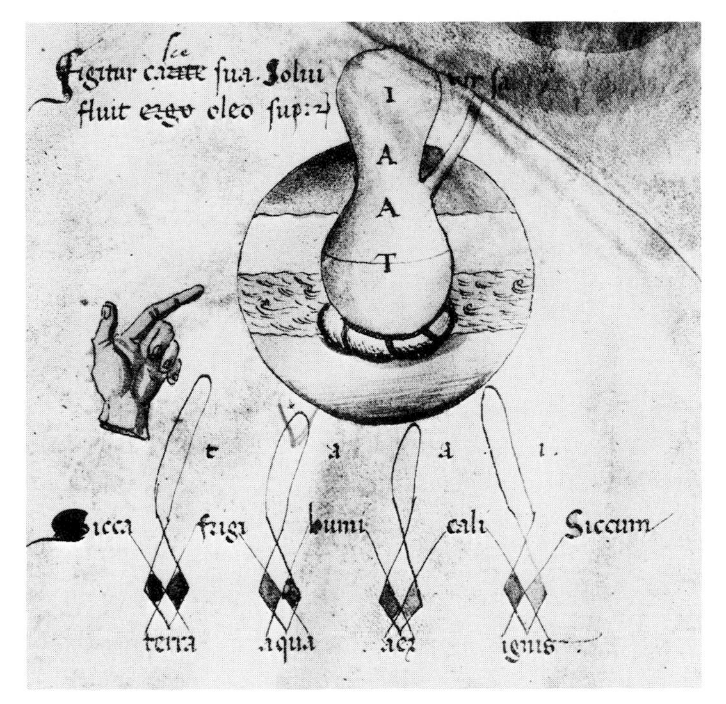

sociated them with their characteristic primary qualities. Moving diagonally, then, from one corner to another immediately reveals the opposition of the elements (air at the upper right, for example, flanked by hot and moist, is easily seen as opposed to earth, cold and dry, at the lower left). This much alone provides what the diagram has to say about the elements. But three additional factors are represented: the cardinal directions, the seasons, and a map of the world. Normally, when such supplementary information was given, the directions and seasons were aligned with the elements (compare Illustration 286) and not forty-five degrees out of phase as in the present instance (north at the bottom, east at the left). This skewing has placed summer *(aestas)*, for example, near air and not next to fire, where it should be. The *O-T* map is also nonstandard in that Asia is not over the *T* but under it. The next diagram, above left, comes from a fifteenth-century manuscript, where it occurs in a work entitled *De terminis naturalibus,* at times ascribed to Thomas Netter of Walden (fl. ca. 1400). It is a circular square of opposition. The primary qualities are announced in circles occupying the corners of the square (heat, dryness, cold, humidity, clockwise from upper left), while the names of the elements are inscribed on the circular bands between these small circles (fire, earth, water, air, clockwise from the top). The diagrammatic features of the square of opposition (see Chapter Seven) dominate the whole: for while it is quite appropriate for the diagonals of the square to indicate that the members of the pairs heat-cold and dryness-humidity are contradictories, it is not only inappropriate, but without any foundation in the theory of elements, to apply the other logical relations represented by a square of opposition (namely, contrary, subcontrary, and subaltern) to the four primary qualities. Nevertheless, this is what the present diagram does. It is an instance of using a visual device initially constructed to serve one doctrine to serve another without any alteration in the device, thus implying the existence of factors within the second doctrine that are not there. Illustration 67 provides another case of "force-fitting" a square of opposition to a particular conception in natural philosophy. But it has also been seen that, by substituting different relations represented by a square of opposition for the original logical ones, one could use such a square as an effective visual tool for the explanation of new—or at least variant—areas of thought (Illustrations 63, 65, and 66). In particular, Illustration 66 shows a more successful application of this type of diagram to the elements. The final picture of the Aristotelian theory of the elements, above right, is taken from a fifteenth-century manuscript of an alchemical miscellany. The names of the four elements are given in the lowest line of text, each one joined to its relevant pairs of primary qualities directly above. The additional information furnished in this figure is the implied importance of the elements in alchemy, provided here by the initial letter of each element's name inscribed in an alchemical vessel: *I(gnis)*, *A(er)*, *A(qua)*, *T(erra)*. The vessel also rests in a circle divided into strips representing the elements.

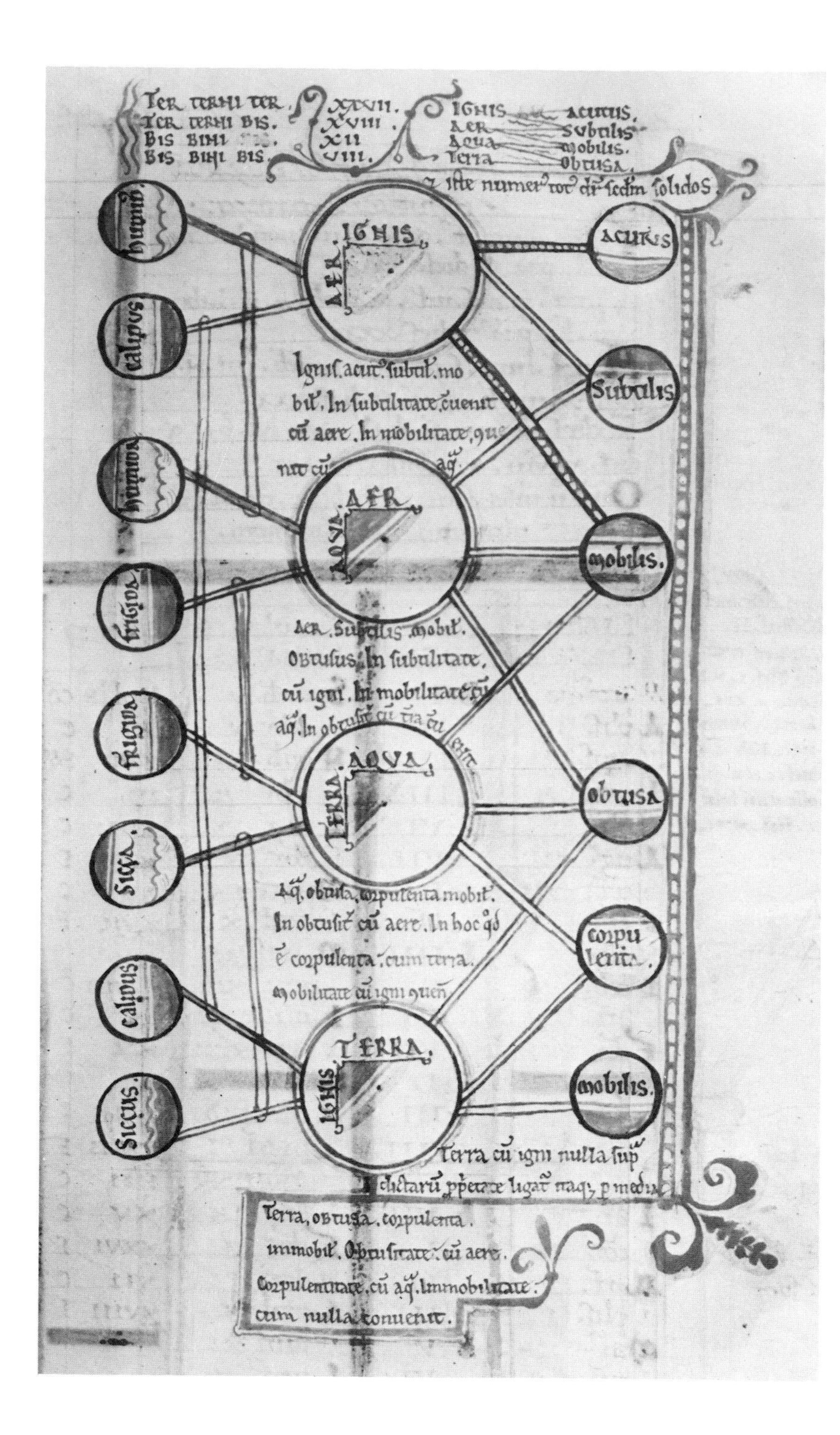

285. Platonic plus Aristotelian. Although a far greater share of medieval diagrams of the elements present either the Platonic or the Aristotelian theory thereof, the penchant to provide alternative views that one finds in so many medieval "textbook" presentations of natural philosophy has provided a fair number of diagrams covering both theories. The first example of such a combination diagram, given here at the left, comes from a twelfth-century miscellany that is a rich source of visual material of all sorts (also in Illustrations 31, 51, 53, 69, 249, 279, and 290). Most of the diagram is devoted to the Platonic theory, the names of the elements appearing horizontally in four central circles, each of these circles in turn joined by straight lines to three appropriate circles at the right bearing the names of the triplet of primary qualities denominating the element in question. As if this were not indication enough of the doctrine involved, brief texts written below each central circle repeat the pertinent triplets and specify which other elements share in members of these triplets. Inscriptions at the top of the diagram provide, first of all, the numerical information about the elements already seen in the series 8, 12, 18, 27 in the diagram in Illustration 282 and in the illuminated initial in Illustration 283 and, secondly, an erroneous matching of the elements with but four of the Platonic primary qualities. The left side of the diagram presents the Aristotelian view. The pairs of primary qualities are given in eight circles at the left, the names of the elements to which they belong written vertically in the central circles. This is rather puzzling, because the same horizontal names provided

(Continued)

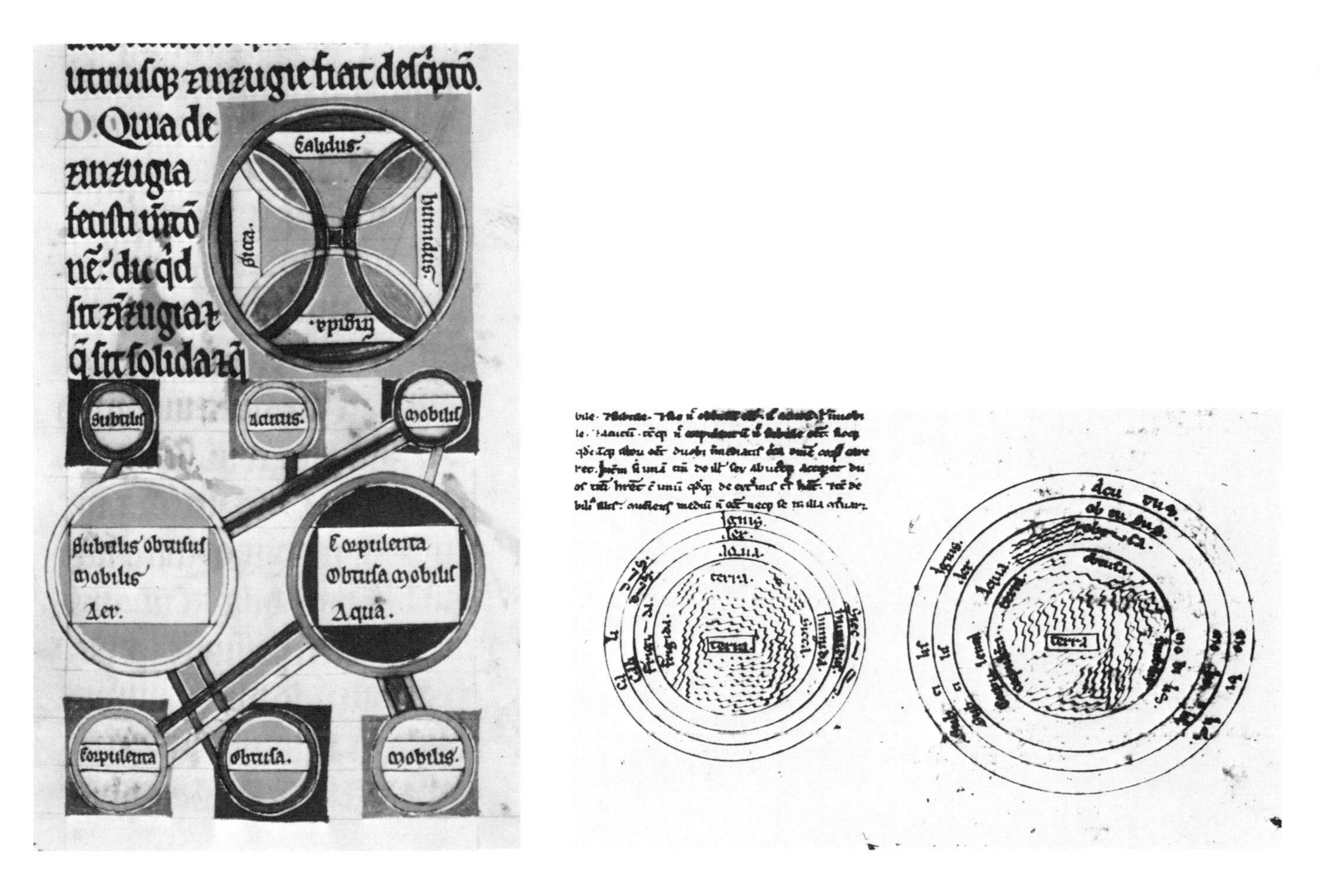

for the Platonic theory could have been used in presenting the Aristotelian pairs of qualities. Only five circles would have been needed to show these pairs, an arrangement that would have been more effective in revealing which elements share which qualities. As it is, the use of eight circles for these qualities has necessitated the less visually revealing ploy of drawing vertical connectors between the "stalks" of the qualities' circles in order to indicate the shared qualities. Clearly, less attention and care have been paid to the Aristotelian part of the diagram. The next two diagrams putting Aristotelian theory together with Platonic are both from thirteenth-century manuscripts of the *Dragmaticon philosophiae* of the twelfth-century scholar William of Conches. The first (at the left above), part of a quite stylized and decorative codex (also in Illustrations 132 and 250), presents its charge in striking reds, greens, blues, yellows, and browns. But its usefulness has been sacrificed for the sake of such colorfulness. The upper circle, representing the Aristotelian theory, contains labels for the four relevant primary qualities, but neglects the names of the elements they characterize. The lower Platonic diagram fares only slightly better. Five of the six primary qualities are given in smaller circles, one *(mobilis)* given twice, another *(acutus)* not relevant to the two elements (*aer* and *aqua*) presented, even though an added line descending from the circle of *acutus* makes it appear as if it is a property of *aer* (thereby erroneously assigning it four primary qualities instead of three). Only *aer* and *aqua* are represented in the diagram because only these two elements are discussed in the text being illustrated. The final set of figures at the right occurs at the same juncture in the *Dragmaticon,* but now the manner of presenting the alternative theories of the elements is not at all standard. For both theories, the elements have been drawn in the form of the shells of a stratified universe, earth at the center, fire at the circumference, each shell carrying, in addition to the name of the pertinent element, an indication of the primary qualities belonging to that element. The Aristotelian two-quality theory is at the left; the Platonic three-quality one at the right.

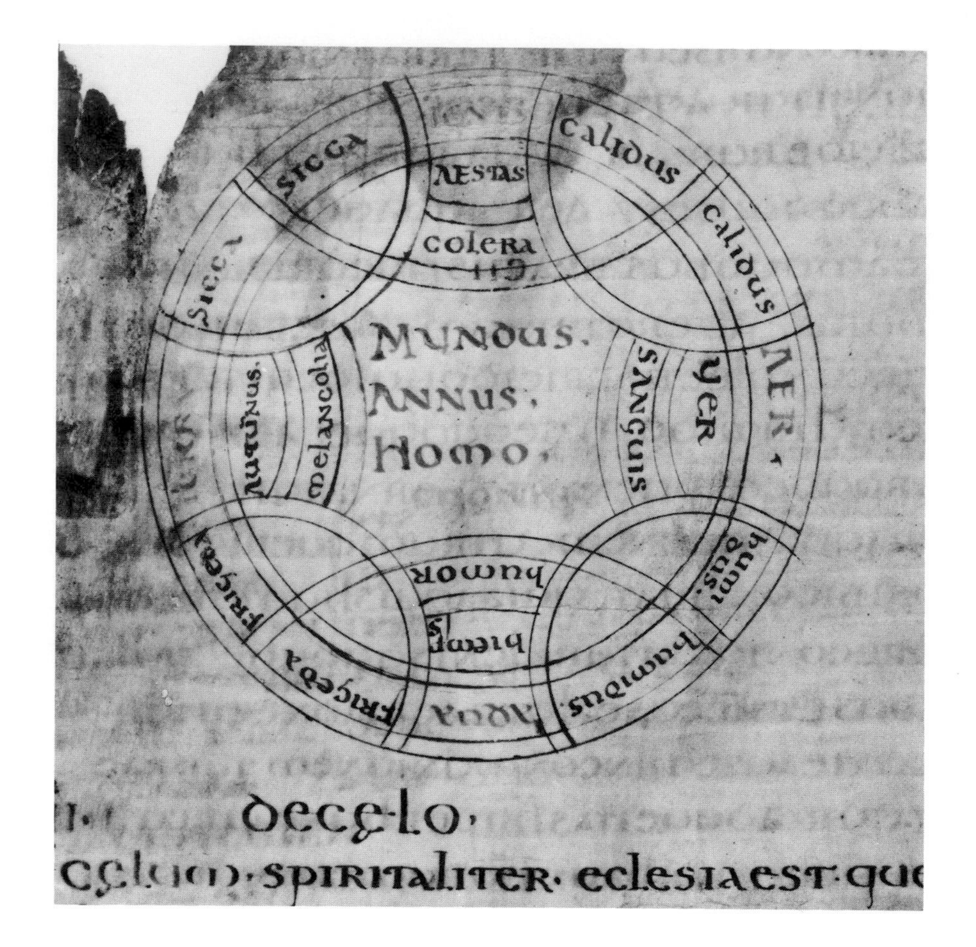

286. Microcosm–Macrocosm. This is the final Isidorean wheel to appear in this volume (others were given in Illustrations 46, 47, 48, 279, and 280). Taken from the eighth-century codex of Isidore's *De natura rerum* used in Illustrations 247 and 280, the theory of the elements it portrays is Aristotelian (Isidore having set forth the Platonic theory in the immediately preceding paragraph by means of his *cybus elementorum:* Illustration 247). He tells us that the theory of Aristotle—whom he does not name—comes from St. Ambrose (fourth century), his reference most likely deriving from the fact that Ambrose had approvingly recounted the theory in his commentary on the first chapter of Genesis. Although the text that the present circular diagram accompanies does not explicitly mention man as a microcosm, this is clearly implied by the "name" assigned the circle: *Mundus, Annus, Homo* (world, year, man). Isidore had already related the seasons to shared pairs of primary qualities in his "wheel of the year" (a copy of which has been provided in Illustration 47). Here, however, in addition to repeating the same information about the seasons (the names of which appear just inside the outermost circular band: *aestas, ver, hiemps, autumnus* = summer, spring, winter, autumn), the primary-quality pairs are also related to the elements at the circumference of the wheel and, in the centermost position, to the four humors (*colera,* for red bile; *sanguis,* blood; *humor,* for phlegm; *melancolia,* black bile). Clearly, the humors are meant to refer to the realm of *Homo,* man as microcosm, while the macrocosm relates to the realms of *Mundus* and *Annus,* duly represented by the four elements and the four seasons, respectively.

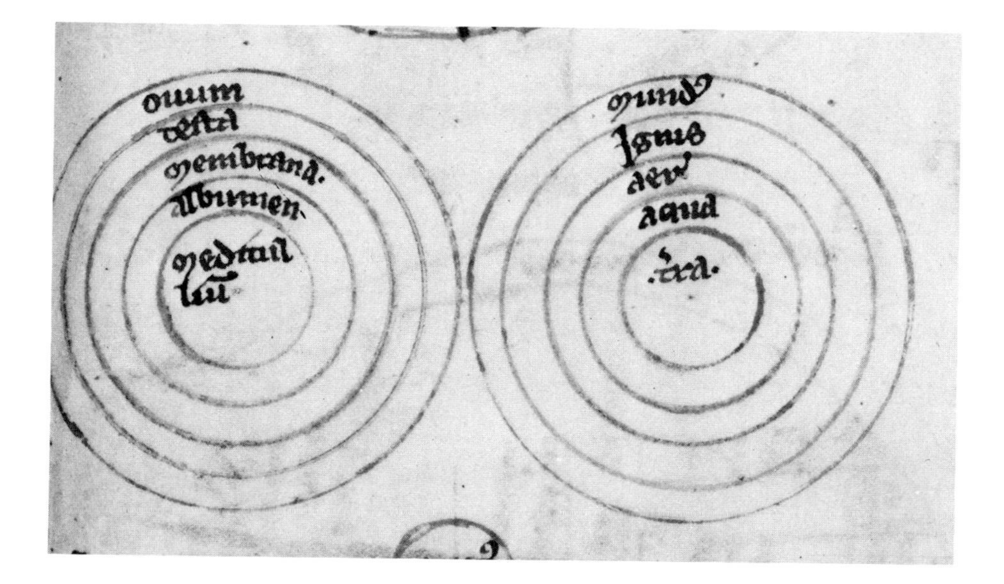

287. The World as Quadripartite. The preceding illustration has made it clear that the four elements were not the only set of four that was fundamental to the fabric of the world within ancient and medieval natural philosophy. The cardinal directions, the seasons, and the bodily humors, together with the temperaments that they determined, were relatively standard fourfold associates of the elements. Samples of yet other aspects of a quadripartite nature are given in the two present figures. The first, above, comes from a set of diagrams appended to a fourteenth-century copy (also seen in Illustrations 250 and 276) of William of Conches's *Dragmaticon.* The two circles represent an egg (at the left) and the world, their outermost circular bands carrying appropriate identification: *ovum, mundus.* The remaining strata within each circle urge a strict correspondence in the constitution of the entities they depict. Thus, just as the stratified universe begins in its outermost sublunar reaches with fire and proceeds through air and water to earth at its center, so an egg begins with its shell *(testa)* and progresses through its membrane and albumen to a central yolk (here named *meditulium*). The second figure, at the right, is taken from an eleventh- or twelfth-century manuscript of an anonymous work, *De quaternario.* In 118 brief chapters (appropriately divided into four books), this anonymous treatise finds fourfoldedness everywhere in the world, both when there are four discriminable entities (such as the elements, humors, sea-

(Continued)

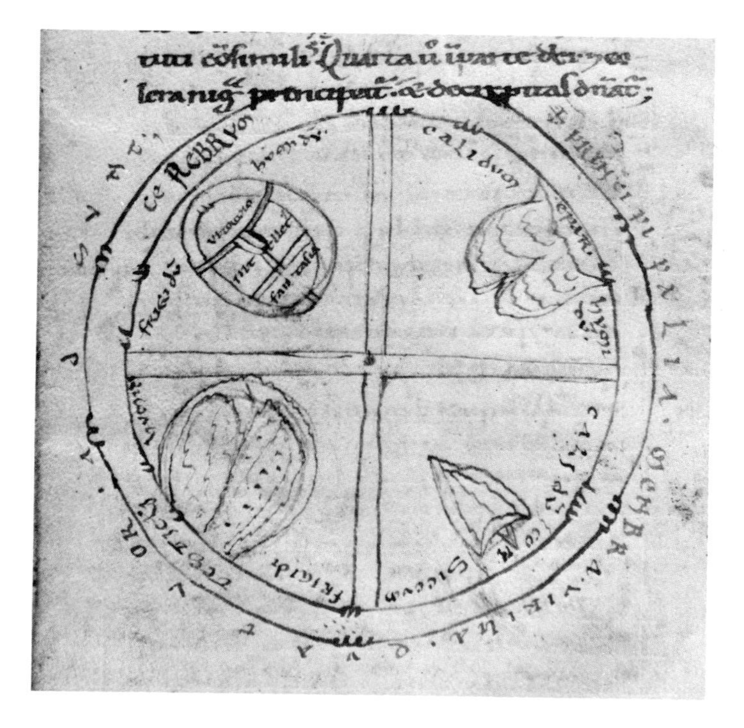

(Continued)

sons, and so on) and when there is some easily effected division into four (such as the division of the zodiacal signs into four groups of three). The final figure in the final chapter of this curious work is reproduced here. The legend surrounding the circle reads: *Principalia membra virilia quatuor adsunt* (there are four principal members in a man). They are, beginning in the upper left quadrant and proceeding clockwise: (1) the brain *(cerebrum)*, characterized by the primary qualities cold and moist and divided into three cells: *memoria, intellectus,* and *fantasia* (compare Illustration 271); (2) the liver *(epar)*, hot and moist; (3) the heart *(cor)*, hot and dry; and (4) the testicles *(testiculus)*, cold and dry (though the diagram has the words cold and moist, missing the correction made in the text). As a whole, then, these four *membra virilia,* like the four elements, share the same paired primary qualities, even though the artist or scribe who drew the figure has not positioned the names of these qualities in a way that would make their sharing visually evident.

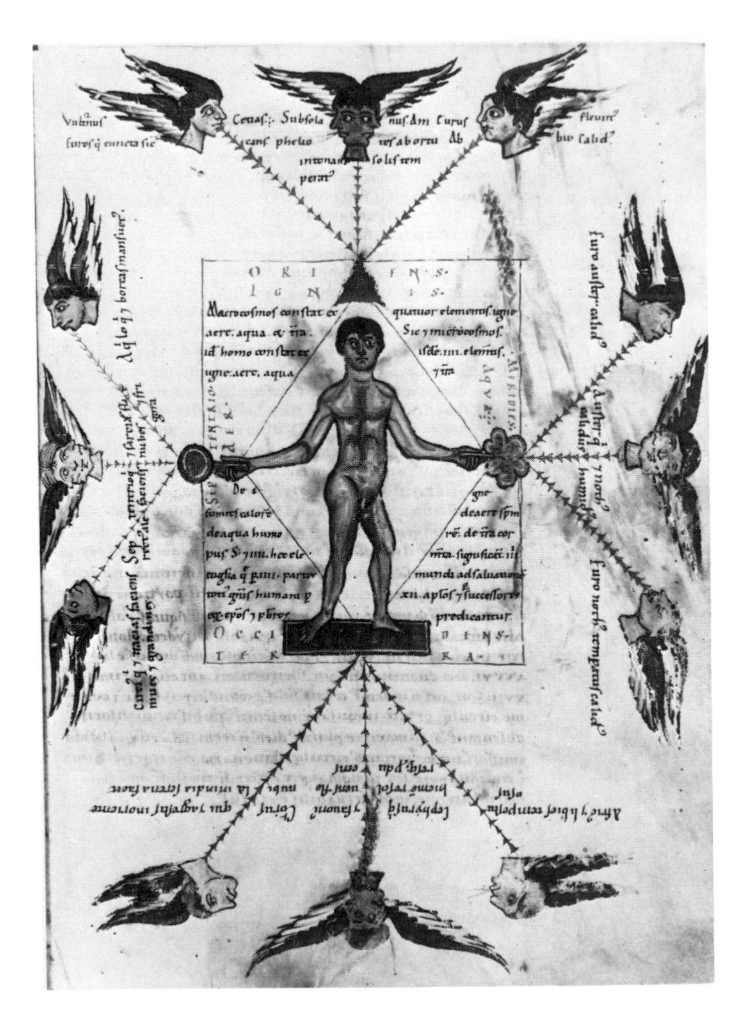

288. Man and the Four Elements. The macrocosm–microcosm relations portrayed in the three pictures given here are pictographically and textually more explicit than the relations of the same type expressed in the previous two illustrations. The first, at the left, comes from the twelfth-century Prüfening monastery miscellany that has already appeared in Illustrations 279 and 281. Personifications and properties of the twelve winds (see Illustration 280) establish a frame for the human figure at the center whose right hand holds a disk representing air and the north and whose left grasps a flower representing water and the south, while his feet at the west stand upon a rectangle representing earth for support, his head surmounted by an eastern triangle depicting fire. The relations between man and the universe implied by this arrangement are set forth explicitly by the text filling the corners of the rectangle that encloses the human figure. Just as the macrocosm consists of the four elements, so the microcosm or man consists of the same four elements. The text occupying the lower corners of the rectangle specifies which microcosmic constituents derive from the elements (heat from fire, spirit from air, humor from water, and body from earth) and goes on to claim that the four elements signify the four gospels, the preaching of which throughout the four corners of the world leads to human salvation. Another twelfth-century Prüfening manuscript (also seen as the earlier codex in Illustration 20), above right, has provided the second

(Continued)

microcosmos picture. Here, even more information is given concerning the relation of man to the universe. Indications of the four elements appear twice. The first instance is in the four corners of the picture's frame (fire at the upper left, followed clockwise by air, earth, and water) where quarter-arc bands contain their names and, in quite small letters, the names of the Platonic quality triplets characterizing them. Spokelike bands running from these four corners to the dominant Christlike figure tell us of the origin of each of the five senses from the elements: vision derives from fire, hearing from the superior air, smell from the inferior air, touch from earth, and taste from water. Indications of the elements are repeated elsewhere on the picture, often associated with other bodily constituents and phenomena. *Aer* and *ignis* appear directly overhead without added comment, but *terra* appears below the central figure with the counsel that, as the feet support the body's weight, so earth supports all things *(ut pedes molem corporis, terra sustenat omnia)*. The name *aer* occurs again on the breast *(pectus)*, for the breath and coughing in the breast are as the winds and thunder in the worldly air *(in quo flatus et tussis ut in aero venti et tonitrua)*. Water belongs to the belly *(venter)* into which all things flow like rivers into the sea *(in quo confluent omnia ut in mare flumina)*. Other properties of the elements, as well as some that have already been announced, are given in hexameters written on the frame of the whole: fire provides warmth and mobility, air furnishes sound or speech, earth gives rise to flesh and weight, water supplies blood and the other humors. Two other spokes forming an equilateral triangle with the base of the picture relate parts of the body to parts of nature: the hairs are like blades of grass, the bones like stones, and the nails like trees. The halo borne by the central figure carries an inscription that likens the head to a celestial sphere *(instar celestis spere)*. This is further confirmed by seven rays issuing from the seven openings in the head, each of them bearing the name of one of the seven planets. The final picture, bottom right, taken from a fourteenth-century manuscript, is much simpler and totally pictorial in the transmission of its macrocosm–microcosm message. It adorns the beginning of Book IV of a French translation of the encyclopedic *De rerum proprietatibus* of the thirteenth-century scholar Bartholomaeus Anglicus (Bartholomew the Englishman). This particular book deals with the properties of the human body and opens with the warning that anyone who wishes to treat this topic must begin with the qualities of the four elements and humors. The miniature above these opening lines presents this admonition visually in its depiction of a human body lying in the midst of a primordial nature that is neatly divided into strata representing the four elements.

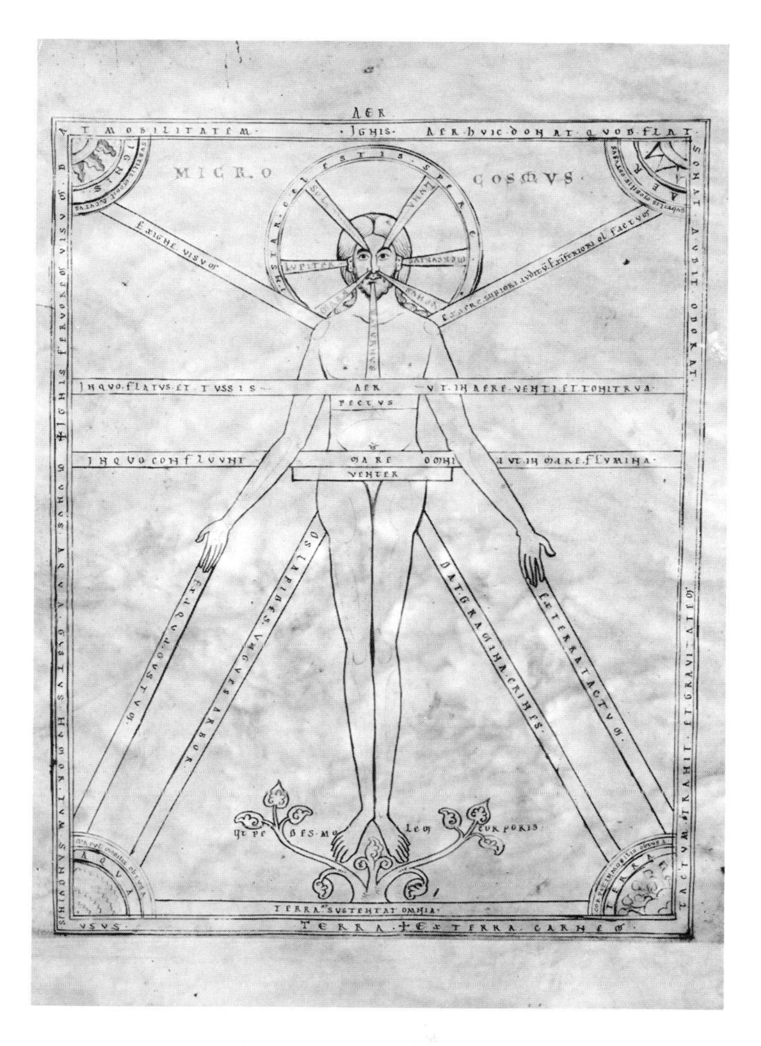

CHAPTER TWENTY-SIX

Encyclopedic Views of Medieval Systems of the World and Its Parts

MEDIEVAL LATIN ENCYCLOPEDIC works devoted to cosmography and the simpler aspects of natural philosophy were, if not astonishingly abundant, at least plentiful enough. The works of Macrobius, Isidore of Seville, the Venerable Bede, William of Conches, Honorius Augustodunensis, Lambert of St. Omer, Michael Scot, and John of Sacrobosco—to mention only those who have provided raw material for the previous pages of this work—provide adequate evidence for that fact.

In contrast, encyclopedic illustration of cosmography and natural philosophy was extraordinarily rare. It is true that the illustrations in some works in these fields were in ways more important and as well known as or even better known than the texts to which they belonged. Isidore of Seville's *De natura rerum* is an excellent example: his schemata of the months, seasons, cardinal directions, planets, terrestrial zones, elements, humors, and winds (see Illustrations 46–48, 247, 279, and 280) seem to have enjoyed a greater circulation than the work they originally accompanied. Similarly, a brief encyclopedic handbook like the anonymous compilation sampled in Illustrations 131, 251, and 280 often devoted as much space to diagrams as to text. But neither of these works had to face the task of depicting as many facets as possible of natural philosophy or cosmography within the compass of one or two diagrams or pictures. Two examples of such an achievement are given as the final two illustrations of this chapter.

Illustration 290, from a twelfth-century manuscript, concentrates on the external world and broaches the human realm only in terms of the so-called ages of man, a limitation congenial to its emphasis on other calendrical factors. Illustration 291 is from a late-thirteenth-century codex, though the ideas it expresses belong to the twelfth century. Microcosmic as well as macrocosmic, its coverage of elements pertinent to the human constitution is more complete.

A quite different kind of encyclopedism is exemplified in the first illustration of this chapter. It is one of the occasional clumps of diagrammatic material found in medieval codices. One or two pages of such collections of schemata are often found covering the flyleaves of a medieval book. In rarer instances, a fair number of folios, either at the end of or within a manuscript, may be devoted to such a collection. At times these collections may be of diagrams belonging to a text found elsewhere in the manuscript, but in other cases their textual origin is nowhere to be found within the codex at hand.

A few of the previous illustrations in this volume have been taken from such collections. For example, the paired diagrams of an egg and the world given in Illustration 287 and the figure of the Platonic *regiones* of the universe in Illustration 276 come from a collection of thirty diagrams clumped after the text of the *Dragmaticon* of William of Conches. These appended figures all relate to items discussed in the text of the *Dragmaticon,* even though the text itself contains numerous other diagrams.

The case of the collection of diagrams selected for Illustration 289 is slightly different. Part of a twelfth-century manuscript, these diagrams too occur at the end of a text that contains diagrams of its own (the lambda figure in Illustration 283 is one such)—namely, a text of Chalcidius' Latin translation of Plato's *Timaeus.* But only a few of the eighteen diagrams constituting this collection truly belong to the *Timaeus.* Most have been imported from the likes of Macrobius, Boethius, or William of Conches, here left to stand on their own without any relevant text.

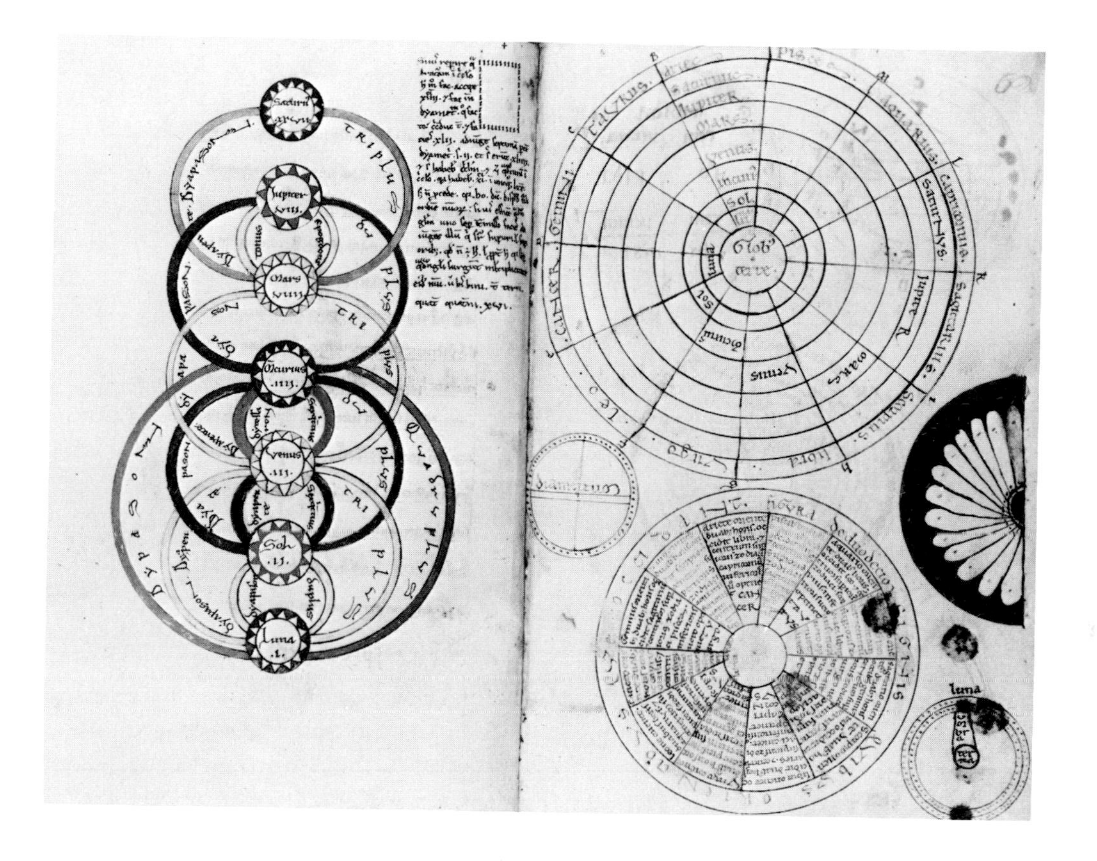

289. An Encyclopedic Spread. Taken from the twelfth-century codex described at the conclusion of this chapter's introduction, the seven manuscript pages reproduced on pages 361 to 363 contain diagrams and figures drawn from a variety of medieval works. Most of them have appeared within, or are related to, earlier illustrations in this volume; a few are new. *Page 1* (page 361, left): Each of the small starred circles running vertically through this diagram bears the name of one of the seven planets, together with one number from the intercalated series of square numbers that Plato used in the first step of the division of the World Soul (Illustration 283). Here these numbers explain the harmony of the spheres, the arcs joining the small central circles accounting for the musical intervals obtaining between the spheres of the heavenly bodies named within those circles. The diagram most likely belongs to those paragraphs in Macrobius' *Commentary on the Dream of Scipio* (II, 2) in which this particular doctrine is discussed. Macrobius, in turn, probably lifted the whole idea from Chalcidius (ch. 96). (The text surrounding the small rectangle in the upper right-hand corner of the page deals with numerical procedures unrelated to the main diagram.) *Page 2* (page 361, right): The large circle occupying the top of the page is nothing more than a diagram of the orbs of the planets surrounding the central sphere of the earth *(globus terrae)*, in turn surrounded by the sphere of the fixed stars, here identified by the twelve names of the zodiacal signs. The large circle below this is a wheel intended to show the correlation of the risings, settings, zeniths, and nadirs of the zodiacal signs. Thus, one sector tells us that during the two hours Aquarius is rising, Leo is setting, Scorpio is passing through the zenith, and Taurus is passing through the nadir. In fact, the whole diagram simply provides an alternative way to show which zodiacal signs stand at 90° to one another. The circular figure in the lower right corner diagrams a lunar eclipse (cf. Illustration 253). The purpose of the small circle at the left seems to be no more than to explain what a diameter is, while the semicircle on the right serves no function at all in its present state. It may have been intended to show which zodiacal signs are visible when a given

(Continued)

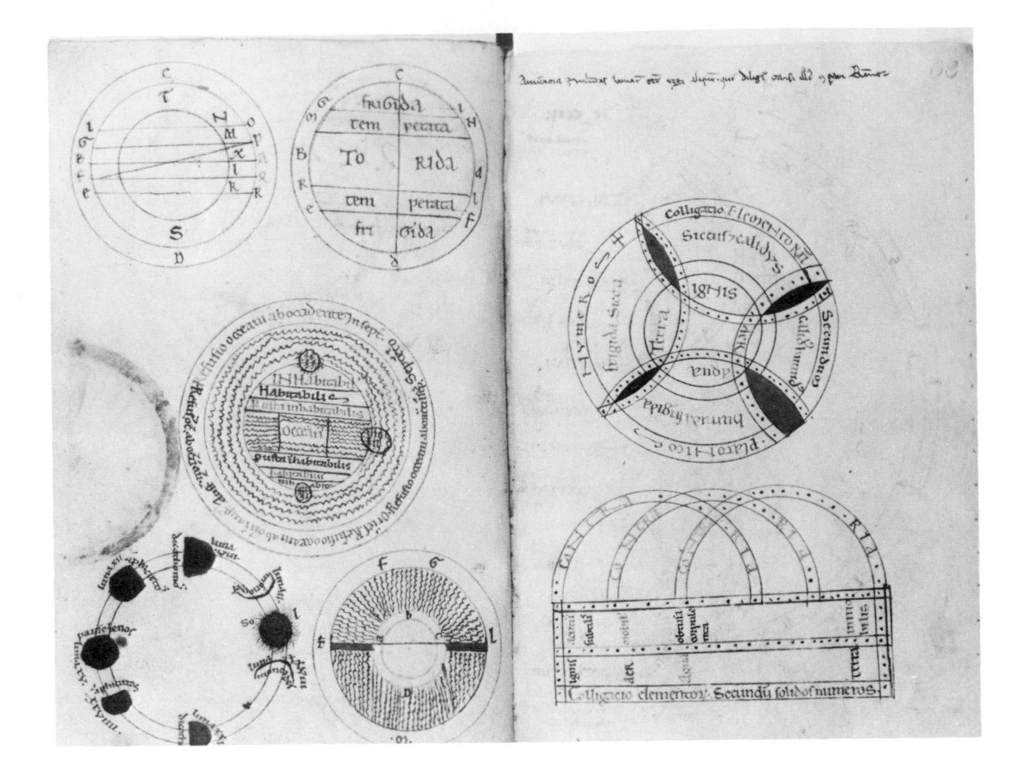

(Continued)
sign is at the zenith. Such semicircular diagrams were used for that purpose in some manuscripts of the *Dragmaticon* of William of Conches. *Page 3* (page 362, left): The top two circles are diagrams of the zones of the earth. Quite unlike the Isidorean wheels of the zones given in Illustration 279, both are Macrobean. That on the right reveals the zones of the earth alone (*Comm. Scip.,* II, 5). That on the left represents Macrobius' attempt (II, 7) to show the correspondence between the heavenly zones mentioned by Vergil and the earthly zones discussed by Cicero. The central diagram is a rather poor attempt at illustrating the Macrobean theory of the tides (cf. the first figure given in Illustration 251). Were it not for the words *refusio oceani* (backflow of the ocean) appearing several times on the circumference of the circle, it would be difficult to guess the purpose of the diagram. As was standard with diagrams of this sort, the habitable and uninhabitable zones of the earth are represented in a central "map." The three small circles riding on the circumference of the central circle are probably intended to depict the rising, zenith, and setting of the sun. The diagram at the lower left is clearly intended to present the phases of the moon (cf. Illustration 234). The figure in the lower right corner is, taken alone, more problematic. Because it is detached from the text it was intended to explain, only acquaintance with a similar diagram associated with the text will allow one to determine its purpose. Comparison with the veritable bevy of similar diagrams in Illustration 248, however, makes it certain that it is intended to represent Macrobius' rainfall theory. That it was held to be meaningful apart from the relevant text in Macrobius can be viewed as evidence of how well known that text must have been to some medieval scholars. *Page 4* (page 362, right): The circular schema given here does not match its announced function. The inscription on the circumference claims that it is a representation of the "connection of the elements according to Platonic numbers" *(colligatio elementorum secundum platonicos numeros);* but the diagram in fact depicts the Aristotelian theory of the elements in terms of shared pairs of primary qualities (cf. Illustration 284). The lower arc figure is not much more successful. Alleged to be a diagram of the *colligatio* of the elements according to solid numbers (compare the numerical relations of the elements exhibited in Illustration 282), it in no way makes the relevant mathematics clear. What is more, its representation of the Platonic quality triplets characterizing the elements (again see Illustration 282) is incomplete and the arcs designed to illustrate the contrary re-
(Continued)

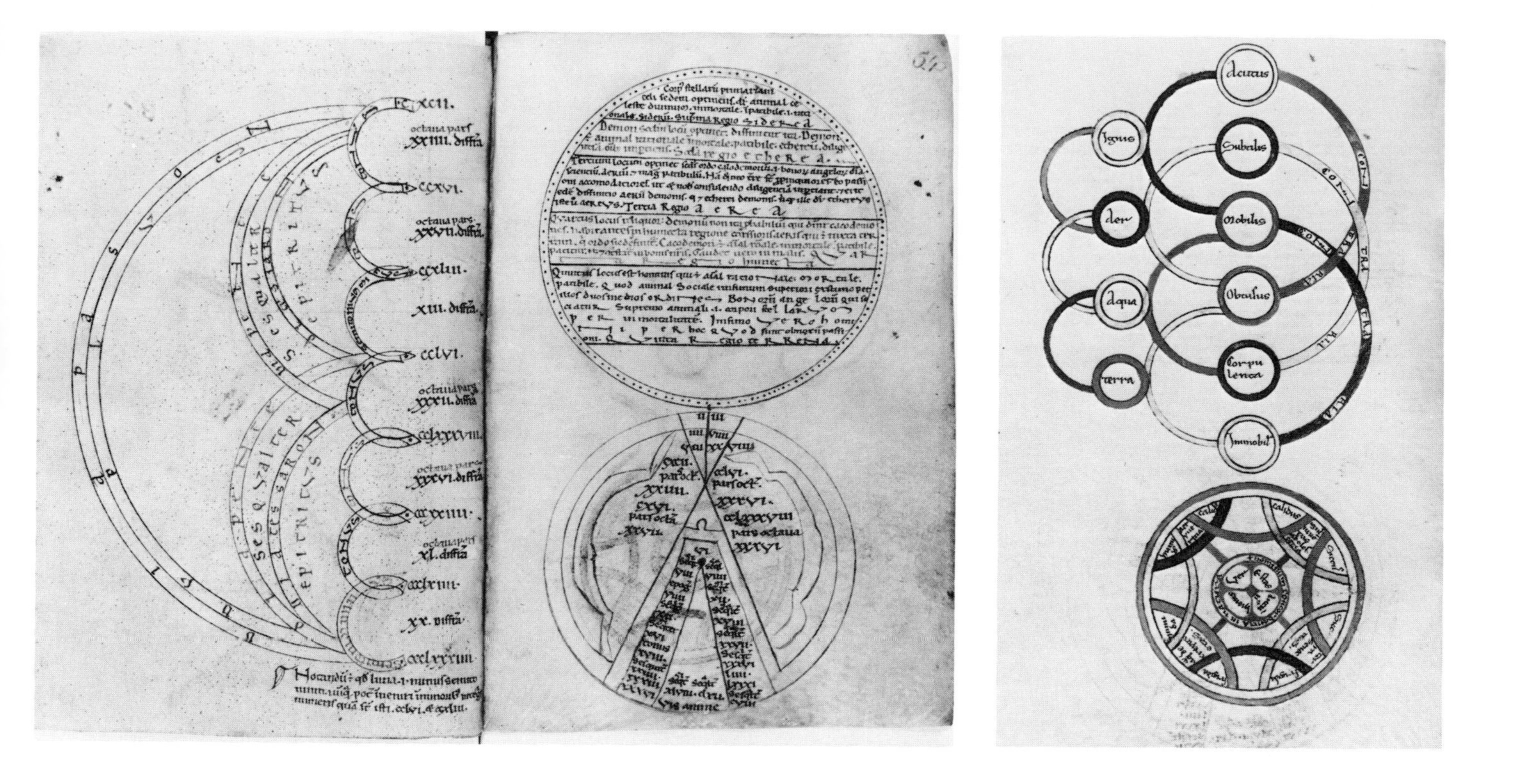

lations between these qualities are nonfunctional. *Page 5* (page 363, left): This whole page is occupied by an arc diagram of musical intervals similar to those seen in numerous instances above (Illustrations 95 and 100–105). In this case, the purpose is to represent the fact that the octave is made up of five whole tones and two minor semitones as well as of an interval of a fifth plus an interval of a fourth. The arcs carry inscriptions indicating these intervals and, in the case of the octave, fifth, and fourth, the ratios corresponding to those intervals. The Roman numerals at the left attempt to represent these ratios in whole numbers, beginning with 192 at the top and going to 384 at the bottom, thus stipulating the double ratio of the octave. The whole numbers given here derive from the series Boethius had used in investigating the fifth and the fourth (*De musica,* I, 18): namely, 192, 216, 243, 256, 288. This series successfully provides the relevant ratios involved, for example, in the interval of a fifth (which consists of three whole tones plus a minor semitone). Thus, the ratio of 256 to 243 is that of a minor semitone, while the remaining ratios given by the series (216:192, 243:216; 288:256) are all equivalent to 9:8 (which is the ratio corresponding to a whole tone). This latter point is established—and duly noted at the left of the diagram—by the fact that the differences between the numbers determining the whole tone ratios are eighth parts of the lower term of each such ratio (hence, 216 − 192 = 24 = an eighth part of 192, etc.) So far—that is, up to 288—the diagram is successful. But in an attempt to have a whole-number diagram for the octave and not merely for the fourth and the fifth, the designer of the diagram has added the succeeding numbers 324, 364, and 384 to the original Boethian series. When he did so, the upper fourth and fifth of the octave were represented by appropriate ratios (that is, 384:288 = 4:3, the sesquitertian ratio of the fourth, and 384:256 = 3:2, the sesquialtern ratio of the fifth). But the difference between 364 and 324 does not yield the eighth part necessary for determining the required whole-tone ratio; nor does the ratio 384:364 correspond to the required minor semitone (something that follows even from the annotation at the bottom of the diagram asserting that the *lima* or minor semitone can never be found in whole numbers less than 256:243). The medieval creator of this diagram did not see that extending Boethius' whole-number series for the fourth and fifth to cover the octave would not work. *Page 6* (page 363, center): The upper circle on this page is an alternative representation of the five Platonic *regiones* of the cosmos already noted in the diagram from William

(Continued)

(Continued)
of Conches given in Illustration 276. Now, however, the *regiones*—also referred to as *loci*—are depicted as strips or belts instead of the circular bands or shells used in the previous diagram, a change in format that has permitted the inclusion of more information about each *regio.* The circular figure below contains three different elements. The first consists of the six Roman numerals at the very top of the figure: they give the two series of square numbers 2, 4, 8 and 3, 9, 27 (erroneously written as 28). The two sets of numerals below this on both sides of the central, tepeelike figure repeat parts of the musical interval diagram occupying the previous page. This part of the diagram is incomplete and not without error, but it seems that the intention was to use one side of the central triangle for the relevant series of the lower fourth of the octave and the other side for the series of the upper fifth. The two numerical series within the central triangle are a cruder version of the series representing the second step in Plato's division of the World Soul already seen in the lambda diagram in Illustration 283. Unlike that earlier, far more elegant diagram, the present lambda adds information about the kinds of ratios (or, in one instance, musical intervals) that obtain between some of the numbers in the double series. (The whole diagram is labeled below as *vis anime* [the power of the soul], an expression most likely derived from Chalcidius' comment on the Platonic text in question.) *Page* 7 (page 363, right): This final page contains neatly crafted schemata giving both the Platonic and the Aristotelian theories of the elements (compare Illustrations 284 and 285). The relevant numbers are absent in the upper Platonic figure and connectors are missing between one member of each triplet of qualities and the element to which it belongs, but otherwise the diagram is effective. The circular Aristotelian diagram below is in the standard form, employing intersecting arcs to indicate the primary qualities shared by the four elements (compare Illustration 286 and the circular figure on the fourth page of the present collection). The inscription on the circumference of the central circle *(hominum concordantia in natura)* reveals that additional macrocosmic-microcosmic information is being furnished. Thus, the elements (clockwise from fire at the upper right through earth, water, and air) are matched with the seasons at the center of the diagram and with the ages of man inscribed just below each element's name (*ignis, iuventus* and *adolescentia; terra, senectus; aqua, decrepita etas; aer, pueritia*).

290. Time, Space, and Matter. It is fitting for one of the final illustrations in this volume to be of a full page from the twelfth-century miscellany so frequently mined before (Illustrations 31, 51, 53, 69, 249, 279, and 285). As the first line of text on this page tells us, this diagram was designed by a monk of Ramsey in Huntingdonshire named Byrhtferth *(Hanc figuram edidit bryhtferde monachus ramesiensis),* who apparently flourished during the reign of King Aethelred (978–1016). Its claim to be a diagram depicting the concord of the months and the elements *(de concordia mensium et elementorum)* is a rather modest description of what it actually covers. Setting aside the elements as the basic material constituents of the world, one might consider the months as an umbrella for all of the various temporal factors represented in the figure (the seasons, the ages of man, the solar and lunar months themselves, and, because of their motion through the heavens, the signs of the zodiac). But this leaves a variety of spatial factors depicted by the diagram unaccounted for: the cardinal directions, the equinoctial and solstitial points, and the winds. All in all, then, the diagram treats of space, time, and matter. Turning to the manner in which elements belonging to each of these areas is represented, one can begin with the central diamond and the cardinal directions indicated at its vertices. The names of the directions (east at the top, north at the left) are given in both Latin *(oriens, meridies, occidens, aquilo)* and transliterated Greek *(anathole, mesembrios, disis, arcton),* the four initial letters of the latter giving rise to the inscription of the name of Adam in large script about the central diamond. The circles occupying the vertices of the larger diamond contain the names of the four elements and of the twelve winds. The sides of the diamonds joining these circles are interrupted by other circles bearing the names of the seasons corresponding to the elements (each season belonging to the element appearing next in a counterclockwise direction), of the dates on which these seasons begin, and of the ages of man (including their span in years) associated with the elements. The names of the pairs of primary qualities shared by both the elements and the seasons are written along the sides of the two diamonds. Turning to the exterior portions of the diagram, the arcs joining the four element circles carry, first, indications of the zodiacal signs, and then the solar and lunar months, both with specifications of the number of days for each. Smaller arcs capping the elements' circles give the dates of the equinoxes and solstices. The center of the diagram contains the Greek letters ($\chi\rho s$) twice as a reference to Christ. The texts surrounding the diagram do no more than explain its contents and function. Given the care with which the diagram appears to have been constructed, it is surprising to see that the correlation of the elements and seasons with the cardinal directions is not that traditionally found. Normally, east is correlated with air and spring, south with fire and summer, west with earth and autumn, and north with water and winter. Here, however, the concomitants of the east have been interchanged with those of the west and those of the north switched with those of the south. The resulting incongruities (for example, hot and dry fire and summer associated with the north) certainly seem striking enough to have made the designer of the diagram aware of the error of his ways. But apparently this did not occur. The connection of *iuventus* with autumn and *senectus* with winter is also nonstandard (cf. the final figure on the final page reproduced in the previous illustration).

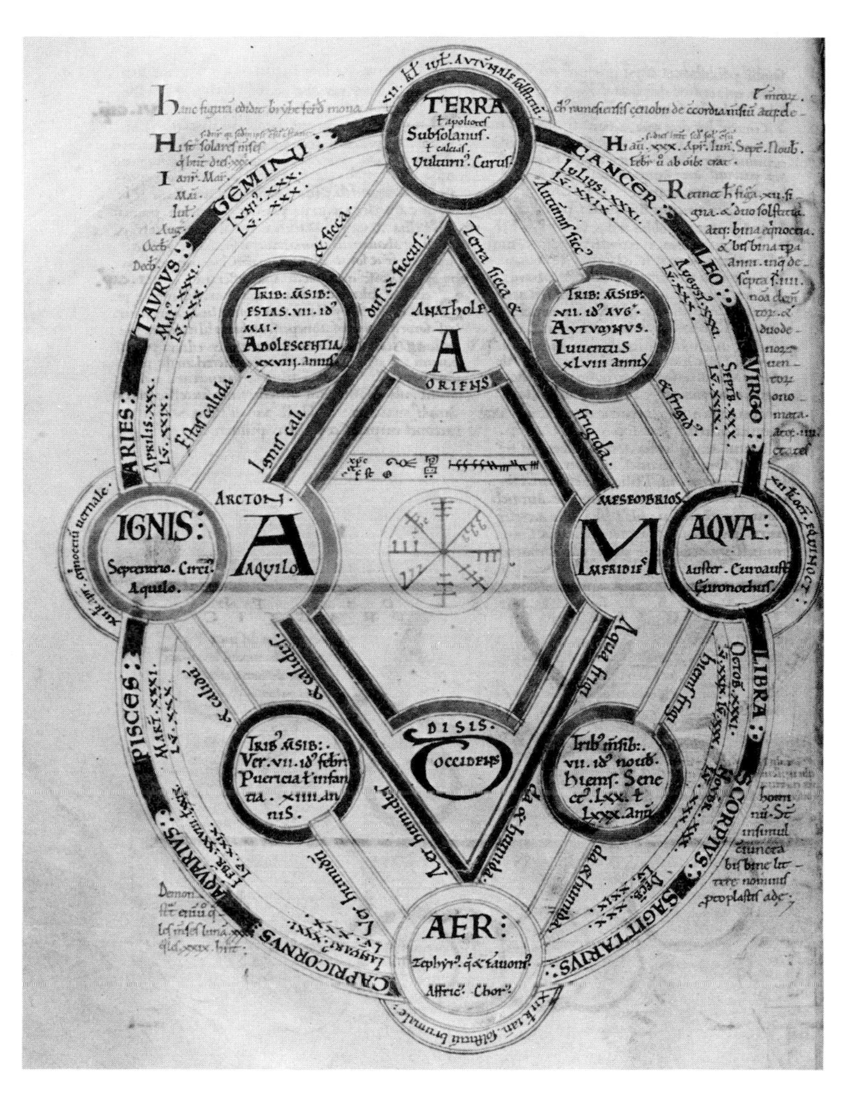

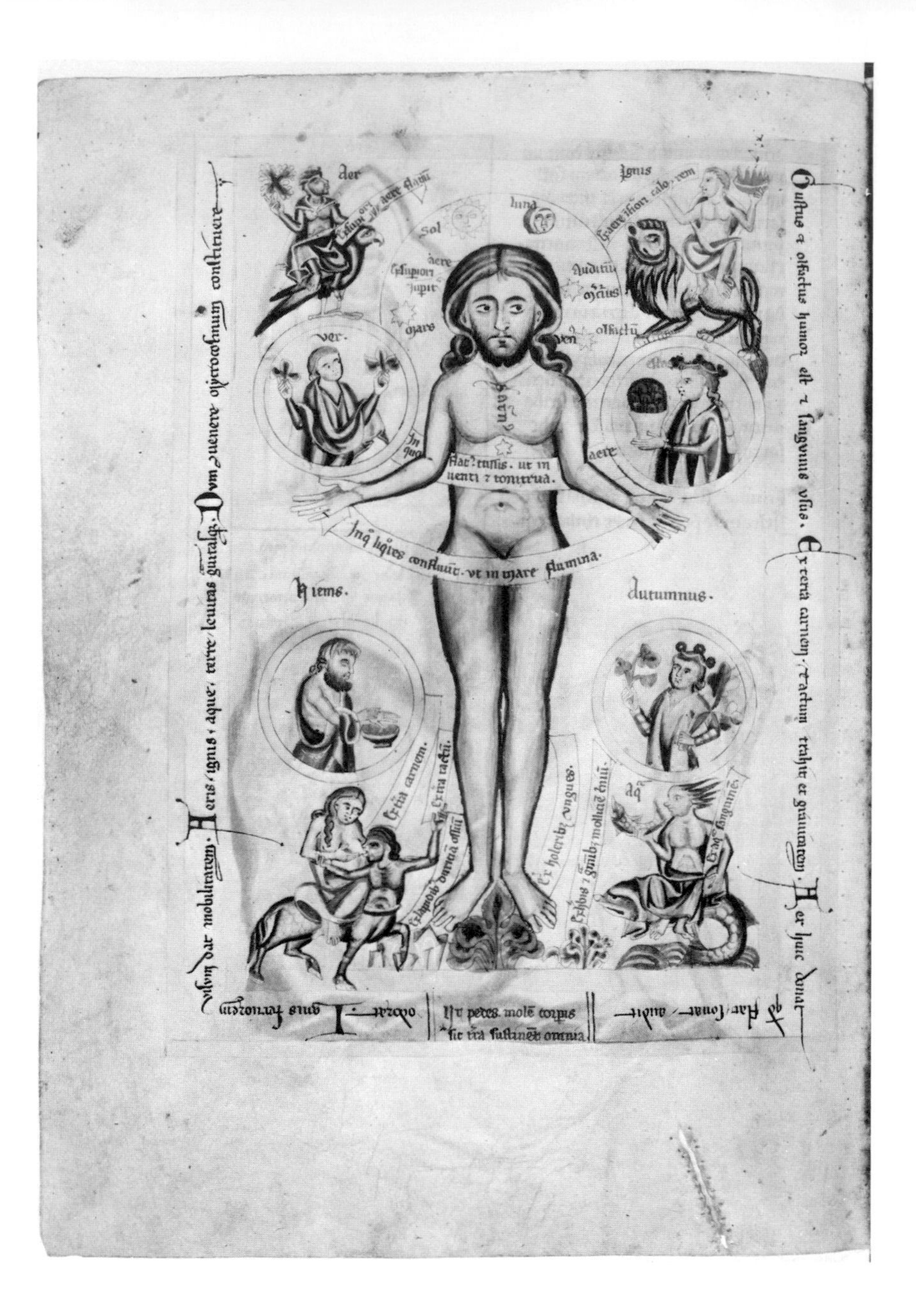

291. A Twelfth-century System Visualized in the Thirteenth Century. The two pictures reproduced here occur on facing pages that fall between a copy of the *De natura rerum* by the thirteenth-century Flemish Dominican encyclopedist Thomas of Cantimpré (see Illustration 19) and one of the *Philosophia mundi* by the twelfth-century French scholar William of Conches found in a manuscript written at Aldersbach in 1295 (also seen in Illustration 253). The picture on the left-hand page should be compared with the figures personifying the elements given in Illustration 281 and with the second picture reproduced in Illustration 288. The two twelfth-century manuscripts in which these latter illustrations occur were both produced at the Benedictine monastery in Prüfening (near Regensburg) not far from Aldersbach (near Passau), so the resemblance many features in the present pictures bear to elements in the earlier productions provides a good example of how scientific illustrations circulated and were reproduced within the confines of a relatively small geographic area. Thus, the personifications of the elements shown here in the four corners of the page are iconographically quite like those given in Illustration 281. The major exception is the figure of water at the lower right, who now rides

(Continued)

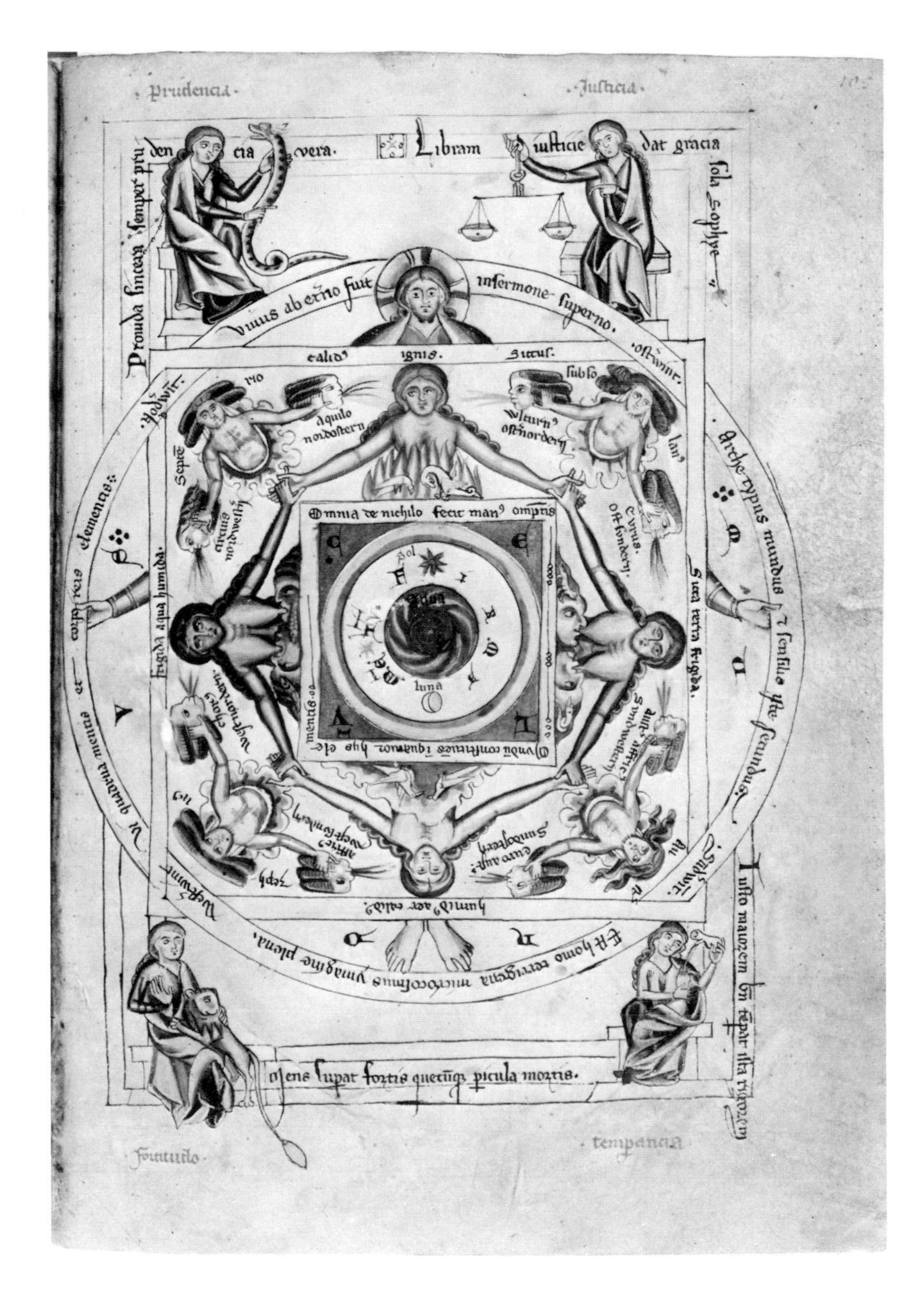

and carries a fish instead of sitting astride a griffin while pouring water from a vessel as in Illustration 281. The similarities with the second microcosm man in Illustration 288 are even more marked. The inscriptions on the scrolls crossing the torso of the central figure are all but identical in both illustrations. So is the information provided by the other scrolls in the picture and within the frame of the whole. The only notable differences are that the nails are now related to vegetables *(ex holeribus)* rather than to trees and that the inferior air is held (inadvertently, it would seem) to give rise to heat rather than to the sense of smell. Further, the left-hand border of the frame contains additional hexameters asserting that when the heaviness and lightness of the four elements come together, a macrocosm is created *(Aeris ignis aque terre levitas gravitasque, Dum convenere microcosmum constituere)*. Four circles enclosing personifications of the seasons provide another feature absent in the microcosm man in Illustration 288. But the halo of the present microcosmic figure contains the seven planets in the same order in which they appear in the twelfth-century illustration, even though their correspondence with the seven openings of the head is not as clearly depicted. The illustration on the right-hand page is at once more

(Continued)

(Continued)
complex and, as far as one can judge, more original. The corners of the outer rectangle are inhabited by personifications of the four cardinal virtues, each accompanied by one of her standard attributes (Prudence with a serpent, Justice with scales, Fortitude with a lion, Temperance with two vases). Christ as Logos stands inside and embraces a circular band containing inscriptions that set forth a hierarchy ranging from the supernal Word through the *mundus archetypus* and man as microcosm to the corporeal elements. A large square, interwoven with the circular band, harbors triplets of the twelve winds in each of its corners. The names of the winds are given in both Latin and (with some exceptions) German (for example, *Subsolanus, Osternwint; Chorus, Westnordern*), thus revealing the Bavarian origin of the illustration. The four elements (from fire at the top, clockwise through earth, air, and water) are depicted by female busts surrounding an inner square. The nourishment of the world by earth and water is apparently the reference behind the faces feeding at the breasts of these two elemental personifications. Animals traditionally associated with one or another of the elements (phoenix and salamander, quadruped and reptile, birds, and fish) appear below the elements, while the names of each element and of its primary qualities are written directly overhead in the frame of the larger square. Unfortunately, the arresting appearance of this depiction is marred by the fact that the proper positions of air and water have been switched. As a result, the elements do not visually represent their opposition (fire, as hot and dry, should stand opposite water, as cold and moist). This improper positioning has also obscured the sharing of primary qualities by the elements, in spite of the fact that this sharing is the apparent reference behind the personified elements joining hands. The inscription on the frame of the innermost square proclaims that "the hand of the Omnipotent had made all things from nothing, constructing the world out of these four elements." The letters in the corners of the smaller square spell out the word 'heaven' *(celum)*, while the letters scattered—together with the sun, the moon, and one star—within the circle contained by this square yield the word *firmamentum*. The innermost circles represent the earth and water, though only the latter is identified by name. Almost all of the conceptions portrayed in this illustration are, as even a brief glance at some of the previous illustrations in this volume will indicate, quite standard. One conception, however, is less commonplace: the hierarchy professed by the verses written on the circular band held by Christ. These date the illustration, it has been suggested, as a thirteenth-century reflection of twelfth-century thinking, a suggestion that is confirmed when one compares the present illustration with the hierarchical elements depicted in the twelfth-century figures given above in Illustrations 274 and 275. Put together with the undeniable fact that the first illustration reproduced here derives from a twelfth-century source, there is hardly any doubt that the whole is a fragment of the legacy of that century.

Bibliography and Guide to Further Reading

Although bibliographies of ancient and medieval science can be found in numerous publications concerned with these periods in the history of science, there are almost no bibliographies of the history and importance of pictorial material within the early history of science. With this in view, the initial part on scientific illustrations in the present bibliography has been made, relatively speaking, more complete than the second covering scientific content. Emphasis in this more limited second half of the bibliography has been, wherever possible, on works in English, on works that can serve as introductions to their subject, and on works that themselves contain substantial bibliographies that can facilitate further reading and research. With one notable exception, editions and translations of primary sources have been kept to an absolute minimum in the second part. This exception—section 7—has been made in order to provide references to translations of those scientific texts whose accompanying illustrations or diagrams have repeatedly furnished pictorial material for this *Album,* the purpose being a partial fulfillment of the ideal of providing bibliographic documentation for every text explained by an illustration reproduced in the book.

I. THE SCIENTIFIC ILLUSTRATION

1. Specific manuscripts

Codices of scientific works that are in some way especially notable for the illustrations they carry have from time to time been reproduced in toto in facsimile. Most frequently these codices have belonged to the area of materia medica. As expected, the Greek Dioscorides (seen above in Illustrations 188, 193, and 202) executed for Princess Anicia Juliana early in the sixth century has had the privilege of such a reproduction; first, in black and white, as *Codex Aniciae Iulianae picturis illustratus* . . . Josef von Karabacek *et al.,* eds. (Leiden: Sijthoff, 1906). More recently, this presentation manuscript has been reproduced in full color as *Codex Vindobonensis med. Gr. 1 der Österreichischen Nationalbibliothek,* 2 vols. (Graz: Akademische Druck, 1965–1970). Two of the earlier manuscripts (neither utilized in the present volume) of the herbal of Pseudo-Apuleius have also appeared in facsimile: F. W. T. Hunger, *The Herbal of Pseudo-Apuleius from the Ninth Century MS in the Abbey of Monte Cassino—Codex Casinensis 97* . . . (Leiden: Brill, 1935); *The Herbal of Apuleius Barbarus, from the Early Twelfth-century manuscript formerly in the Abbey of Bury St. Edmunds (MS Bodley 130),* described by Robert T. Gunther (Oxford: Roxburghe Club, 1925). Other illustrations from various codices of Pseudo-Apuleius have been reproduced together with the critical edition of the text by Ernest Howald and Henry E. Sigerist, *Antonii Musae de Herba vettonica liber, Pseudo-Apulei Herbarius, Anonymi de Taxone liber, Sexti Placiti Liber medicinae ex animalibus,* in Corpus medicorum latinorum, IV (Leipzig: Teubner, 1927). Another facsimile of a Pseudo-Apuleius codex is *Medicina antiqua; Libri quattuor medicinae; Codex Vindobonensis 93 der Österreichischen Nationalbibliothek* (Graz: Akademische Druck, 1972). Similarly, the miniatures in the tenth-century manuscript of Apollonius of Citium seen in Illustration 168 have been reproduced as a supplementary insert to Apollonius Citiensis, *In Hippocratis de articulis commentarius,* J. Kollesch and F. Kudlien, eds., German trans. by J. Kollesch and D. Nickel, in the Corpus medicorum graecorum, XI, 1, 1 (Berlin: Akademie Verlag, 1935). Note should also be taken of the *Tacuinum sanitatis* facsimile cited in the picture credit to Illustration 270. Outside the field of medicine, the sixth-century *agrimensores* codex used in Illustration 146 has recently appeared in facsimile as *Codex Arcerianus A,* H. Butzmann, ed., in Codices graeci et latini photographice depicti XXII (Leiden: Sijthoff, 1970). Yet another manuscript reproduction is that of the autograph of Lambert of St. Omer's *Liber floridus* (see Illustration 251): appearing in Ghent (E. Story-Scientia, 1968), it was followed by Albert Derolez, ed., *Liber floridus Colloquium* (Ghent: E. Story-Scientia, 1973).

A number of manuscripts have been utilized repeatedly in this volume, but none of them has appeared in facsimile. Most have received not much more attention than the usual manuscript-catalog description. A few, however, have been given slightly more extensive notice. Thus, MS 17 of St. John's College, Oxford (seen in Illustrations 31, 51, 53, 69, 249, 279, 285, and 290) has been the subject of discussion by Charles Singer: "A Review of the Medical Literature of the Dark Ages, with a New Text of About 1110," in *Proceedings of the Royal Society of Medicine,* 10 (1917), pp. 107–160. A relatively detailed analysis of MS W.73 (Illustrations 50, 131, 251, and 280) of the Walters Art Gallery in Baltimore has been given by Harry Bober: "An Illustrated Medieval Schoolbook of Bede's *De natura rerum,*" in *Journal of the Walters Art Gallery,* 19–20 (1956–1957), pp. 65–97. Although the collective manuscript Burney 275 of grammar, rhetoric, logic, arithmetic, geometry, music, and astronomy (sampled in Illustrations 44, 59, 60, 62, 104, 178, 241, and 252) was presented to Jean, Duke of Berry, by Pope Clement VII in 1387, it has received only passing notice by those historians of art who have dealt with the duke's illuminated codices. References to texts and translations relevant to other frequently mined manuscripts are given below in section 7.

2. Manuscript illustration in general and the iconography of science

Among those scholars of the illustrated ancient and medieval book who have in recent years treated aspects of their field that are of particular importance for the history of science, two are preeminent: Fritz Saxl and Kurt Weitzmann. Saxl's *Lectures,* 2 vols. (London: Warburg Institute, 1957) contains chapters on science and art in general as well as on such more specific topics as astrology, belief in the stars, and microcosm–macrocosm conceptions in the Middle Ages. Further discussion and reproduction of illustrations dealing with astrological, astronomical, and related topics are to be found in the first three volumes of his *Verzeichnis astrologischer und mythologischer illustrierter Handschriften des lateinischen Mittelalters,* vols. 1 and 2 in Sitzungsberichte der Heidelberger Akademie der Wissenschaften, phil. hist. Klasse, Abh. 6–7 (1915) and Abh. 2 (1925–1926), vol. 3 (London: Warburg Institute, 1953). Note should also be taken of Saxl's early "Beiträge zu einer Geschichte der Planetendarstellung im Orient und im Okzident," in *Der Islam,* 3 (1912), pp. 151–177 and his later "A Spiritual Encyclopaedia of the Later Middle Ages," in *Journal of the Warburg and Courtauld Institutes,* 5 (1942), pp. 82–142. Kurt Weitzmann's most comprehensive treatment of scientific illustrations can be found in the opening chapter on scientific and didactic treatises in his *Ancient Book Illumination* (Cambridge, Mass.: Harvard Univ. Press, 1959). His provocative, if somewhat controversial, view of the papyrus roll ancestry of manuscript illustration is also of interest: *Illustrations in Roll and Codex: A Study of the Origin and Method of Text Illustration,* rev. repr. (Princeton, N.J.: Princeton Univ. Press, 1969). For another of Weitzmann's important contributions, see section 3.

Guides for the symbols and iconography frequently found in medieval scientific works are: Karl Künstle, *Ikonographie der christlichen Kunst* (Freiburg im Breisgau: Herder, 1928); *Lexikon der christlichen Ikonographie,* Engelbert Kirschbaum, ed., 8 vols. (Rom: Herder, 1968–1976); and *Reallexikon zur deutschen Kunstgeschichte,* Otto Schmitt *et al.,* eds. (Stuttgart: Metzlersche Verlagsbuchhandlung, 1937–). The iconographic elements involved in the personification of the liberal arts (see Illustrations 172–180) are the subject of Paolo d'Ancona, "Le rappresentazioni allegoriche delle Arti Liberali nel medioevo e nel Rinascimento," in *L'Arte,* 5 (1902), pp. 137–155, 211–228, 269–289, 370–385; compare d'Ancona's *L'Uomo e le sue opere* (Florence: Società editrice "La Voce," 1923); Adolf Katzenellenbogen, "The Representation of the Seven Liberal Arts," in Marshall Clagett, Gaines Post, and Robert Reynolds, eds., *Twelfth-century Europe and the Foundations of Modern Society* (Madison: Univ. of Wisconsin Press, 1961), pp. 39–55; and Philippe Verdier, "L'Iconographie des arts libéraux dans l'art du moyen âge jusqu'à la fin du quinzième siècle," in *Arts libéraux et philosophie au moyen âge: Actes du quatrième Congrès international de philosophie médiévale* (Paris: Vrin, 1969), pp. 305–355. The iconography of music, as well as a good deal more of the diagrammatic and pictorial material relative to the medieval book-history of this liberal art, are excellently presented in Joseph Smits van Waesberghe's *Musikerziehung. Lehre und Theorie der Musik im Mittelalter* (Leipzig: Deutscher Verlag für Musik, 1969).

Manuals or model books (see Illustrations 16–18, 207) for the production of manuscript illustrations are the subject of R. W. Scheller, *A Survey of Medieval Model Books* (Haarlem: Erven F. Bohn, 1963). Various aspects of the mnemonic illustration or picture (see Illustrations 71–76) are treated by Ludwig Volkmann, "Ars memorativa," in *Jahrbuch der kunsthistorischen Sammlungen . . . in Wien,* n.s. 3 (1929), pp. 111–200.

3. Arabic scientific illustrations

Although treating manuscript illumination in general, Richart Ettinghausen, in *Arab Painting* (Geneva: Albert Skira, 1962), devotes a great deal of attention to scientific works (on cosmography, medicine, zoology, constellations, technology, and so on). Specifically concerned with science are: A. I. Sabra, "The Scientific Enterprise: Islamic Contributions to the Development of Science," in Bernard Lewis, ed., *The World of Islam: Faith, People, Culture* (London: Thames and Hudson, 1976), pp. 181–200; A. I.

Sabra, S. K. Harmarneh, and Donald R. Hill, chapters on exact sciences, life sciences, and mechanical technology in John R. Hayes, ed., *The Genius of Arab Civilization: Source of Renaissance* (New York: N.Y.U. Press, 1975), pp. 121–189; and Seyyed Hossein Nasr, *Islamic Science: An Illustrated Study,* with photographs by Roland Michaud (Westerham, Kent: World of Islam Festival Publ., 1976). That the accompanying pictorial material as well as the content of much Arabic science derives from Greek forbears is argued by Kurt Weitzmann, "The Greek Sources of Islamic Scientific Illustrations," in George C. Miles, ed., *Archaeologica orientalia in memoriam Ernst Herzfeld* (Locust Valley, N.Y.: Augustin, 1952), pp. 244–266; this article is reprinted in Weitzmann's *Studies in Classical and Byzantine Manuscript Illumination* (Chicago: Univ. of Chicago Press, 1971), pp. 20–44.

4. Cosmography and astronomy

The picturing of the medieval Christian view of the cosmos is the concern of Jurgis Baltrusaitis in *Cosmographie chrétienne dans l'art du moyen âge* (Paris: Gazette des Beaux-arts, 1939), as well as of Ellen J. Beer in *Die Rose der Kathedrale von Lausanne und der kosmologische Bilderkreis des Mittelalters* (Bern: Bentcli Verlag, 1952). The depiction of the medieval cosmos as tied to several specific texts is incisively analyzed in the two articles by Marie-Thérèse d'Alverny cited in the picture credits to Illustrations 274 and 275. The creation (compare Illustration 273) is treated by André Grabar in *Les voies de la création en iconographie chrétienne: Antiquité et moyen âge* (Paris: Flammarion, 1979).

Astronomical and astrological illustrations are treated in almost all of the material by Fritz Saxl that has been cited in section 2 above. Similar concerns are exhibited in the multi-authored (Raymond Klibansky, Erwin Panofsky, and Fritz Saxl) *Saturn and Melancholy: Studies in the History of Natural Philosophy, Religion, and Art* (London: Thomas Nelson, 1964). In addition to these works, the representation of constellations is treated in the relatively old, but still valuable work of Georg Thiele, *Antike Himmelsbilder* (Berlin: Weidmann, 1898). The basic Arabic work whose codices present excellent depictions of the stars (see Illustrations 227–228) is the subject of Joseph M. Upton, "A Manuscript of 'The Book of Fixed Stars' by ᶜAbd ar-Raḥmān as-Ṣūfī," in *Metropolitan Museum Studies,* 4 (1933), pp. 179–197, as well as of the article by Emmy Wellesz cited in the picture credit to Illustration 228. Line drawings reproducing manuscript figures are included in H. C. F. C. Schjellerup, ed. and trans., *Description des étoiles fixes . . . par l'astronome persan Abd-al-Rahman al-Sûfi* (St. Petersburg: Commissionaires de l'Académie impériale des sciences, 1874). Important work on astronomical diagrams can be found in the recent articles of Bruce Eastwood: "The Diagram *Spera celestis* in the *Hortus deliciarum:* A Confused Amalgam from the Astronomies of Pliny and Martianus Capella," in *Annali dell'Istituto e Museo di storia della scienza di Firenze,* 6 (1981), pp. 177–186; "Origins and Contents of the Leiden Planetary Configuration (MS Voss. Q.79, fol. 93v), an Artistic Astronomical Schema of the Early Middle Ages," in *Viator,* 14 (1983), pp. 3–40; and "Characteristics of the Plinian Astronomical Diagrams in a Bodleian Palimpsest, Ms. D'Orville 95, ff. 25–38," in *Sudhoff's Archiv,* 67 (1983), pp. 2–12. The subject of the last of these articles bears directly on Illustration 249.

5. Flora and fauna

The classic article with which to begin one's study of the illustrated herbal is Charles Singer's "The Herbal in Antiquity," in *Journal of Hellenic Studies,* 47 (1927), pp. 1–52. Some of the more important herbals discussed by Singer can be viewed in more complete and more elegant formats in the facsimile editions cited above in section 1. Sample illustrations (mostly in color) of yet others can be seen in Wilfred Blunt and Sandra Raphael, *The Illustrated Herbal* (New York: Thames and Hudson, 1979). Herbal illustrations as well as the depiction of animals are central in the important article by Otto Pächt, "Early Italian Nature Studies and the Early Calendar Landscape," in *Journal of the Warburg and Courtauld Institutes,* 13 (1950), pp. 13–47. The topic of the stylized versus realistic portrayal of plants on Gothic capitals has been given classic treatment by Denise Jalabert in "La flore gothique, ses origines, son evolution du XIIᵉ au XVᵉ siècle," *Bulletin monumentale,* 91 (1932), pp. 181–246. The same topic has been discussed further by Lottlisa Behling, *Die Pflanzenwelt der mittelalterlichen Kathedralen* (Cologne and Graz: Böhlau, 1964), who earlier carried a similar issue into the area of painting in her *Die Pflanze in der mittelalterlichen Tafelmalerei* (Weimar: Böhlau, 1957; 2nd ed. Cologne and Graz: Böhlau, 1967).

A quite comprehensive treatment of animal representation can be found in Francis Klingender, *Animals in Art and Thought to the End of the Middle Ages,* Evelyn Antal and John Harthan, eds. (Cambridge, Mass.: M.I.T. Press, 1971). Relatively exhaustive treatment of the same problem of animal depiction within narrower confines is the recent work of Zoltán Kádár, *Survivals of Greek Zoological Illuminations in Byzantine Manuscripts,* Timothy Wilkinson and Miklós Kretzoi, trans. (Budapest: Akadémiai Kiado, 1978). Three-dimensional animal representation is the subject of Victor H. Debidour, *Le bestiaire scuplté du moyen âge en France* (Paris: Arthaud, 1961). Fundamental for the study of the illustrations of the medieval Latin bestiary is M. R. James, *The Bestiary . . .* (Oxford: Roxburghe Club, 1928). This work is important not only for the illustrations

Scribner's, 1970–1980). More complete bio-bibliographical information from the time of Homer through the fourteenth century frequently can be had from George Sarton's *Introduction to the History of Science,* 3 vols. in 5 (Baltimore, Md.: Williams and Wilkins, 1927–1948). Two general histories of science covering antiquity and the Middle Ages are: Eduard J. Dijksterhuis, *The Mechanization of the World Picture,* C. Dikshoorn, trans. (Oxford: Clarendon, 1961) and René Taton, ed., *Ancient and Medieval Science from the Beginnings to 1450,* A. J. Pomerans, trans. (New York: Basic Books, 1963). An absolutely brilliant treatment of its subject is Owsei Temkin, *Galenism: Rise and Decline of a Medical Philosophy* (Ithaca, N.Y.: Cornell Univ. Press, 1973). In spite of its apparently limiting title, the following is still the most adequate work covering the technicalities of astrology for any period: Auguste Bouché-Leclercq, *L'astrologie grecque* (Paris: Leroux, 1899; repr. Brussels: Culture et Civilisation, 1963). Similarly, the following encyclopedia article goes well beyond antiquity, especially in terms of its coverage of medieval depictions of the zodiac: Hans Georg Gundel, "Zodiakos. Der Tierkreis in der antiken Literatur und Kunst," in *Pauly's Realencyclopädie der classischen Altertumswissenschaft,* X.A, cols. 461–710. Eric J. Holmyard's *Alchemy* (Harmondsworth: Penguin, 1957) is still the most adequate brief treatment of its subject. William Kneale and Martha Kneale's *The Development of Logic* (Oxford: Clarendon, 1962) is the best general history of logic, but information about the technical aspects of logic involved in the diagrams reproduced in the present volume (Illustrations 12, 33, 43, 44, 55, and 59–62) is more easily available in a work like H. W. B. Joseph's *An Introduction to Logic,* 2nd ed. (Oxford: Clarendon, 1916).

9. Ancient science

Translations (unfortunately often suffering because of their brevity) from a good selection of primary sources are given in Morris R. Cohen and I. E. Drabkin, eds., *A Source Book in Greek Science* (New York: McGraw-Hill, 1948; repr. Cambridge, Mass.: Harvard Univ. Press, 1958). An introductory treatment of its subject is Marshall Clagett, *Greek Science in Antiquity* (New York: Abelard-Schuman, 1955), the second half of which deals with the legacy of Greek science through the early Middle Ages. G. E. R. Lloyd has produced two more recent introductory volumes, *Early Greek Science: Thales to Aristotle* (London: Chatto and Windus, 1970) and *Greek Science After Aristotle* (London: Chatto and Windus, 1973). These can be supplemented with Lloyd's more advanced *Magic, Reason and Experience: Studies in the Origins and Development of Greek Science* (Cambridge: Cambridge Univ. Press, 1979) and *Science, Folklore and Ideology: Studies in the Life Sciences in Ancient Greece* (Cambridge: Cambridge Univ. Press, 1983). Important bibliographies are found in all of these volumes. Surprisingly thorough surveys of medicine and mathematics and applied science in the period in question are given in Peter M. Fraser, *Ptolemaic Alexandria,* 3 vols. (Oxford: Clarendon, 1972). The most adequate and easily available comprehensive work on Greek mathematics is Sir Thomas L. Heath, *A History of Greek Mathematics,* 2 vols. (Oxford: Clarendon, 1921). Two works that cover Egyptian and Babylonian mathematics as well as Greek are Otto Neugebauer, *The Exact Sciences in Antiquity,* 2nd ed. (Providence, R.I.: Brown Univ. Press, 1957) and Bartel L. van de Waerden, *Science Awakening,* Arnold Dresden, trans. (Groningen: Noordhoff, 1954). Otto Neugebauer's *A History of Ancient Mathematical Astronomy,* 3 vols. (Berlin, etc.: Springer-Verlag, 1975) is a masterly treatment of its topic. Covering Egyptian and Babylonian as well as Greek material, in addition to an incisive analysis of the mathematics of astronomy, this work contains important remarks about the nature and significance of astronomical diagrams, especially as they are found in the manuscript sources themselves. The third volume also contains an excellent bibliography of the subject. Notable for present purposes due to the number of plates reproducing pictorial material relevant to decans, star clocks, constellations, and cosmological schemes is Otto Neugebauer and Richard A. Parker, *Egyptian Astronomical Texts,* 3 vols. (Providence, R.I.: Brown Univ. Press, 1969). Eustace D. Phillips' *Greek Medicine* (London: Thames and Hudson, 1973) is a reasonable introduction to its subject. It should be supplemented by the work of Temkin on Galenism listed in section 8 of this bibliography, by the work of Fraser just cited, by Wesley D. Smith's *The Hippocratic Tradition* (Ithaca, N.Y.: Cornell Univ. Press, 1979), and by the very valuable collection of articles in Owsei Temkin and C. Lilian Temkin, eds., *Ancient Medicine: Selected Papers of Ludwig Edelstein,* C. Lilian Temkin, trans. (Baltimore, Md.: Johns Hopkins Press, 1967). The immensely important topic of the role of Aristotle within ancient science might best be approached by reading Friedrich Solmsen, *Aristotle's System of the Physical World: A Comparison with His Predecessors* (Ithaca, N.Y.: Cornell Univ. Press, 1960) or, at a more elementary level, G. E. R. Lloyd, *Aristotle: The Growth and Structure of His Thought* (Cambridge: Cambridge Univ. Press, 1968). Alternatively, there are a number of quite provocative and valuable articles collected by Jonathan Barnes, Malcolm Schofield, and Richard Sorabji as *Articles on Aristotle,* 4 vols. (London: Duckworth, 1975–1979), especially in vol. 1: *Science,* vol. 2: *Metaphysics,* and vol. 4: *Psychology and Aesthetics.* All four of the volumes contain extremely useful bibliographies. For Greek philosophy in general, the standard work in English is now W. K. C. Guthrie, *A History of Greek Philosophy,* 6 vols. (Cambridge: Cambridge Univ. Press, 1962–1981). Comprehensive, and containing a good deal of important bibliographic information, the quality and usefulness of the later volumes do not match those of the earlier ones. The sixth and final volume is devoted to Aristotle, so this work must be supplemented by others for the remainder of the history of ancient philosophy. Reasonable beginning guides in this regard are A. A. Long's *Hellenistic Philosophy: Stoics, Epicureans, Sceptics* (London: Duckworth, 1974) and R. T. Wallis' *Neoplatonism* (London: Duckworth, 1972).

10. Arabic science

There are no general histories of science—introductory or advanced, no matter—in this area to match those that exist for Greek and medieval Latin science. In lieu of such works, the text of the article by A. I. Sabra on "The Scientific Enterprise" in Islam that has been cited in section 3 above is the best introduction to the whole. To this, one might add the chapters on science and philosophy in the two completely different editions of Sir Thomas Arnold and Alfred Guillaume, eds., *The Legacy of Islam* (Oxford: Clarendon, 1931), 2nd ed. rev. by Joseph Schacht and C. E. Bosworth (Oxford: Clarendon, 1974). For individuals of importance within the history of Arabic science one can consult the *Dictionary of Scientific Biography* and Sarton's *Introduction to the History of Science* (both cited in section 8). Translations of brief selected texts, again carrying the disadvantages of being pulled from their contexts, can be found in Seyyed Hossein Nasr's *Science and Civilization in Islam* (Cambridge, Mass.: Harvard Univ. Press, 1968). For the exact sciences, a nucleus of important articles is collected in E. S. Kennedy *et al., Studies in the Islamic Exact Sciences* (Syracuse, N.Y.: Syracuse Univ. Press, 1983). Further references can be found in David A. King, "The Exact Sciences in Medieval Islam: Some Remarks on the Present State of Research," in *Middle East Studies Association Bulletin,* 4 (1980), pp. 10–26. For mathematics in particular see Adolf P. I͡Uschkevich, *Les mathématiques arabes (VIIIᵉ–XVᵉ siècles)*, M. Cazenave and K. Jaouiche, trans. (Paris: Vrin, 1976), which is a translation (with corrections and additions) of the section on Arabic mathematics of a work that appeared earlier in a German translation as *Geschichte der Mathematik im Mittelalter* (Basel: Pfalz Verlag, 1964). A brief introduction to Arabic medicine is provided by Manfred Ullman in *Islamic Medicine* (Edinburgh: Edinburgh Univ. Press, 1978). The exceptionally rich Arabic literature on ophthalmology (compare Illustrations 65 and 213) is treated by Julius Hirschberg in *Geschichte der Augenheilkunde,* 2nd ed., II and III (Leipzig: Engelmann, 1908). For Arabic philosophy see Majid Fakhry, *A History of Islamic Philosophy* (New York: Columbia Univ. Press, 1970) and Tjitze J. de Boer, *The History of Philosophy in Islam,* Edward R. Jones, trans. (London: Luzac, 1903) which, in spite of its age, is still an intelligent introduction to the subject. An introduction to the extremely important issue of the impact of Greek learning on Arabic philosophy and science is provided by Richard Walzer, *Greek into Arabic: Essays on Islamic Philosophy* (Oxford: Bruno Cassner, 1962); Francis E. Peters, *Aristotle and the Arabs: The Aristotelian Tradition in Islam* (New York: N.Y.U. Press, 1968), which contains a useful bibliographic appendix; and ᶜAbdurraḥmān Badawi, *La transmission de la philosophie grecque au monde arabe* (Paris: Vrin, 1968). To these volumes, one should add Franz Rosenthal's *The Classical Heritage in Islam,* Emil Marmorstein and Jenny Marmorstein, trans. (Berkeley: Univ. of California Press, 1975), which consists of relevant translated passages from primary sources in philosophy and, especially, the sciences; and the partial translation of the classic (1915) article of Ignaz Goldziher, "The Attitude of Orthodox Islam Toward the 'Ancient Sciences,'" in *Studies on Islam,* Merlin L. Swartz, ed. and trans. (New York and Oxford: Oxford Univ. Press, 1981), pp. 185–215.

11. Medieval Latin science

Two works provide easy access to the field. First, Edward Grant, ed., *A Source Book in Medieval Science* (Cambridge, Mass.: Harvard Univ. Press, 1974), consisting of annotated translations, with introductions, of passages from relevant primary sources, in most instances fortunately not as brief and "out of context" as is frequently the case in works of this sort. Secondly, David C. Lindberg, ed., *Science in the Middle Ages* (Chicago: Univ. of Chicago Press, 1978), which contains chapters on each of the various "sciences," as well as on such topics as the transmission of Greek and Arabic learning to the Latin West, the philosophical setting of science, and the significance of universities for the scientific enterprise. Another introductory work is Alistair C. Crombie's *Medieval and Early Modern Science,* 2nd ed., 2 vols. (Garden City, N.Y.: Doubleday, 1959), while the relevant Roman background and some of the earlier Middle Ages are treated in William H. Stahl, *Roman Science* (Madison: Univ. of Wisconsin Press, 1962). The two central classic works of Pierre Duhem, who in a sense started the serious study of medieval science as a whole, are *Études sur Léonard de Vinci: Ceux qu'il a lus et ceux qui l'ont lu,* 3 vols. (Paris: Hermann, 1906–1913; repr. Paris: F. de Nobele, 1955) and *Le système du monde,* 10 vols. (Paris: Hermann, 1913–1959). Corrections to the exaggerated claims made by Duhem on behalf of medieval science can be found in the works of Marshall Clagett and Anneliese Maier. Clagett's most fundamental contribution in this regard is his *The Science of Mechanics in the Middle Ages* (Madison: Univ. of Wisconsin Press, 1959). For Anneliese Maier, one must turn to her *Studien zur Naturphilosophie der Spätscholastik,* 5 vols. (Rome: Edizione di Storia e Letteratura, 1955–1968). Selected chapters from these five volumes have recently appeared in English translation as *On the Threshold of Exact Science,* Steven D. Sargent, ed. and trans. (Philadelphia: Univ. of Pennsylvania Press, 1982). For mathematics in the Latin West, a brief survey of salient elements can be found in the German edition of IUschkevich cited in the previous section, but the best single source is still Moritz Cantor, *Vorlesungen über Geschichte der Mathematik,* vol. 1, 3rd ed. (Leipzig: Teubner, 1907), and vol. 2, 2nd ed. (Leipzig: Teubner, 1900). Another work which, in spite of the restriction indicated by its title, provides a good idea of the nature of medieval Latin mathematics is Marshall Clagett, *Archimedes in the Middle Ages,* 4 vols. to date (Madison: Univ. of Wisconsin Press, 1964; Philadelphia: American Philosophical Society, 1976–1980). Emmanuel Poulle, in *Les instruments astronomiques du moyen âge* (Oxford: Museum of the History of Science, 1969), 12 pp., explains in outline the structure and function of the basic medieval instruments, each explanation accompanied by a photograph of a surviving original, while the content and nature of medieval texts is the subject of the same author's *Les sources astronomiques (Textes, tables, instruments)* in Typologie des sources du moyen âge occidental, XXXIX (Turnhout: Brepols, 1981). There is hitherto no satisfactory English account of medieval optics in general, but a partial history of that involved is furnished by David C. Lindberg, *Theories of Vision from al-Kindi to Kepler* (Chicago: Univ. of Chicago Press, 1976). Robert Halleux's *Les textes alchimiques,* in Typologie des sources du moyen âge occidental, XXXII (Turnhout: Brepols, 1979) is an excellent guide to the literature, and is most valuable in the specification it provides of which medieval texts can be found in the various early modern collective publications of alchemical writings; in its discussion of the adequacy of early printed alchemical lexica; and especially in the critical evaluation it furnishes of the great amount of unreliable material with which any historian of alchemy must deal. Two works—the first far more elementary than the second—that yield an adequate picture of medieval Latin medicine in general (even though they tell their story in settings limited to but parts of the European scene) are: Charles H. Talbot, *Medicine in Medieval England* (London: Oldbourne, 1967) and Nancy G. Siraisi, *Taddeo Alderotti and His Pupils: Two Generations of Italian Medical Learning* (Princeton: Princeton Univ. Press, 1981), the latter of which contains a bibliography that is valuable for further exploration of the field. The older work of Charles Singer, *From Magic to Science: Essays on the Scientific Twilight* (New York: Boni and Liveright, 1928; repr. New York: Dover, 1958) treats medicine and materia medica in the earlier Middle Ages (as well as Hildegard of Bingen, compare Illustration 277). The works already listed in sections 1 and 5 on herbals, plants, and animals will furnish a great deal of information concerning the content of the segments of the history of medieval science to which they belong, but the account they provide can be extended by the following four works in the same area: Hermann Fischer, *Mittelalterliche Pflanzenkunde* (Munich: Verlag der Münchener Drucke, 1929); Agnes Arber, *Herbals: Their Origin and Evolution, a Chapter in the History of Botany 1470–1670,* 2nd ed. (Cambridge: Cambridge Univ. Press, 1938); T. H. White, ed. and trans., *The Book of Beasts . . . from a Latin Bestiary of the Twelfth Century* (London: Jonathan Cape, 1954); and Florence McCulloch, *Mediaeval Latin and French Bestiaries* (Chapel Hill: Univ. of North Carolina Press, 1962). Details concerning the twelfth-century Neoplatonic view of nature, of which aspects have been depicted in Illustrations 274, 282, and 283, can be found in publications by three scholars whose contributions to the field in question have been especially outstanding: Marie-Dominique Chenu, *Nature, Man, and Society in the Twelfth Century,* selected, ed., and trans. by Jerome Taylor and Lester K. Little (Chicago: Univ. of Chicago Press, 1968); Tullio Gregory, *Anima mundi: La filosofia di Guglielmo di Conches e la Scuola di Chartres* (Florence: Sansoni, 1955); and Edouard Jeauneau, *"Lectio philosophorum": Recherches sur l'École de Chartres* (Amsterdam: Hakkert, 1973). Julius R. Weinberg's *A Short History of Medieval Philosophy* (Princeton: Princeton Univ. Press, 1964) is one of the best introductions to the history of medieval philosophy as a whole, while further study will find excellent guides in A. H. Armstrong's *The Cambridge History of Later Greek and Early Medieval Philosophy* (Cambridge: Cambridge Univ. Press, 1967) and Norman Kretzmann, Anthony Kenny, and Jan Pinborg, eds., *The Cambridge History of Later Medieval Philosophy . . . 1100–1600* (Cambridge: Cambridge Univ. Press, 1982), which contains an extensive bibliography. Finally, good introductions to the problem of the transmission of Greek and Arabic philosophy and science to the Latin West are provided by the chapter specifically concerned with this problem in David Lindberg's *Science in the Middle Ages* (cited in section 8), by the chapter by Bernard Dod on Aristoteles Latinus in the just-mentioned *Cambridge History of Later Medieval Philosophy,* and especially by Marie-Thérèse d'Alverny, "Translations and Translators," in Robert L. Benson and Giles Constable with Carol D. Lanham, eds., *Renaissance and Renewal in the Twelfth Century* (Cambridge, Mass.: Harvard Univ. Press, 1982), pp. 421–462. Detailed episodes in this "transmission story" are, inter alia, incisively documented in Charles H. Haskins, *Studies in the History of Mediaeval Science,* 2nd ed. (Cambridge, Mass.: Harvard Univ. Press, 1927).

Picture Sources and Credits

In addition to the identification and acknowledgment of sources, the following list contains a number of bibliographic references that provide further information relative to the manuscript at hand or to some particular aspect of the picture reproduced. Lack of space has prevented the listing of references to modern editions and translations of texts explained by the relevant pictures and diagrams. A list of abbreviations for frequently cited manuscript collections is provided below.

ABBREVIATIONS

Balt WAG	Walters Art Gallery, Baltimore, Maryland
Bamb	Staatliche Bibliothek, Bamberg, Federal Republic of Germany
Bas	Öffentliche Bibliothek der Universität, Basel, Switzerland
Bern	Burgerbibliothek, Bern, Switzerland
BL	British Library, London, England
BL Add	———, Additional MSS
BL Ar	———, Arundel MSS
BL Burn	———, Burney MSS
BL Cott	———, Cotton MSS
BL Egert	———, Egerton MSS
BL Harl	———, Harley MSS
BL Roy	———, Royal MSS
BL Slo	———, Sloane MSS
BN	Bibliothèque nationale, Paris, France
BN lat	———, fonds latin
BN n.a.l.	———, nouvelles acquisitions latines
BN fr.	———, fonds français
BN graec	———, fonds grec
BN arab	———, fonds arabe
Bol Univ	Biblioteca Universitaria, Bologna, Italy
Camb Fitz	Fitzwilliam Museum, Cambridge, England
CU	University Library, Cambridge, England
Fl Laur	Biblioteca Medicea Laurenziana, Florence, Italy
Lamb Pal	Lambeth Palace, London, England
Leid	Bibliotheek der Rijksuniversiteit, Leiden, The Netherlands
Mi Ambr	Biblioteca Ambrosiana, Milan, Italy
Mu	Bayerische Staatsbibliothek, Munich, Federal Republic of Germany
Mu CLM	———, codices latini
Mu CGM	———, codices germanici
Mu Arab	———, codices arabici
Nap	Biblioteca Nazionale, Naples, Italy

Ox Bl	Bodleian Library, Oxford, England
Ox Bl Ash	______, Ashmolean MSS
Ox Bl Auct	______, Auctarium MSS
Ox Bl Bodl	______, Bodley MSS
Ox Bl Can Class Lat	______, Canonici Latin Classical MSS
Ox Bl Can Misc	______, Canonici Miscellaneous MSS
Ox Bl Digby	______, Digby MSS
Ox Bl Douce	______, Douce MSS
Ox Bl Laud Misc	______, Laud Miscellaneous MSS
Ox Bl Or	______, D'Orville MSS
Ox Bl Rawl	______, Rawlinson MSS
Ox CCC	Corpus Christi College, Oxford, England
Ox SJ	St. John's College, Oxford, England
Vat	Biblioteca Apostolica Vaticana, Vatican City
Vat lat	______, Fondo Vaticano Latino
Vat graec	______, Fondo Vaticano Greco
Vat Barb lat	______, Fondo Barberini Latino
Vat Chigi	______, Fondo Chigi
Vat Ottob lat	______, Fondo Ottoboniano Latino
Vat Pal lat	______, Fondo Palatino Latino
Vat Reg lat	______, Fondo Reginense Latino
Vat Urb lat	______, Fondo Urbinate Latino
VE	Biblioteca Nazionale Marciana, Venice, Lat. MSS
VE f.a.	______, fondo antico
VI	Österreichische Nationalbibliothek, Vienna, Latin MSS

Frontispiece MS Ox Bl Can Class Lat 257, fol. lv.
1. Universitní Knihovna, Prague: MS Kap. A. XXI, fol. 133r. Courtesy of the State Library of the Czech Socialist Republic—University Library, Prague.
2. Mastaba of Ka-ni-nesut, Giza, Egypt. Photograph courtesy of the Österreichische Akademie der Wissenschaften, Vienna.
3. Library of the Süleymaniye Mosque, Istanbul: MS Esad Efendi 3638, fol. 3v.
4. MS Ox Bl Bodl 602, fol. 36r.
5. Bibliothèque municipale, Épinal, France: MS 73, fol. lr.
6. MS BN lat 16141, fol. 25r.
7. MS BN lat 17155, fol. 327v.
8. Wellcome Historical Medical Library, London, England: MS Or 10a, fol. 1v. By courtesy of the Wellcome Trustees.
9. MS Vat lat 2159, fol. 85r.
10. MS Vat Chigi F.VII.159, fol. 236v.
The sixth-century original from which this miniature derives is reproduced in Kurt Weitzmann, *Late Antique and Early Christian Book Illumination* (1977), pl. 17.
11. MS BN lat 7330, fols. 36v, 41v.
12. MS BN lat 14716, fol. 17r.
13. MS Vat Chigi E.VI.197, fol. 135r.
14. MS Vat Pal lat 1581, fol. 70r.
15. MS Ox Bl Can Class Lat 257, fols. 15v–16r.
16. MS VI 507, fol. 7v. This manuscript has been analyzed by P. J. H. Vermeeren, *Über den Kodex 507 der Oesterreichischen Nationalbibliothek* (1956).
17. Knihovna Metropolitní Kapituli, Prague: MS M.100 (1459), fol. 53r.
18. Houghton Library, Harvard University: MS Typ 101 H. This brief and unusual codex has been described by Samuel Ives and Hellmut Lehmann-Haupt, *An English 13th-century Bestiary* (H. P. Kraus, 1942). Reproduced courtesy of Mr. Philip Hofer and H. P. Kraus.
19. MSS Leid BPL 144, fol. 38v; VI 13440, fol. 4r.
20. MSS Mu CLM 13002, fol. 3r; Mu CLM 17403, fol. 4v. An analysis of CLM 13002 can be found in Albert Böckler, *Die Regensburg-Prüfeninger Buchmalerei* (1924), pp. 20–29.
21. MS BL Ar 83, fol. 129r.
22. MS Ox Bl Digby 190, fol. 181v.
23. MS BL Harl 4348, fol. 26v.
24. MS BL Cott Nero D.IV, fol. 14v.
25. MS BN lat 7377B, fols. 126v–127r.
26. MS Mu CLM 10268, fol. 107r.
27. MS BN lat 8500, fol. 34v.
28. MS Ox Bl Add. C.144, fol. 64v.
29. MS BL Egert 935, fol. 36r.
30. MS BN lat 8500, fol. 33v.
31. MS Ox SJ 17, fol. 7r.

32. MS Bas F.II.8, fol. 45r.
33. Vienna, Dominikanerbibliothek, MS 160/130, fol. 26v.
34. MS BN n.a.l. 1401, fols. 40r–40v.
35. MS Nap XIII.G.38, fol. 5r.
36. MS Mu CLM 13046, fol. 39v.
37. MS Bern 36, fol. 57v.
38. MS Bern 263, fol. 14r.
39. Pierpont Morgan Library, New York, MS Glazer 37, leaf 1.
40. MS BN lat 3473, fol. 80v.
41. MS BN lat 7877, fol. 79v.
42. MS BN lat 15125, fol. 48v.
43. MS BL Roy 8.A.XVIII, fol. 3v.
44. MS BL Burn 275, fol. 166r.
45. MS BL Ar 83, fol. 130v.
46. MS BN lat 4860, fol. 99v.
47. MS BN lat 4860, fol. 100r.
48. MSS BN lat 4860, fol. 103v; BN lat 6649, fol. 16r.
49. MS Mu CLM 14353, fol. 75r.
50. MS Balt WAG W.73, fol. 9r, courtesy of the Trustees of the Walters Art Gallery.
51. MS Ox SJ 17, fol. 27r.
52. MS Mu Arab 454, fol. 42v.
53. MS Ox SJ 17, fol. 81v.
54. MS Vat lat 2056, fol. 55v.
55. MS BN lat 14716, fol. 62r.
56. MS Ox Can Misc 26, fols. 3v–4r.
57. MS BL Cott Tiberius C.I, fol. 5v.
58. MS BL Ar 83, fol. 126r.
59. MS BL Burn 275, fol. 279v.
60. MS BL Burn 275, fol. 193r.
61. MS BN lat 14716, fol. 17v; Juan Celaya, *Expositio . . . in primum tractatum Summularum Magistri Petri Hispani* (1525). Phot. Bibl. nat. Paris.
62. MS BL Burn 275, fol. 287r.
63. MS BN lat 7361, fol. 59v.
64. MS BN fr. 1082, fol. 53v.
65. National Library, Cairo, Egypt: MS Ṭibb Taymūr 100, p. 7.
66. MS Ox Bl Digby 119, fol. 147r.
67. MS Mu CLM 26838, fol. 202v. The doctrine of maxima and minima represented in the lower entries in the square of opposition in this Illustration is not treated in the text of Burley to which the square is appended, but it is the central subject of an anonymous tract that directly follows in the manuscript. Once again, the ultimate source involved was Aristotle, specifically his decision to define a power in terms of the maximum it can accomplish (*De caelo,* I, ch. 11). The lower inscription in all four of the corner circles is the same: distance or weight *(distantia vel pondus)*. It refers to what were the paradigm cases of the maxima and minima literature, namely, the setting of limits to someone's active ability or power to lift different weights and the limiting of the active power of someone's vision either by the objects one is able to see at a given distance or by the distances over which one is able to see a given object. Something more of how maxima and minima function as limits relative to weight and distance is provided when the relevant explanatory phrases next to the circle in this square are examined. Beginning at the upper left, the maximum weight Socrates *can* lift (in medieval terms, a *maximum quod sic*) is defined as that which he can lift but than which he can lift no greater. Diagonally opposite, the definition of the minimum weight Socrates *cannot* lift (in medieval terms, a *minimum quod non*) is given as that which he cannot lift, and than which any equal or greater he cannot lift, but which, given any lesser weight, there is always a weight greater than that lesser weight which he can lift. The rather involved definitions of these maxima and minima aside, two things about these limit notions are made clear by the square of opposition. Thus, as the label on the diagonal joining them explicitly says, *maxima quod sic* and *minima quod non* are contradictories. This means that only one of them can be used to define or limit a given active power—in the case of the example used in the square, Socrates' power of lifting. The other two circular elements in the square represent *minima quod sic* (at the upper right) and *maxima quod non* (at the lower left), definitions of these terms and the relevant contradictory relations being given as before. Finally, the square points out that *maxima quod sic* (upper left) are contrary to *maxima quod non* (lower left) although, as in the case of the square's treatment of first and last instants, the subaltern and subcontrary "connectors" lack relevance.
68. Musée du Louvre, Paris: Letronne papyrus I.2.325.
69. MS Ox SJ 17, fol. 14r.
70. MS BN lat 16533, fol. 66v.
71. MS Mu CGM 4413, fols. 160v–161r.
72. Thomas Murner, *Logica memorativa* (1509). By permission of the Houghton Library, Harvard University.
73. MS Mu CLM 697, fols. 150v–151r.
74. MS BN lat 3352B, fol. 2r.
75. MS Ox Bl Digby 56, fol. 165v.
76. MS Mi Ambr D.75 inf, fol. 6r.
77. MS BN lat 7207, fol. 211r.
78. MS VE sm. 299, fol. 6r.
79. MS Leid Voss Chem Q.51, fol. 1v.
80. MS Mu CLM 10268, fol. 47v.
81. MS BN graec 2180, fol. 108r.
82. MS Ox Bl Auct F.3.13, fol. 112r.
83. British Museum, London: leather roll, Acc. 10250.
84. Columbia University, New York: cuneiform tablet, Plimpton 322.
85. MS Ox Bl Can Misc 554, fol. 112v.
86. MS Ox Bl Auct F.1.9, fol. 32v.

87. MS Leid Or. 185, fol. 42v.
88. MS BN graec 2389, fols. 68v–69r.
89. MS BN lat 14738, fol. 9v.
90. MS Ox Bl Digby 176, fol. 95r.
91. MS BN arab 2426, fols. 173v–174r.
92. MS Mu CLM 10691, fols. 13v–14r.
93. Crawford Library, Royal Observatory, Edinburgh, Scotland, perpetual calendar, MS 5–7; [Johann Lichtenberg], *Pronosticatio in latino* (1492), by permission of the Houghton Library, Harvard University.
94. MS Ox Bl Can Misc 554, fol. 127v.
95. MS Lamb Pal 67, fol. 62v. By courtesy of His Grace the Archbishop of Canterbury and the Trustees of Lambeth Palace Library.
96. MS BN lat 16644, fol. 56v.
97. MSS Bamb H.J.IV.12, fol. 28r; Bas F.II.33, fol. 74v.
98. MS Bas F.II.33, fol. 83r.
99. MS Ox Bl Auct F.5.28, fol. 16r.
100. MS Lamb Pal 67, fols. 103r, 104v. By courtesy of His Grace the Archbishop of Canterbury and the Trustees of Lambeth Palace Library.
101. MSS Ox Bl Digby 190, fol. 195v; Mu CLM 10268, fol. 40v.
102. MS BN lat 7361, fol. 82r.
103. MSS BN lat 7202, fol. 48r; BN lat 7361, fol. 101v.
104. MSS Lamb Pal 67, fol. 151v, by courtesy of His Grace the Archbishop of Canterbury and the Trustees of Lambeth Palace Library; BL Burn 275, fol. 382v.
105. MSS BL Harl 5237, fols. 35v–36r; VI 51, fol. 2r.
106. British Museum, Rhind Papyrus, Acc. 10057, recto.
107. British Museum, cuneiform tablet, Acc. 15285.
108. MS Ox Bl Douce 125, pp. 30–31; Bibliothèque Ste. Geneviève, Paris: MS 2200, fol. 154v.
109. MSS BN lat 10257, fol. 12v; Vat Ottob lat 1862, fol. 42r; Mu CLM 13021, fol. 167r.
110. MSS BL Roy. 15.A.XXVII, fol. 11v; BL Slo 285, fol. 49v.
111. MS Ox Bl Digby 98, fol. 131v.
112. MS Nap VIII.C.22, fol. 8v.
113. MS Vat Reg lat 1268, fol. 21v.
114. MS Bern A.50, fol. 198v.
115. MS Vat Reg lat 1268, fols. 10v, 5v.
116. MSS BN arab 2457, fol. 115v; Aya Sofya, Istanbul, 2762, unfoliated; Leid Or. 14(I), p. 117.
117. MS Vat graec 190, fol. 213v.
118. MS BL Add 34018, fol. 54v.
119. MSS BL Harl 2686, fol. 36r; BL Add 15603, fol. 32r; BN lat 8500, fol. 41r.
120. MSS Mu CLM 23511, fol. 24r; Vat graec 190, fol. 227r; Ox Bl Auct F.5.28, fol. 4r; BL Ar 84, fol. 94r.
121. MS Ox Bl Or 301, fol. 354r. Diagrams from Thomas L. Heath, *The Thirteen Books of Euclid's Elements* (1908), vol. 3, pp. 490 and 487.
122. MSS Vat graec 190, fol. 245r; Vat Reg lat 1268, fol. 133r; VE f.a. 332, fol. 222v. Diagrams from Thomas L. Heath, *The Thirteen Books of Euclid's Elements* (1908), vol. 3, pp. 502 and 499.
123. *The Elements of Geometrie of the most auncient Philosopher Euclide . . . translated into the Englishe toung by H. Billingsley,* London (1570), fols. 331r and 332r. From copy in Houghton Library, Harvard University.
124. MSS Ox Bl Auct F.5.28, fol. 48v; Bas F.II.33, fol. 116r.
125. MS BN graec 2389, fol. 75v.
126. MS Mu Arab 464, fol. 13r.
127. MSS Leid Or. 905, fol. 43r; CU Add. 3589, fol. 155r.
128. MSS Ox Bl Digby 215, fol. 5v; BN lat 7333, fol. 59v.
129. MS CU Kk.I.I, fol. 219v.
130. MSS Ox Bl Digby 40, fol. 46r; Ox Bl Digby 20, fol. 1v.
131. MSS Balt WAG W.73, fol. 5r, courtesy of the Trustees of the Walters Art Gallery; Ox Bl Auct F.3.25, fol. 1v.
132. MSS BL Ar 377, fol. 101v; BL Roy 12.F.X, fols. 14v, 15r; BL Slo 2424, fol. 20v.
133. MSS Ox Bl Auct F.5.28, fol. 27r; BL Slo 2156, fol. 38r.
134. MSS Ox CCC 150, fols. 35v, 36r, courtesy of the President and Fellows of Corpus Christi College, Oxford; Ox Bl Ash 424, fol. 77r.
135. MSS Ox Bl Ash 424, fol. 76v; Aya Sofya, Istanbul, 2451, fol. 43v.
136. MS Aya Sofya, Istanbul, 2451, fol. 9r.
137. MS Aya Sofya, Istanbul, 2451, fol. 33v.
138. MS Bas F.1V.30, fols. 21r, 34r.
139. MS Ox Bl Auct F.5.28, fol. 110r.
140. MSS BN lat 8680A, fol. 8v; BL Harl 13, fol. 138v.
141. MS Ox Bl Auct F.5.28, fol. 133r; Wissenschaftliche Allgemeinbibliothek, Erfurt, German Democratic Republic, MS Amploniana Quarto 325, fol. 191r.
142. MSS BL Slo 2156, fol. 164r; Mu CLM 26838, fol. 77r; Ox Bl Can Misc 177, fol. 16r.
143. MSS Bl Slo 2156, fol. 191r; Vat Chigi E.VI.197, fol. 146r.
144. MSS BN lat 6558, fol. 6r; Vat Chigi E.VI.197, fol. 137v.
145. MSS Vat lat 4445, fol. 61v; Vat lat 4448, fol. 22r.
146. Herzoglichen Bibliothek, Wolfenbüttel, Federal Republic of Germany, MS 36, 23, Aug. 2°, fol. 62v; Bern MS 299, fol. 4r. Other *agrimensores* illustrations from the Wolfenbüttel Codex Arcerianus, as well as from other MSS, can be found in O. A. W. Dilke, "Illustrations from Roman Surveyor's Manuals," *Imago Mundi,* 21 (1967), pp. 9–29, and in his *The Roman Land Surveyors: An Introduction to the Agrimensores* (1971). For a facsimile reproduction of the Codex Arcerianus, see the Bibliography, sect. 1.
147. MS Bern 87, fol. 16r.

148. MS Vat Barb lat 92, fol. 8r.
149. MS Mu CLM 13021, fol. 193v; Ox Bl Auct F.5.28, fol. 135r.
150. MSS Ox Bl Douce 276, fol. 55v; tombstone of Hugh Libergiers in Rheims Cathedral. See the discussion of the square depicted on this tombstone in B. C. Morgan, *Canonic Design in English Medieval Architecture* (1961), pp. 45–69, and in Guy Beaujouan, "Réflexions sur les rapports entre théorie et pratique au moyen âge," in J. Murdoch and E. Sylla, eds., *The Cultural Context of Medieval Learning* (1975), pp. 459–461.
151. Library of the Seraglio, Istanbul, codex Palatii Veteris 1, fol. 47v; MS Ox Bl Can. Ital. 197, fol 61r.
152. MSS Ox Bl Ash 1522, fol. 72r; BN lat 7192, fol. 33r.
153. MS BN fr. 19093, fol. 20r. This "Sketchbook" has been edited by Richard Hahnloser, *Villard de Honnecourt. Kritische Gesamtausgabe* (1935). There is an English trans. by Theodore Bowie, *The Sketchbook of Villard de Honnecourt* (1959).
154. MS BN lat 8686, fol. 10v.
155. University Library, Istanbul, MS FY 1404, no folio.
156. Bas relief from Ostia Museum, Ostia, Italy, photograph courtesy of the Metropolitan Museum of Art, gift of Ernest and Beata M. Brummer, 1948, in memory of Joseph Brummer; MSS Ox Bl Arab. d. 138, fol. 2v; BN lat 6823, fol. 2r.
157. MS Mu CLM 12, fol. 3r.
158. MS BL Cott. Claudius E.IV, fol. 201r.
159. MS Vat lat 2094, fol. 1r.
160. MS Bol Univ 3632, fol. 210r.
161. MS Ox Bl Ash 304, fol. 2v; BL Add 10302, fol. 32v.
162. MS VE graec. Z.388, frontispiece; BN fr. 565, fol. 1r.
163. MS BL Roy 20.B.XX, fol. 3r.
164. Eton College Library, Eton, England: MS 204, fols. 1v–2r.
165. Glasgow University Library, MS Hunter 9, fol. 30v: discussion of illustrations in this MS can be found in E. C. Streeter and Charles Singer, "Fifteenth-century Miniatures of Extramural Dissections," in *Essays on the History of Medicine Presented to Professor Karl Sudhoff,* ed. Charles Singer and Henry E. Sigerist (1924). MS BL Harl 4425, fol. 59r.
166. Balt WAG, Dioscorides MS (Arabic), courtesy of the Trustees of the Walters Art Gallery. On this MS see Hugo Buchthal, "Early Islamic Miniatures from Baghdad," *Journal of the Walters Art Gallery,* 5 (1942), pp. 18–39.
167. MSS BL Slo 1975, fol. 93r; BN suppl. turc 693, fol. 106r. On the latter MS, see Pierre Huard and Mirko Drazen Grmek, *Le premier manuscrit chirurgical Turc redigé par Charaf ed-Din (1465)* (1960).
168. MSS Ox Bl Rawl C.328, fol. 4r; Fl Laur Plut LXXIV, 7, fol. 204v (for facsimiles from this MS see the ed. of Apollonius of Citium in the Bibliography, sect. 1).
169. MS Nap. XIII.G.38, pars III, fol. 70r.
170. MS BN lat 8500, fol. 54v; engraving by the Master of the Amsterdam Cabinet. On the legend depicted by this engraving and others see the article of George Sarton, "Aristotle and Phyllis," *Isis,* 14 (1930), pp. 8–19.
171. Erasmus, *Encomium moriae* (1515), with marginal drawings by Holbein (on which see Betty Radice, "Holbein's Marginal Illustrations in Praise of Folly," *Erasmus in English,* no. 7 [1975], pp. 7–17): photograph from H. A. Schmid's ed. (1931); Holbein broadsheet, *Hercules Germanicus:* from the Dept. of Prints and Drawings of the Zentralbibliothek, Zurich, Switzerland.
172. MS Bamb HJ.IV.12, fol. 9v.
173. [Strasbourg MS lost]: new ed. of this lost *Hortus Deliciarum* by Rosalie Green has appeared in 2 vols. (1979).
174. Chartres Cathedral, west facade, right portal, archivolt. Photograph courtesy of James Austin.
175. Spanish Chapel, S. Maria Novella, Florence: Andrea Bonaiuti Fresco. Photograph courtesy of Princeton University Press.
176. MS Mi Ambr. B.42 inf, fol. 1r. The all but identical iconography in the *Canzone delle virtù e delle scienze* of Bartolomeo di Bartoli has been reproduced in facsimile by Leon Dorez (1904).
177. MS BN lat 8500, fols. 33r, 38r, 41v.
178. MS BL Burn 275, fols. 293r, 359v.
179. MS BL Add 15692, fol. 29v; Gregor Reisch, *Margarita philosophica* (1517): by permission of the Houghton Library, Harvard University.
180. Thomas Murner, *Logica memorativa* (1509): by permission of the Houghton Library, Harvard University.
181. MS Ox Bl Can Class Lat 257, no folio.
182. MS BL Add 10302, fol. 37r.
183. MSS BN graec 2243, fol. 10v; BN fr. 218, fol. 111r.
184. Naples, Museo Nazionale, Roman mosaic depicting Plato's Academy.
185. MS BN arab 5847, fol. 5v.
186. Archives nationale, Paris: MS MM 406, fol. 10v. On this MS, see Astrik L. Gabriel, *Student Life in Ave Maria College, Medieval Paris: History and Chartulary of the College* (1955).
187. MSS BN fr. 574, fol. 27r; Biblioteca Trivulziana, Milan, MS 2167, fol. 13v.
188. VI, MS Med. graec. 1, fol. 6v.
189. MSS Fl Laur Plut 82, 1, fol. 2v; BN fr. 565, fol. 23r.
190. MS BL Cott Nero D.I, fol. 23v. For the earlier MS from which this miniature is derived (as noted in caption), see M. R. James *et al., Illustrations to the Life of St. Alban in Trinity College, Dublin, MS E i 40* (1924).
191. MSS BL Slo 1975, fol. 48r; OX Bl Bodl 764, fol. 10v; BN lat 5543, fol. 167v.
192. Wellcome Historical Medical Library, London,

England: Johnson Papyrus. By courtesy of the Wellcome Trustees.
193. MSS VI med. graec. 1, fol. 98r; Mu CLM 337, fol. 66r; Leid Voss Lat Q.9, fol. 60v.
194. MSS Ox Bl. Bodl Arab d. 138, fol. 126r; Landesbibliothek, Kassel, Federal Republic of Germany: 2° Phys 10, fol. 34v; Ox Bl Ash 1462, fol. 67r; VE VI.59, fol. 53r.
195. MSS Fl Laur Plut LXXIII, 16, fol. 44r; BL Harl 1585, fol. 17r.
196. MS BN suppl. graec 247, fol. 16v.
197. MSS BL Egert 2020, fol. 161v (for indication of the earlier MS from which this illustration was taken, see Otto Pächt, "Early Italian Nature Studies . . . ," *Journal of the Warburg and Courtauld Institutes,* 13 [1950], pp. 13–47); Öffentliche Studienbibliothek, Salzburg, Austria: MS V.C.H.166, fol. 173r.
198. Gothic Capitals in Notre Dame, Paris; Naumburger Dom. Photographs from Art Reference Bureau and Bildarchiv Foto Marburg.
199. Bibliotheek der Rijksuniversiteit, Utrecht: MS 32, fol. 1r; Trinity College Library, Cambridge, England: MS R.17.1, fol. 5v.
200. MS Smyrna B.8, fol. 52r (destroyed, reproduced from Josef Strzygowski, *Der Bilderkreis des griechischen Physiologus* [1899], pl. VI, which contains reproductions of other miniatures from this lost MS); MS Bern 318, fol. 15r (this MS has been reproduced in Christoph von Steiger and Otto Homburger, *Physiologus Bernensis. Voll-Faksimilie-Ausgabe des Codex Bongarsianus 318 der Burgerbibliothek, Bern* [1964]).
201. MSS Ox Bl Laud Misc 247, fol. 158v; Ox Bl Bodl 764, fol. 96r; Ox Bl Ash 1511, fol. 80v.
202. MSS Ox Bl Bodl 764, fol. 56v; VI med graec 1, fol. 483v; BN fr. 12400, fol. 30v.
203. MSS Trinity College, Dublin, 57, fol. 191v; BL Harl 1585, fol. 66v; BN fr. 19093, fol. 24v (see note in credit for Illustration 153); Mu Arab 464, fol. 177r; BN arab 5036, fol. 136v.
204. MS Vat lat 2761, fol. 36v.
205. MS Mi Ambr AF Ar.D.140 inf, fol. 9r. This manuscript is treated in Oscar Löfgren, *Ambrosian Fragments of an Illuminated Manuscript Containing the Zoology of al-Jāḥiẓ, Uppsala Universitets Arsskrift* (1946), no. 5.
206. MSS Ox Bl Bodl 602, fol. 3v; Ox Bl Ash 1511, fol. 22v; Conrad Gesner, *Historiae animalium* (1551), fol. 978.
207. MS BL Harl 2799, fol. 243r.
208. MSS BN fr. 1377, fol. 27r; Mu Arab 464, fol. 59v.
209. MSS Ox Bl Ash 399, fol. 18r; India Office Library, London: MS Persian 2296, unfoliated: photograph courtesy of the India Office Library and Records.
210. Kungliga Biblioteket, Stockholm: MS X.118, no fol. (scroll): this MS has been published in facsimile, ed. and trans. by D'Arcy Power, *De arte Physicali et de Cirurgia of Master John Arderne* (1922); MS BN lat 16169, fol. 59v.
211. Biblioteca Universitaria, Pisa: MS Roncioni 99, fol. 2r; MS Ox Bl Ash 399, fol. 23r.
212. MS Ox Bl Ash 399, fol. 13v; *Primus (liber) Canonis Avicenne principis cum explanatione Jacobi de partibus . . .* (1498), no foliation: photograph courtesy of the Yale Medical Library, Yale University.
213. National Library, Cairo: MS Ṭibb Taymur 100, no folio given.
214. MS Ox Bl Digby 77, fol. 7r.
215. MS BL Slo 981, fol. 68r.
216. MS Ox Bl Ash 424, fol. 65v; Witelo (Vitellionis), *Optica* (1535), fol. 55r, by permission of Houghton Library, Harvard University; Witelo, *Optica* (1572), p. 87: by permission of Houghton Library, Harvard University; Andreas Vesalius, *De corporis humani fabrica* (1543), p. 643: photograph courtesy of the New York Academy of Medicine Library.
217. Tomb of Senmut at Deir-el-Bahri, Luxor: Ceiling detail; Tomb of Seti I (No. 17) in Valley of Kings, Luxor: Ceiling of Hall K, detail. Photographs courtesy of the Metropolitan Museum of Art.
218. Temple of Hathor at Dendera, ceiling (now in Louvre, Paris); Coffin lid of Sotor, Luxor: British Museum, no. 6705.
219. Berlin Museen, Berlin: cuneiform tablet VAT 7851, front side.
220. Musée du Louvre, Paris: Letronne papyrus I.2.325.
221. Museo Nazionale, Naples: Farnese globe. Photograph from Georg Thiele, *Antike Himmelsbilder* (1898), pl. VI, courtesy of the Marquand Library of Art and Archaeology, Princeton University.
222. MS Vat graec 1291, fol. 2v (for which, see Franz Boll, *Beiträge zur Überlieferungsgeschichte der griechischen Astrologie und Astronomie* [1899], 1); Stiftsbibliothek, Sankt Gallen, Switzerland: MS 902, p. 76.
223. MS Vat Barb lat 76, fol. 3r. Another medieval diagram of the heavens utilizing a projection similar to this diagram has been noted and analyzed in J. D. North, "Monasticism and the First Mechanical Clocks," in J. T. Fraser *et al.,* eds., *The Study of Time,* II (1975), pp. 386–388.
224. Albrecht Dürer, woodcut: Bartsch no. 151: photograph from Willi Kurth, *Complete Woodcuts of Albert Dürer* (1927); MS VI 5415, fol. 168r.
225. San Lorenzo, Florence: Sagrestia Vecchia, altar ceiling, photograph from *Mitteilungen des Kunsthistorischen Instituts in Florenz* (1912). The most recent discussion of this ceiling is: Alessandro Parronchi, *Il cielo notturno della sacrestia vecchia di S. Lorenzo* and Francesco Gurrieri, *Considerazioni sulla "sphaera" celeste della sacrestia vecchia di S. Lorenzo,* publ. together (Florence: Biblioteca Medicea Laurenziana, n.d.).
226. MS BL Harl 647, fol. 10v.
227. MS BN arab 5036, fols. 118r, 118v.

228. MSS Leid Voss Lat Q.79, fol. 30v; Ox Bl Marsh 144, p. 167 (this MS is analyzed by Emmy Wellesz, "An Early al-Ṣūfī MS in the Bodleian Library in Oxford," *Ars Orientalis,* 3 [1959], pp. 1–26); VI 5318, fols. 26v–27r.
229. MSS BL Add 23770, fol. 18v; Vat Reg lat 1283, fol. 4v.
230. MSS Vat Reg lat 1283, fol. 9v; BN lat 7330, fol. 10r.
231. MS Mu CLM 10268, fol. 85r.
232. MS BN lat 7330, fols. 54v, 55r.
233. MS Ox CCC 157, p. 380, photograph courtesy of the President and Fellows of Corpus Christi College, Oxford.
234. MSS Mu CLM 14353, fol. 77r; BN fr 565, fol. 19v; BL Oriental 8349, fol. 72v.
235. Bayeux Tapestry, courtesy of Photographie Giraudon; MSS Ox Bl Bodl 614, fol. 34r; Camb Fitz 167, fols. 88v–89r, courtesy of the Fitzwilliam Museum, Cambridge.
236. Museum of History of Science, Oxford, England: brass *quadrans vetus;* MS Ox Bl Ash 1522, fol. 72r.
237. Adler Planetarium, Chicago: Astrolabe M-27 (front and back); modern diagrams from J. D. North, "The Astrolabe," *Scientific American,* 230 (1974), pp. 97, 104; an easily available and quite adequate explanation of the astrolabe can be found in this article.
238. MSS Oxf. Bl Auct F.5.28, fol. 99v; BN lat 7293A, fol. 15r; BN lat 7412, fol. 23v; Ox Bl Ash 1522, fol. 83v.
239. MS Vat lat 8174, fol. 137r.
240. MS Ox CCC 144, fol. 80r, photograph courtesy of the President and Fellows of Corpus Christi College, Oxford. J. D. North, *Richard of Wallingford: An Edition of His Writings with Introductions, English translations and commentary* (1976), contains a full explanation of the construction and utilization of this instrument.
241. Bibliothèque de l'Arsenal, Paris: MS 1186, fol. 1v; MS BL Burn 275, fol. 390v.
242. MS Mu CLM 10269, fol. 130r.
243. MSS BN lat 6514, fols. 68r, 69v; BN lat 7147, fol. 1v.
244. MSS Vat lat 4467, fol. 21r (reproductions of drawings of surgical instruments from MSS of the Arabic text of Abulcasis can be seen in Abulcasis, *On Surgery and Instruments: A Definitive Edition of the Arabic Text with English Translation and Commentary,* by M. S. Spink and G. L. Lewis (1973); VE VII.32, fol. 38v.
245. MSS Mu CLM 337, fol. 11r; Vat Chigi F.VII.159, fol. 224v.
246. MS BN lat 8865, fol. 56v.
247. MSS BN lat 6413, fol. 4v; BL Cott Tiberius C.I, fol. 6v; BN lat 6649, fol. 9r; Vat Fondo Rossiano 247, fol. 60r; Ox Bl Auct F.2.20, fol. 5v.
248. MSS Ox Bl Or 77, fol. 88v; BL Add 11943, fol. 29v; Ox Bl Selden Supra 25, fol. 199v; BL Egert 2976, fols. 49v, 50r; BN lat 6415, fol. 63v; BL Harl 4794, fol. 46v.
249. MSS Bern 347, fol. 25r; Vat Reg lat 123, fol. 169v; BL Roy 13.A.XI, fol. 143v; Ox SJ 17, fol. 38r. Reproduction of another MS copy of this diagram, and discussion thereof, can be found in Harriet Pratt Lattin, "The Eleventh Century MS Munich 14436: Its Contribution to the History of Coordinates, of Logic, of German Studies in France," *Isis,* 38 (1948), pp. 205–225.
250. MSS BL Egert 2976, fol. 20r; Ox BL Digby 107, fol. 20v; BL Roy 12.F.X, fol. 20v.
251. MSS Bibliotheek van de Rijksuniversiteit, Ghent, Belgium 92, fol. 225r; Balt WAG W.73, fol. 8v, courtesy of the Trustees of the Walters Art Gallery; Mu CLM 10268, fol. 33r.
252. MSS BL Burn 275, fol. 459v; Ox Bl Digby 40, fol. 14r.
253. MSS Mu Arab 464, fol. 11v; Mu CLM 2655, fols. 115v–116r; BL Roy 19.C.I, fol. 40v; Bibliothèque municipale, Troyes, France, 62, fol. 105v.
254. MSS BN fr. 1082, fols. 150v, 8v; Nap XIII.G.38, pars II, fol. 113v; Vat Pal lat 1389, fol. 114v.
255. MSS Vat Chigi E.VI.197, fol. 142v; Biblioteka Jagiellónska, Krakow, 748, fol. 49r; BN fr. 1082, fol. 55v.
256. MSS Biblioteca Universitaria, Pavia, Italy, Aldini 314, p. 151; Vat lat 3064, fol. 84r.
257. MSS BN graec. 2327, fol. 279r; BN lat 7147, fol. 81v.
258. MSS BL Add 4913, fol. 17r; CU Ii.3.12, fol. 61v.
259. MS BN lat 11229, fols, 37v, 31r. The collection of texts to which these miniatures belong, which was made by Johannes de Ketham and printed in 1491 and later dates, has appeared in facsimile as vol. 1 of the series *Monumenta medica,* Henry E. Sigerist, ed.: *Facsiculus medicinae of Johannes de Ketham Alemanus,* Charles Singer, ed. and trans. (1924); Singer also edited a reproduction of the 1493 Venice ed. in Italian as vol. 2 of the *Monumenta medica.*
260. Trinity College Library, Cambridge, England, MS 922 (R.14.52), unfoliated, by courtesy of the Master and Fellows of Trinity College, Cambridge; Ox Bl Ash 391, fol. 10r.
261. University Library, Edinburgh, Scotland, MS 170 (D.b.VI.1), fol. 111v; Camb Fitz MS Fr. 298, vol. 81r, courtesy of the Fitzwilliam Museum, Cambridge.
262. MSS Mu CLM 161, fol. 39v; BN lat 16169, fol. 134r.
263. MSS BL Harl 4751, fol. 40r; BL Cott Tiberius C.I, fol. 7v. Reproduction and discussion of other versions of the latter predictive, circular diagram can be found in Henry E. Sigerist, "The *Sphere of Life and Death* in Early Medieval Manuscripts," *Bulletin of the History of Medicine,* 11 (1942), pp. 292–303; and E. Wickersheimer, "Figures Médico-astrologiques des IXe, X^e et XIe siècles," *Janus,* 19 (1914), pp. 157–177.
264. MS BL Slo 1975, fol. 92v.
265. MSS BL Egert 2572, fol. 50r; BL Harl 3719, fols. 158v 159r; BL Egert 2572, fol. 61r; BL Add 29301, fol. 44v.
266. MSS BN lat 7028, fol. 154r; BN lat 11229, fol. 45r; Ox Bl Ash 391(V), fol. 9r; Camb Fitz 167, fol. 102r, courtesy of the Fitzwilliam Museum, Cambridge.

267. Wellcome Institute, London, by courtesy of the Wellcome Trustees. Description, Latin text, and trans. of this medical girdle book are given in Charles H. Talbot, "A Medieval Physician's Vade Mecum," *Journal of the History of Medicine and Allied Sciences,* 16 (1961), pp. 213–233.
268. Universitní Knihovna, Prague, MS Raudnitz IV.fc.29, fol. 98r, courtesy of the State Library of the Czech Socialist Republic; Kongelige Bibliotek, Copenhagen, MS Kgl. Saml. 84b, fol. 3r.
269. MSS BL Add 29301, fol. 25r; Bol Univ graec. 3632, fol. 383r.
270. MSS Fl Laur Pal lat 586, fol. 17v; Vl n.s. 2644, fol. 21r. The latter MS has been published in facsimile as *Tacuinum Sanitatis in Medicina,* ed. and German trans. by Franz Unterkircher, English trans. by Heide Saxer and Charles H. Talbot (1968).
271. MSS BL Add 22553, fol. 1v; Ox Bl Can Misc 366, fol. 1v; Universitní Knihovna, Prague, IV.F.18, fol. 143v, courtesy of the State Library of the Czech Socialist Republic; Ox Bl Ash 399, fols. 22v–23r. Discussion and reproduction of these and other brain diagrams can be found in Edwin Clarke and Kenneth Dewhurst, *An Illustrated History of Brain Function* (1972).
272. Deir el-Bahri, papyrus, photograph courtesy of Princeton University Press; Tomb of Seti I, Valley of the Kings, Luxor, photograph courtesy of Phaidon Press.
273. MSS Vl 2554, fol. 1r; Ox Bl Bodl 270b, fol. 4r; Ox Bl Ash 1511, fol. 6v.
274. MS BN lat 6734, fols. 3v, 1v. These miniatures are analyzed in detail in Marie-Thérèse d'Alverny, "Le cosmos symbolique du XIIe siècle," *Archives d'histoire doctrinale et littéraire du moyen âge,* 28 (1953), pp. 31–81.
275. MSS BN lat 6734, fol. 2v; BN lat 3236A, fol. 90r; Mu CLM 26838, fol. 80r. For the first of these three figures, see the article referred to in the previous note; for the second, see Marie-Thérèse d'Alverny, "Les pérégrinations de l'âme dans l'autre monde d'après un anonyme de la fin du XIIe siècle," *Archives d'histoire doctrinale et litteraire du moyen âge,* 13 (1940–1942), pp. 239–299.
276. MSS BL Ar 83, fol. 123r; Ox Bl Digby 107, fol. 51r; BL Roy 19.C.I, fol. 34v.
277. MSS Mu CLM 10268, fol. 22r; Nassauische Landesbibliothek, Wiesbaden, Federal Republic of Germany, B, fol. 14r.
278. MSS Lamb Pal 67, fol. 111v, by courtesy of His Grace the Archbishop of Canterbury and the Trustees of Lambeth Palace Library; Mu CLM 210, fol. 123r; BN fr. 1082, fol. 127 v.
279. MSS BN lat 6649, fol. 8v; Vl 12600, fol. 75r; Ox SJ 17, fol. 40r; Ox Bl Laud Misc 356, fol. 120r; BL Oriental 8349, fol. 65v.
280. MSS BN lat 6413, fol. 27v; Balt WAG W.73, fol. 1v, courtesy of the Trustees of the Walters Art Gallery; Mu CLM 210, fol. 139r; Ox Bl Douce 117, fol. 6r.
281. MSS Vl 12600, fol. 30r; Nap XIII.G.38, pars II, fol. 62r.
282. MS BN lat 7361, fol. 51v; Anagni Cathedral, Italy: crypt fresco, photography from Léon Pressouyre, "Le cosmos platonicien de la cathédrale d'Anagni," in *Mélanges d'archéologie et d'histoire,* 78 (1966), pp. 551–593.
283. Bibliothèque royale, Brussels, MS Roy. II.1639, fol. 6v: this miniature is analyzed in Harry Bober, "In principio: Creation Before Time," in Millard Meiss, ed., *De artibus opuscula XL: Essays in Honor of Erwin Panofsky,* 2 vols. (1961), vol. 1, pp. 13–28; Vat Reg lat 123, fol. 129r; Ox Bl Digby 23, fol. 25v.
284. MSS Mu 210, fol. 132v; Ox Bl Digby 75, fol. 129r; Fl Laur Ashburnam 1166, fol. 8r.
285. MSS Ox SJ 17, fol. 13r; BL Roy 12.F.X, fol. 2r; BL Add 18210, fol. 69r.
286. MS BN lat 6413, fol. 5v.
287. MSS Ox Bl Digby 107, fol. 52r; Gonville and Caius College, Cambridge, England, MS 428/428, fol. 50r, courtesy of the Master and Fellows of Gonville and Caius College, Cambridge.
288. MSS Vl 12600, fol. 29r; Mu CLM 13002, fol. 7v (cf. reference in picture credit 20); Bibliothèque royale, Brussels, Roy. 9094, fol. 45r.
289. MS Ox Bl Digby 23, fols. 51v–54v.
290. MS Ox SJ 17, fol. 7v.
291. MS Mu CLM 2655, fols. 104r–105r.

Index

The numbered references in this index refer either to textual material, which is designated by page numbers in lightface, or caption material, which is designated by illustration numbers in **boldface.** It is suggested that, in addition to the page and illustration number references given here for any particular subject, the reader also consult the Picture Sources and Credits section for any illustration listed, which may provide some additional information.